Gallium Nitride and Related Materials II

MATERIALS RESEARCH SOCIETY
SYMPOSIUM PROCEEDINGS VOLUME 468

Gallium Nitride and Related Materials II

Symposium held April 1–4, 1997, San Francisco, California, U.S.A.

EDITORS:

C.R. Abernathy
University of Florida
Gainesville, Florida, U.S.A.

H. Amano
Meijo University
Nagoya, Japan

J.C. Zolper
Sandia National Laboratories
Albuquerque, New Mexico, U.S.A.

MATERIALS
RESEARCH
SOCIETY

PITTSBURGH, PENNSYLVANIA

This work was supported in part by Army Research Office under Grant Number ARO:DAAG55-97-1-0125. The views, opinions, and/or findings contained in this report are those of the author(s) and should not be construed as an official Department of the Army position, policy, or decision, unless so designated by other documentation.

CAMBRIDGE
UNIVERSITY PRESS

32 Avenue of the Americas, New York NY 10013-2473, USA

Cambridge University Press is part of the University of Cambridge.

It furthers the University's mission by disseminating knowledge in the pursuit of education, learning and research at the highest international levels of excellence.

www.cambridge.org
Information on this title: www.cambridge.org/9781558993723

CODEN: MRSPDH

A catalogue record for this publication is available from the British Library

Library of Congress Cataloguing in Publication data

Gallium nitride and related materials II : symposium held April 1–4, 1997,
    San Francisco, California, U.S.A. / editors, C.R. Abernathy, H. Amano,
    J.C. Zolper
        p. cm—(Materials Research Society symposium proceedings ; v. 468)
        Includes bibliographical references and index.
        ISBN 1-55899-375-X
        1.   Gallium nitride—Congresses.   2.   Semiconductors—Materials—
    Congresses.   3.   Electroluminescent devices—Materials—Congresses.
    4.   Lasers—Materials—Congresses.   5.   Epitaxy—Congresses.
    I.   Abernathy, C.R.   II.   Amano, H.   III.   Zolper, J.C.   IV.   Series: Materials
    Research Society symposium proceedings ; v. 468.
TK7871.15.G33G35   1997   97-13305
621.3815´2—dc21   CIP

ISBN 978-1-558-99372-3 Hardback

CONTENTS

PART I: GROWTH AND DOPING

*Invited Paper

PART III: <u>CHARACTERIZATION</u>

*Invited Paper

*Invited Paper

PART V: DEVICE PERFORMANCE AND DESIGN

*Invited Paper

PREFACE

This symposium, one in a continuing series of MRS symposia dedicated to III-nitrides, focused on recent developments in GaN, InN, AlN and their alloys that are now finding application in short-wavelength lasers (~400 nm, cw at room temperature) and high-power electronics (2.8 W/mm at GHz). This proceedings is an accurate representation of the meeting in that approximately 70% of the presented papers are included.

The most controversial topic of the meeting involved the recent report of enhanced conductivity in p-type material through co-doping of oxygen and Mg. A model used to explain this behavior elicited several minutes of lively discussion. The topic of In-segregation and quantum dot formation came up throughout the meeting and is also an area of controversy. Convincing evidence was presented showing inhomogeneous emission from the edge of InGaN quantum well structures. Reports that In composition modulations in the quantum well give rise to quantum dots and their potentially positive impact on laser performance was mentioned by several speakers, though theoretical calculations reported at the meeting suggest that readsorption may cause an increase in the threshold current. While much of the symposium reported on advances in material preparation and understanding of defect issues, similar advances in material and device processing were also reported.

Because of the strong attendance and diversity of topics covered, we believe this proceedings represents an accurate and informative picture of the present status of III-nitride science and technology.

C.R. Abernathy
H. Amano
J.C. Zolper

April 1997

ACKNOWLEDGMENTS

We would like to thank the U.S. Army Research Office for its generous financial support of the symposium.

We would also like to thank our invited speakers who contributed greatly to the success of the meeting.

T. Azuhata	S.J. Rosner
W.W. Chow	R.J. Shul
S.P. DenBaars	M. Shur
K. Kim	M. Suzuki
R.J. Molnar	S. Tanaka
S.J. Pearton	M. van Schilfgaarde
F.A. Ponce	

We are also grateful to the MRS staff for their assistance in preparation of this volume. Finally, we wish to commend and congratulate the authors for the high quality of their presentations and papers, and for the timely submission of their manuscripts.

MATERIALS RESEARCH SOCIETY SYMPOSIUM PROCEEDINGS

MATERIALS RESEARCH SOCIETY SYMPOSIUM PROCEEDINGS

Prior Materials Research Society Symposium Proceedings available by contacting Materials Research Society

Part I

Growth and Doping

IMPURITY CONTAMINATION OF GaN EPITAXIAL FILMS FROM THE SAPPHIRE, SiC AND ZnO SUBSTRATES

Galina Popovici *, Wook Kim *, Andrei Botchkarev *, Haipeng Tang * James Solomon ** and Hadis Morkoç *+

* University of University of Illinois at Urbana-Champaign, Coordinated Science Laboratory, 1101 West Springfield Avenue, Urbana, IL 61801
** Dayton Research Institute, Dayton, OH 45469-0167
+ On leave at Wright Laboratory Wright Patterson AFB under a URRP program funded by AFOSR

ABSTRACT

Likely contamination of GaN films by impurities emanating from Al_2O_3, SiC and ZnO substrates during growth has been studied by secondary ion mass spectrometry analysis. The highly defective interfacial region allows impurities to incorporate more readily as compared to the equilibrium solubility in a perfect crystal at a given temperature as evidenced by increased impurity levels in that region as detected by SIMS. The SIMS measurements in GaN layers grown on SiC and sapphire showed large amounts of Si and O, respectively, within a region wider than the highly disordered interfacial region pointing to the possibility of impurity diffusion at growth temperatures. The qualitative trend observed is fairly clear and significant. These observations underscores the necessity for developing GaN and/or AlN substrates.

INTRODUCTION

Wide band gap semiconductor materials extend the applications of semiconductors outside the realm of classical semiconductors such as Si and GaAs. [1-3] The large band gap, high thermal conductivity, and chemical inertness of nitrides pave the way for high power/temperature operation and light emission in green, blue and ultraviolet region of the spectrum. Of particular interest is the steady development of GaN technology in the last years, culminating in the demonstration of efficient green and blue light emitting diodes, and violet lasers.[4]

In spite of the rapid development many problems remain. The lack of an ideal substrate presents a major problem in GaN growth. Because of the high decomposition nitrogen pressure of GaN, no conventional crystal growth method can be employed to obtain bulk GaN crystals. Despite steady progress, GaN crystals grown from Ga solution under high pressure are not yet being produced in quantities needed, causing the growth efforts to heteroepitaxy on a variety of substrates, such as sapphire (Al_2O_3) [5], 6H-SiC [6], Si [7], GaAs [8], ZnO [9] and others, among which sapphire is the most widely employed one. Despite the poor lattice and thermal match, the best material grown today is on sapphire substrates. The preference towards sapphire substrates can be attributed to their ease of availability, hexagonal symmetry, and minimal pre-growth cleaning requirements. Sapphire is stable at high temperatures ($\sim 1000\ °C$) required for epitaxial growth by vapor phase techniques. SiC (mismatch -3.4%) and ZnO (mismatch 2.0%) appear promising due to better lattice match compared to sapphire (mismatch

-13%). ZnO has the best lattice match, but it decomposes at typical growth temperatures employed. Use of substrates other than GaN presents not only the inconvenience of the lattice and thermal mismatch, but also a possibility of unintentional contamination from the substrate during growth. While the influence of thermal and lattice mismatch from different substrates on GaN crystal structure and defect content have been studied extensively, possible contamination of GaN films by substrates during growth has not gotten much attention In this paper, contamination of GaN films by impurity out-diffusion from Al_2O_3, SiC and ZnO substrates during growth will be discussed. We show that the substrate may be a source of considerable contamination of the grown layer.

EXPERIMENTAL

GaN films were grown by reactive molecular beam (RMBE) details of which have been described elsewhere.[10] The temperature of the ammonia injector was kept at 300 C. Two tandem Nanochem ammonia purifiers were utilized to reduce the oxygen level in ammonia. Basal plane sapphire substrates were degreased with organic solvents, and etched in a hot solution of H_2SO_4 and H_3PO_4 (H_2SO_4: H_3PO_4 = 3:1) for 20 min. They were then rinsed with deionized(DI) water and blow dried with filtered nitrogen. Molten indium was then used to mount the sapphire substrates on Si templates mounted on molybdenum blocks with high purity carbon screws. Once in the growth chamber, nitridation was performed by exposing the sapphire substrate to an ammonia flux of 16 sccm for 1 min. at 850 °C. This nitridation condition was chosen as it provides the smoothest sapphire surface morphology. Growth temperature for both AlN buffer layer and GaN film was 800 °C. The thickness of the AlN buffer layers was 60 ~ 80 nm. The growth rate employed was about 0.7 μm/hr.

The depth profile of impurities along the growth direction was obtained by secondary ion mass spectroscopy(SIMS) performed using a CAMECA IMS-4f double focusing ion microanalyzer configured for cesium primary ion bombardment. The quantification of elemental concentration for this instrument was done by using relative sensitivity factors (RSFs) derived from the analysis of a GaN standard with known doses of impurity implants. The SIMS data near the surface (tens of nm) are in the limits of equilibration distance and should be disregarded. In addition, analyses were performed with 12kV oxygen and 10kV cesium with a quadrupole based instrument. No standards were available for quantification for this instrument.

RESULT AND DISCUSSIONS

Fig. 1 shows the SIMS depth profiles of impurities for the GaN layer grown on SiC. AlN buffer layer was used. One can see that Si is present near the interfacial layer, up to approximately 1/3 of the thickness, diffusing from the substrate. Moreover, concentration of oxygen and hydrogen is larger in this region as compared to the rest of the layer. O incorporates in the film from two sources: from the buffer layer and from the gas phase. H enters from the gas phase since ammonia is used as the nitrogen source. Al, Si and Ga profiles for the same sample, measured with the quadrupole instrument are shown in Fig. 2. These data are not quantitative, but a qualitative trend is clearly seen: Again, one can see that the concentrations of Al and Si are larger near the interface.

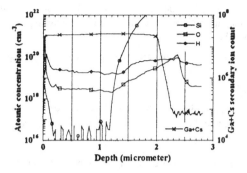 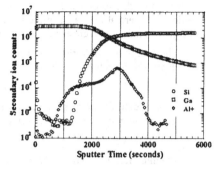

Fig.1. The SIMS profile for Ga, Si, O, and H for the sample Grown on 6H-SiC. SIMS profiles were measured with a CAMECA IMS-4f double focusing ion microanalyzer. The data for Si are not quantitative.

Fig. 2. The SIMS profile for Ga, Si, and Al for the same sample as in Fig. 1 measured with a quadrupole based instrument. The data are not quantitative.

Table I. Growth condition and the contamination depth of O, Si and Zn from Al_2O_3, SiC and ZnO substrates in the GaN films All samples were grown on AlN buffer layer.

Sample #	Substrate	$T_{substrate}$ (°C)	Growth time (h)	Sample thickness (μm)	Measur. Impurity	Contam. depth (μm)
5465	Sapphire	800	2	2.5	Al	0.6
5536	Sapphire	793	2	0.8	O	0.30
5553	Sapphire	795	6	3.0	O	1.0
5554	Sapphire	800	6	3.0	O	1.0
5602	Sapphire	800	5	3.0	Al O	1.0 1.0
5550	SiC	810	6	2.0	Si Al	0.9 0.6
5518	ZnO	800	2	1.05	Zn O	0.2 0.2

It is well known that the presence of defects especially extended defects like dislocations, grain and twin boundaries, stacking faults and boundaries of the inversion domains enhance diffusion of impurities. The TEM studies of GaN films show that the region near the interface has the highest defect density compared to the region near the surface.[11-12] Therefore, one can expect that the amount of the impurities near the interface to be higher than it is permited by the solubility limit of the perfect crystal.

The depth of contaminated layer depends on growth time as one can see from the Table I. From Table I it can be seen that the larger growth times result in larger impurities depths. The

5

latter can be due to diffusion, since the thickness of the defective near-interface layer is set by lattice and thermal mismatch of the layer and substrate and should not significantly depend on the growth time, if the growth conditions are the same.

The results on contamination of the GaN layer from the substrate points to two stringent problems which hinder III-N research:

1. Lack of GaN substrates for the growth of GaN based devices.

2. Frequently in the existing literature, different deep levels are assigned to native defects, without taking into account contaminants, which could be responsible for deep levels as well.

CONCLUSIONS

The interfacial GaN layer near the substrate is contaminated during growth by the impurities from the substrate. The thickness of the contaminated layer is larger for longer growths.

ACKNOWLEDGMENTS
The research is funded by grants from ONR(contract #N00014-95-1-0635, #N00014-89-J-1780), AFOSR(contract #F49620-95-1-0298),under the supervision of Mr. M. Yoder, and Drs. G. L. Witt, Y. S. Park, and C. E. C. Wood.

REFERENCES

1. H. Morkoç Progress and Prospects of Group-III Nitride Semiconductors, International Symposium on blue Lasers and Light Emitting Diodes, Chiba Univ., Japan, March 5-7, 1996, p. 23-29.
2. S. N. Mohammad, A. Salvador, and H. Morkoç, Emerging Gallium Nitride Devices, Proc. IEEE v.83, 1306 (1995).
3. H. Morkoç, S. Strite, G. B. Gao, M. E. Lin, B. Sverdlov, and M. Burns, Large-band-gap SiC, III-V nitrides, and II-VI ZnSe-based Semiconductor Devices Technologies, J. Appl. Phys. 76, 1363 (1994).
4. I. Akasaki and H. Amano, Current Status of III-V Nitride Research, International Symposium on blue Lasers and Light Emitting Diodes, Chiba Univ., Japan, March 5-7, 1996, p. 11-16.
5. T. Lei, K. F. Ludwig, Jr. , and T. Moustakas, J. Appl. Phys. 74, 4430 (1993)
6. M. E. Lin, S. Strite, A. Agarwal, A. Salvador, G. L. Zhou, N. Teraguchi, A. Rocket and H. Morkoç, Appl. Phys. Lett. 62, 702 (1993)
7. A. Ohtani, K. S. Stevens, and R. Beresford, MRS Symp. Proc. v.339, p.471-476.
8. J. W. Yang, J. N. Kuznia, Q. C. Chen, M. A. Khan, T. George, M. De Graef, and M. Mahajan, Appl. Phys. Lett. 67, 3759 (1995)
9. F. Hamdani, A. Botchkarev, W. Kim, H. Morkoç M. Yeadon, J. M. Gibson, S.-C. Y. Tsen, D. J. Smith, D. C. Reynolds, D. C. Look, K. Evans and C. W. Litton, Appl. Phys. Lett. 70, 467 (1997)
10. W. Kim, Ö. Aktas, A. E. Botchkarev, A. Salvador, S. N. Mohammad and H. Morkoç, J. Appl. Phys. 79, 7657 (1996).
11. Q. Zhu, A. Botchkarev, W. Kim, O. Aktas, A. Salvador, B. Sverdlov, H. Morkoc, S.-C.-Y. Tsen, and D. J. Smith, Appl. Phys. Lett, 68, 1141 (1996)
12. Z. Liliental - Weber, S. Ruvimov, T. Suski, J. W. Ager III, W. Swider, Y. Chen, Ch. Kisielowski, J. Washburn, I. Akasaki, H. Amano, C. Kuo, and W. Imler, MRS Symp. Proc. v. 423 (MRS, Pittsburg, PA, 1996 p. 487.

RELIABLE, REPRODUCIBLE AND EFFICIENT MOCVD OF III-NITRIDES IN PRODUCTION SCALE REACTORS

B. Wachtendorf, R. Beccard, D. Schmitz and H. Jürgensen
AIXTRON GmbH, Kackertstr. 15-17, D-52072 Aachen, Germany
O. Schön, M. Heuken
Institut für Halbleitertechnik I, RWTH Aachen, , Sommerfeldstr. 24, D-52074 Aachen, Germany
E. Woelk
AIXTRON Inc., 1569 Barclay Blvd., Buffalo Grove, IL 60089, U.S.A.

ABSTRACT

In this paper we present a class of MOCVD reactors with loading capacities up to seven 2" wafers, designed for the mass production of LED structures.

Our processes yield device quality GaN with excellent PL uniformities better than 1 nm across a 2" wafer and thickness uniformities typically better than 2%.

We also present full 2" wafer mapping data, High Resolution Photoluminescence Wafer Scanning and sheet resistivity mapping, revealing the excellent composition uniformity of the nitride compounds InGaN and AlGaN. As well we will show sheet resistivity uniformity for Si-doped GaN and Mg-doped GaN.

INTRODUCTION

The worldwide demand for Ultra-High-Brightness blue and green LEDs has driven the development of MOCVD for Al-Ga-In-N alloy systems towards efficient multiwafer technology. We developed MOCVD reactors with loading capacities up to seven 2" wafers. The layers grown in these machines show excellent reproducibility from wafer to wafer and from run to run. The reactor design produces material with abrupt interfaces, even when using different substrates like Al$_2$O$_3$, SiC, Si. As a standard tool for characterization we use non-destructive wafer topography, which is used to optimize the process for uniformity and yield.

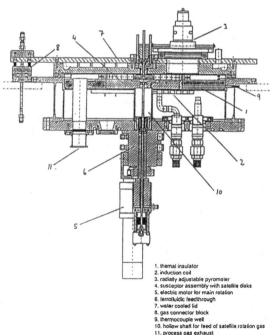

1. thermal insulator
2. induction coil
3. radially adjustable pyrometer
4. susceptor assembly with satellite disks
5. electric motor for main rotation
6. ferrofluidic feedthrough
7. water cooled lid
8. gas connector block
9. thermocouple well
10. hollow shaft for feed of satellite rotation gas
11. process gas exhaust

Fig. 1: cross section of the AIX2000HT Planetary Reactor

MOCVD REACTOR SYSTEMS

Key features of all the reactors are: Flexibility in the choice of the carrier gas in each single step of the structures, extremely low thermal mass allowing quick adjustment of different growth temperatures for each layer and real two-flow injection of the group III and group V reactants to minimize undesired prereactions. This also allows the process to be easily adapted from one machine to another.

Fig. 1 shows the cross section of the AIX2000 HT Planetary Reactor. The susceptor is heated by induction from the RF-coil which can be adjusted in height for the achievement of uniform temperature profiles. The

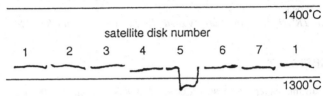

Fig. 2: temperature measurement of a planetary disk

low thermal mass of the susceptor allows steep cooling and heating ramps of about 5°C/s for the different process-steps. Like all AIXTRON reactors, the AIX2000HT is equipped with Gas Foil Rotation technology for the rotation of each substrate in addition to the rotation of the complete susceptor.

As it is well known that the nitride process is most sensitive to temperature fluctuations, special care was taken to design the coil in such a way that temperature gradients in the satellites are avoided. A pyrometer, which is mounted at the top of the reactor allows the continuous monitoring of the temperature through a purged viewport which can be used also for other in-situ measurement techniques like reflectivity based growth-rate determination. Fig. 2 shows a temperature measurement of a rotating planetary disk, with one SiC-wafer inside as a marker. This wafer is approx. 25°C colder as the empty satellites, so we can conclude that the temperature profile is better than 1°C within a 2" wafer.

The MOCVD process is also designed to ensure maximum reliability and reproducibility. In particular, the initial deposition steps, which are commonly known to have a great influence on the layer quality have been optimized.

RESULTS

The standard tool for characterization is non-destructive wafer topography. It combines RT-PL-mapping, sheet resistivity mapping and also mapping of the layer thickness by white-light reflectrometry.

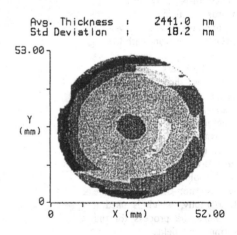

Fig.3: Thickness map of a GaN layer

This method has the following advantages:

- The characterization of the wafer is non-destructive, therefore it is not necessary to sacrifice wafers for characterization.
- The measurement is fast. No contacts or metallization are needed for thickness measurements or sheet resistivity mapping, which correspond to the doping profile. This means it is possible to receive fast feedback before device processing.
- The measurement is full automated and easy to use.

Our standard process for the growth of undoped GaN starts with a desorption step at elevated temperatures. After this a low-temperature GaN-nucleation layer is grown which will be annealed at high temperatures again. The GaN layer is grown on this nucleation layer. Depending on the parameters during nucleation

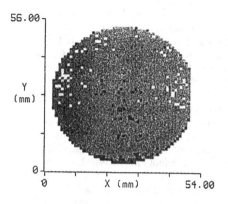

Fig. 4: Peak wavelength map of undoped GaN

and growth the material quality can be adjusted in a way that either background carrier concentrations below $1 \times 10^{17} cm^{-3}$ are achieved or the layers are highly resistive.

Using the method of wafer mapping for characterization, we optimized the process in term of homogeneity and composition. Fig. 3 shows a thickness map of a very uniform GaN-layer. The standard deviation is below 0.8%, which is one of the best results ever reported for a full 2"-wafer. With an edge exclusion of 3mm the standard deviation is even better than 0.5%. The RT-PL map of this layer shows that the emission is very uniform across the full 2"-wafer (Fig. 4), the standard deviation of the peak wavelength is only 0.07nm, mostly due to noise.

Our processes yield doping uniformities better than 5% for Si-doping and better than 20% for Mg-doping. This can be verified with Hall-measurements, but this is a destructive method. It also takes some time to provide contacts, especially to the p-type material. For this reason we perform wafer mappings of sheet resistivity, which is directly related to the

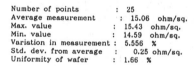

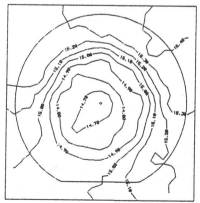

Fig. 5: sheet resistivity map of a Si:GaN-layer

doping level, taking the thickness uniformity into account. Fig. 5. shows the sheet resistivity map of a Si-doped GaN-layer. The standard deviation of the sheet resistivity is better than 2%. If we assume that the electron mobility is the same across the wafer, we can conclude that the doping level shows a uniformity better than 2% on the full wafer. The corresponding PL-Map shows a peak wavelength uniformity better than 0.25nm (Fig. 6).

For Mg-doping the sheet resistivity map shows a uniformity better than 20% after thermal activation of the holes (Fig. 7). This value is not as good as that for Si-doping because of interface effects between the GaN-buffer and the doped film and because of the more difficult thermal activation process of the holes.

InGaN MQW serves to adjust the wavelength of an LED. So the accurate adjustment of the concentration and PL-intensity distribution of InGaN are of major importance. Fig. 8 shows the peak wavelength distribution of a bulk InGaN-layer. The ultra high homogeneity in composition is reflected in the small standard deviation of the peak wavelength less than 0.75nm. The PL-intensity map of the same layer shows an intensity distribution in the range of 10% (Fig. 9). Both values fully meet the requirements of LED mass production.

The same applies to AlGaN-layers.The production grade uniformity of the Al-GaN-composition is shown in Fig. 10, where the PL peak wavelength map is shown. The standard deviation of the average wavelength is of about 0.25nm, which meets all the requirements for the mass production of devices.

CONCLUSION

A new class of MOCVD reactors for the mass production of GaN-based optoelectronic devices was presented. We also presented uniformity data for doped and undoped GaN-layers as well as for the ternary alloys InGaN and AlGaN. It was proven that the material quality is optimized for the production standards. The advantages of non-destructive wafer topography for statistical process control and also for process optimization were shown.

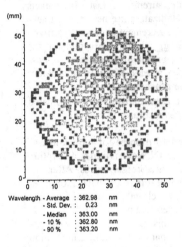

Wavelength - Average : 362.98 nm
       - Std. Dev. : 0.23 nm
       - Median : 363.00 nm
       - 10 % : 362.80 nm
       - 90 % : 363.20 nm

Fig. 6: RT-PL peak wavelength of a Si:GaN-layer

Number of points	: 25
Average measurement	:. 7871. ohm/sq.
Max. value	: 9878. ohm/sq.
Min. value	: 5824. ohm/sq.
Variation in measurement	: 51.501 %
Std. dev. from average	: 1260.23 ohm/sq.
Uniformity of wafer	: 16.01 %

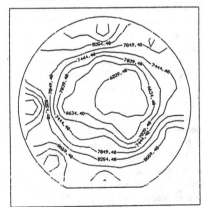

Fig. 7: Sheet resistivity map of a Mg-doped GaN layer

ACKNOWLEDGMENT

The authors acknowledge the work of W. Michel at the AIXTRON Application Lab, Aachen, Germany, as well as the contribution of H. Obloh, P. Schlotter, U. Kaufmann and N. Herres of the Fraunhofer Institute for Applied Solid State Physics, Freiburg, Germany.

```
Avg. Wavelength :    461.58  nm
Std Deviation   :      0.68  nm
```

```
Avg. Intensity :    1270.5  CU
Std Deviation  :     127.1  CU
```

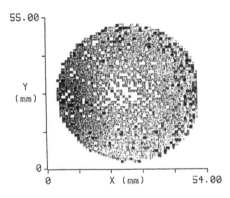

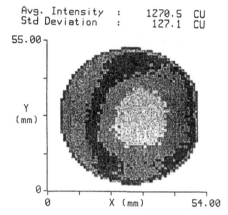

Fig. 8: peak wavelenght of an InGaN-layer

Fig. 9: PL-intensity distribution of an InGaN-layer

```
Avg. Wavelength :    340.31  nm
Std Deviation   :      0.25  nm
```

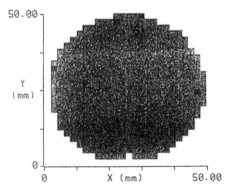

Fig. 10: RT-PL map of the peak wavelength of a Al-
GaN layer

11

Growth and Characterization of In-Based Nitride Compounds and their Double Heterostructures

V. A. Joshkin[*], J. C. Roberts[*], E. L. Piner[**], M. K. Behbehani[**], F. G. McIntosh[*], L. Wang[***], S. Lin[***], I. Shmagin[*], S. Krishnankutty[*], R. M. Kolbas[*], N. A. El-Masry[**] and S. M. Bedair[*]

[*] Dept. of ECE, Campus Box 7911, N. C. State University, Raleigh, NC 27695
[**] Dept. of MSE, Campus Box 7907, N. C. State University, Raleigh, NC 27695
[***] Sandia National Laboratories, P. O. Box 5800, Alburquerque, NM 87185

ABSTRACT

We report on the growth and characterization of InGaN bulk films and AlGaN/InGaN/AlGaN double heterostructures (DHs). Good quality bulk InGaN films have been grown by metalorganic chemical vapor deposition (MOCVD) with up to 40% InN as characterized by x-ray diffraction. The effect of hydrogen in the growth ambient on the InN% incorporation in the InGaN films is presented. Photoluminescence (PL) spectra of AlGaN/InGaN/AlGaN DHs exhibit emission wavelengths from the violet through yellow depending on the growth conditions of the active InGaN layer. The PL spectra are fairly broad both at room temperature and 20 K, and could be a result of native defects or impurity related transitions. We also observed a linear dependence between the PL intensity and excitation power density in the $0.001 W/cm^2$ to $10 MW/cm^2$ range. Time resolved PL of one of these DHs suggest a recombination lifetime on the order of 520 ps.

INTRODUCTION

InGaN device quality films and their related heterostructures play a critical role in the development of nitride devices. Unfortunately, a limited amount of informative research regarding the epitaxial growth of InGaN and related quantum well (QW) structures exists, and can be attributed to several fundamental problems associated with the growth of In-based compounds. For example, the high equilibrium vapor pressure required during growth to prevent the dissociation of the In-N bond represents a significant problem in the growth of In-based nitride compounds. While MOCVD growth of GaN at 1000 °C results in sufficient decomposition of NH_3 at the growing nitride surface[1], such a high growth temperature cannot be used with In-based compounds due to the weak In-N bonds as well as the possibility of In atoms desorbing from the growing surfaces. There is also the potential of phase separation in this materials system, especially for films having high In concentration.[2]

AlGaN/InGaN/AlGaN double heterostructures (DHs) currently used in optoelectronic device structures suffer from the fairly high lattice mismatch between these ternary alloys. For example, for green emission light emitting diodes (LEDs), the thickness of the InGaN well should not exceed ~30 Å i.e., less than the critical thickness to suppress the generation of misfit dislocations as based on Mathew's model.[3] Imposing such strict thickness tolerances on thin InGaN active layers will make production scale-up of LEDs more difficult. Furthermore, high quality interfaces

Mat. Res. Soc. Symp. Proc. Vol. 468 © 1997 Materials Research Society

between the InGaN wells and the GaN or AlGaN barrier layers are difficult to achieve. Poor interfaces can result from poor nucleation (3D instead of 2D), surface reconstruction, segregation and reaction of In at the interfaces and incompatible growth temperatures. In addition, the lattice mismatch between (AlGa)N barrier layers and the InGaN active layer can influence the InN% in the InGaN alloy.[4] In this paper we will review our current progress in the area of In-based nitride compounds and their DHs. We have investigated photoluminescence (PL) over the temperature range 20 < T < 300K, performed time resolved spectroscopy, and investigated the dependence of PL intensity on excitation power density of blue and green emitting DHs.

EXPERIMENT

Our nitride research activities use a versatile growth system with a unique susceptor that is capable of operating in either Atomic Layer Epitaxy (ALE) or conventional metalorganic chemical vapor deposition (MOCVD) growth mode. In ALE growth mode, the rotating part of the graphite susceptor first exposes the substrate to the column III precursors, followed by 180° rotation of the substrate to column V precursors. The reactants are physically separated by injecting them through windows on opposite sides of a stationary susceptor. The exposure time of the substrate to each reactant, as well as the time per growth cycle are controlled by the rotational speed and by the pause times under the reactant gas streams[5]. Since the cycle times can be very short, on the order of one second, this design does not suffer from the slow growth rates common with conventional vent/run ALE configurations. In the conventional MOCVD growth mode, no rotation takes place and the substrate is kept fixed under a mixed flow of column III and V precursors. Both of these growth modes have been used to successfully deposit high quality GaN, InGaN, and AlGaInN epitaxial films.

Films were grown by MOCVD in an atmospheric pressure vertical reactor on (0001) sapphire substrates. Source gases used were ammonia (NH_3), trimethylgallium (TMG, -10 °C), trimethylaluminum (TMA, +18 °C) and ethyldimethylindium (EdMIn, +10 °C); N_2 was used as the carrier gas. Sapphire substrates were annealed at 1050 °C in N_2 and NH_3 for 15 minutes and 1 minute respectively, followed by the ALE growth of an AlN buffer layer at 700 °C and 100 Torr. Prior to the growth of "bulk" InGaN films, an AlGaN graded to GaN prelayer was deposited at 950 °C and 750 Torr for 15 min. and 10 min. respectively, with a 2.5 min. grading period in between. The InGaN layers were grown for 1 h by MOCVD in the temperature range 710-780 °C. The quality and InN percent of the films were characterized by θ-2θ x-ray diffraction (θ-2θ XRD) and x-ray rocking curve (XRC).

For the growth of DHs, a 0.1 μm thick $Al_{0.09}Ga_{0.91}N$ cladding layer was grown on the AlN buffer layer at 950 °C and 750 Torr, followed by a graded In_xGa_{1-x} N layer deposited while the temperature was ramped down from 800 °C to 750 °C. Active InGaN layers were then grown for a period of one (sample B) or 2.5 (sample A) minutes at 750 °C, followed by a ~500 Å $Al_{0.09}Ga_{0.91}N$ upper cladding layer grown at 950 °C to complete the DH. The thickness of InGaN active layers in these DHs were not measured directly, but were extrapolated using data by TEM from thicker InGaN films in similar DHs. In this study the InGaN active layer thickness can be estimated to be in the 30-50 Å range, however an exact value is not currently available. Another difficulty encountered is the lack of good uniformity, resulting in a variation of PL peak emission of about 280 meV over a sample area of about 1 cm². This variation could be a result of the variation of indium mole fraction over the sample area.

RESULTS

The InN percent in MOCVD and ALE grown InGaN was found to be significantly influenced by the amount of hydrogen flowing into the reactor. The temperature ranges for this study are 710 - 780 °C for MOCVD. For a given set of growth conditions, an increase of up to 25% InN in InGaN, as determined by x-ray diffraction, can be achieved by reducing the hydrogen flow from 100 to 0 sccm.

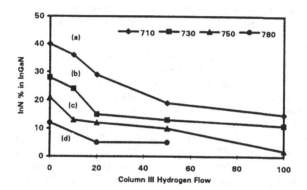

Figure 1. InN percent in InGaN vs. hydrogen flow at the growth temperature
(a) 710, (b) 730, (c) 750 and (d) 780 °C.

Figure 1 plots the InN percent, as determined by θ-2θ XRD, in InGaN as a function of the hydrogen flow injected with the nitrogen carrier gas. All of the samples were grown by MOCVD using 1 sccm of TMGa (-10 °C), and 90 sccm of EDMIn (+10 °C) in 5 slm of nitrogen carrier gas and 5 slm of ammonia at the temperatures and hydrogen flows indicated. The figure indicates that the InN percent in the films drops significantly as the hydrogen flow increases. This trend occurs at all four temperatures investigated with very good consistency. The general trend of Figure 1 shows a fairly rapid decrease in the InN percent as the hydrogen flow is increased from 0 to 20 sccm, followed by a more gradual decrease upon further increased hydrogen flow from 20 to 100 sccm. All films are single crystalline, good quality, and without indium metal as-grown, as indicated by θ-2θ XRD.

In an effort to determine the effect on the InN percent due to the hydrogen

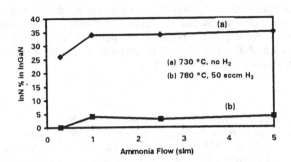

Figure 2. InN percent in InGaN vs. ammonia flow at (a) 730 °C and no hydrogen, and (B) 780 °C and 50 sccm hydrogen.

generated during the decomposition of ammonia, a series of experiments was performed in which the overall ammonia flow was varied while keeping all other parameters constant. The results of that study are shown in Figure 2 in which the ammonia flow was varied from 0.3 to 5 slm while using nitrogen as a make-up gas to keep the overall column V flow constant at 5 slm. The samples represented by curve (a) in Figure 3 were grown at 730 °C with no hydrogen while those of curve (b) were grown at 780 °C with 50 sccm hydrogen. Both were deposited by MOCVD with the same TMGa and EDMIn flows as those in Figure 1. Both curves show little change in InN percent for ammonia flows greater than 1 slm indicating a sufficient supply of reactive nitrogen species at the growing surface with little effect from the hydrogen being generated by the decomposition of ammonia. This is an indication that the decomposition of ammonia into nitrogen and hydrogen at these temperatures is extremely low. If the decomposition were occurring at a higher rate, the increased hydrogen being generated from the ammonia would result in a decrease in the InN percent similar to that observed in Figure 1. Using a ±1 InN% error associated with the points on curve (a) of Figure 2 and comparing this number to the zero hydrogen point on curve (b) of Figure 1 indicates that 5 sccm of hydrogen, at most, is being generated from the ammonia. Assuming 5 slm of ammonia, 15 sccm of liberated hydrogen would result from 0.1% decomposition of the ammonia at the temperature range studied. Therefore, the data from Figures 1 and 2 suggest an ammonia decomposition rate of less than 0.1%.

PL spectra of a series of AlGaN/InGaN/AlGaN heterostructures have been measured using the 325 nm output of a He-Cd laser as the photoexcitation source. Figure 3 shows this series of room temperature PL spectra for these DHs in which the growth conditions for these films were held constant except the growth time of the active InGaN layer was steadily decreased from (a) to (d). The AlGaN barrier layers were grown with approximately 10% AlN. The graph shows emission from the active

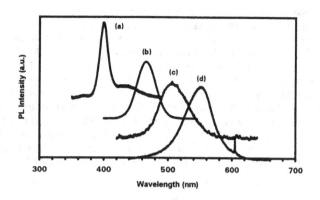

Figure 3. RT PL of MOCVD grown AlGaN/InGaN/AlGaN DHs.

layer corresponding to 400, 456, 505, and 550 nm for (a), (b), (c) and (d), respectively. The FWHMs are 115, 189, 241, and 211 meV, respectively which are comparable to those reported by Nichia obtained from their electroluminescence studies. Although absolute thickness measurements for these samples were not available, TEM data from other similar samples coupled with appropriate scaling of the growth times suggest thicknesses on the order of 25 to 250 Å. The active layers of these films were too thin to obtain x-ray diffraction to verify the InN% in the well regions.

Two other AlGaN/InGaN/AlGaN DHs with blue (sample A) and green (sample

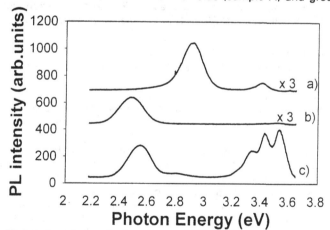

Figure 4. PL spectra of sample A and sample B at different temperatures: (a) sample A at RT, (b) sample B at RT; (c) sample B at 20K.

B) emissions were studied in further detail. Figure 4 shows the PL spectrum of these two DHs, where a He-Cd laser beam (325 nm) was used as the photoexcitation source. Figures 4a and 4b show the spectrum of samples A and B at room temperature (RT), while Figure 4c shows the PL spectrum at 20K of sample B. In order to study the effect of excitation power density on the PL emission two experimental setups were used. The first one uses a He-Cd laser where the excitation power density could be varied from 10^{-3} W/cm^2 to 2W/cm^2. The second setup uses the third harmonic (280 nm) of a modelocked Ti:Sapphire laser, operating at 840 nm with a pulse duration of 250 fs and repetition rate of 76 MHz, with excitation power levels in the 0.1 MW/cm^2 to 10 MW/cm^2 range. Figure 5 shows RT PL spectra at excitation levels of about 10 MW/cm^2 of the two structures sample A (Figure 5, curve a) and B (Figure 5, curve b). Figures 6 and 7 show the variation of the PL intensity

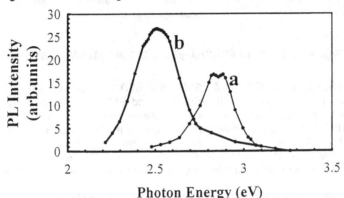

Figure 5. PL spectra of DHs measured at RT with a high excitation power density (10MW/cm^2): a) sample A, b) sample B.

with pumping power density for these two samples. Figures 6 and 7 are for the He-Cd cw laser set up (low excitation density) and pulsed laser set up (high excitation density) respectively.

To determine the life time of these optical transitions, excitation correlation spectroscopy was used. In this technique, an ultrafast optical pulse train was divided into two by a beam splitter, with one pulse train was delayed with respect to the other. After the two beams were focused to the same spot on the sample the correlation signal at the sum frequency yields the carrier lifetime. In these experiments the second harmonics of an 800 nm laser pulse generated from a tunable Ti:Sapphire laser (pulse duration of about 200 fs) was used as the excitation source. The time resolution of this setup is determined by the pulse duration of the laser, 200 fs. Figure 7 shows the dynamic response at 40K for peak luminescence of 2.7 eV for a DH emitting in the blue region. From this time resolved PL, the decay can be fitted to the well known relation $I(t) = I_0 \exp(-t/\tau)$ yielding τ, the recombination lifetime of about 520 ps. To gain insight of the recombination mechanism at high excitation levels we estimated the radiative quantum efficiency η. We estimate η to be very small, on the order of 10^{-3} to 10^{-2}, as determined by measuring the input laser beam flux and the PL emission measured at a given solid angle. η can be written as:

$$\eta = 1/(1 + \tau_r/\tau_{nr}), \qquad (1)$$

where τ_r is time constant of radiative recombination, and τ_{nr} is time constant of nonradiative recombination. For τ we can write:

$$1/\tau = 1/\tau_r + 1/\tau_{nr}. \qquad (2)$$

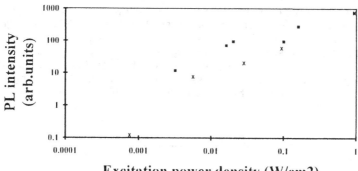

Figure 6. Relation between moderate excitation power densities (He-Cd laser, 325 nm) and PL intensity of samples A (■'s) and B (x's) at 300K.

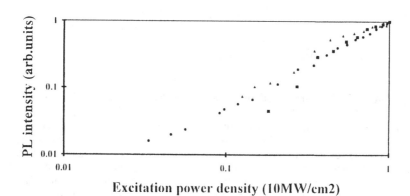

Figure 7. Relation between higher excitation power density and PL intensity of sample A @ 300K (■'s), sample B @ 300K (●'s) and sample B @ 77K (▲'s).

According to these equations, $\tau_{nr} << \tau_r$ and our decay time is apparently determined by nonradiative recombination. Nonradiative processes could be associated with the expected high density of defects in these DHs.

The above PL data represent preliminary efforts to optically characterize AlGaN/InGaN DHs grown on ALE AlN buffer layers. As seen in Figure 4, the PL spectra are fairly broad for both structures. At RT the full width at half maximum (FWHM) for samples A (blue emitting) and B (green emitting) are 230 meV and 220 meV respectively, while at 20K the FWHM for sample B is about 200 meV. These line widths are comparable to that reported by Nakamura et. al. for their electroluminescence (EL) spectrum of green LEDs based on similar DHs. It should be mentioned in an early paper, for the same group they claimed that the PL and EL spectrum line widths were comparable. Even though our DHs are not grown on optimized, thick (4 μm) cladding layers like the Nichia structures, our PL line widths are comparable to their structures. From Figures 4 and 5, the line width increases with the excitation power density. Also at very high excitation power, the blue emission (Figure 5, curve a) consists of several overlapping spectra peaking at 2.8 eV and 2.9 eV. Thus it is possible that two radiative recombination processes exist, with

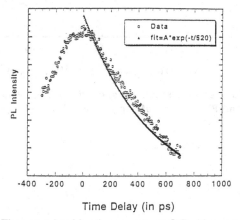

Figure 8. Time-resolved luminescence at 2.7 eV measured at 40K.

peak emission spectra spaced by about 0.1 eV in energy. The nature and origin of these defects are not yet identified. From Figure 5, curve b, the green emission also seems to consist of several peaks. These peaks are not as well resolved as those of sample A, but they can be used to support the argument that the PL spectra of these DHs could result from several transitions that might be attributed to defects. It should also be noted from Figure 4c that the line width did not significantly decrease for the low temperature PL measurement performed at 20K. This could also be evidence that the observed emissions are defect related rather than pure band to band recombination. It is also possible that the InGaN/AlGaN interfaces are not abrupt and that the grading of the In composition could add to the broadening of these emission spectra. The linear relationship, shown in Figures 6 and 7, between the excitation power density and the resulting PL signal, can imply that band edge transitions are taking place. However, when these data are considered within the context of the spectra shown in Figures 4 and 5, the nature of the emission spectra could also be due to defect related energy levels with fairly high occupancies.

CONCLUSIONS

In summary, flowing small amounts of hydrogen during the MOCVD growth of InGaN films has been shown to have a profound effect on the InN percent in III-nitride compounds. An increase of up to 25 percent InN in InGaN, depending on the growth temperature, can ben obtained by reducing the hydrogen flow from 100 to 0 sccm. The amount of hydrogen being generated by the ammonia at these temperatures does not appreciably change the InN percent in these films, indicating a very low decomposition rate of the available ammonia. Preliminary data about the optical properties of blue and green AlGaN/InGaN DHs has also been presented. Although exact identification of the optical transitions in these structures from preliminary data would be premature, it seems unlikely that the observed emission spectra are pure band-to-band in character. Further work is under way to elucidate the exact mechanisms of these transitions and to relate them to the bulk properties of InGaN films.

ACKNOWLEDGMENTS

This work is supported by the ONR, University Research Initiative (URI), ARPA optoelectronics program, and ARO.

REFERENCES

[1] K. Doverspike, L. Rowland, D. K. Gaskill and J. A. Freitas, J. Electron. Mat. **24** (1995) 269.
[2] R. Singh, D. Doppalapudi and T. D. Moustakas, to be published in Appl. Phys. Lett.
[3] L. W. Mathews and A. E. Blokeslee, J. Crystal Growth **27** (1974) 118.
[4] Y. Kawaguchi, M. Shimizu, K. Hiramatsu and N. Sawaki, to be published in MRS Fall 1996 Symp. Proc.
[5] J. R. Gong, S. Nakamura, S. M. Bedair and N. A. El-Masry, J. Electron. Mat. 21 (1992) 965.

MOVPE GROWTH AND CHARACTERIZATION OF Al$_x$Ga$_{1-x}$N LAYERS ON SAPPHIRE

S.CLUR[1], O.BRIOT[1], J.L.ROUVIERE[3], A.ANDENET[1], Y-M.LE VAILLANT[1], B.GIL[1], R.L.AULOMBARD[1], J.F.DEMANGEOT[2], J.FRANDON[2], M.RENUCCI[2]

[1]GES, CC074, Université MONTPELLIER II, 34095 Montpellier Cedex 5, FRANCE
[2]Laboratoire de Physique des Solides, Université P.Sabatier, 118 rte de Narbonne 31062 Toulouse Cedex FRANCE
[3]CEA-Grenoble, DRFMC/SP2M, 17 rue des Martyrs, 38054 Grenoble Cedex 9, FRANCE

ABSTRACT

AlGaN is an important material for the realization of nitride heterostructures, involved in most device designs. We have studied the growth of this alloy using low pressure MOVPE (76 Torr), and using triethyl-gallium (TEGa), trimethyl-aluminum (TMAl) and ammonia (NH$_3$) as precursors. First the solid -gas aluminum segregation was studied in order to calibrate the incorporation of Al in the solid phase. We found that aluminum is more readily incorporated than gallium in the solid, leading to an apparent Al segregation coefficient greater than unity. A simple kinetic model is used to fit the experimental data. Scanning electron microscopy has been used to investigate the morphology of the samples through the whole range of Al content (x = 0 to 1), and we observe a clear evolution of the surface features versus aluminum concentration : at low Al contents, small (below 1 µm) hexagonal holes are observed while at high Al, acicular features are observed, with a sudden transition between those morphologies around x = 0.5. Transmission electron microscopy was used to analyze the crystalline structure of these samples. Finally, the samples were studied by low temperature (2K) reflectivity and Raman spectroscopy. We report the evolution of the optical quality of samples (x < 0.4) versus Al content, as evaluated from the broadening of the observed excitonic transitions in the 2K reflectivity.

INTRODUCTION

Most device structures require confining layers, either for photons, like in lasers where cladding layers are deposited to ensure optical guiding, or to confine carriers (electrons and/or holes) in optical devices or in transistors. In the nitride system, the AlGaN alloy forms a solid solution over the entire range of composition and has a bandgap scaling between 3.4 and 6.2 eV. Moreover, the lattice mismatch between AlN and GaN is "only" about 2%. As a result, AlGaN has been successfully employed in many devices like lasers[1] or transistors[2].
From the few available data in the literature, it appears that the crystal quality of AlGaN rapidly degrades with increasing Al content. Fortunately, due to the large bandgap of AlN, confinement can be obtained even for low Al content. In this paper, we have studied the MOVPE growth of the alloy in the whole range of composition, and we demonstrate that at high Al contents, the layers are highly defective, they include grains oriented along the ($10\bar{1}1$) axis which have grown from hexagonal grain facets present in the buffer layer. Our interpretation is that such misoriented planes are grown at high Al contents due to the low surface mobility of aluminum, while C-faces would grow preferentially when there are mobile adatoms on the surface (Ga rich alloys).

EXPERIMENTAL

The AlGaN layers were grown by low pressure MOVPE (76 Torr) using triethylgallium, trimethylaluminum and ammonia as precursors. The layers were deposited onto (0001)

Mat. Res. Soc. Symp. Proc. Vol. 468 © 1997 Materials Research Society

sapphire substrate, on top of 500Å AlN buffer layers deposited at 800°C. For all the AlGaN samples, the growth temperature was 980°C, equal to the optimum growth temperature for GaN epilayers, in our system. The V/III molar ratio was kept constant to 10,000 by keeping the total group III molar flow constant, changing only the Al/Ga ratio. The Al composition was determined by energy dispersive analysis of x-ray (EDAX), making the proper corrections for atomic number, absorption and fluorescence. Cross-sections for TEM were prepared using the standard procedures : mechanical polishing and Argon ion milling. TEM observations were realized on a JEOL4000EX electron microscope. Raman spectra were excited with the 488 nm line of an Ar$^+$ ion laser, for which the layers and the substrate are both transparent, in backscattering geometry along or perpendicular to the c axis. All the experiments were performed at room temperature using a Dilor spectrometer, which allows conventional as well as micro measurements. Spectra from the edge of the layers were taken in the micro-Raman configuration, when the incident light is focused onto a 1 micrometer diameter spot; the directions of incident and scattered light are then both normal to the z-axis and parallel to an x' axis in the basal plane of the crystal. Reflectance experiments were carried out at 2K, in pumped liquid helium.

RESULTS AND DISCUSSION

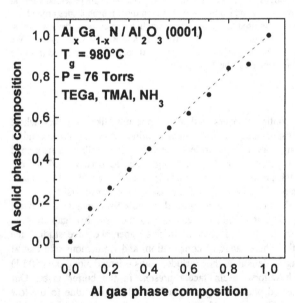

Figure 1: Solid phase composition versus gas phase composition. The circles represent the experimental data while the dashed line corresponds to the modeling.

First, we studied the incorporation of aluminum in the solid AlGaN during the MOVPE growth. Few elements concerning this point are available in the literature: Koide[3] demonstrated that the AlGaN alloy could be grown at atmospheric pressure over the entire range of composition, provided that care is taken to avoid premature reactions between the precursors. However, their data shows a noticeable dispersion which prevent detailed analysis of the incorporation mechanism. Wickenden et al.[4] reported data obtained from low pressure growth, free from scattering due to premature reactions, but their data is limited to Al contents below 0.4 . In figure 1, we report our own low pressure MOVPE results concerning the Al content in the solid phase versus the gas phase composition. We modeled the experimental data in the following way: there are experimental evidences supporting the fact that the growth is limited by mass-transport of

24

group III elements through the gas phase[5], so we assume that the solid composition will be determined by the relative Al and Ga fluxes reaching the interface by diffusion. The details concerning this model will be reported elsewhere. It must be noted that there is a significant bowing in the data of figure 1. In our model, this behavior is related to the value of the ratio of the TEGa to TMAl diffusion coefficient, which is equal to 0.82 in our case. Next, we investigated the structural quality of our samples. Figure 2 displays scanning electron microscope images of the sample surfaces, for Al contents from x=0.16 to x=0.85.

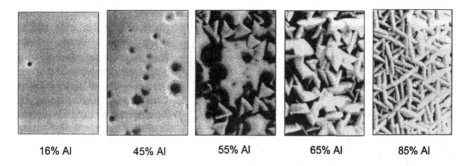

| 16% Al | 45% Al | 55% Al | 65% Al | 85% Al |

Figure 2: SEM images of Al$_x$Ga$_{1-x}$N alloys, for x ranging from 0.16 to 0.85. Each picture is 2μm wide.

We observe a clear evolution of the surface features versus aluminum concentration : at low Al contents, small (below 1 μm) hexagonal holes are observed while at high Al, acicular features are observed, with a sudden transition between those morphologies around x = 0.5. The reasons of these modifications of the morphologies will be given below, from the TEM results. Four Al$_x$Ga$_{1-x}$N samples with different Al compositions (x = 0.16, x = 0.45, x = 0.55, x = 0.85) have been observed using Transmission Electron Microscopy. The two last samples greatly differ from the two first ones. The samples with a low Al concentration looks like traditional GaN layers. They are monocrystalline with a Ga polarity (ref.[6,7,8]) and contain a high density of dislocations (about 10^{10} / cm^2) and many nanopipes (ref.[9]). The samples with an Al content higher than 0.55 contain two kinds of textures : the planes parallel to the initial (0001) sapphire surface plane can be the (0001) plane or the pyramidal {10$\bar{1}$1} planes. These two textures clearly appear in the diffraction pattern of Fig 3b. The angular width of these textures is about ± 4°. Due to the symmetry of the wurtzite material, the {10$\bar{1}$1} texture has six variants. These six variants can be characterized by the orientation of the c and a = [2$\bar{1}\bar{1}$0] axis with respect to the texture of the (0001) AlN buffer layer which is defined by : AlN (0001) [2$\bar{1}\bar{1}$0] // Al$_2$O$_3$ (0001) [10$\bar{1}$0]. The six different orientations of the grains are :

AlGaN (0001) [2$\bar{1}\bar{1}$0] // AlN (01$\bar{1}$1) [2$\bar{1}\bar{1}$0]

AlGaN (0001) [2$\bar{1}\bar{1}$0] // AlN (0$\bar{1}$11) [2$\bar{1}\bar{1}$0]

AlGaN (0001) [2$\bar{1}\bar{1}$0] // AlN (10$\bar{1}$1) [1$\bar{2}$10]

AlGaN (0001) [2$\bar{1}\bar{1}$0] // AlN ($\bar{1}$011) [$\bar{1}$2$\bar{1}$0]

AlGaN (0001) [2$\bar{1}\bar{1}$0] // AlN (1$\bar{1}$01) [$\bar{1}$1$\bar{2}$0]

AlGaN (0001) [2$\bar{1}\bar{1}$0] // AlN ($\bar{1}$101) [1$\bar{1}$20]

Two of these variants can be clearly seen on a HREM picture (Fig 3a) taken along the a=[$2\bar{1}\bar{1}0$] direction. The four other variants cannot be resolved on this HREM image, but are present due to the symmetry of the crystal. It appears that at high Al concentration, the {$10\bar{1}1$} is favored. Such a texture has already been reported in GaN layers grown on (0001) sapphire[10]. But in this case, the {$10\bar{1}1$} textured grains were very small and appeared only in the buffer layer. In our case, HREM images reveal that the {$10\bar{1}1$} textures start growing at once on the AlN buffer layer (not on the sapphire surface) and that they propagate throughout the layer. These grains are responsible of the rough surfaces observed by SEM (Fig 2) : {$10\bar{1}1$}

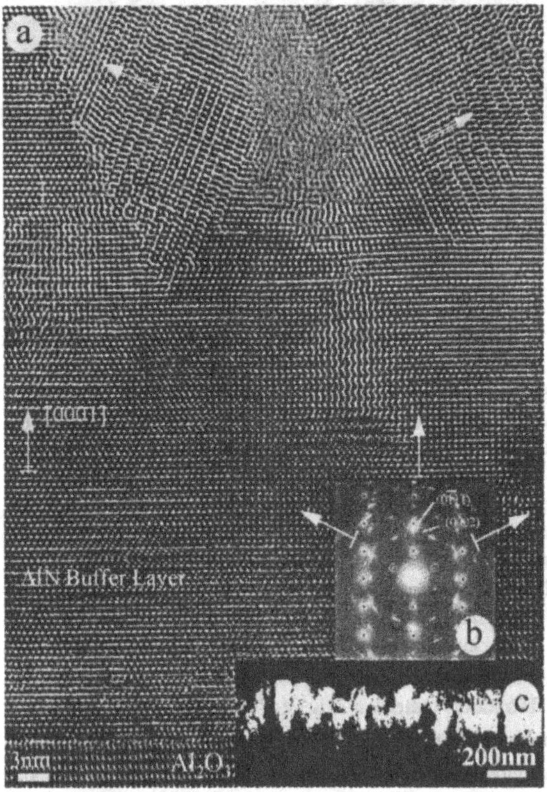

Figure 3: Transmission Electron Microscopy of an $Al_{0.85}Ga_{0.15}N$ sample: a) High resolution image, b) Electron diffraction and c) low resolution image showing the grains shapes.

grains are higher than (0001) grains and forms grains elongated along the {$2\bar{1}\bar{1}0$} directions. For x = 0.85, the average width of these grains is 75nm; their length cannot be determined on the TEM cross-section. Since the buffer texture is the same in all the samples, we have to correlate the fact that the {$10\bar{1}1$} textures appear in the epilayers with the increasing aluminum content. Our hypothesis is that the low surface mobility at this particular growth temperature (which is optimum for GaN, but apparently not for AlGaN), prevent the re-arrangement of adatoms, which would favor the growth of C-faces. In order to test this, we have grown AlN epilayers at increasing temperatures, and we report the x-ray diffraction data concerning these layers in figure 4. It can be clearly seen in this figure that the crystal quality of AlN increases with increasing growth temperature. This data strongly support our hypothesis concerning AlGaN, and we may anticipate that the growth temperature has to be raised when the Al content increases. Figure 5-a shows Raman spectra recorded in the x'(y'y')x' geometry, for various aluminum molar fraction x in the solid solutions. According to the selection rules only the A1(TO) and E2 phonons of wurtzite structure are allowed in this case. Their frequency variation can be easily followed from the well-known GaN phonon frequencies [11] up to x = 0.55. For higher x values, the corresponding peaks are wider and their intensity becomes lower, then rendering the evolution of the modes not so easy to follow. The decreasing intensity of spectra may be correlated the drastic change of microscopic structure of the layers, revealed by scanning electron microscopy for x > 0.5. The A1(LO), E1(LO), E1(TO) phonons were identified in the spectra using other scattering geometries. Measured phonon frequencies are plotted in Figure 5-b, as a function of x. The values for x = 1 were taken from the study of MacNeil et al. on AlN single crystals [12]. A one mode behavior is clearly observed for the Raman active polar modes of A1 and E1 symmetry. For the (non polar) E2 mode, the situation is more complicated. The peak, clearly evidenced for x = 0.85 (see Fig. 5-a) and located near 660 cm^{-1} close to the E2 mode frequency in pure AlN, is tentatively assigned to the E2 symmetry. This mode can be hardly followed for lower aluminum molar fraction down to x = 0.65.

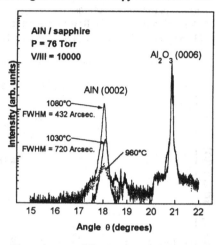

Figure 4: x-ray diffraction linewidths for AlN epilayers grown onto (0001) sapphire at different growth temperatures.

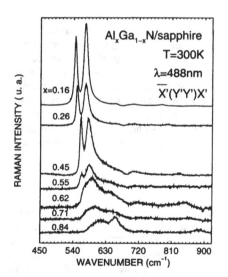

Figure 5a: Room temperature Raman spectra from Al$_x$Ga$_{1-x}$N layers in the x'(y'y')x' scattering configuration.

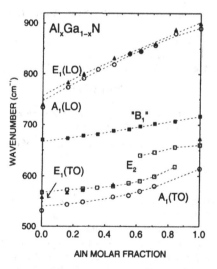

Figure 5b: Variation of measured phonon frequencies of Al$_x$Ga$_{1-x}$N layers as a function of the aluminum molar fraction.

On the other hand, a second branch originating from the E2 mode of GaN is also observed at lower frequencies for x < 0.72. Then the non polar mode seems to display a two mode behavior in the solid solutions.

A dip is also found in the Raman spectra from the alloys: its frequency varies linearly between 673 and 704 cm^{-1}, when x increases from 0.16 to 0.85, and may be correlated with the location of the B1 "silent" vibrational mode previously calculated for wurtzite GaN and AlN pure crystals [13,14].

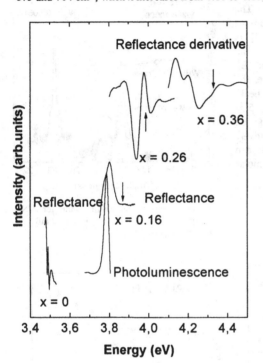

The optical properties of the AlGaN layers are reported in Figure 6. It has not been possible to measure the optical properties for alloys with aluminum content above x=0.4. For samples with lower Al content, the energy position of the transitions in the reflectance and reflectance derivative spectra were obtained from a fit using a local approximation of the dielectric constant. We observe a rapid degradation of the optical quality when increasing the Al content. An important Stokes-shift is already present between photoluminescence and reflectivity at x = 0.16. Due to limitation of our laser (325 nm), we were not able to pump the photoluminescence of samples with higher Al contents. Structures could only be detected in the derivative of the reflectance spectra for x above 0.16. Above x = 0.36, we have not been able to detect any structure in the reflectance spectra or their derivatives, due to the response of our grating.

Figure 6: 2K optical properties (reflectance, reflectance derivative, photoluminescence) of AlGaN

CONCLUSION

We have successfully grown AlGaN alloy on the whole range of composition, and modeled the Al incorporation from the gas phase to the solid, taking diffusion effects into account. The structural quality of the alloy samples was investigated by SEM and TEM, and we found that misoriented grains are grown at high Al contents. These misorientations originate in the buffer layer and are not "flattened" at high Al content, because Al has a low surface diffusion at the growth temperature of 980°C. We suggest that the growth temperature should be raised accordingly with increasing aluminum fraction in the samples. Raman spectroscopy, photoluminescence and reflectance experiments have been performed and demonstrate that the crystalline quality degrades continuously with increasing Al content.

ACKNOWLEDGMENTS

This work has been supported by the DRET/DGA and by THOMSON-CSF-LCR.

REFERENCES

[1] S.Nakamura, M.Senoh, S.Nagahama, N.Iwasa, T.Yamada, T.Matsushita, Y.Sugimoto and H.Kiyoku Appl.Phys.Lett.70,7 (1997) 868

[2] M.A.Khan, Q.Chen, C.J.Sun, J.W.Yang, M.Blasingame, M.S.Shur and H.Park Appl.Phys.Lett.68,4 (1996) 514

[3] Y.Koide, H.Itoh, N.Sawaki, I.Akasaki and M.Hashimoto J.Electrochem.Soc. 133,9 (1986) 1956

[4] D.Wickenden, C.Bargeron, W.Bryden, J.Miragliotta and T.Kistenmacher Appl.Phys.Lett.65,16 (1994) 2024

[5] M.A.Khan, R.Skogman, R.Shulze and M.Gershenzon Appl.Phys.Lett.43,5 (1983) 492

[6] J.L.Rouvière, M.Arlery, R.Niebuhr, K.H.Bachem and O.Briot MRS Internet J.Nitride Semicond.Res.1 (1996) 33

[7] B.Daudin, J.L.Rouvière and M.Arlery Appl.Phys.Lett.69,17 (1996) 2480

[8] F.A.Ponce, D.P.Bour, W.T.Young, M.Saunders and J.W.Steeds Appl.Phys.Lett. 69,3 (1996) 337

[9] W Qian, M Skowronski, K Doverspike, LB Rowland, DK Gaskill Journal of Crystal Growth 151 (1995) 396

[10] S.Christiansen, M.Albrecht, W.Dorsch, H.P.Strunk, C.Zanotti-Fregonara, G.Salviati, A.Pelzmann, M.Mayer, M. Kamp, K.J.Ebeling. MRS Internet J. Nitride Semicond. Res. 1, (1996) 19

[11] F. Demangeot, J. Frandon, M.A. Renucci, B. Beaumont and P. Gibart, Proc. 9th SIMC Conf.(IEEE), Toulouse (1996)

[12] L.E, MacNeil, M. Grimsditch and R. H. French, J. Am. Ceramic Soc. 76, 1132 (1993)

[13] T. Azuhata, T. Matsunaga, K. Shimada, K. Yoshida, T. Sota, K. Suzuki and S. Nakamura, Physica B 219 - 220, 493 (1996)

[14] I. Gorzcyca, N.E. Christensen, E. L. Peltzer y Blanca and C.O. Rodriguez, Phys. Rev. B 51, 11936 (1995)

GROWTH OF TERNARY SILICON CARBON NITRIDE AS A NEW WIDE BAND GAP MATERIAL

L. C. Chen[1], C. K. Chen[1], D. M. Bhusari[2], K. H. Chen[1,2], S. L. Wei[1], Y. F. Chen[3], Y. C. Jong[3], D. Y. Lin[4], C. F. Li[4], and Y. S. Huang[4]

[1] Center for Condensed Matter Sciences, National Taiwan University, Taipei, Taiwan
[2] Institute of Atomic and Molecular Sciences, Academia Sinica, Taipei, Taiwan
[3] Physics Department, National Taiwan University, Taipei, Taiwan
[4] Department of Electronic Engineering, National Taiwan Institute of Technology, Taipei, Taiwan

ABSTRACT

Growth of pure crystalline carbon nitride (c-CN) with crystal sizes large enough to enable measurement of its properties has not been achieved so far. We report here that incorporation of silicon in the growth of CN can promote formation of large, well faceted crystallites. Crystalline thin films of SiCN have been grown by microwave plasma-enhanced chemical vapor deposition using CH_4, N_2, and SiH_4 gases. Auger electron spectroscopy, scanning electron microscopies, and X-ray diffraction spectroscopy have been employed to characterize the composition, the morphology and the structure of the films. The new crystalline ternary compound $(C; Si)_x N_y$ exhibits hexagonal structure and consists of a network wherein the Si and C are believed to be substitutional elements. While the N content of the compound is about 35%, the extent of Si substitution varies from crystal to crystal. In some crystals, the Si content can be as low as 10%. Optical properties of the SiCN compounds have been studied by photoluminescence (PL) and piezoreflectance (PzR) spectroscopies. From the PzR measurement, we determine the band gap of the new crystals to be around 3.8 eV at room temperature. From the PL measurement, it is found that the compounds have a strong subband-gap emission centered around 2.8 eV at room temperature, which can be attributed to the effect of defects containing in the crystals.

INTRODUCTION

The search of blue laser diodes has been continued for about three decades. To achieve blue light emitting, a large direct band-gap (> 2.5 eV) is necessary. Research and development works on materials with such property have been flourished in three groups of materials: SiC, II-VI compounds, and III-V nitrides (AlN, GaN and InN). SiC has produced blue light-emitting diodes (LEDs) with rather low efficiency, while II-VI compounds still suffer from long term operation capability. On the other hand, GaN shows great progress for use in blue LEDs. Further, ternary III-V semiconductor compound such as InGaN is useful because its band-gap can be varied from 1.95 to 3.4 eV depending on the indium mole fraction. Despite the superior performance of III-V nitrides for applications in LEDs, challenges remain for applications in laser diodes. Hence it would be tantamount to look for opportunities in other new wide band-gap materials. In the present paper we report the growth of a novel IV-V nitride (specifically, ternary SiCN) and show that the optical properties of this compound are promising.

Early unpublished patent disclosure by Sung and theoretical studies by Liu and Cohen suggest that a tetrahedral compound of carbon and nitrogen, preferably in the β-C_3N_4 structure that is isomorphic to β-Si_3N_4, might have bulk modulus superior to that of diamond [1]. C_3N_4 in different structure forms other than β-C_3N_4 [2-4] as well as structure with less nitrogen-to-carbon ratio such as C_4N_3 [5] are also proposed to have high bulk modulus. The prediction of such a revolutionary material can be intuitively understood since the C-N bond exhibits low ionicity, and both carbon and nitrogen have small atomic size as well as sufficiently high coordination number in the covalently bonded crystalline C-N solid (c-CN). In addition to high bulk modulus, c-CN is also expected to possess several other properties such as wide band-gap, high thermal conductivity, high strength, high decomposition temperature and excellent resistance to corrosion and wear. With all these predicted properties, c-CN once realized is useful for high performance opto-electronic as well as engineering applications.

Following such high expectation, synthesis of the hypothetical c-CN has attracted a great deal of interest recently [6-12]. Besides academic interest in proving the theoretical prediction, the growth of high quality c-CN film is also essential before its many promising applications can be realized. However, until now carbon nitride films reported in the literature are mainly amorphous or polymeric with the nitrogen content generally below 40%. Only a few laboratories have reported synthesizing very tiny crystals with a size typically between nanometer and submicron. Further, these crystals are embedded in an amorphous matrix, wherein volume of the crystalline phase accounts for only about 5% of the total volume of the deposited film [9]. Under such circumstance, it is already difficult even for mere identification of the c-CN. Consequently, most of the predicted properties of c-CN such as exceptional hardness and wide band-gap have not yet been verified. It is thus of significant importance to grow large sized crystals or a continuous polycrystalline film to enable property measurements. Some progress towards this direction started from the synthesis of large (several tens of microns), well-faceted Si-containing c-CN crystals by microwave plasma enhanced chemical vapor deposition (MWCVD) [13-15]. In our previous works, silicon was incorporated to the growth of c-CN primarily by chemical reactions between the Si substrate and gasseous species in the plasma at high temperature. In the present work, silane is added as a source gas for MWCVD. In particular, the effects of the substrate temperature and the silane content of the gas mixtures on the film growth are investigated. Some preliminary studies on the optical properties of the resultant films are also presented in this paper.

EXPERIMENTAL

An AsTeX 5kW microwave reactor was used to synthesize carbon nitride films. Most of the films were deposited on Si (100) wafers after standard HF acid deoxidization and distilled-water cleaning. In some of the deposition runs non-Si substrates such as sapphire and AlN were also used. Separate substrate heating can be controlled by a resistive heater. A thermal couple placed underneath the Mo substrate holder was used to monitor the substrate holder temperature. In addition, a two-color pyrometer was also used to monitor the substrate temperature. During CVD process, the top surface of the substrate is usually at a higher temperature than the substrate holder. The "substrate temperature" used hereafter is thus referred to that of the pyrometer reading. A mixture of semiconductor grade CH_4, N_2, H_2 and SiH_4 (5% SiH_4 in N_2 dilution) gases in various proportions was used as source gas. Typical flow rates were 80 sccm for H_2, 20 sccm for CH_4, 80 sccm for N_2 while the flow rate for SiH_4 was varied between 0-10 sccm. The chamber pressure was maintained at 30 Torr and the microwave power was kept at 1.5 kW. The range of substrate temperature was 400-1200 °C.

Film morphology was examined by a conventional scanning electron microscopy (Hitachi S - 800 SEM). Film crystallinity was determined by a Rigaku D/MAX-3C X-ray diffractometer using Cu K_α radiation at 1.54 Å. Two theta scan was taken from 12 to 80 degree while theta was fixed at 5 degree. Chemical composition of the individual crystals was determined from the scanning Auger depth profile data recorded on a Perkin Elmer Scanning Auger Nanoprobe system (SAN 760). An electron beam of energy 5 keV and Ar ion beam of energy 4 kV were employed for depth profile measurements.

Photoluminescence (PL) spectra were obtained at temperatures from 10 K to 300 K using a He-Cd laser operating at 325 nm. The direct band gap of the new compound was determined by piezoreflectance (PzR) spectroscopy at temperatures from 15 K to 500 K. A detailed description of the PzR apparatus has been reported elsewhere [16].

RESULTS AND DISCUSSIONS

Both the morphology and microstructure of the films are a strong function of the substrate temperature as well as the SiH_4 content in the gas mixture. Up to a deposition temperature of about 550 °C and a SiH_4 flow rate of about 4 sccm only broad bumps were observed in x-ray diffraction (XRD) spectra, indicating that the coatings were amorphous. Above 800°C and 8 sccm SiH_4, the coatings were completely crystalline. It should be emphasized that some amount of SiH_4 is needed for the formation of crystallites. For instance, at a substrate temperature of 800°C, no crystallite can be observed without SiH_4 addition while deposition of micron-sized, well-faceted crystallites were observed when SiH_4 flow rate was greater than 6 sccm. Further, the higher the SiH_4 content, the lower the deposition temperature was needed for crystallite formation. When 10 sccm of SiH_4 was used, aggregation of fine grains with a size of about a few nm started to form once the substrate temperature was kept above 600 °C. Figures 1 and 2 show the SEM micrographs of a nearly continuous film deposited at 800°C and with 10 sccm of SiH_4 viewed normal to and 45° to the wafer surface, respectively. The film appears transparent and consists of randomly oriented rod-shaped crystals. Most of the crystals exhibit a hexagonal cross section. Typical rod length to width ratio is about 5-10, suggesting that growth rate for these crystals was apparently larger along their c-axis than along their a-axis.

Figure 1   Figure 2

Typical SEM image of the SiCN film viewed normal to (Fig.1) and 45° to (Fig. 2) the wafer surface.

It may be mentioned here that the incorporation of silicon is critical in the growth of carbon nitride based material. The role of Si is believed to catalyze the chemical reactions in the plasma that eventually lead to the formation of crystalline C-N network. Furthermore, Si may also stablize the resulting crystallite. In our previous works, silicon was incorporated primarily by chemical reactions between the Si substrate and gasseous species in the plasma at temperatures higher than 1000 °C. With SiH_4 addition, the deposition temperature for crystallite formation can be reduced by a few hundreds of degree C.

To determine the chemical composition of individual crystals, scanning Auger spectroscopy with depth profiling and with a probe size smaller than the crystal dimension is used. A typical Auger spectrum taken on the as-deposited crystal surface indicated the presence of Si, C, N and O (Figure 3, top). The corresponding Auger spectrum taken from the same spot after sputtering at 4 kV for 6 minutes is also shown in Figure 3 (bottom). The atomic concentration of each element with sensitivity correction is indicated in the Auger spectrum. The senstivity factors for Si, C, N and O were taken to be 0.28, 0.14, 0.23 and 0.4, respectively. While nitrogen remains unchanged, significant reduction of carbon was observed after sputtering. Depth profile studies indicate that the carbon content reaches a constant level after the initial reduction upon sputtering, suggesting that part of the C signal was due to surface adsorption. For the specific crystal shown in Figures 3, the composition is $Si_{20.6}C_{40.5}N_{34.4}O_{4.5}$ and $Si_{37}C_{22.5}N_{35.9}O_{4.6}$, respectively, before and after sputtering. We believe that the oxygen is present predominantly as impurity. It is observed that the chemical composition varies considerably even for the different crystals on the same substrate. While average [N] can be seen to be about 35 at.%, [Si] and [C] can vary over a range of a few tens of at.%. In some crystals, the Si content can be as low as 10 at.%. Remarkably, although the chemical composition of the crystals varies widely, their morphology remains unchanged. We propose the existence of a rich phase-space for the Si-C-N system, presumably with a similar crystalline structure wherein the Si and C are substitutional elements occupying similar sites in the network. The new SiCN phase studied in the present report can be conveniently expressed as $(C; Si)_2N$.

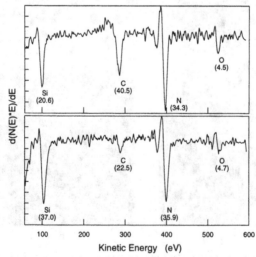

Figure 3   Typical Auger spectrum taken from the as-received SiCN crystal (top) and the same area after sputtering for 6 min at 4 kV (bottom).

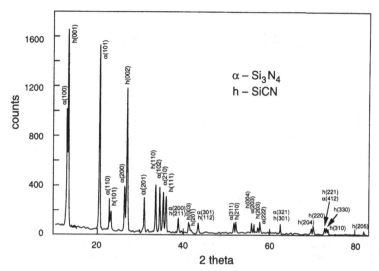

Figure 4   A typical XRD spectrum of the film. Contributions from the α-Si$_3$N$_4$ and hexagonal-SiCN are indicated.

Figure 4 shows a typical XRD spectrum of the crystalline film such as the one shown in Figures 1 and 2. Some of the diffraction peaks can be indexed to the α-Si$_3$N$_4$ phase whereas most of the other peaks are indexed tentatively to the new hexagonal-SiCN phase. A couple of peaks can be fitted simultaneously to both phases are also indicated. It is noted that the relative intensities of the α-Si$_3$N$_4$ peaks are in good agreement with those reported in the X-ray powder file, suggesting a randomly-oriented polycrystalline film which is consistent with the SEM investigation. The lattice parameters, derived from the XRD spectrum and suggested by TEM diffraction patterns, of the new h-SiCN phase were determined te be a = 5.4 Å and c = 6.7 Å. For films deposited with a SiH$_4$ flow rate greater than 10 sccm and at substrate temperature higher than 950 °C, XRD spectra were dominated by α-Si$_3$N$_4$.

Typical PzR spectra of the SiCN film measured at several temperatures between 20-500 K is shown in Figure 5. The values of the band gap and the broadening parameter of direct band-to-band transitions can be obtained by a detailed experimental lineshape fit to Cohen's expression [17]. The obtained values of band gap are indicated by arrows in the figure. At 300 K, a band gap value of 3.81 eV was determined. As the measurement temperature decreases, the band gap value only increases slightly, quite typical of high band gap materials such as diamond [18]. Detailed analyses of the temperature dependence are to be published elsewhere [19].

The PL spectra excited by a HeCd laser working at 325 nm have been performed at temperatures from 10 K to 300 K. Typical PL spectra for four temperatures are displayed in Figure 6. An intense broad emission centered around 2.8 eV was observed. Notice that the intensity of this emission increases initially then decreases as the measurement temperature increases. Because the photon energy of the emission is much smaller than the band gap value 3.81 eV, it can be attributed to the transition originated from deep defects. However, the origin of the defects and the temperature dependence of the emission intensity still needs to be clarified.

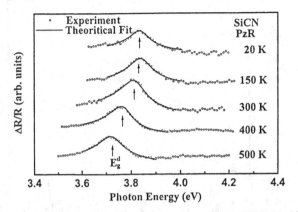

Figure 5　The experimental PzR spectra (dotted curves) of SiCN at 20, 150, 300, 400 and 500 K, respectively. The solid curves are least-squares fits to the first derivative of a Lorentzian profile, which yields the direct band-to-band transition energy indicated by the arrow.

Figure 6　The experimental PL spectra of SiCN at 10, 100, 200 and 300 K, respectively.

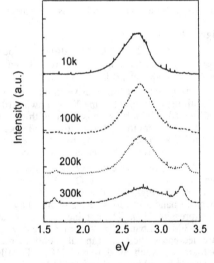

SUMMARY

In conclusion, we report here the process of a novel SiCN compound containing about 35 at.% of N and exhibiting hexagonal structure wherein the Si and C are substitutional elements in the network. Si has been observed to play an important catalytic type of role in promoting the growth of CN based crystals. We also demonstrated that this new compound has a band gap of about 3.8 eV and a deep defect emission around 2.8 eV.

ACKNOWLEDGMENT

The authors would like to thank Dr T. J. Chuang for helpful discussions. The project is supported by the fund from the NSC, Taiwan under contract No. NSC-86-2113-M-002-036.

REFERENCES

1. A. Y. Liu and M. L. Cohen, Science **245**, 841 (1989).
2. A. Y. Liu and R. M. Wentzcovitch, Phys. Rev. B **50**, 10362 (1994).
3. J. Ortega and O. F. Sankey, Phys. Rev. B **51**, 2624 (1995).
4. Y. Guo and W. A. Goddard III, Chem. Phys. Lett. **237**, 72 (1995).
5. J. V. Badding and D. C. Nesting, Chem. Mater. 8, 535 (1996).
6. C. Niu, Y. Z. Lu, and C. M. Lieber, Science **261**, 334 (1993).
7. D. Li, Y. W. Chung, M. S. Wong, and W. D. Sproul, J. Appl. Phys. **74**, 219 (1993).
8. S. Kumar and T. L. Tansley, Solid State Commun. **88**, 803 (1993).
9. K. M. Yu, M. L. Cohen, E. E. Haller, W. L. Hansen, A. Y. Liu, and I. C. Wu, Phys. Rev. B **49**, 5034 (1994).
10 D. Marton, K. J. Boyd, A. H. Al-Bayati, S. S. Todorov, and J. W. Rabalais, Phys. Rev. Lett. **73**, 118 (1994).
11. Z. M. Ren, Y. C. Du, Z. F. Ying, Y. X. Qiu, X. X. Xiong, J. D. Wu, and F. M. Li, Appl. Phys. Lett. **65**, 1361 (1994).
12. J. P. Riviere, D. Texier, J. Delafond, M. Jaouen, E. L. Mathe, and J. Chaumond, Mater. Lett. **22**, 115 (1995).
13. L. C. Chen, C. Y. Yang, D. M. Bhusari, K. H. Chen, M. C. Lin, J. C. Lin, and T. J. Chuang, Diamond and Related Materials 5, 514 (1996).
14. D. M. Bhusari, C, K, Chen, K. H. Chen, T. J. Chuang, L. C. Chen, and M. C. Lin, J. Mat. Res. 12, in press (1997).
15. L. C. Chen, D. M. Bhusari, C. Y. Yang, K. H. Chen, T. J. Chuang, M. C. Lin, C. K. Chen, and Y. F. Huang, Thin Solid Film, in press (1997).
16. C. F. Li, D. Y. Lin, Y. S. Huang, Y. F. Chen, and K. K. Tiong, J. Appl. Phys. **81**, 400 (1997).
17. M. L. Cohen, Phys. Scr. **T1**, 5 (1982).
18. Y. P. Varshni, Physica 34, 149 (1967).
19. D. Y. Lin, C. F. Li, Y. S. Huang, Y. C. Jong, Y. F. Chen, L. C. Chen, C. K. Chen, K. H. Chen, and D. M. Bhusari (submitted to Phys. Rev. B).

NEW PRECURSOR ROUTES TO NANOCRYSTALLINE CUBIC/HEXAGONAL GALLIUM NITRIDE, GaN

R. L. WELLS*, J. F. JANIK‡*, W. L. GLADFELTER#, J. L. COFFER†, M. A. JOHNSON†, and B. D. STEFFEY†

* Department of Chemistry, Duke University, Durham, NC 27708-0346
‡ on leave from the University of Mining and Metallurgy, Krakow, Poland
# Department of Chemistry, University of Minnesota, Minneapolis, MN 55455
† Department of Chemistry, Texas Christian University, Ft. Worth, TX 76129

ABSTRACT

Two precursor routes culminating in bulk nanocrystalline gallium nitride materials are reported, with emphasis on the materials' XRD/crystalline features and photoluminescence (PL). First, the new polymeric gallium imide, $\{Ga(NH)_{3/2}\}_n$, can be converted to nanocrystalline, cubic/hexagonal GaN ranging in color from yellow to light gray. Second, a new route to gallazane, $[H_2GaNH_2]_x$, from the combination of $LiGaH_4$ and NH_4X (X = Cl, Br) in Et_2O is shown to result in a material that slowly converts to a polymeric solid via H_2 and NH_3 elimination-condensation pathways. Both the gallazane and the polymeric solid are pyrolyzed to dark gray nanocrystalline, phase-inhomogeneous GaN as above. Specific variations in the pyrolysis conditions enable some control over the particle nanosize and a degree of crystalline phase-inhomogeneity of the materials. These nanophase GaN materials have also been characterized by room temperature photoluminescence (PL) measurements. In general, the observed emission spectra are strongly dependent on pyrolysis temperature and typically exhibit weak defect yellow-green emission. While the as-prepared GaN does not exhibit band-edge PL, a brief hydrofluoric acid etch yields nanophase GaN exhibiting an intense blue-emitting PL spectrum with an emission maximum near 420 nm.

INTRODUCTION

The broad direct bandgap semiconductor gallium nitride, GaN, has two common crystalline polymorphs - the thermodynamically stable hexagonal or wurzite form and the metastable cubic or zinc blende form [1]. Most of the vapor deposition processes as well as bulk precursor routes result in the hexagonal GaN, a variety compatible with the Group 13 nitrides, AlN or InN, with which it can form solid solutions of intermediate semiconductor/electronic properties.

In recent years, however, there has been an increasing number of reports demonstrating the formation of cubic GaN from precursor routes under standard laboratory conditions. Gladfelter and coworkers [2] reported that the pyrolysis of the cyclotrigallazane, $[H_2GaNH_2]_3$, (obtained from the reaction between $H_3Ga\cdot NMe_3$ and NH_3 [3]), resulted in the cubic/hexagonal variety of GaN best described as a mixture of cubic and hexagonal close-packed layers. Wells and Janik [4] found that the conversion of the new gallium imide precursor, $\{Ga(NH)_{3/2}\}_n$, yielded similar cubic/hexagonal GaN. In a related precursor chemistry, Gonsalves et al. [5] described the pyrolysis of $[Ga(NMe_2)_3]_2$ under NH_3 which led to GaN labeled as cubic with stacking faults. However, based on model calculations, most of these results can be interpreted as having varying amounts of order in the stacking direction [2b, c]. For the purpose of this report, we will use interchangeably "cubic/hexagonal" or "phase-inhomogeneous" to denote such GaN phases.

In regard to the luminescent spectral behavior of the GaN materials, generally there are four types of possible scenarios anticipated: first - intrinsic bandgap photoluminescence, emitting in the blue region with a maximum near 410 nm, second - broad defect photoluminescence, known to emit in the yellow region [6], third - non-radiative states, lowering the observed quantum yield of band edge and/or defect photoluminescence, and fourth - any possible combination of the above three scenarios.

Mat. Res. Soc. Symp. Proc. Vol. 468 © 1997 Materials Research Society

EXPERIMENTAL

Synthetic Procedures

The preparation of the polymeric gallium imide, $\{Ga(NH)_{3/2}\}_n$, was accomplished according to the published method [4]. Before use, the imide was conditioned under ammonia for one day.

The gallazane, $[H_2GaNH_2]_x$, was synthesized from the new reaction system, $LiGaH_4/NH_4X$ (X = Cl, Br) in Et_2O, to be described elsewhere [7]. The optimal preparation used the *in situ* formed $LiGaH_4$ (not isolated from the ether-liquor) in combination with solid NH_4Cl at O °C, wherein the ether insoluble LiCl could be conveniently separated from the initially soluble gallazane. If, on the other hand, one isolated $LiGaH_4$ by evacuation, the compound partially decomposed forming an insoluble gray slurry upon redissolution in fresh Et_2O and contaminating the products. The isolated white gallazane decomposed at room temperature with the evolution of H_2 and NH_3, and converted slowly to a gray polymeric solid, elimination-condensation product, especially upon evacuation. Also, a several day long stirring of the $LiGaH_4/NH_4Br$ system resulted in the spontaneous precipitation of yet another polymeric solid similar to the one just described. The as-prepared gallazane, as well as the polymeric solids obtained from it after definite reaction and evacuation times, were immediately used in all subsequent pyrolyses. Due to the thermal instability of both the $LiGaH_4$ and $[H_2GaNH_2]_x$, the preparation details (*in situ* or isolated materials) and handling times of the by-products as well as evacuation appeared to influence the properties of the final GaN materials.

For pyrolysis under an ammonia flow, liquid NH_3 over sodium was slowly evaporated into the furnace-heated pyrolysis tube that contained a precursor in a quartz boat, and left the system through a bubbler. A decomposition of the gallium imide was also performed in boiling N,N,N',N'-tetramethyl-1,6-hexanediamine (bp 210 °C). In this experiment, an amine suspension of the imide was placed in a flask equipped with a condenser and was refluxed under argon for 48 hours. Upon completion, the supernatant was removed and the solid evacuated overnight yielding a yellow grayish product.

Characterization Methods

XRD data were collected using oil coated samples on a Phillips XRD 3000 diffractometer. Elemental analyses were provided by E + R Microanalytical Laboratory, Corona, NY. Room temperature photoluminescence (PL) data were recorded using a SPEX Flurolog-2 instrument equipped with a double emission monochromator and a R928 photomultiplier tube. Excitation was provided by a 450 W Xe lamp whose output was focused into a 0.22 m monochromator to provide wavelength selection. PL spectra for some samples were also recorded by excitation with the 325 nm line of a Linconix HeCd laser with an average power of 7 mW.

Organization and Characterization of Samples

The gallium imide derived GaN samples have the following labels for the listed pyrolysis conditions: A - reflux in the amine, bp 210 °C, 48 h; B - 600 °C, vacuum, 3 h; C - 500 °C, NH_3, 4 h; D - 600 °C, NH_3, 4 h. These four samples were used for the XRD determinations. Some of them were also used in the photoluminescence (PL) study. An additional sample was labeled E - 300 °C, NH_3, 4 h.

The gallazane derived samples of GaN have the following designations (in parentheses, preparative route to gallazane): 1 - pyrolysis of the polymeric solid/gallazane obtained upon 20 h evacuation of gallazane (*in situ* $LiGaH_4/NH_4Br$) - 600 °C, vacuum, 4 h; 2 - pyrolysis of the precipitated (60 h) polymeric solid (evacuated $LiGaH_4/NH_4Br$) - 600 °C, NH_3, 3 h; 3 - pyrolysis of the polymeric solid/gallazane obtained upon 5 h evacuation of gallazane (evacuated $LiGaH_4/NH_4Br$) - 600 °C, NH_3, 3 h; 4 - pyrolysis of the precipitated (72 h) polymeric solid (*in situ* $LiGaH_4/NH_4Br$) - 600 °C, NH_3, 4 h; 5 - pyrolysis of the as-prepared gallazane (*in situ* $LiGaH_4/NH_4Cl$) - 600 °C, NH_3, 5 h. Samples 1-5 were used for the XRD determinations and sample 4 was additionally included in the photoluminescence (PL) measurements.

EA: C: Ga, 81.02; N, 15.87; C, 0.22; H, 0.54; Ga/N = 1.03/1.00; 3: Ga, 83.06; N, 16.54; H, 0.05; C < 0.3; Ga/N = 1.05/1.00; 5: Ga, 86.31; N, 13.49; C, H < 0.1%; Ga/N = 1.29/1.00.

RESULTS AND DISCUSSION

Two different GaN precursors used in this study, i.e. gallium imide and gallazane (*via* a new synthetic route from LiGaH$_4$/NH$_4$X (X = Cl, Br) in Et$_2$O), were pyrolytically converted to a range of similar phase-inhomogeneous GaN materials. The difference between the chemical and physical properties of the product formed from the reaction of LiGaH$_4$ with NH$_4$X (designated gallazane) and that of previously characterized cyclotrigallazane is noteworthy. Cyclotrigallazane produced directly from the reaction of H$_3$Ga·NMe$_3$ and NH$_3$ is an insoluble, white crystalline solid that has been fully characterized. It can be slowly sublimed under vacuum at 80 °C without decomposition, and thermogravimetric analyses established a weight loss only above 140 °C. In contrast, the material produced by the double hydrogen elimination from LiGaH$_4$ plus NH$_4$X exhibits solubility in ether and a greater thermal instability. This new synthetic route may produce material having the formula [H$_2$GaNH$_2$]$_x$ where x does not exclusively equal 3. It is notable that some of the syntheses of borazane [H$_2$BNH$_2$]$_x$ yield a family of compounds related by different values of x.

Figure 1 illustrates some of the XRD spectra obtained for the gallium imide derived materials.

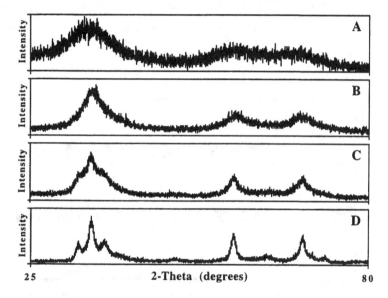

Figure 1. XRD powder patterns for GaN obtained by pyrolysis of gallium imide: **A** - reflux in N, N, N', N' - tetramethyl - 2, 6 - hexanediamine, bp 210 °C, 48 h; **B** - 600 °C, vacuum, 3h; **C** - 500 °C, NH$_3$, 4 h; **D** - 600 °C, NH$_3$, 4 h.

None of the samples shows a typical diffraction pattern for the hexagonal GaN (JCPDS file 2-1078) with the characteristic triplet feature in the 2-theta 30° to 40° range for the (100), (002), and (101) diffractions at (in parentheses, intensity) 32.4° (70), 34.6° (50), and 37.0° (100), respectively, which, approximately and with caution, can be used to compare with those of the phase-inhomogeneous GaN samples. For example, the spectra for samples **C** and **D** show the triplet nature but with the intensities different from the relevant hexagonal pattern. The high intensity middle peak in the triplet may correspond to either a significant cubic phase in the admixture with some hexagonal phase or the whole pattern may result from statistically scrambled layers of cubic and hexagonal GaN in a certain ratio [2a, b]. Alternatively, the pattern can be accounted for by assuming different quantities of stacking faults in either the cubic or hexagonal varieties, or by different cubic or hexagonal polymorphs [2c]. Although the phase character of these solids seems to be varied and remains unknown, it is clear that it is not purely hexagonal or cubic but rather of an intermediate and complex origin.

Sample **A** was obtained at the lowest temperature of 210 °C after reflux in the amine. It shows three diffraction halos in the positions that are consistent with the onset of GaN crystallinity. The remaining diffractograms confirm the influence of the applied pyrolysis conditions on both the nanocrystallinity and phase character of the samples. The average crystallite size estimated from the Scherrer equation is 13 nm for sample **D** and smaller for other samples. Samples **B** and **D** were both heated at 600 °C; however, the first one (pyrolyzed under vacuum) displays very broad features and, hence, contains small crystallites while the second one (pyrolyzed under NH_3) is relatively much sharper and consists of bigger crystallites. In addition to the application of vacuum or a NH_3 atmosphere, the pyrolysis temperature is also found to influence the particle nanosize. This is evident from comparison of the patterns for samples **C** and **D** (pyrolysis under NH_3); the increase of the temperature from 500 °C (sample **C**) to 600 °C (sample **D**) results in a noticeable sharpening of the peaks and, thus, increased particle size.

Figure 2 shows the XRD results for the gallazane derived GaN materials. In this case, all the pyrolyses were performed at 600 °C. Similarly, as in the relevant gallium imide route case, the pyrolysis under vacuum yields smaller average nanoparticles than the pyrolysis under NH_3 (compare sample **1** with all the remaining samples). In this case, however, the diffractions are broader than for related sample **B** which implies this precursor route affords smaller nanoparticles of GaN than the gallium imide route under similar pyrolysis conditions. The Scherrer estimation gives for sample **3** the largest in the series average particle size of 13 nm, coinciding well with the particle sizes approximated above for the gallium imide derived GaN. Also, sample **5** obtained from the as-prepared gallazane shows a pattern almost identical with sample **3**.

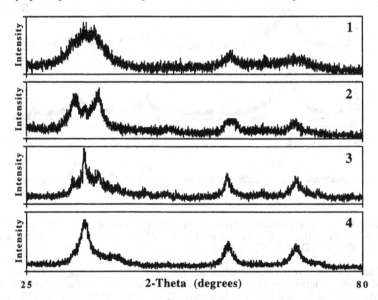

Figure 2. XRD powder patterns for gallazane derived GaN. Pyrolysis conditions for: 1 - 600 °C, vacuum, 4 h; 2 - 600 °C, NH_3, 3 h; 3 - 600 °C, NH_3, 3 h; 4 - 600 °C, NH_3, 4 h.

Samples **2-5** result from almost identical pyrolysis conditions but display dramatically contrasting diffractograms. For example, the pattern for sample **2** fits well with the pattern for pure hexagonal GaN (JCPDS file 2-1078). On the other hand, the pattern for sample **4** matches reasonably well the theoretically calculated pattern for cubic GaN [2b], and samples **3** and **5** show an intermediate phase form comparable with the previously discussed sample **D**. The pyrolysis conditions being virtually the same, other factor(s) in the gallazane precursor route must be responsible for the observed span in the GaN varieties. We believe that it is the nature of the precursors utilized in this system. As mentioned above, the gallazane as well as the polymeric solids obtained from it are all thermally unstable. It is apparently this feature, which is the function of the preparative factors such as the type of the polymeric precursor, handling times, and extent of evacuation, that imprints in the chemical character of the precursors and, upon pyrolysis, is further transformed into the specific crystalline variety. For example, the elemental analyses indicate that there is a tendency for gallium enrichment in the final materials (see EA for samples **3** and **5**).

Representative room-temperature PL spectra for GaN prepared by pyrolysis of the gallium imide are illustrated in Figure 3. Such measurements are useful in this context as a means of assessing the presence of defect or non-radiative sites in the solid(s). However, spectroscopic distinctions between cubic and hexagonal forms are possible only *via* analysis of excitonic luminescence at low temperature. Nanophase GaN prepared by the pyrolysis of the gallium imide precursor at or above a temperature of 500 °C either *in vacuo* or in an NH_3 atmosphere will typically produce samples exhibiting defect emission and a relatively low quantum efficiency. This is exemplified by the PL spectrum of GaN prepared by the pyrolysis of gallium imide at 600 °C under vacuum (sample **B**). Such weak PL is attributed to the presence of non-radiative defects of either interfacial or interior character. However, much stronger PL (albeit defect emission) can be observed when the pyrolysis temperature is lowered or the reaction medium is changed.

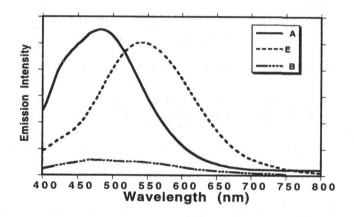

Figure 3. Room temperature PL spectra for GaN samples prepared by pyrolysis of gallium imide: **A** (solid line) - reflux in N,N,N',N'-tetramethyl-2,6-hexanediamine, bp 210 °C, 48 h; **B** (dot-dashed line) - 600 °C, vacuum, 3 h; **E** (dashed line) - 300 °C, NH_3, 4 h.

For example, when the gallium imide precursor is pyrolyzed in an ammonia atmosphere at a lower temperature of 300 °C, a more intense defect PL spectrum with a broad peak maximum near 560 nm is observed (sample **E**). It is found that a vacuum anneal of this type of GaN at 500 °C overnight results in diminution of the intensity of this band, consistent with its characterization as defect luminescence. Carrying out the reaction in a liquid phase *via* pyrolysis of the gallium imide in the relatively high boiling amine solvent N,N,N',N'-tetramethyl-2,6-hexanediamine (bp 210 °C) for 48 hours yields GaN which emits strongly in the blue region of the spectrum (430-480 nm); however, some of this broad blue light originates from the pyrolyzed amine coating present on the semiconductor surface (sample **A**).

Given the presence of defects in many of these as-prepared GaN nanophase materials, finding a suitable physico-chemical method to eliminate the defects and produce blue band-edge luminescent GaN is mandatory. Previous accounts regarding bulk epitaxially grown GaN have demonstrated that HF and HNO_3 have some utility as etchants for removing oxide and

43

hydrocarbon impurities from GaN [8]. Hence we decided to evaluate the efficacy of these acids on the photoluminescence of nanophase GaN prepared by the routes described in this paper. For these experiments, gallazane derived GaN sample 4 (pyrolysis at 600 °C, NH_3, 4 h) was exposed to solutions of 35% HNO_3 or 48% HF. It is found that a very brief etch (20 seconds) using 48% HF followed by a water and an ethanol wash yields GaN exhibiting an intense blue-emitting PL spectrum (Figure 4).

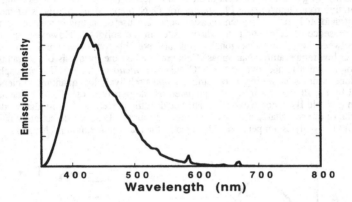

Figure 4. Room temperature PL spectrum of gallazane derived GaN sample 4 exposed to a brief HF etch.

At present, details of the mechanism of improved luminescence for these GaN nanophase samples are not clear. However, it is likely that the removal of oxide-related non-radiative centers by the acid at the nanoparticle surface plays a key role in these observations.

ACKNOWLEDGMENT. R. L. W. thanks the Office of Naval Research for its financial support.

REFERENCES

1. For example see: (a) S. Strite and H. Markoç, J. Vac. Sci. Technol. B **10**, p. 1237 (1992). (b) D. A. Neumayer and J. G. Ekerdt, Chem. Mat. **8**, p. 9 (1996).

2. (a) J.-W. Hwang, S. A. Hanson, D. Britton, J. F. Evans, K. F. Jensen, W. L. Gladfelter, Chem. Mat. **2**, p. 342 (1990). (b) J.-W. Hwang, J. P. Campbell, J. Kozubowski, S. A. Hanson, J. F. Evans, W. L. Gladfelter, Chem. Mat. **7**, p. 517 (1995). (c) W. L. Gladfelter and J. P. Campbell, personal communications.

3. A. Storr, J. Chem. Soc. (A), p. 2605 (1968).

4. J. F. Janik and R. L. Wells, Chem. Mat. **8**, p. 2708 (1996).

5. K. E. Gonsalves, G. Carlson, S. P. Rangarajan, M. Benaissa, M. J.-Yacamán, J. Mater. Chem. **6**, p. 1451 (1996).

6. (a) T. Ogino, M.Aoki, Jpn. J. Appl. Phys. **19**, p. 2395 (1980). (b) E. Glaser, T. Kennedy, K. Doverspike, L. Rowland, D. Gaskill, J. Freitas, M. Asif Khan, D. Olson, J. Kuznia, D. Wickenden, Phys. Rev. B **51**, p. 13326 (1995).

7. J. F. Janik and R. L. Wells, Inorg. Chem., submitted.

8. L. Smith, S. King, R. Nemanich, R. Davis, J. Electron. Mater. **25**, p. 805 (1996).

TOWARD GROWING III-V CLUSTERS WITH METALORGANIC PRECURSORS

A. DEMCHUK, J. PORTER, and B. KOPLITZ
Department of Chemistry, Tulane University, New Orleans, LA 70118-5698

ABSTRACT

The present work reports on the formation of GaN-containing clusters from metalorganic precursors by combining pulsed laser photolysis and pulsed nozzle methods. Ammonia (NH_3) and triethylgallium $(C_2H_5)_3Ga$ (TEG) or trimethylgallium $(CH_3)_3Ga$ (TMG) with He, Ar, or N_2 as the carrier gas are introduced into a high vacuum chamber via a specialized dual pulsed nozzle source. The light from an ArF excimer laser (193 nm, 23 ns FWHM) is focused into the mixing and reaction region of the nozzle source, and the products are then mass analyzed with a quadrupole mass spectrometer. Efficient laser-assisted growth of $(GaN)_x$-containing clusters is shown with this technique.

INTRODUCTION

Metalorganic chemical vapor deposition (MOCVD) is a common technique for growing GaN films. In this method, triethylgallium $(C_2H_5)_3Ga$ (TEG) or trimethylgallium $(CH_3)_3Ga$ (TMG) and ammonia (NH_3) are often used as Ga and N sources, respectively [1]. Incorporation of carbon atoms during high temperature MOCVD process is a major obstacle for obtaining high quality films [2]. In our work, we are developing new approaches to studying photo-assisted GaN growth from metalorganic precursors by using UV pulsed laser photolysis and pulsed nozzle methods. Figure 1 illustrates the physical processes involving the formation of clusters using this technique.

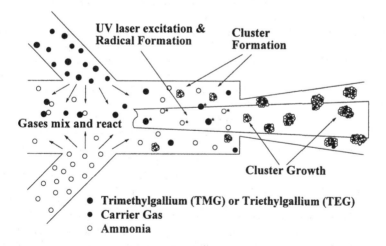

Fig. 1 Schematic of laser-assisted growth of GaN-containing clusters from metalorganic precursors in a dual pulsed nozzle source.

EXPERIMENTAL

The experimental apparatus consists of a high vacuum chamber (base pressure ~ 10^{-6} Torr) equipped with a quadrupole mass spectrometer and a specialized dual pulsed nozzle source, shown in Fig. 2.

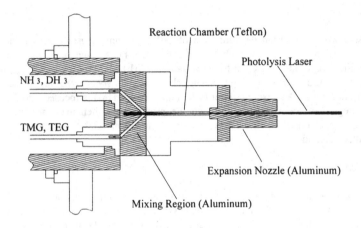

Fig. 2 Schematic of dual pulsed nozzle source.

Triethylgallium $(C_2H_5)_3Ga$ (TEG) or trimethylgallium $(CH_3)_3Ga$ (TMG) (99+% purity, Strem Chemical) is used with He, Ar or N_2 (99%, Air Products) as the carrier gas. Procedurally, the vapor of TEG or TMG was fed into a stainless steel sample chamber by bubbling the carrier gas through a temperature control stainless steel bubbler to a sample pressure varied between 760-2200 Torr. The TMG and TEG source bottles were maintained at 25 °C and 100 °C, respectively, resulting in TMG and TEG vapor pressures of ~ 220 and ~ 190 Torr [3]. In experiments with TEG, the sample chamber walls were heated to 100 °C by an electrical tape heater to prevent TEG condensation. This temperature is not high enough to cause any significant dissociation of the TEG, which begins to occur in the 250-300 °C temperature range [4].

The vapors of TMG or TEG and ammonia, NH_3 (99.99% purity, Air Products) were introduced into the high vacuum chamber via the nozzle assembly shown above (Fig. 2). Two sample gas pulses (TMG or TEG and NH_3) were simultaneously injected into the mixing/reaction region of the nozzle. The opening time of the pulsed valve was adjusted to be 0.2-0.3 ms. In order to label the products and to distinguish between possible reaction pathways, one can use deuterated ammonia, ND_3 (99% purity, Cambridge Isotope Lab.).

UV light from an ArF excimer laser (Lambda Physik LEXtra-200, 193 nm, 23 ns FWHM) operating at constant energy was mildly focused with a quartz lens (f=250 mm) into the mixing/reaction region of the nozzle through a 0.7 mm diameter pin hole such that its direction was perpendicular to the mass spectrometer axis and propagated toward the sample molecular beam axis. The incident laser energy was varied between 0.1-20.0 mJ per pulse. The delay time (~5 ms) between the laser pulse and gas sample injection was adjusted to obtain a maximum product signal. The repetition rate of sample gas injection/laser irradiation pulses was 10 Hz.

Products were mass analyzed with a UTI 100C quadrupole mass spectrometer, and mass spectra were recorded on a digital oscilloscope (LeCroy 9370), typically with 500 pulses averaged to obtain a suitable signal/noise ratio. The data were then transferred and stored on a computer.

RESULTS AND DISCUSSION

Figures 3 and 4 contain typical electron impact (EI) mass spectra that show the adduct formation due to the gas phase reaction involving TMG and NH_3 (Fig. 3) and TEG and NH_3 (Fig. 4) when using our laser-assisted cluster growth technique.

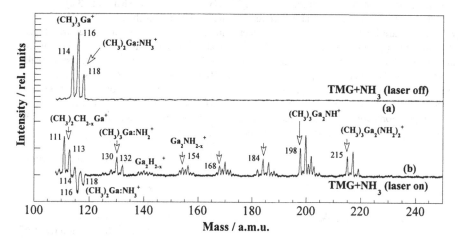

Fig. 3 EI mass spectra of clusters produced in the dual pulsed nozzle during the expansion of TMG and NH_3. The laser intensity was ~ 5 mJ/pulse.

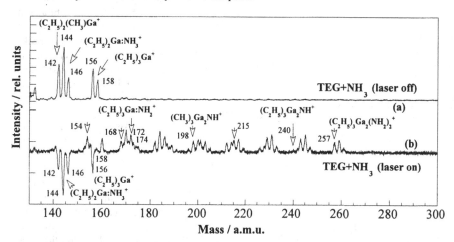

Fig. 4 EI mass spectra of clusters produced in the dual pulsed nozzle during the expansion of TEG and NH_3. The laser intensity was ~ 5 mJ/pulse.

During the mixing and expansion of TMG or TEG and NH_3, the adducts $(CH_3)_2Ga:NH_3$ (Fig. 3 a) or $(C_2H_5)_2Ga:NH_3$ (Fig. 4 a), respectively, are detected even without laser irradiation. They have the proper gallium isotopic relation (relative abundances of the two isotopes of gallium: ^{69}Ga, 60.4%; ^{71}Ga, 39.6% [5]). During UV laser irradiation of such gas mixtures, efficient cluster formation takes place as illustrated by Figure 3(b) and Figure 4(b). These spectra were created by subtracting the spectra measured separately with the "laser on" and the "laser off." It is clearly seen that laser photolysis of the $(CH_3)_3Ga$ parent molecule and the $(CH_3)_2Ga:NH_3$ adduct (or $(C_2H_5)_3Ga$ and the $(C_2H_5)_2Ga:NH_3$ adduct) result in a decrease in the intensity of these peaks and the formation of deprotonated species, e.g. $(CH_3)_2CH_{2-x}Ga$ in the case of TMG (Fig 3 a).

At the same time, during laser irradiation new adducts such as $(CH_3)_3Ga:NH_2$ or $(C_2H_5)_3Ga:NH_2$ are formed as well as GaN-cluster features containing two gallium atoms (Fig. 3 b and 4 b). The spectrum reveals a series of intense peaks that can be assigned to the following mixture of clusters: R_2Ga_2NH and $R_3Ga_2(NH_2)_2$ or $R_3Ga_2(NH_3)NH$ and less intensive peaks attributed to $R_3Ga_2(NH_2)(NH)_2$; $R_4Ga_2(NH_2)NH$; $R_3HGa_2(NH_2)NH$; $R_2Ga_2(NH_3)NH_2$ and $[GaNH(NH_2)]_2$ clusters and their fragments, where R is CH_3 or C_2H_5, respectively. Note that these clusters may be formed as a result of dissociation of clusters with higher masses due to electron impact ionization. Note also that the detection of species with masses higher than 300 a.m.u. is currently not possible with our experimental apparatus.

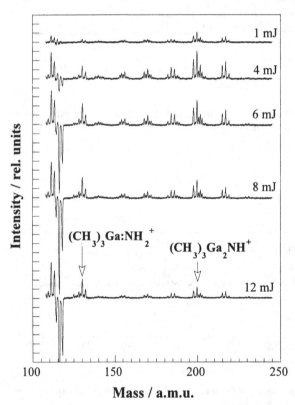

The intensity of the parent TMG and adduct $(CH_3)_2Ga:NH_3$ peaks typically decrease with increasing laser fluence as a result of photolysis (Fig 5). Simultaneously, the intensity of the deprotonated TMG species increases. Likewise, the intensity of the $(CH_3)_3Ga:NH_2$ peak increases, but it saturates at ~ 10 mJ/pulse. In contrast, the intensity of GaN-clusters peaks containing two Ga atoms changes nonmonotonously. The maximum in these cluster peaks is reached at 4-5 mJ/pulse with a subsequent drop off at higher power. Analogous behavior in cluster formation is observed in the case of TEG.

← **Fig. 5** EI mass spectra of GaN-containing clusters produced in a dual pulsed nozzle during expansion of TMG and NH_3 with UV laser irradiation at different photolysis laser energies.

Changing the sample pressure of NH_3 and TMG or TEG changes the intensity relation of some cluster peaks. Typically the intensity of the cluster peaks with masses of 168, 170, 172 a.m.u. and 184, 186, 188 a.m.u. increase when the pressure of the NH_3 exceeds the pressure of TEG (or TMG). These peaks can be attributed to the formation of $[GaNH(NH_2)]_2$ clusters and their fragments, in particular $[GaNH]_2$ and $[GaNH]_2NH_2$. Note that if instead of helium, argon or nitrogen is used as the carrier gas for TMG or TEG under otherwise identical conditions, the mass spectrum does not change.

In the following discussion, we focus on the possible photochemistry that takes place in such gas mixtures and give a description of what may be happening in such excited gas systems. It is known from several studies [6] that NH_3 molecules resonantly absorb 193 nm photons (NH_3 absorption cross-section, $\sigma = 9\pm3\cdot10^{-18}$ cm^{-2} at λ=193 nm [7]) and predissociate on a femtosecond time scale [8, 9] to produce NH_2 radicals. The primary photolysis step of TMG or TEG is the excitation of the parent molecule [10] ($\sigma_{TMG} = 20\pm6\cdot10^{-18}$ cm^{-2} and $\sigma_{TEG} = 7.4\pm1.6\cdot10^{-18}$ cm^{-2} at λ=193 nm [7]) with successive release of a methyl or ethyl group [11, 12] on a nanosecond time scale [13]. At least two 193 nm (6.4 eV) photons are necessary to obtain $(CH_3)_3Ga^+$ or $(C_2H_5)_3Ga^+$ ions [14], which occurs in competition with the dissociation channel. The photolysis reaction of $(CH_3)_2Ga:NH_3$ and $(C_2H_5)_3Ga:NH_3$ adducts can be described analogously. Thus, the laser photodissociation and photoionization of TMG or TEG and NH_3 and their adduct molecules result in the formation of different radicals and ions that can be involved in cluster growth. Cluster formation and growth might occur by gas phase reaction between radicals as well as parent molecules and radicals.

The most probable reaction in the post laser excitation processes is the reaction between the NH_2 radical and TMG or TEG. This reaction results in the formation of a Lewis acid-Lewis base adduct, such as $(CH_3)_3Ga:NH_2$ or $(C_2H_5)_3Ga:NH_2$. Although the $(CH_3)_3Ga:NH_3$ adduct can be formed during the high temperature MOCVD reaction between TMG and NH_3 [15, 16], $(CH_3)_3Ga:NH_2$ and $(C_2H_5)_3Ga:NH_2$ are clearly detected by electron impact ionization.

It is remarkable that the laser energy dependence for the formation of such adducts show typical behavior for the saturation processes attributed to the photodissociation of NH_3 and NH_2 radical formation. For clusters containing two Ga atoms, this measurement shows nonmonotonous behavior in the clusters formation that could be attributed to competition between the photoionization and photodissociation processes taking place during photolysis of TMG and the $(CH_3)_2Ga:NH_3$ adduct [17]. It is suggested that TMG or TEG and their adduct radicals play an important role in the formation of these clusters. Studies at higher mass will reveal to what extent "larger" clusters are being formed.

Finally, the relative pressure of ammonia and TMG or TEG has a strong influence on cluster formation. Higher ammonia pressures lead to formation of lower molecular weight GaN clusters that contain relatively less carbon (not shown). Further experimental investigation of cluster formation by this technique can shed additional light on the chemistry associated with III-V cluster growth when metalorganic precursors are involved.

CONCLUSIONS

A new cluster-generation method that combines UV pulsed laser photolysis and a dual pulsed expansion nozzle is described for the formation of III-V clusters from metalorganic precursors. GaN-containing clusters formed due to laser-assisted growth using TMG or TEG and NH_3 precursors have been identified. The laser energy and the vapor pressure of the gases have a significant influence on cluster formation. In additional to GaN, other important semiconductors growth systems can be targeted with this method.

ACKNOWLEDGMENTS

This project was supported by the Department of Energy, the National Science Foundation through the Center for Photoinduced Processes at Tulane University, and the State of Louisiana via its LEQSF program.

REFERENCES

1. R.D. Dupuis, A.L. Holmes, P.A. Grudowski, K.G. Fertitta, F.A. Ponce, Mat. Res. Soc. Symp. Proc. **395**, 183 (1996).

2. J. Kouvetakis, J. McMurran, D.B. Beach, D.J. Smith, Mat. Res. Soc. Symp. Proc. **395**, 79 (1996).

3. G.E. Coates, M.L.H.Green, K.Wade, Organometallic Compounds, V.1: The Main Group Elements: by G.E. Coates and K.Wade, (Methuen & Co Ltd, London, 1967) p. 343.

4. I. Qureshi, P.K. Ajmera and S. Felps, Mater. Letters **14**, 107 (1992).

5. Organometallic Compounds of Aluminum, Gallium, Indium and Thallium. Ed. by A. McKillop, J.D. Smith and I.J. Worral, (Chapman and Hall Ltd., New York, 1985) p.113.

6. H. Okabe, Photochemistry of Small Molecules, (Wiley Inc., New York, 1978) pp. 269-272.

7. D. Bauerle, Laser Processing and Chemistry, 2nd ed. (Springer-Verlag, Berlin, Heidelberg, New York, 1996), pp. 583-587.

8. L. Ziegler, J. Chem. Phys. **82**, 664 (1985).

9. H. Shinohara, J. Chem. Phys. **79**, 1732 (1983).

10. A. Sato, Y. Tanaka, M. Tsunekawa, M. Kobayashi, H. Sato, J. Phys. Chem. **97**, 8458 (1993).

11. Th. Beuermann and M. Stuke, Appl. Phys. B **49**, 145 (1989).

12. Y. Zhang, Th. Beuermann, and M. Stuke, Appl. Phys. B **48**, 97 (1989).

13. S.A. Mitchell, P.A. Hackett, D.M. Rayner, M.R. Humphries, J. Chem. Phys. **83**, 5028 (1985).

14. S.A. Mitchell, and P.A. Hackett, J. Chem. Phys. **79**, 4815 (1983).

15. M.J. Almond, C.E. Jenkins, D.A. Rice, J. Organometallic Chem. **439**, 251 (1992).

16. M.J. Almond, G.B. Drew, C.E. Jenkins, D.A. Rice, J. Chem. Soc. Dalton Trans. **1**, 5 (1992).

17. J.L. Brum, S. Deshmukh, and B. Koplitz, J. Chem. Phys. **93**, 7946 (1990).

OPTIMIZATION OF III-N BASED DEVICES GROWN BY RF ATOMIC NITROGEN PLASMA USING IN-SITU CATHODOLUMINESCENCE

J. M. Van Hove, P.P Chow, J.J. Klaassen, R. Hickman II, A.M.Wowchak, D.R. Croswell, and C. Polley.
SVT Associates, 7620 Executive Dr. Eden Prairie, MN 55344,
jvanhove@svta.com

ABSTRACT

In-situ cathodoluminescence (CL) is presented as a technique to optimize GaN, and AlGaN films deposited by MBE using an RF plasma as a source of reactive nitrogen. Excitation of the MBE grown nitride films is conveniently achieved in the preparation chamber using an Auger electron gun. The photoemission is monitored through a side port and dispersed with a 1/8 m monochromator with a typical resolution of 3 nm. The in-situ CL spectra of AlGaN and GaN films provides quick determination of both material composition, doping, and quality from the position and width of the band edge emission. The use of CL for the assessment of material composition in the growth of nitride materials is extremely beneficial since the complementary technique of RHEED oscillations is not routinely observed for these systems. The determination of material quality using CL has been used to optimize growth conditions for GaN PIN junction photovoltaic detectors on (0001) sapphire. Detectors having peak responsivity of 0.175 A/W at the GaN band edge of 365 nm and a UV to visible rejection ratio of greater than 10^5 have been fabricated. The high rejection ratio is accredited to the reduction of the yellow defect levels in the MBE grown material. Material optimization using in-situ CL for growth of AlGaN MODFETs having drain currents of 425 ma/mm and g_m of 66 mS/mm is discussed.

Introduction

Interest in the group III (Al, Ga, In) nitrides and their ternary and quartenary alloys has increased dramatically with the recent progress in developing blue semiconductor lasers. Other devices such as high temperature, high power electronics and solar blind UV detectors are being developed in this wide band gap material system. Recent progress has been made in molecular beam epitaxy (MBE) of these materials. One advantage MBE has over other growth techniques is the ability to monitor the growth process *in-situ*. Reflection high energy electron diffraction (RHEED) is now a commonly used *in-situ* technique for calibration of composition and growth rates for most III-V materials. Unfortunately, RHEED oscillations have only been observed under limited conditions for nitride growth. This makes film composition difficult to determine quickly *in-situ*. In this work, we present the use of *in-situ* cathodoluminescence (CL) for determining film composition, optical quality, and doping levels of AlGaN films.

Cathodoluminescence is an optical technique which uses an electron beam to excite the film. The resulting emission provides material information similar to that obtained by photoluminescence, such as the position of the band edge and the

51

mid gap energy levels [1]. This technique has been used by Rouleau and Park to monitor *in-situ* the blue/green CL emission from MBE grown ZnSe films [2]. A significant advantage of *in-situ* CL is that it can be accomplished using a standard RHEED gun present in most MBE systems. Dispersion of the RHEED streak fluorescence through a simple monochromator /detector allows the band gap of the deposited material to be determined. From this measurement, the composition of $Al_xGa_{1-x}N$ and $In_xGa_{1-x}N$ films can be determined. The optical quality of the deposited film can also be evaluated from the FWHM of the band-edge emission, and qualitative assessment of the doping level can be made from the structure of the CL emission.

Experiment
In-situ cathodoluminescence was done in the analysis chamber of the MBE system. CL excitation was done with the Auger electron gun at 2 to 8 KeV. The emitted light was measured with a 1/8 meter (3 nm resolution) monochromator and PMT detector. Similar measurements have been reported by us using the RHEED gun in the growth system and a telescoping lens/detector system [3].

An atomic RF plasma source developed specifically for the growth of MBE nitride was used for this work and has been described previously [4]. The basic concept is to create a plasma of nitrogen with the use of a RF field. RF energy (200 to 550 W) is fed into the gun through a water cooled copper coil. A pyrolitic boron nitride (PBN) tube with a changeable nozzle is centered between the RF coils. Nitrogen is introduced to the tube with a leak valve and a plasma is created within the tube. Flow rates of 2 to 3 sccm of nitrogen is enough to produce growth rates around 1 μm/hr. Elemental Ga and Al supplied from effusion cells were used for the group III elements. Mg doping was done using a conventional effusion cell while Si doping was done using a compact e-beam source. Sapphire (0001) was used as the substrate with a low temperature AlN buffer layer grown prior to GaN deposition. Growth temperatures ranged between 750 and 900 °C and a Ga rich III-V flux ratio was used.

Devices were processed on 2" sapphire substrates and used a low pressure $SiCl_4$ based RIE for recessing. Refractory Ti/Mo/Au contacts were used for the N ohmics and Ni/Au contacts for P type ohmics. Ti/Pt/Au was used to form the gate of the MODFET.

Results
Figure 1 shows the measured spectral response of a GaN PIN diode. Details about the growth, processing and test of similar PIN diodes have been reported elsewhere. [5,6]. The peak responsivity of 0.175 A/W occurs at 360 nm. The visible rejection with respect to the peak responsivity at 360 nm is 5 orders of magnitude. The visible rejection was attributed to the absence of yellow defect states in the MBE grown material. The maximum peak responsivity at unity external quantum efficiency was calculated to be 0.23 A/W including a 19% reflection loss at the air-GaN interface; therefore the internal quantum efficiency from the device was about 76%. The in-situ CL was used to optimize the undoped

material to remove the "yellow" defect as well as improve the quality of the P type GaN. Further improvements are needed in the P type material to sharpen the response curve.

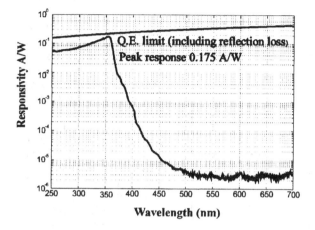

Figure 1. The spectral response of a GaN photovoltaic diode with a peak responsivity of 0.175 A/W and almost 5 orders of magnitude rejection of radiation in the visible and near IR.

Figure 2 shows the CL spectrum from a 1 μm thick layer of GaN on sapphire. These scans were taken at a beam voltage of 4 KeV in the analysis chamber of the MBE system. The CL intensity is plotted on a log scale to highlight the yellow emission level which is present in most GaN films [7]. The log scale reveals considerable information lost in traditional linear plots. Figure 2 (A) shows the room temperature CL emission from an unintentionally doped highly resistive GaN film. This film shows very little yellow emission around 550 nm but has a sharp band edge peak with a FWHM of 66 meV. Figure 2(B) shows the room temperature CL spectrum from a GaN film deposited under non-ideal conditions. The extremely large yellow emission peak centered around 550 nm is possibly due to the presence of oxygen and carbon in the GaN film. Similar yellow emission levels are observed under low growth temperatures, high N/Ga flux ratios, and on poor quality AlN buffer layers. Using the *in-situ* CL scans, both the yellow defect emission level and the FWHM of the GaN band edge can be measured quickly and growth conditions can subsequently be adjusted. The reduction of the yellow emission level has proved to be important in optimizing the UV/visible rejection ratio of the UV PIN photodiodes and in reducing the background carrier concentration needed for AlGaN FET isolation.

The *in-situ* CL scans proved extremely useful for optimizing the Mg doped P type GaN needed for the PIN UV detectors. Figure 3 shows the CL spectrums from three Mg doped P type GaN layers grown on sapphire. The measured Hall concentration is given for each curve. The band edge emission is still present for each doping level but a strong peak centered around 390 nm increases in intensity as the doping level increases. These measurements can be done in-situ and different doping levels and growth conditions tried to optimize the material properties.

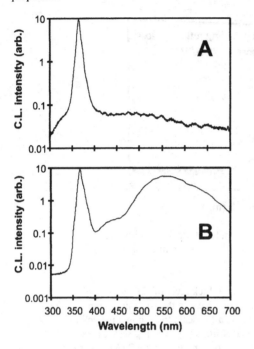

Figure 2. *In-situ* CL spectrum from 1 μm of GaN on sapphire taken after growth at two different growth conditions. The CL intensity is plotted on a log scale to highlight the yellow emission level. (A). Room temperature CL emission from high resistivity GaN showing very little yellow emission at 550 nm. The FWHM of the band edge peak is 66 meV. (B) Room temperature CL from GaN deposited in an unoptimized MBE system. Extremely intense yellow emission centered around 550 nm is observed.

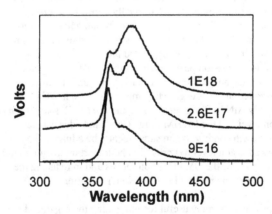

Figure 3. *In-situ* CL spectrum from several Mg doped P type GaN films on sapphire. The measured Hall carrier concentration are given for each curve.

A major advantage of *in-situ* CL is that determination of the composition of AlGaN films by measuring the band gap of the deposited material. From this measurement, the Al content can be determined [8]. Figure 4 shows the in-situ CL spectrum from a AlGaN MODFET structure taken at 2 KeV. Emission from the 300 Å layer of AlGaN is visible at 338 nm as well as the underlying undoped GaN layer at 362 nm. CL emission has been observed from $Al_x Ga_{1-x} N$ for $x = 0$ to 1. CL is extremely useful for scanning high Al content films where deep UV lasers are costly for PL measurements. Figure 5 shows the measured drain-source current Ids vs drain-source voltage Vds for various gate-source voltages for a 75 μm gate width MODFET. Measured extrinsic transconductance g_m was 66 mS/mm and maximum drain current was 425 mA/mm on this device. Optimization of the growth of this device was done with the in-situ CL to determine the Al concentration of the device and the proper growth conditions to give narrow, intense CL emission from the AlGaN. The insulating GaN buffer used similar growth conditions as that used for the PIN detector to reduce the visible response. This involved reducing the yellow defect while narrowing the FWHM of the GaN peak.

Conclusions
Reliable analysis techniques are necessary for the continued advancement of group III nitride technology. *In-situ* CL is presented as a valuable technique for determining film composition, optical quality, and doping levels of MBE grown nitride films. It is a straightforward technique which utilizes existing RHEED and/or Auger equipment for a new purpose, and can be readily retrofitted to older MBE systems. Device results demonstrated the techniques usefulness for optimizing performance.

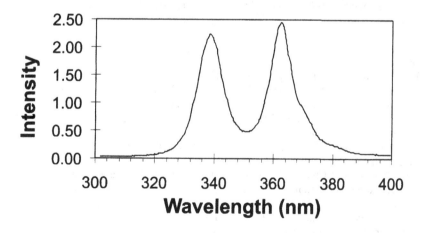

Figure 4. In-situ CL spectrum taken at 2 KeV from an AlGaN MODFET layer showing the AlGaN layer at 338 nm and the GaN buffer at 362 nm.

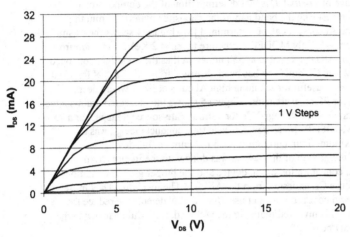

AlGaN HEMT (75 µm - 1 µm gate)

1 V Steps

Figure 5. Drain-source current Ids vs. drain-source voltage Vds for various gate-source voltages for a 75 µm gate width AlGaN MODFET. Measured extrinsic transconductance g_m was 66 mS/mm and maximum drain current was 425 mA/mm on this device.

Acknowledgments

This work was supported in part by NASA Contract NAS5-38054 (D.A. Mott) , BMDO Contract NOOO14-96-C-0251, monitored by C. Wood ONR, and BMDO Contract DASG60-96-C-0120, monitored by M. McCurry, Army Space Command.

References

[1]. B. G. Yacobi, D.B. Holt, J. Appl. Phys. **59**, R1 (1986).
[2]. C.M. Rouleau and R.M Park, Appl. Phys. Lett. **60**, 2723 (1992).
[3]. J.M. Van Hove, P.P. Chow, A.M. Wowchak, J.J. Klaassen, M.F. Rosamond and D.R. Croswell, 1996 Molecular Beam Epitaxy Conference, accepted for publication in J. Crystal Growth **175** (1997).
[4]. J.M. Van Hove, G.J. Cosimini, E. Nelson, A.M. Wowchak, P.P Chow, J. Crystal Growth **150,** 908 (1995).
[5]. J.M. Van Hove, P.P. Chow, R. Hickman, A.M. Wowchak, J.J. Klaassen, C.J. Polley, accepted for publication in Proceedings of the Fall 1996 MRS Nitride Conference.
[6]. J.M. Van Hove, R. Hickman, J.J. Klaassen, P.P. Chow P.P. Ruden, Accepted for publication in Appl. Phys. Lett. **70**, April 28, 1997.
[7]. J. Neugebauer, C. G. Van de Walle, Appl. Phys. Lett. **69**, 503 (1996).
[8]. S. Yoshida, S. Misawa, and S. Gonda, J. App. Phys **53**, 6844 (1982).

$In_xGa_{(1-x)}N$ ALLOYS AS ELECTRONIC MATERIALS

O.K.SEMCHINOVA *, S.E.ALEXANDROV**, H.NEFF***, D.UFFMANN*.
* LFI University of Hannover, Schneiderberg 32, 30167 Hannover, Germany.
** Technical University St-Petersburg, Polytechnicheskaya 26, Russia.
*** TZN GmbH, Neuensothriether Str.18-20, 29345 Unterlüss, Germany.

ABSTRACT

In this work, we present growth of $In_xGa_{(1-x)}N$ films by CVD technique and their optical characterization. Experimental results indicate that these films are promising materials for semiconductor device applications. We focus on solar cells and present preliminary experimental data on prototype devices.

INTRODUCTION

Gallium nitride and its alloys with Al and In are important candidate materials for blue-UV lasers, short wavelength radiation detectors and high-temperature electronics. This is evident from several impressive device achievements in the last few years [1-2].But the problem to fabricate stable and reliable devices, based on Al(Ga,In)N layers intimately is associated with the need of lattice-matched substrates [3]. Less attention is dedicated to indium nitride and its alloys. The primary reason is a smaller band gap (~ 1.9 eV), and more difficulties for film growth [4].

The development of high quality photodiodes and solar cells is a demanding materials problem. Large scale utilization of solar cells for solar power conversion rather relies on low cost , large area film deposition technologies. This includes low cost substrates and sufficient device lifetime, to establish a positive net power generation rate. Thin film flat panel solar cell quantum efficiencies should exceed 10% to be competitive. Semiconductor films, deposited on low quality substrates commonly grow polycrystalline or even amorphous, depending on the deposition conditions. Under these circumstances, cell performance is controlled by electrically active interfacial defects, located primarily at interfaces between grain boundaries. They introduce high concentration of mid gap states and recombination centers reducing both, carrier mobility and lifetime. For the direct gap materials, like InP, GaAs and the group III - Nitrides, these adverse effects are compensated partly by a very short optical absorption length of the order of $1\mu m$ or less.[5].

From this point of view, investigations of $In_xGa_{(1-x)}N$ polycrystalline films deposited by CVD technique on p-type Si substrates for solar cell application is of scientific and commercial interest. The materials combination has been selected since it should provide a very broad spectral response within $1eV < h\nu < 3eV$. The short wavelength contribution at $h\nu > 1.5$ eV is important in the northern hemisphere, where usually rather large concentrations of water vapor are present in the upper atmosphere. This is illustrated in Fig.1. showing the theoretically achievable conversion efficiency η_{max}, as a function of the optical band gap E_g. [5].

For airmass 1 (AM 1)condition at zero concentration of water vapor (w=0) the optimum E_g is set at 1.5 eV. For higher concentration of water vapor (w=1) , a further shoulder appears at E_g =1.8 eV [6]. Therefore, use of a pure n-InN front window layer would be the material of choice, but commonly requires a very high doping level to establish a low series resistence, thus shifting the band gap to slightly lower value, known as the Burstein shift [5].

57

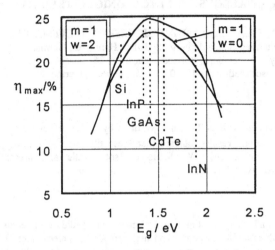

Fig 1. Calculated solar power conversion efficiency η versus band gap energy E_g.

EXPERIMENTAL

In this work, we present growth and characterization of n- $In_xGa_{(1-x)}N$ (x=0.20) films deposited on p-Si substrates by CVD technique.

The deposition process of the films based on pyrolysis of the complex of indium and gallium trichloride with ammonia $InCl_3NH_3$ and $GaCl_3NH_3$. In order to minimize any possible influence on the process chemistry, and also because some compounds, whose formation is to be expected in the Ga-Cl-N-H system, display strong absorption bands in the UV region [7], direct "in situ" ultraviolet spectroscopic analysis of the gas phase during the CVD process has been carried out. A schematic drawing of the experimental apparatus used for these investigations is shown in Fig. 2.

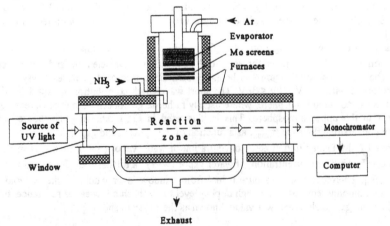

Fig 2. Schematic diagram of the experimental apparatus.

A specially designed two zone hot wall model reactor was placed between the source of UV light (deuterium lamp, 30W) and the monochromator. The reactor has been equipped with quartz windows permitting the recording of transmittance spectra in the wavelength region 200-350nm. The intensity of the monochromatic light was measured with a (FEU-100) photomultiplier. The vapor of the precursor ($InCl_3NH_3$) from the evaporator was provided with a carrier gas (Ar 5.2, electronic grade) to the reaction zone. A system of molybdenum screens, placed between the evaporation and the reaction zone, protected the evaporator from overheating. Both zones were separately heated with resistance furnaces. The experimental equipment is described in more detail in [7].

The complex of indiumtrichloride with ammonia $InCl_3NH_3$ was previously synthesized from indium (99,99%) and NH_4Cl (electronic grade). The partial pressure of the initial complex can be varied from 20 to 3000Pa by adjusting the evaporation temperature, kept at 360Pa during the experiments. The typical UV spectra of the gas phase, formed during pyrolysis at different temperatures (400,600,800,1000°C), is shown in Fig.3.

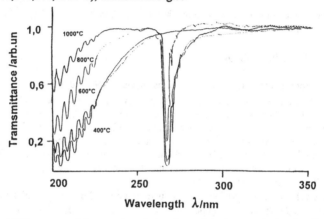

Fig 3. The typical UV spectra of the gas phase being formed during pyrolysis of $InCl_3NH_3$.

Analysis of the transmittance spectra of the gas phase demonstrates that molecules of NH_3, $InCl_2$ and $InCl$ are present in the reaction gas mixture as taken from the position of absorption band maxima. The procedure is well described in [7]. Also it was shown that the relative concentration of the molecules mentioned above are dependent on the temperature during pyrolysis. To examine trends in the variation of concentrations of these compounds with temperature in the reaction zone, the integral absorption coefficient has been calculated which is proportional to the concentration of the absorbing species. Typical relation of the integral absorption intensity for $InCl_3$, NH_3 and $InCl$ with temperature in the reaction zone is presented in Fig.4.

The data obtained in the present paper are similar to those for the deposition process of the GaN films based on the pyrolysis of the $GaCl_3NH_3$ [11]. This fact allows to succeed in the deposition of $In_xGa_{(1-x)}N$ films by CVD technique using In(Ga) Cl_3NH_3 system.

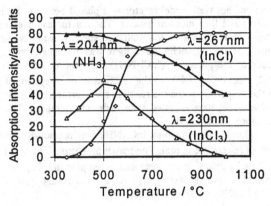

Fig 4. Typical relation of the integral absorbtion intensity for InCl₃, InCl and NH₃ with temperature in the reaction zone.

RESULTS AND DISCUSSION

During the experiments, formation of the films $In_xGa_{(1-x)}N$ (x=0,20) starts at 630°C (substrate temperature) but the optimized deposition process varies between 650°C and 750°C. The growth rate of the films was 10nm/min at 700°C and 5.6nm/min at 650°C. After deposition a visual examination of the surfaces showed that the films deposited on Si as well as on glass substrates are quite smooth, without any visible inclusions. The films deposited on glass substrates exhibited variations of the refraction index between 2.05 and 2.1. The variation in the thickness of deposited films was small and the thickness of the top layer was near 100nm , the resistivity was around 0.9×10^3 Ohm-cm.

In this work we also report preliminary results on the photodiode characteristic of n - In $_xGa_{(1-x)}N$ /p-Si heterojunctions. Metal contacts were established by chromium evaporation forming stripes at the InGaN - top layer. A rather high series resistance > 500 Ohm has been observed. Fig.5 (a,b) shows optical absorption spectra of $In_{0.8}Ga_{0.2}N$ films deposited on glass substrates.

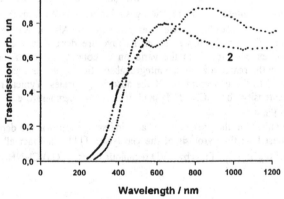

Fig.5. Optical absorption spectra of $In_{08}Ga_{02}N$ films: 1 - 220 nm, 2 - 190 nm thick.

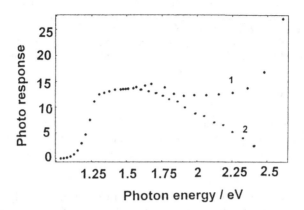

Fig 6. Spectral response of the p-Si/In$_{08}$Ga$_{02}$N heterojunction (1) and Si homojunction (2).

Fig.6. illustrates the spectral response of the heterojunction, within the photon energy range 0.5 eV < hv < 2.5 eV. The lower cut-off around 1eV arises from electron-hole pair generation within the silicon base material. At hv > 2.0 eV the response again increases, due to photoabsorption in the front window In$_{0.8}$Ga$_{0.2}$N layer. The maximum responsivity achieved is approx. 130mV/W and sensitivity depends on the chopper frequency of the incident radiation. For comparison the typical photoresponse pattern of a Si-homojunction is shown, that is gradually decreasing at hv >1.5 eV.

This is illustrated in Fig.7., where the photoresponse as a function of chopper frequency of the increasing radiation is plotted at a wavelength of 633 nm. Typically, a very weak signal is observed under dc-conditions that gradually increases. At higher values, a degrading response is found, due to the high electrical time constant and RC-product of the device. The strong frequency dependence is understood on the basis of the high recombination rate of minority carriers within the junction accounting for very small life times and, hence, a low μT product, defining the minority carrier diffusion length.

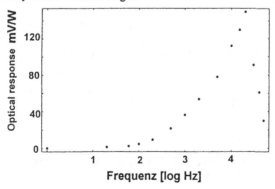

Fig 7. Photoresponse as a function of chopper frequency.

With regard to solar cell performance, these findings represent a serious obstacle, towards further utilization, since at dc-operation, the optical response virtually disappears. Scanning electron microscopy already has indicated that the grain size is below 0.5 μm. Therefore the impurity level would be unacceptably high. Grain size should be of the order of 100μm or more to drastically lower the recombination rate. Solutions are either thermal treatment and postanneal. Also outdiffusion of impurities in a suited chemical environment could be favorable which also could lead to further defect passivation.

CONCLUSIONS

The investigations demonstrate the possibility of the deposition of $In_xGa_{(1-x)}N$ films by CVD technique using $In(Ga)Cl_3NH_3$ system. It was suggested that CVD of $In_xGa_{(1-x)}N$ layers based on pyrolysis of $In(Ga)Cl_3NH_3$ complex is determined by competition of the irreversible reactions.

$In Cl_3NH_3$ ----> $InN + 3HCl$ (1a) $InN + HCl$ ----> $InCl + 1/2\ H_2 + 1/2\ N_2$ (2a)

$GaCl_3NH_3$ ----> $GaN + 3HCl$ (1b) $GaN + HCl$ -----> $GaCl + 1/2H_2 + 1/2\ N_2$ (2b)

The growth of $In_xGa_{(1-x)}N$ films occurs only because of the higher rate of reaction (1) in comparison with that of the reaction (2) under the same conditions.

The possibility of using $In_xGa_{(1-x)}N$ films in the solar cells is demonstrated. The research and development of In(Ga)N - based solar cells are in the initial stages. Further improvements in technology and material quality should lead to the development of much better devices. Ohmic contact formation to the III-Nitrides must be improved and this problem is presently under intense investigation. The use of different substrates and the investigation of n-type In(Ga)N / p-polySi may be of great interest. We hope that the technological task will be established in near future. The full potential of In(Ga)N - based devices is still to be discovered.

REFERENCES

1. S.Nakamura, MRS Bull. 22, 2, 29 (1997).

2. M.S.Shur, M.A. Khan , MRS Bull. 22, 44(1997).

3. T.Egawa, H.Ishikawa, T.Jimbo, M.Umeno, Appl.Phys.Lett.,69, 830 (1996).

4. H.Morkoc, S.Strite, G.B.Gao, M.E.Lin, B.Swerdlov and M.Burns, J.Appl.Phys. 76 (3), 1363 (1994).

5. K.J.Bachmann in Current Topics in Materials Science, edited by E.Kaldis (North-Holland publishing Company 1979) p 482.

6. J.J.Loferski, J.Appl.Phys. 27, 510 (1961).

7. S.E.Alexandrov, A.Y.Kovalgin and D.M.Krasovitsky, J. de Physigue IV, colloque C5, supplement au Journal de Physique II, 5, 1995, p c5-183.

SURFACE PREPARATION AND GROWTH CONDITION DEPENDENCE OF CUBIC GaN LAYER ON (001) GaAs BY HYDRIDE VAPOR PHASE EPITAXY

H. TSUCHIYA, K. SUNABA, S. YONEMURA, T. SUEMASU and F. HASEGAWA
Institute of Materials Science, University of Tsukuba, 1-1-1 Tennohdai, Tsukuba, Ibaraki 305, Japan

ABSTRACT

GaN buffer layers and thick GaN layers were grown on (001) GaAs substrates by hydride vapor phase epitaxy. The ratio of cubic to hexagonal components in the grown layer was estimated from the ratio of the integrated X-ray diffraction intensities of the cubic (002) plane and hexagonal (1011) planes measured by w scan. The optimum growth conditions were thermal cleaning at 600°C, growth temperature of 500°C and thickness of 30 nm for the buffer layer, and the V/III ratio of 300 for thick GaN growth at 800°C. Cubic component in the layer grown with those conditions was more than 85% and strong cubic photoluminescence emission was observed at 377 nm (3.28 eV).

INTRODUCTION

GaN is an attractive material for optoelectronic devices in the region of blue to ultraviolet light and high temperature electronic device applications. Usually, the GaN layers with the best optical and electrical properties have been grown on basal plane sapphire [1,2]. On the other hand, it has been reported that cubic GaN can be grown on substrates with cubic structure, such as (001) GaAs [3,4] and (001) Si [5]. Compared to a sapphire substrate, a (001) GaAs substrate is attractive for GaN laser diodes (LDs) because the cleavage along (110) planes are possible by growing cubic crystals, which form mirrors in the LDs. Furthermore, GaAs substrate is easy to etch by aqua regia and to make ohmic contact on. It is known, however, that the hexagonal phase can coexist with the cubic phase in the layers on (001) GaAs substrates depending on the growth conditions [6-8].

Hydride vapor phase epitaxy (HVPE) has been widely used as a growth method for high-quality compound semiconductors at high growth rate [9]. In this case of hexagonal GaN, Hiramatsu et al. obtained very thick (~1200 μm) layers on sapphire substrates at a high growth rate (30-70 μm/h) by HVPE [10]. Recently, Miura et al. have reported that nominally cubic GaN was obtained using GaN buffer layer using TEGa, HCl and NH₃ [11].

It has been well known that most of hexagonal component in the GaN layer grown on a (001) GaAs substrate has a crystal structure whose c-axes are parallel to the GaAs <111> direction. In this case, the volume of hexagonal GaN is difficult to be estimated by a θ-2θ scan because this hexagonal GaN layer does not have any lattice plane paralleled to the GaAs (001) plane.

In this paper , we reported that the ratio of cubic to hexagonal components in GaN layers grown on (001) GaAs can be estimated by the ω scan. Furthermore, we estimated the dependence of the hexagonal component on the growth condition by this method. It was found that the cubic/hexagonal ratio greatly depended on the thermal cleaning temperature and the V/III ratio during the growth. It was the demonstrated that a thick GaN layer whose cubic component was more than 85% was obtained under the optimized growth condition and strong cubic photoluminescence (PL) emission was observed.

EXPERIMENTS

The GaN layers were grown in a conventional HVPE system using H₂ as a carrier gas. Metal Ga and NH₃ were used as the gallium and nitrogen sources, respectively. HCl gas from a gas cylinder (10% HCl diluted with N₂) was supplied to the Ga source path to produce GaCl. Before being inserted into the growth zone, (001) GaAs substrates were etched in a etchant of H₂SO₄:H₂O₂:H₂O = 5:1:1 at 60°C for 2 min and dipped into HF solution for 15 min. The GaAs substrates were submitted to thermal cleaning in H₂ ambient prior to the growth to remove the surface oxide. To obtain the optimum thermal cleaning, the temperature was changed from 550°C to 650°C. After being cooled down to the desired substrate temperature (~ 500°C), the GaAs substrate was kept in an NH$_3$ ambient for 15 minutes, and HCl supply was started to grow the buffer layer. The thickness of the GaN buffer layer was about 30 nm. Then the substrate temperature was raised to the growth temperature of 800°C in a flow of H$_2$ and NH$_3$. Subsequently, GaCl was supplied to grow a thick GaN layer (with growth rate of about 2μm/hour in this case. In order to determine the optimum growth conditions, the V/III ratio was varied between 200 and 600.

The measurement of XRD (X-ray diffraction) reciprocal space mapping of GaN layer was carried out with a X'Pert-MRD (Phillips). The ratio of cubic to hexagonal components in the GaN layers grown on (001) GaAs was estimated from the ratio of the integrated XRD intensities of the cubic (002) and hexagonal (10$\bar{1}$1) planes measured by ω scan. Hexagonal GaN layers whose c-axes are oriented to the four equivalent GaAs <111> directions may be grown on GaAs or cubic GaN (001) surfaces. The XRD pattern of each grown layer was taken by a ω scan in the [110] and [1$\bar{1}$0] directions and the four integrated XRD intensities from the hexagonal (10$\bar{1}$1) planes were summed up.

Luminescent property of the GaN layers was studied by PL at 77K using a He-Cd laser at 325 nm as an excitation source.

RESULTS AND DISCUSSION

Figure 1 show typical XRD reciprocal space mapping of GaN layer grown by HVPE. Besides the (002) reflection of the GaAs substrate, the (002) reflection of cubic GaN is visible in this scan. The additional diffraction feature detected from the (10$\bar{1}$1) reflection of hexagonal GaN domains oriented such that [0001]//[111]. From this result, it is confirmed that the hexagonal phase is introduced in the cubic phase in GaN layers on (001) GaAs substrates. These results correspond to the previous report by many researchers [6,7,12].

However, the estimation of the ratio of cubic to hexagonal components in the grown GaN layer is not so simple as it appears, because XRD peak intensities usually change for different lattice planes. Furthermore, There is no report on the XRD intensity of cubic compared with hexagonal GaN because a high-quality thick cubic GaN layer has not yet been obtained. So, the theoretical XRD intensities of cubic and hexagonal GaN were calculated. The integrated XRD intensity I is given by

$$I = I_0 \times |F|^2 \times M \times L \times V \times N \times e^{-2M}$$

where I_0, F, M, L, V, N, e^{-2M} are the incident X-ray intensity, structure factor, multiplicity factor, volume, Lorentz polarization factor, a number of unit cell per unit volume and temperature factor, respectively. Table I shows the relative XRD intensity from one plane for several GaN planes. It was found that the integrated XRD intensity from the hexagonal (10$\bar{1}$1) plane was nearly equal to that from the cubic (002) plane. Therefore, the ratio of cubic to hexagonal components in grown GaN layers can be estimated from the ratio of the integrated XRD intensity from the cubic (002) plane to the summation of those from the hexagonal (10$\bar{1}$1) planes measured by the ω scan. Figure

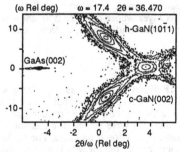

(ω Rel deg)   ω = 17.4 2θ = 36.470

Fig. 1 XRD reciprocal space mapping.

Table I Theoretical XRD intensities of cubic and hexagonal GaN.

Structure	Peak	θ(deg.)	I($\times 10^4$)	I/I$_{c\text{-}(002)}$
Cubic	(111)	17.2	18.8	2.35
	(002)	20.0	7.98	1
	(220)	28.9	4.86	0.61
	(311)	34.5	2.10	0.26
Hexagonal	(10$\bar{1}$0)	16.1	7.6	0.95
	(0002)	17.3	18.5	2.32
	(10$\bar{1}$1)	18.4	8.47	1.06
	(10$\bar{1}$2)	24.2	1.63	0.20

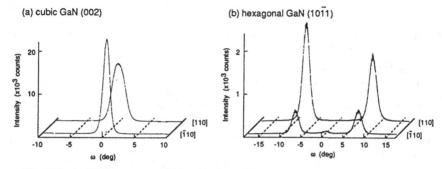

(a) cubic GaN (002)

(b) hexagonal GaN (10$\bar{1}$1)

Fig. 2 Typical XRD patterns of (a) cubic GaN (002) and (b) hexagonal GaN (10$\bar{1}$1)
peaks measured by the ω scan.

2 show typical XRD patterns of cubic GaN (002) and hexagonal GaN (1011) peaks measured by
ω scan. We defined ω as the difference between the incident angle and the reflection angle into the
(001) GaAs substrate for a certain detection position. From the (002) diffraction peak of cubic
GaN obtained by ω measurement, it is found that a GaN layer has a cubic structure with the [001]
direction perpendicular to the GaAs substrate as shown in Fig. 2(a). On the other hand, the
hexagonal (10$\bar{1}$1) planes are tilted from the GaAs (001) plane about 7° toward the <111> direction
as shown in Fig. 2(b). It was found that the c-axis of hexagonal GaN in the grown layer was
parallel to the <111> axis of the GaAs substrate, taking into account that the angles between the
GaAs (001) and GaAs (111) planes and between the hexagonal (0002) and hexagonal (10$\bar{1}$1)
planes were 54.7° and 61.9°, respectively.

Figure 3 shows the integrated XRD intensities as a function of the V/III ratio. The V/III ratio
was changed from 200 to 600 by changing the NH$_3$ flow rate at the growth temperature of 800°C.
When the obtained GaN layer was measured by θ-2θ scan, there was no peak corresponding to
hexagonal GaN, and only the peak corresponding to the (002) diffraction of cubic GaN was
observed. The four curves in Fig. 3 shows the XRD intensities of the cubic (002) plane and the
hexagonal (10$\bar{1}$1) planes with two different incident azimuths of X-ray. When the V/III ratio was
600, the integrated XRD intensity of the hexagonal (10$\bar{1}$1) plane for an incidence azimuth of [110]
was the largest and the others were negligible. This result indicates that a hexagonal GaN layer
whose c-axes are oriented in the GaAs <111>B direction is dominantly grown on (001) GaAs,
even though only an XRD peak of cubic GaN is observed by the θ-2θ scan. On the other hand, the
integrated XRD intensity of cubic GaN increases, and that of hexagonal GaN decreases sharply

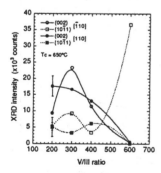

Fig. 3 Dependence of the integrated XRD
intensities on the V/III ratio.

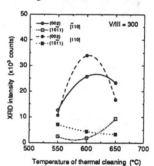

Fig. 4 Dependence of the integrated XRD
intensities on the thermal cleaning
tenperature.

with a decrease in the V/III ratio. This tendency is the same as that obtained by MBE and MOVPE growth of GaN [6,13]. When the V/III ratio was 300, the largest integrated XRD intensity of cubic GaN was obtained. In this case, the integrated XRD intensity of cubic GaN was about twice as large as that of hexagonal GaN. When the V/III ratio was 200, however, the integrated XRD intensity of cubic GaN was reduced. From these results, it was found that cubic GaN can be dominantly grown on (001) GaAs using the optimized V/III ratio even in HVPE.

From TEM measurements, it is known that the formation of hexagonal GaN or stacking faults in GaN grown on (001) GaAs originates from the GaN/GaAs interface [12]. Furthermore, the surface of GaAs frequently has facets of {111} planes. We suspected that these facets might be formed during the thermal cleaning. In order to suppress the formation of facets, therefore, GaN layers were grown at various thermal cleaning temperatures. Figure 4 shows dependence of the integrated XRD intensity measured by the ω scan on the thermal cleaning temperature. When the thermal cleaning was carried out at 600°C for 10 minutes in an H_2 atmosphere, the integrated XRD intensity of cubic GaN was about ten times as large as that of hexagonal GaN. Furthermore, the surface of the GaN layer obtained with the thermal cleaning at 600°C and V/III ratio of 300 (Fig. 5(c)) was much smoother than those obtained with other growth conditions (Figs. 5(a), (b)). This result indicates that the surface morphology of the GaAs substrate prior to the growth greatly influences the surface and quality of thick GaN layers.

From these results, it was concluded that the ratio of cubic to hexagonal components in the grown GaN layers greatly depended on the V/III ratio during growth and the surface morphology of the GaAs substrate prior to the buffer layer growth. Under optimized growth conditions, it was found that a cubic dominant GaN layer can be grown on (001) GaAs.

In order to clarify the luminescent property of the GaN layer, the PL spectra were measured at 77K. When the GaN layer was grown with the thermal cleaning at 650°C and V/III ratio of 600, a large broad emission near 500 nm in addition to the 362 nm ultraviolet (UV) emission was observed as shown in Fig. 6(a). The emission intensity itself was small, as can be imagined from the noise of spectrum.

The emission at around 500 nm almost disappeared for the sample grown with the thermal cleaning at 650°C and V/III ratio of 300 as shown in Fig. 6(b). Furthermore, The UV emission peak was narrower and the peak wavelength (371 nm) was longer than those of the sample grown with the thermal cleaning at 650°C and V/III ratio of 600.

When the GaN layer was grown with the thermal cleaning at 600°C and V/III ratio of 300, the spectrum exhibited a strong emission at 377 nm (3.28 eV) and no emission was seen in the visible spectral region and corresponding to hexagonal GaN. This peak agrees well with band edge emission of cubic GaN layer reported by Okumura et al [14]. These PL results also indicate that cubic dominant GaN can be grown by optimizing the surface preparation and growth conditions.

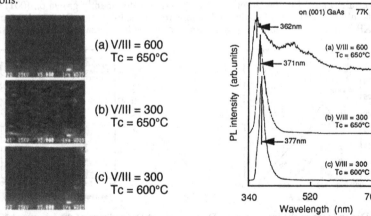

(a) V/III = 600
Tc = 650°C

(b) V/III = 300
Tc = 650°C

(c) V/III = 300
Tc = 600°C

Fig.5 Surface images of the GaN layers obtained at various growth conditions.

Fig.6 PL spectra of the GaN layers obtained at various growth conditions.

CONCLUSION

Thick GaN layers were grown by HVPE on (001) GaAs substrates with GaN buffer layers grown under various growth conditions. The cubic to hexagonal component ratio of the grown layer was investigated by XRD measurements using the ω scan. The optimum growth conditions to obtain cubic-dominant GaN were found to be a thermal cleaning temperature of 600°C and a V/III ratio of 300. With these growth conditions, a GaN layer whose cubic component was about 85% was grown on (001) GaAs. Furthermore, a strong emission at 377 nm (3.28 eV) without any emission in the visible region indicated high-quality of the cubic GaN.

ACKNOWLEDGEMENT

The authors would like to express their sincere thanks to Mr. K. Saito of Philips Japan, LTD. for measurement of XRD reciprocal space mapping, Mr. T. Takahashi of Electrotechnical Laboratory for his help with thickness measurement of the GaN buffer layer by ellipsometry, and to Dr. S.H. Cho and Professor K. Akimoto of our institute for the PL measurement.

REFERENCES

1. S. Nakamura, M. Senoh, S. Nagahama, N. Iwasa, T. Yamada, T. Matsushita, H. Kiyoku and Y. Sugimoto, Japan. J. Appl. Phys., **35**, L74 (1996).
2. M.A. Kahn, M.S. Shur, J.N. Kuznia, Q. Chen, J. Burn and W. Schaff, Appl. Phys. Lett., **66**, 1083 (1995).
3. M.Mizuta, S.Fujieda, Y.Matsumoto and T. Kawamura, Japan. J. Appl. Phys., **25**, L945 (1986).
4. H. Tsuchiya, K. Sunaba, S. Yonemura, T. Suemasu and F. Hasegawa, Japan. J. Appl. Phys., **36**, L1 (1997).
5. T. Lei, M. Fancilli, R.J. Monar, T.D. Moustakas, R.J. Graham and J. Scanlon, Appl. Phys. Lett., **59**, 944 (1991).
6. O. Brandt, H. Yang, B. Jenichen, Y. Suzuki, L. Daweritz and K.H. Ploog, Phys. Rev. B, **52**, R2253 (1995).
7. A.A. Yamaguchi, T. Manako, A. Sakai, H. Sunakawa, A. Kimura, M. Nido and A. Usui, Japan. J. Appl. Phys., **35**, L873 (1996).
8. A. Nakadaira and H. Tanaka, *Proc. Int. Symp. Blue laser & Light Emitting Diodes*, Chiba, Japan 1996 (Ohmsha Ltd., Tokyo, 1996) pp. 90.
9. J.J. Tietjen and J.A. Amick, J. Electrochem. Soc., **133**, 724 (1966).
10. K. Hiramatsu, T. Detchprohm and I. Akasaki, Japan. J. Appl. Phys., **32**, 1528 (1993).
11. Y. Miura, N. Takahashi, A. Koukitu and H. Seki, Japan. J. Appl. Phys., **35**, 546 (1996).
12. N. Kuwano, Y. Nagatomo, K.Kobayashi, K. Oki, S. Miyoshi, H. Yaguchi, K. Onabe and Y. Shiraki, Japan. J. Appl. Phys., **33**, 18 (1994).
13. M.Sato, J. Appl. Phys.,**78**, 2123 (1995).
14. H. Okumura, S. Yoshida and T. Okahisa, Appl. Phys. Lett., **64**, 2997 (1994).

GROWTH OF GaN THIN FILMS ON SAPPHIRE SUBSTRATE BY LOW PRESSURE MOCVD

M. Ishida, T. Hashimoto, T. Takayama, O. Imafuji, M. Yuri, A. Yoshikawa, K. Itoh,
Y. Terakoshi*, T. Sugino*, and J. Shirafuji*
Electronics Research Lab., Matsushita Electronics Corporation, Osaka 569, Japan,
masa@erl.mec.mei.co.jp
*Department of Electrical Engineering, Osaka Univ., Osaka 565, Japan

ABSTRACT

High quality GaN films are grown on sapphire(0001) substrates by low pressure MOCVD using TMG and NH_3 as source materials. Effects of surface nitridation and buffer layer thickness on the quality of over-grown GaN films are investigated. It is revealed by atomic force microscope (AFM) observations that surface roughness of the annealed buffer layers strongly depends on the nitridation time. Dislocation density and surface morphology of the high temperature GaN layer depend on the buffer layer thickness. It is found that sufficient surface nitridation of sapphire makes the buffer layer just prior to the high temperature growth very smooth, which is essential to obtain flat thick-GaN on it. It is also found that thickness of the buffer layer largely influences the dislocation density in the over-grown thick GaN. In order to obtain good surface morphology and low dislocation density at the same time, both nitridation time and buffer layer thickness must be optimized.

INTRODUCTION

GaN and related compounds are promising materials for short wavelength light emitting diodes (LEDs) [1], laser diodes(LDs) [2,3], and high-temperature/high-speed electronic devices [4], because of their large band-gaps, high thermal conductivity, high saturation carrier velocity, and strong chemical bonding. However, high equilibrium pressure of nitrogen over GaN surface at growth temperatures makes it difficult to obtain bulk single crystal [5]. Therefore, heteroepitaxial growth on various substrates has been investigated.

Sapphire is one of the best substrates for GaN growth and many attempts to grow GaN epitaxially on sapphire have been made. Amano et al. and Nakamura reported that GaN epitaxial layers can be obtained on a thin AlN or GaN buffer layer grown at a low temperature by MOCVD, in spite of the large mismatch in lattice constants and thermal expansion coefficients between GaN and sapphire [6,7]. Uchida et al. found that crystalline quality of GaN depends on nitridated layers formed on sapphire [8]. Wu et al. reported on structural and morphological evolution of GaN grown on GaN buffer layers at high temperatures [9]. However, detailed mechanisms of these processes are only partially understood. Also, the role of the nitridated layer and buffer layer in low pressure MOCVD is still not clear since these previous attempts were made by atmospheric pressure MOCVD.

In this paper, we report the dependence of the structural and optical properties of GaN on the nitridation time and the buffer layer thickness in low pressure MOCVD. The surface morphology of low temperature buffer layer and over-grown high temperature GaN are examined for various nitridation time and buffer layer thickness, through which growth mechanisms and optimum conditions are discussed.

EXPERIMENTAL

GaN films were grown on c-plane sapphire (α-Al$_2$O$_3$(0001)) by low pressure MOCVD. Trimethyl gallium(TMG) and ammonia(NH$_3$) were used as source materials and H$_2$ was used as a carrier gas. The substrates were cleaned in organic solvents and loaded in a horizontal MOCVD reactor. The substrate was preheated in H$_2$ ambient for 10min at 1000°C. Then, the substrate was nitridated at 1000°C in NH$_3$/H$_2$ ambient. The nitridation time was varied from 5min to 30min. After the nitridation of the substrate, a GaN buffer layer was deposited at 600°C. The growth time of the GaN buffer layer was varied from 5min to 30min. We estimated the thickness of the buffer layer assuming the growth rate of 120nm/hour, which is obtained separately by growing a thick buffer layer (~0.5μm) under the same condition. An epitaxial GaN film was grown on the buffer layer for 1 hour (~1μm) at 1000°C with a V/III ratio of 5500. The pressure was maintained at 6.67×10^3 Pa(50Torr) throughout the growth. Growth sequence and flow rates of the source gases are shown in Fig. 1.

Surface morphology of GaN was observed with Nomarski microscope. Photoluminescence spectra were recorded at room temperature using front excitation and front emission configuration. All samples were excited by the 325nm line of a He-Cd laser. In order to investigate the initial stage of GaN growth, surface morphology of the buffer layer itself was studied by atomic force microscopy (AFM). Dislocation density in the thick GaN films was estimated by cross-sectional transmission electron microscopy (TEM) observation.

	Thermal Cleaning	Nitridation	Buffer layer growth	High temp. growth
TMG Flow rate			15.3μmol/min	30.7μmol/min
NH$_3$ Flow rate			170 mmol/min	
H$_2$ Flow rate	9 slm			
Substrate temp.	1000°C	1000°C	600°C	1000°C
Pressure	6.67×10^3Pa (50Torr)			
Time	15min	5~30min	5~30min	60min

Fig. 1. Growth sequence and flow patterns of the gases. Nitridation time and buffer layer growth time are varied. The growth rate of the buffer layer is 120nm/hour.

RESULTS AND DISCUSSION

Figure 2 shows the surface morphology of the high-temperature-grown GaN films with various nitridation time and buffer layer thickness. The films grown with 30min nitridation and a 40nm buffer layer(Fig. 2(d)) or 60nm buffer layer(Fig. 2(e)) exhibit very smooth surface, while those grown with shorter nitridation time or a thinner buffer layer show 3D growth (Fig. 2 (a),(b),(c))

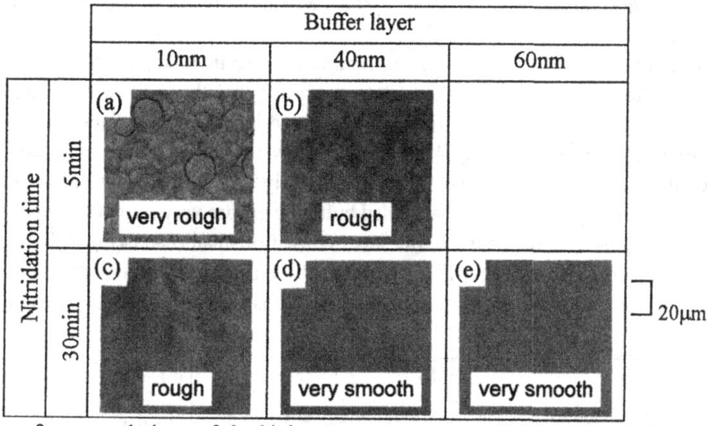

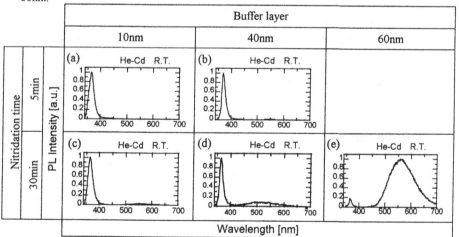

Fig. 2. The surface morphology of the high-temperature-grown GaN: (a) nitridation 5min, buffer layer 10nm; (b) nitridation 5min, buffer layer 40nm; (c) nitridation 30min, buffer layer 10nm; (d) nitridation 30min, buffer layer 40nm;and (e) nitridation 30min, buffer layer 60nm.

Fig. 3. Room-temperature photoluminescense spectra of GaN: (a) nitridation 5min, buffer layer 10nm; (b) nitridation 5min, buffer layer 40nm; (c) nitridation 30min, buffer layer 10nm; (d) nitridation 30min, buffer layer 40nm; and (e) nitridation 30min, buffer layer 60nm.

Figure 3 shows the photoluminescence spectra from the GaN films corresponding to those in Fig. 2. Strong near band-edge emission around 370nm is observed from the films with 10nm or 40nm buffer layers (Fig. 3 (a),(b),(c),(d)). However, the spectrum from the film grown with 30min nitridation and a 60nm buffer layer(Fig. 3(e)) is dominated by deep level emission around 550nm though the surface morphology is very smooth. This means that the GaN films with strong near band-edge emission do not always exhibit a good surface morphology. It is suggested that the strong near band-edge emission from the GaN films with rough surface morphology reflects good crystalline quality of each of the columns or grains in the GaN films.

In order to study the effects of the nitridation process in detail, the surface of the buffer layer as deposited and just prior to the high temperature growth(i.e. after annealing at 1000°C for 5min) with different nitridation time was observed by AFM. The thickness of the buffer layer is 10nm for all samples. Fig. 4 and Fig. 5 show the AFM images of as-deposited and annealed surface of the GaN buffer layer, respectively. Without nitridation of sapphire, the surface of both as-deposited and the annealed buffer layers are very rough(Fig. 4(a), Fig. 5(a)). For 10min nitridation, as-deposited buffer layer consist of small islands(Fig. 4(b)). It is clearly seen, by comparing Fig. 4(b) and Fig. 5(b), that the buffer layer experiences reconstruction during annealing, resulting in a smaller number of islands with larger sizes. For 30min

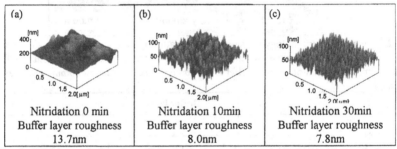

Fig. 4. AFM images of surface of the buffer layer as deposited: (a) nitridation 0min; (b) nitridation 10min; and (c) nitridation 30min.

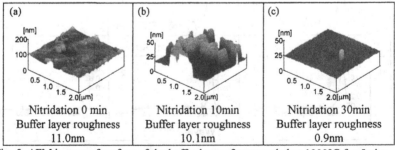

Fig. 5. AFM images of surface of the buffer layer after annealed at 1000°C for 5min; i.e. the surface just before the high temperature growth: (a) nitridation 0min; (b) nitridation 10min; and (c) nitridation 30min.

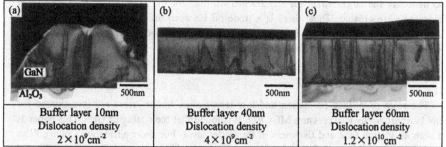

Buffer layer 10nm	Buffer layer 40nm	Buffer layer 60nm
Dislocation density	Dislocation density	Dislocation density
$2 \times 10^9 cm^{-2}$	$4 \times 10^9 cm^{-2}$	$1.2 \times 10^{10} cm^{-2}$

Fig. 6. Cross-sectional TEM images of the GaN films: (a) buffer layer 10nm, (b) buffer layer 40nm, (c) buffer layer 60nm. Densities are counted for the dislocations which reach the flat part of the surface.

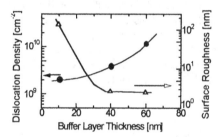

Fig. 7. Dislocation density and surface roughness of high temperature GaN dependence on buffer layer thickness.

nitridation, the islands of as-deposited GaN buffer layer(Fig. 4(c)) become much smaller than Fig. 4(b) and the surface of the annealed buffer layer (Fig.5(c)) becomes very smooth. This clearly demonstrates that the surface of the buffer layer is strongly influenced by nitridation and that sufficient surface nitridation(1000°C, 30min in the present experiments) is essential to obtain a flat buffer layer just prior to the high temperature growth. Uchida et al. reported that nitridation at 1050°C for 3min creates a uniform nitridated layer and they obtained a flat GaN film on such a flat buffer layer [8]. We believe that difference in nitridation time between the present results and their report comes from the difference in pressure, temperature, or flow conditions of NH_3/H_2 gas.

Cross-sectional TEM observation was performed to evaluate the dislocation density in the GaN films. Fig. 6 shows the TEM images of GaN films with different buffer thickness. The nitridation time is chosen to be 30min for all samples, which is needed to obtain a flat buffer layer as discussed above. In case of Fig. 6(a), we count the dislocations which reach the flat part of the surface. Surface roughness is also measured by a surface profiler. Fig. 7 shows the dependence of the dislocation density and the surface roughness on the buffer layer thickness. As the buffer layer becomes thinner, the dislocation density decreases, and the surface becomes rougher. We believe that the buffer layer thickness determines the nucleation density. When a buffer layer is thin, the nucleation density becomes low and GaN film favors 3D growth, which leads to a rough surface. When a buffer layer is thick, on the other hand, the nucleation density becomes high and GaN films favors 2D growth. However, in order to accommodate the larger strain for the 2D growth than for the 3D growth, high density of dislocations are created. Thus, there is a trade-off between dislocation density and surface morphology. In order to achieve low dislocation density and smooth surface at the same time, the buffer layer thickness must be controlled around 40nm.

CONCLUSIONS

The roles of surface nitridation and low temperature buffer layers are investigated for GaN films grown by low pressure MOCVD. It is found that the quality of GaN films depends on both nitridation time and thickness of the buffer layer. For short nitridation (5min), films generally exhibit 3D growth from which strong band-edge PL is observed. For long nitridation (30min), the surface morphology is significantly improved especially when the buffer layer is thick(40 or 60nm). The role of surface nitridation on the structural properties of buffer layer itself is examined separately, which revealed that sufficient nitridation (30min) is necessary to obtain a flat surface of the buffer layer. Thus, sufficient nitridation and a thick buffer layer are important from the viewpoint of surface morphology. From the viewpoint of dislocation density in the films, on the other hand, it is found that a thinner buffer layer is favorable. Thus, there is optimum buffer layer thickness from the viewpoint of the surface morphology, photoluminescence spectra, and dislocation density.

ACKNOWLEDGEMENTS

The authors would like to thank Dr. T. Baba and Mr. T. Ueda for their helpful discussions, Dr. Y. Yabuuchi of Matsushita Technoresearch Inc. for his cooperation in TEM observation and useful discussions. The authors are also grateful to Dr. G. Kano for his continuing supports and encouragement to this work.

REFERENCES

1. S. Nakamura, M. Senoh, N. Iwasa, S. Nagahama, T. Yamada, and T. Mukai, Jpn. J. Appl. Phys. **34**, L1332 (1995).
2. S. Nakamura, M. Senoh, S. Nagahama, N. Iwasa, T. Yamada, T. Matsushita, Y. Sugimoto, and H. Kiyoku, Appl. Phys. Lett. **69**, 4056 (1996).
3. K. Itaya, M. Onomura, J. Nishio, L. Sugiura, S. Saito, M. Suzuki, J. Rennie, S. Nunoue, M. Yamamoto, H. Fujimoto, Y. Kokubun, Y. Ohba, G. Hatakoshi, and M. Ishikawa, Jpn. J. Appl. Phys. **35**, L1315 (1996).
4. M. A. Khan, M. S. Shur, J. N. Kuznia, Q. Chen, J. Burm, and W. J. Schaff, Appl. Phys. Lett. **66**, 1083 (1995).
5. T. Matsuoka, Photoluminescence of InN and InGaN, in Properties of Group III Nitrides, edited by J. H. Eddgar, (INSPEC IEE, London, 1994), p. 231.
6. H. Amano, N. Sawaki, I. Akasaki, and Y. Toyoda, Appl. Phys Lett. **48**, 353 (1986).
7. S. Nakamura, Jpn. J. Appl. Phys. **30**, L1705 (1991).
8. K. Uchida, A. Watanabe, F. Yano, M. Kouguchi, T. Tanaka, and S. Minagawa, Solid State Electron. **41**, 135 (1997).
9. X. H. Wu, P. Fini, S. Keller, E. J. Tarsa, B. Heying, U. K. Mishra, S. P. DenBaars, and J. S. Speck, Jpn. J. Appl. Phys. **35**, L1648 (1996).

MBE-GROWTH OF STRAIN ENGINEERED GaN THIN FILMS UTILIZING A SURFACTANT

R. Klockenbrink,*,1 Y. Kim, * M. S.H. Leung,* C. Kisielowski,* J. Krüger,*
Sudhir G.S.,* M. Rubin,** and E. R. Weber*,**

* Department of Materials Science and Mineral Engineering, University of California,
Berkeley, CA 94720
** Materials Science Division, Lawrence Berkeley National Laboratory, Berkeley, CA 94720
1 ralfkl@uclink4.berkeley.edu

ABSTRACT

GaN films were grown on sapphire substrates at temperatures below 725 °C utilizing a Constricted Glow Discharge plasma source. A three dimensional growth mode is observed at such low growth temperatures resulting in films that are composed of individual but oriented grains. The strain that originates from the growth on the lattice mismatched substrate with a different thermal expansion coefficient is utilized to influence the thin film growth. The strain can be largely altered by the growth of suitable buffer layers. Thereby, optical and structural film properties can be engineered. It is argued that the surface diffusion of Ga ad-atoms is affected by engineering the strain. Alternatively, surface diffusion can be influenced by surfactants. It is demonstrated that the use of bismuth as a surfactant allows to modify the surface morphology of the GaN films that reflects the size of the grains in the films. The results suggest that a substantial increase of the oriented grain sizes in the films is possible while maintaining a low growth temperature.

INTRODUCTION

Group-III nitrides are the most promising materials for optoelectronic light sources [1, 2] that can emit light from the ultraviolet to the visible spectral range and as well for high frequency power devices [3]. Usually, GaN films are grown at temperatures that are low compared with its melting point ($T_{growth} < 0.5\ T_m$). In contrast to MOCVD, a MBE growth process can exploit deviations from thermodynamic equilibrium to a larger extend. This may help to increase doping levels or to grow high quality AlN/InN/GaN quantum well structures. However, due to the lower growth temperature (typical: MBE growth at 1000 K; MOCVD growth at 1300 K) the films grown by MBE usually exhibit a three dimensional growth mode with grain sizes that are limited by the growth temperature and by strain. This determines the structural quality of the thin films [4].

Recent progress in understanding GaN thin film growth revealed that strain is one of the key issues that determines the growth and physical properties of the thin films [4-8]. Strain can largely be altered by the growth of a buffer layer [6]. In fact, the active utilization of the buffer-layer growth-temperature [7], of its thickness and of the III/V flux ratio [8] open the possibility to strain engineer desired film properties such as the surface morphology, the size of the grains in the films [4], or the optical film properties [9]. It was argued that the surface diffusion of the Ga ad-atoms and the stoichiometry of the GaN films are affected by strain [4, 6]. Nevertheless, GaN thin film growth at MBE growth temperatures remained three dimensional and there are several options that can be explored to improve the structural film quality by enhancing the surface diffusion length of the Ga ad-atoms during growth and, thereby, the size of the grains in the films.

Obviously, the MBE growth temperatures can be increased. This is possible because nitrogen sources with large fluxes are now available such as our Constricted Glow Discharge (CGD) plasma source (demonstrated growth rate: 1.4 μm/h) [10]. Using several of such sources allows to overcome the film decomposition rate even at growth temperatures above 800 °C. However, this approach puts limits on the exploration of the low temperature growth. It is for this reason that we choose to perform experiments with surfactants in order to enhance surface diffusion while maintaining a growth temperature of 725 °C.

This contribution focuses on the utilization of bismuth (Bi) as a surfactant for the growth of GaN by MBE. Bi seems to be a promising choice for at least two reasons. First, Bi is isoelectronic with N but of extremely different size (covalent radii: N=0.07 nm, Bi=0.15 nm). Thus, it is unlikely that large amounts of Bi will be incorporated into the growing film. On the surface, however, Bi will substitute for N and, thereby, alter the bonding to the diffusing Ga ad-atoms. Second, GaN was grown from Bi solutions.

EXPERIMENTS

GaN layers are grown using a rebuilt Riber 1000 MBE system. Knudsen cells are used to evaporate pure Ga (99.9999%), Mg (99.99%), and Bi (99.9999%) while the activated nitrogen is produced by the CGD plasma source with pure nitrogen gas (99.9999%) along with a Millipore nitrogen purifier. Some details of the source design are given elsewhere [10]. A dc voltage generates a glow discharge that is constricted to an area in the plasma chamber close to the gas exit. It is the pressure difference between the plasma chamber and the MBE growth chamber that extracts the activated nitrogen species with energies less than 5 eV. A liquid nitrogen cryoshroud is used during growth to obtain a base pressure in the chamber of ~ 2 x 10^{-10} Torr. A thin titanium (Ti) layer on the back of the 10 x 11 mm^2 c-plane sapphire substrate absorbs the heat radiated from the tungsten (W) filament heater. The temperature of the substrate is monitored with a pyrometer.

The substrates are degreased by boiling in acetone and ethyl alcohol for 5 minutes each and blown dry with nitrogen. After degassing in the load lock for 30 min at 500 °C, they are transferred into the growth chamber. The substrates are then heated up to 700 °C for thermal desorption of surface contaminants. At this temperature, they are exposed to activated nitrogen for 10 minutes. Subsequently, a thin low temperature GaN buffer layer (~ 250 Å) is deposited on the substrate. Its particular thickness was determined and optimized by intentionally engineering the strain [7, 8]. Finally, the main epitaxial layer is grown on the buffer layer during 4 hours. Typical grown conditions are: Ga source temperature: 1210 K; nitrogen flow rate: 5 - 80 sccm; buffer-layer growth-temperature: 773 K; main-layer growth-temperature: 1000 K. The Bi source temperature was varied in the range of 250 to 550 °C; some among these samples were in addition Mg doped. During the growth, the nitrogen partial-pressure in the chamber is the range of 10^{-5} to 10^{-2} Torr.

RESULTS AND DISCUSSION

In figure 1 we present atomic-force-microscopy (AFM) images of nominally undoped samples grown without Bi and with different amounts of Bi (T_{Bi} = 350 to 550 °C); each depicted area is 2 x 2 μm^2. We assume that the observed features on the surface reflect the size of the grains in the films. The sample in Fig. 1a) that is grown without any bismuth exhibits small grains with irregular boundaries. From previous investigations it is known that these grains can be disconnected [4]. Therefore, a drastically reduced lateral Hall mobility is observed. A small amount of Bi makes the grains coalesce (Fig. 1 b) and the Hall mobility increases from 6 to 73 cm^2/Vs (cf. Table 1). The background n-type carrier concentration was unchanged and exceeded 10^{18} cm^{-3}. This rather high n-type carrier concentration is probably caused by oxygen contamination of the growth chamber which was opened before these runs. A further increase of the Bi temperature to 450 °C (Fig. 1 c) and to 550 °C (Fig. 1 d), respectively, results in an enlargement of the grains. However, a poorer grain coalescence is obtained and the Hall mobility decreases.

Magnesium (Mg) doped GaN thin films exhibit grain sizes that compare well with those of the unintentionally doped n-type films of figure 1a). Unexpectedly, the size of the surface features increase drastically to about 10 μm if the Bi surfactant is used (figure 2; T_{Bi} = 350 °C and T_{Mg} = 280 °C). Similar effects are observed on films that are grown with different Bi source temperatures. In these particular runs the background impurity concentration is lower and the intrinsic n-doping did not exceed 10^{17} cm^{-3}. Thus, an influence of the impurity background concentration on the enhanced surface diffusion cannot be excluded.

Our previous experiments suggested that the feature sizes that can be observed on the films relate to the size of oriented grains which form the GaN thin films. This grain size limitation was attributed to a temperature and strain dependence of the Ga surface diffusion coefficient [4]. Following these arguments, the Bi surfactant seems to alter the Ga surface diffusion coefficient on the GaN (0001) faces, too. If we assume that surface diffusion only occurs during the growth of a double layer of Ga and N (thickness: 0.26 nm), a surface *diffusion length* $x \sim (D\tau_o)^{-1/2}$ can be estimated where D is the surface diffusion coefficient and τ_o is the time that is required to grow the double layer. Since, both, the growth rate and the grain size can be extracted from the experiments, we can estimate Ga surface diffusion coefficients. They are shown in figure 3. Open circles are taken from reference 4 and they depict the temperature and stress dependence of the surface diffusion coefficient. In addition, we present the value obtained from the sample shown in the microphotography of Fig. 4 (solid circle). It is seen that the utilization of Bi as a surfactant leads to a further increase of the diffusion coefficient. We estimate a surface diffusion coefficient that we would have expected for a growth temperature of 985 °C even though the sample is grown at 725 °C. Thus, the use of a bismuth surfactant is beneficial in several respects. First, it allows to extend the growth of GaN thin films with a desired grain size to lower growth temperatures. This is important if deviations from a thermodynamic equilibrium should be exploited. Second the use of a surfactant provides an independent way of tuning the growth to values that result in a film with coalesced grains.

At present, details of the growth mechanism using surfactants are not well understood. We expect that the presence of impurities in the growth chamber must affect the experimental results. Also, a dependence of the feature size on the film thickness is subject of current investigations. Nevertheless, we give experimental evidence on an impact of Bi on the thin film growth if this semi-metal is used as a surfactant.

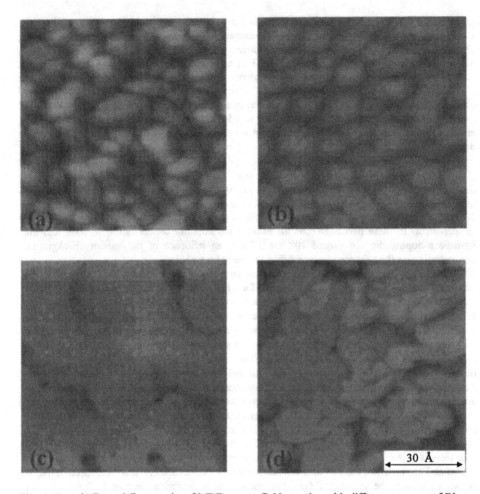

Fig. 1: Atomic Force Micrographs of MBE-grown GaN samples with different amount of Bi as surfactant: a) no Bi, b) $T_{Bi} = 350$ °C, c) $T_{Bi} = 450$ °C, and d) $T_{Bi} = 550$ °C, respectively. The size of the depicted areas is 2×2 μm^2.

Table 1: Carrier mobility of GaN samples grown with and without surfactant in dependence on the bismuth temperature.

T_{Bi} (°C)	-	350	450	550
μ (cm^2/Vs)	6	73	10	3

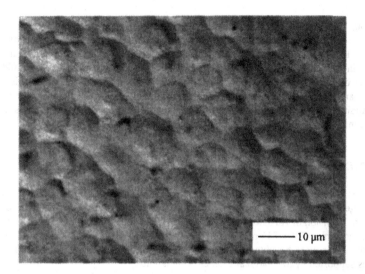

Fig. 3:
Surface image of a MBE GaN sample grown with T_{Bi} = 350 °C and T_{Mg} = 280 °C taken by optical microscopy.

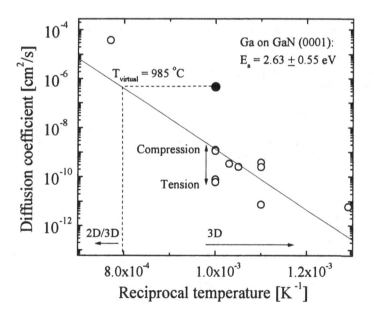

Fig. 4:
Surface diffusion coefficients of the sample grown with Bi (solid circle) in comparison with samples grown without any surfactant (open circles) in dependence on the reciprocal temperature.

CONCLUSIONS

In conclusion, GaN thin films that exhibit surface features of very different sizes were grown at 725 °C. These sizes can actively be determined by engineering the strain in the layers and by using surfactants. It seems likely that the observed surface features reflect the sizes of the grains in the films. In this case, the use of surfactants as well as the presence strain would influence the surface diffusion of Ga ad-atoms. Therefore, the result suggests that the surface diffusion coefficient can be varied by more than 4 orders of magnitude at a growth temperature as low as 725 °C. In addition, surfactants can be used to determine the coalescence of the grains in GaN thin films. This influences the lateral Hall mobility of the films.

ACKNOWLEDGMENTS

A research scholarship provided by the German Science Foundation (DFG) to R. K. is gratefully acknowledged. This work was supported by the Office of Energy Research, Office of Basic Energy Sciences, Division of Advanced Energy Projects (BES-AEP) and by the Laboratory Technology Transfer Program (ER-LTT) of the U.S. Department of Energy under Contract No. DE-AC03-76SF00098. This work benefits from the use of U.C. Berkeley's Integrated Materials Laboratory, which is supported by the National Science Foundation.

REFERENCES

1. S. Nakamura, M. Senoh, N. Iwasa, S. Nagahama, Jpn. J. Appl. Phys. **34**, L797 (1995).
2. S. Nakamura, M. Senoh, S. Nagahama, N. Iwasa, T. Yamada, T. Matsushita, H. Kiyoku, Y. Sugimoto, Jpn. J. Appl. Phys. **35**, L217 (1996).
3. M. A. Kahn, A. Bhattarai, J. N. Kuzina, D. T. Olsen, Appl. Phys. Lett. **63**, 1214 (1993).
4. H. Fujii, C. Kisielowski, J. Krüger, R. Klockenbrink, M. S.H. Leung, Sudhir G.S., Hyunchul Sohn, Mike Rubin, E.R. Weber, submitted to J. Appl. Phys., 1997; and Mat. Res. Soc. Symp. Vol. **449**, 1997.
5. M.S.H. Leung, R. Klockenbrink, C. Kisielowski, H. Fujii, J. Krüger, Sudhir G.S., A. Anders, Z. Liliental-Weber, M. Rubin, and E.R. Weber, Mat. Res. Soc. Symp. Vol. **449**, 221 (1997).
6. C. Kisielowski, J. Krüger, S. Ruvimov, T. Suski, J.W. Ager III, E. Jones, Z. Liliental-Weber, H. Fujii, M. Rubin, E.R. Weber, M.D. Bremser, and R.F. Davis; Phys.Rev.BII **54**, 17745 (1996).
7. R. Klockenbrink, Y. Kim, M. S.H. Leung, C. Kisielowski, J. Krüger, Sudhir G.S., M. Rubin, E.R. Weber, to be published.
8. C. Kisielowski, J. Krüger, M. S.H. Leung, R. Klockenbrink, H. Fujii, T. Suski, Sudhir G.S., J.W. Ager III., M. Rubin and E.R. Weber, Proceedings of the International Conference on the Physics of Semiconductors (ICPS), Berlin 1996 (World Scientific, Singapore) 513 (1996).
9. J. Krüger, C. Kisielowski, R. Klockenbrink, Sudhir G.S. Y. Kim, M. Rubin, E.R. Weber, these Proceedings.
10. A. Anders and S. Anders, Plasma Sources Sci. Technol. **4**, 571 (1995).

QUASI-THERMODYNAMIC ANALYSIS OF METALORGANIC VAPOR PHASE EPITAXY OF GaN

Shukun Duan *, Dacheng Lu **
*National Integrated Optoelectronics Laboratory, Institute of Semiconductors, The Chinese Academy of Sciences, Beijing 100083, CHINA, skduan@red.semi.ac.cn
**Laboratory of Semiconductor Materials of Sciences, Institute of Semiconductors, The Chinese Academy of Sciences, Beijing 100083, CHINA, dclu@red.semi.ac.cn

ABSTRACT

A thermodynamic analysis of GaN grown by MOVPE has been proposed based on quasi-thermodynamic equilibrium established on the solid-vapor interface. Phase diagrams for the MOVPE growth of GaN using TEGa and NH_3 has been calculated. The phase diagram is consists of four phases regions: the region for single condensed phase of GaN, the region for double condensed phase of GaN (s) +Ga(l), the etching region with Ga droplets and the etching region without Ga droplets. The effect of growth temperature, reactor pressure, content of carrier gas, deposition ratio of NH_3 and V/III ratio upon growth of GaN using MOVPE has been studied.

INTRODUCTION

Epitaxial GaN and its alloy films have recently attracted much interest due to their optoelectronic applications in the blue/ultraviolet region and for their high temperature stability. High-brightness blue and green light emitting diodes (LEDs) have already been made commercially available.[1]. Very recently, the observation of the room-temperature operation of GaN-based laser diodes[2] have been reported. MOVPE is the most commonly used technology for the growth of GaN.

In this paper we establish a quasi-thermodynamic model for MOVPE growth of GaN using trimethylgallium (TMGa) and NH_3 as source materials. In this model, solid GaN phase is in equilibrium with gas phase, which contains Ga, NH_3, N_2, H_2, CH_4 and inert gas. When equilibrium partial pressures of these species are known, MOVPE phase diagrams of GaN can be established. High quality of GaN can only grow inside the single condensed GaN phase region. The concept of MOVPE phase diagram was introduced by Stringfellow [3] and was used to describe the single phase region of some III-V and II-VI compounds grown by MOVPE [4,5].

MODEL AND CALCULATION PROCEDURE

The MOVPE phase diagrams based on thermodynamic equilibrium established at the solid-vapor interface have been developed for many III-V and II-VI compound systems[3-5]. According to the experimental results[6,7], we assume that TMGa is decomposed irreversibly near the vapor-solid interface when the temperature is higher than 550 °C. Thus, we have

$$Ga(CH_3)(g)+H_2(g)=Ga(g)+3CH_4(g). \tag{1}$$

Based on thermodynamic calculations, NH_3 is almost fully decomposed to N_2 and H_2 when temperature is higher than 300 °C. However without catalyst, the decomposition rate of ammonia is very low[8]. In order to calculate the influence of decomposition of ammonia, we introduce decomposition fraction of NH_3, η,

$$NH_3(g)=(1\eta)NH_3+\eta/2N_2(g)+3\eta/2H_2(g). \tag{2}$$

Due to the strong inert behavior of N_2, we ignore the reaction between N_2 and Ga in the temperature range of 550-1200 °C. Thus the reaction to form GaN at the interface can be written

$$Ga(g)+NH_3(g)=GaN(s)+3/2H_2(g). \tag{3}$$

These assumptions make this model to be a quasi-equilibrium model.
The equilibrium equation for reaction (3) is as follows:

$$a_{GaN} P_{H_2}^{3/2} / (P_{Ga} P_{NH_3} P_{tot}) = K_1 \tag{4}$$

where P_i is the equilibrium partial pressure of i, K_1 is the equilibrium constant of the reaction, and a_{GaN} is the activity of solid GaN which is equal to 1. The change of the total number of molecules created by the reactions, mainly by the decomposition of ammonia, can not be neglected. In this case, we use molar quantity to express the mass conservation constraints of element N and H, instead of partial pressure,

$$n_{NH_3}^o = n_{GaN} + n_{NH_3} + 2n_{N_2} \tag{5}$$

where n_i^o is the input molar quantity of i and $n_{GaN} = n_{TMG}^o - n_{Ga}$.

$$2n_{H_2}^o + 3n_{NH_3}^o = 2n_{H_2} + 3n_{NH_3} + n_{CH_4} . \tag{6}$$

The partial pressure of species i can be obtained as follows:

$$P_i = n_i (P_{tot} / n_{tot}) \tag{7}$$

where n_{tot} is the total molar quantity in the gas phase,

$$n_{tot} = n_{NH_3} + n_{N_2} + n_{H_2} + n_{Ga} + n_{CH_4} + n_{IG} . \tag{8}$$

The total pressure of the reactor is expressed as follows:

$$P_{tot} = P_{NH_3} + P_{N_2} + P_{H_2} + P_{Ga} + P_{CH_4} + P_{IG} \tag{9}$$

where P_{IG} is the partial pressure of inert gas, $P_{CH_4} = 3P_{TMG}^o$.
The calculations are performed by solving eqs.(4)-(9). The partial pressures can be obtained for a given set of growth temperature, reactor pressure and n_i^o. To calculate the phase diagram for the MOVPE growth of GaN, the following conditions which define the existence region for the single condensed phase of GaN is needed.
1. When partial pressure of Ga in the gas phase exceeds the vapor pressure of liquid Ga saturated with N, double condensed phase region, GaN(s)+Ga(l), is formed. To determine the boundary between GaN(s) and GaN(s)+Ga(l) phase region, we have $P_{Ga} = P_{Ga}^M$, where P_{Ga}^M stands

for the vapor pressure of liquid metal gallium saturated with nitrogen. In the often used growth temperature zone, one can assume that $P_{Ga}^{M} = P_{Ga}^{o}$, where P_{Ga}^{o} represents the vapor pressure of pure gallium [9]. Hereafter this boundary is called Ga forming line.

2. When input TMGa molar quantity is equal to the gallium molar quantity in the gas phase, the growth rate of GaN should be equal to zero. To determine the boundary between the growth region and the etching region, we have $n_{Ga} = n_{TMG}^{o}$ .Hereafter this boundary is called etching line.

3. Nitrogen is a gas. Thus, there is no group V condensed phase in the phase diagram. Under a given input partial pressure of TMGa, there exists a maximum value of V/III ratio, R_{max}. This value can be expressed as follows:

$$P_{TMG}^{o} = P_{tot}/(R_{max}+1) \qquad (10)$$

where R=$n_{NH3}{}^{o}$/$n_{TMG}{}^{o}$. This equation determines the upper boundary of single solid phase of GaN. Hereafter this boundary is called limit line.

Under a set of growth conditions and the limitations discussed above, the phase diagrams for the MOVPE growth of GaN can be obtained. We mainly focus on the boundaries of single solid GaN phase in this report, because GaN epilayers can only grow in this region.

RESULTS AND DISCUSSION

Fig.1 shows the relationship between equilibrium partial pressure of species in the gas phase and the input partial pressure of TMGa at 1050 °C, 1 atm, η=0 and H$_2$ as a carrier gas. In Fig. 1, P_{TMG}^{o} which is kept constant (dashed line) and P_{Ga}^{M} (dotted line) are also shown. One can see that equilibrium partial pressure of Ga decreases with increasing input V/III ratio. The curve of equilibrium partial pressure of Ga crosses the curve of P_{Ga}^{M} at lower V/III ratio. When P_{Ga} is larger than P_{Ga}^{M}, gallium droplets can deposit on the growing surface with GaN together. In the model presented here, $n_{Ga} = n_{TMG}^{o}$, instead of $P_{Ga} = P_{TMG}^{o}$, is used to determine the etching line. However the influence of the formation of gallium droplets is not included in this model. Therefore only the boundaries of the single solid GaN phase region can be calculated precisely. In order to describe the boundary of phase region which contains liquid gallium phase, Eq.(5) should be as follows:

$$n_{NH_3}^{o} = n_{NH_3} + 2n_{N_2} + n_{TMG}^{o} - n_{Ga} - n_{Ga}^{l} \qquad (11)$$

where n_{Ga}^{l} is the molar quantity of liquid gallium droplets. Because n_{Ga}^{l} is a small quantity compared with $n_{NH_3}^{o}$ and n_{NH_3}, our model can be used to roughly estimated the boundaries of phase region which has liquid Ga phase.

Shown in Fig. 2 is the phase diagram for the MOVPE growth of GaN using H$_2$ as a carrier gas at 1050 °C, 1 atm and η=0. The upper phase boundary (thick solid line) for single solid GaN phase (GaN(s)) is described by eq. (12). There is no meaning beyond this boundary. The whole phase diagram is divided into four regions (A, B, C and D) by the etching line (dotted line) and Ga droplets forming line (fine solid line). Region A is a single solid GaN phase region. Region B stands for co-deposition zone of GaN(s) and Ga(l). In the regions C and D, GaN, if exist, would be etched off instead of deposition. The region C is different from D in leaving Ga droplets on

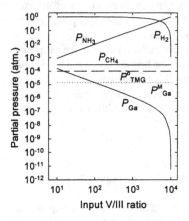

Fig. 1. Equilibrium partial pressure as a function of input V/III ratio for equal input partial pressure of TMGa using H_2 carrier gas. The growth temperature, reactor pressure, P_{TMGa} and η are 1050 °C, 1 atm, 1e-4 atm, and 0, respectively.

the surface during etching.

The effect of composition of a H_2/IG mixture carrier gas on the boundaries of single solid GaN phase is shown in Fig. 3. The parameter F is the molar fraction of H_2 relative to inert gas in the carrier gas,

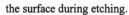

$F = P^o_{H_2} / \left(P^o_{H_2} + P^o_{IG} \right)$. One can see that the single solid GaN phase expands with decreasing F.

A multi-temperature MOVPE diagram of GaN is given in Fig. 4, in which only the boundaries of single solid GaN phase are shown. The etching line and the Ga forming line move toward to the right when growth temperature (T_g) increases., However, T_g has no effect on the limit line. Therefore the GaN(s) phase region (space) becomes narrower when T_g increasing. Considering that the decomposition fraction of NH_3 is a

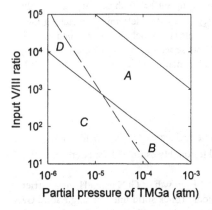

Fig. 2. Phase diagram for the MOVPE growth of GaN at 1050 °C and 1 atm. The carrier gas and η are H_2 and 0, respectively.

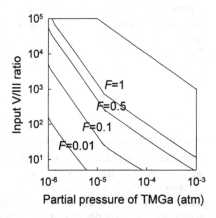

Fig. 3. Phase diagram for the MOVPE growth of GaN using H_2/IG mixture as a carrier gas. (T_g=1050 °C and P_{tot}=1 atm)

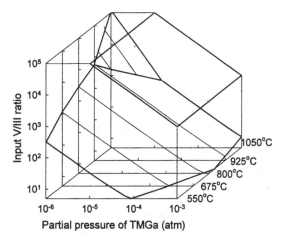

Input V/III ratio

10^5

10^4

10^3

10^2

10^1

10^{-6} 10^{-5} 10^{-4} 10^{-3}

Partial pressure of TMGa (atm)

1050°C
925°C
800°C
675°C
550°C

Fig. 4. Multi-temperature phase diagram for the MOVPE growth of GaN. The carrier gas, reactor pressure and η are H_2, 1 atm and 0, respectively.

function of temperature, the phase region of GaN(s) will be more narrower than that shown in Fig. 4. Clearly, high input V/III ratio is needed when high growth temperature is chosen to improve the crystalline quality of GaN.

The decomposition of ammonia should affect the MOVPE phase diagram of GaN, since GaN is formed by the reaction of Ga with NH_3. To describe this influence, we calculated the dependence of the Ga forming line on the V/III ratio as a function of η at two constant molar flow rates of TMGa. These results are shown in Fig. 5. The V/III ratio for the formation of Ga droplets increases with increasing η, especially when η is larger than 0.6. Many growth parameters affect the decomposition fraction of ammonia, such as growth temperature, reactor pressure, total gas flow rate and composition of mixture carrier gas. Reactor design also plays an important role in reducing the thermo-decomposition of ammonia.

Fig. 6 shows the influence of reactor pressure on the phase diagram. It can be found that the single solid GaN phase expands when the reactor pressure drops from 1 atm to 0.1 atm.

CONCLUSIONS

The quasi-thermodynamic analysis of the MOVPE growth of GaN using TMGa and ammonia has been carried out. MOVPE phase diagrams of GaN have been proposed. The phase diagram for the MOVPE growth of GaN is divided into four regions by the gallium forming line and etching line: single solid GaN phase, GaN(s)+Ga(l) double condensed phase and two etching regions. The special features of this phase diagram are: no group V element condensed phase exists and there are two etching regions. Epitaxial growth of GaN can only grow in the single condensed GaN phase. High input V/III ratio is needed to avoid the formation of gallium droplets on the growing surface. The single condensed GaN phase expands with decreasing growth temperature, reactor pressure, decomposition fraction of NH_3, and fraction of H_2 in a H_2-IG mixture carrier gas.

ACKNOWLEDGMENTS

The authors would thank High Technology Research by the State Science and Technology Commission of China and Hewlett-Packard Company for the financial support and Professor A. Koukitu for useful information.

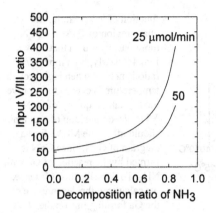

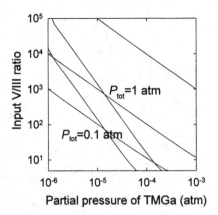

Fig. 5. Effect of the decomposition fraction of NH_3 (η) on the input V/III ratio on the gallium forming line (using H_2 carrier gas, T_g=1050 °C, P_{tot}=1 atm).

Fig. 6. Phase diagram for the MOVPE growth of GaN at the reactor pressure 1 and 0.1 atm. (T_g=1050 °C. P_{tot}=1 atm, η=0).

REFERENCES

1. S. Nakamura, T. Mukai and M. Senoh, Appl. Phys. Lett. **64**, p.1687 (1994).

2. S. Nakamura, M. Senoh, S. Nagahama, N. Iwasa, T. Yamada, T. Matsushita, H. Kiyoku, and Y. Sugimoto, Jpn. J. Appl. Phys. **35**, L74 (1996); Jpn. J. Appl. Phys. **35**, p. L217 (1996).

3. G. B. Stringfellow, J. Cryst. Growth, **70**, p. 33 (1984).

4. D. -C. Lu, X. Liu, D. Wang and L. Lin, J. Cryst. Growth, **124,** pp. 383-388 (1992).

5. Shu-kun Duan and Da-Cheng Lu, Proc. of the 1st Symp. on Blue Laser and Light Emitting Diodes, p. 332 (1996); J. Crystal Growth, **170**, p. 514-517 (1997).

6. P. W. Lee, T. R. Omstead, D. R. Mckenna and K. F. Jensen, J. Cryst. Growth, **85**, p.165 (1987).

7. M. J. Cherng, H. R. Jen, C. A. Larsen and G.B. Stringfellow, J. Cryst. Growth, **77**, p. 408 (1986).

8. J.J. Tietjen, and , Solid State Technology, **15**, p.42 (1972)

9. R. E Honing and D.A. Kramer, RCA Rev. **30**, p.285 (1996).

ALUMINUM NITRIDE THIN FILMS GROWN BY
PLASMA-ASSISTED PULSED LASER DEPOSITION ON Si SUBSTRATES

M. OKAMOTO, T. OGAWA, Y. MORI, and T. SASAKI
Department of Electrical Engineering, Osaka University, 2-1 Yamada-Oka, Suita, Osaka 565,
Japan

ABSTRACT

The smooth and highly oriented AlN films were obtained using pulsed laser deposition from sintered AlN target in a nitrogen ambient. The XRD investigation revealed that highly oriented AlN thin films along the c-axis (AlN (0002)) normal to the substrate were obtained both on Si(111) and on Si(100) substrates. The (0002) x-ray peak width became narrower with increasing substrate temperature. The CL investigation showed that AlN films at high laser energy density (E_d) indicated CL peak at shorter wavelength (306nm) than that at low E_d (394nm). N/Al atomic ratio in AlN films grown at high E_d also increased as comparison with the films grown at low E_d.

INTRODUCTION

Currently, wide band gap semiconductors, such as diamond, c-BN, and AlN, are of the great interest for various applications including light emitting device in the ultraviolet, electron emitting device, and high temperature device. However there is substantial difficulty in the growth of diamond and c-BN because their cubic crystal structure is metastable at low temperature and low pressure. On the other hand, it is easier to grow AlN because its wurtzite crystal structure is stable phase. Furthermore, its capability of n- and p-type doping makes AlN attractive for wide band gap semiconductors[1].

AlN films have been grown by various methods, such as MOCVD[2-4], reactive sputtering[5], plasma-assisted MBE[6], and pulsed laser deposition (PLD)[7-11]. We have employed PLD, because of its low-temperature growth, less contamination of impurities, and nonequibrium process. We had examined the influence of the growth atmosphere on AlN films grown on Si(100) by PLD[12-13]. We found that a nitrogen contained ambient was effective for increasing N/Al composition ratio. In this paper, we investigated difference in AlN films crystallinity between on Si(100) and Si(111), and effect of laser energy density (E_d) on the optical property of AlN films.

EXPERIMENT

The deposition chamber was evacuated to a base pressure of 10^{-8} Torr order by a turbomolecular pump. The fourth harmonics of a Nd: YAG laser ($\lambda = 266$ nm) was employed to ablate the target. The repetition rate was 10 Hz and the nominal pulse width was 3-5 ns. The incident laser beam was focused by a lens into a laser energy density (E_d) of ranging from 1 to 7

J/cm^2 on the target surface. Sintering AlN was used as the target material. The AlN target was rotated at 60 rpm. The Si substrate was located 3 cm away from and parallel to the target surface. Before it was introduced into the deposition chamber, the substrate was cleaned with acetone in an ultrasonic bath to remove organic impurities, and then was cleaned by etching with a 25% HF solution to remove the surface oxide layer. The Si substrate was heated to 950 °C. An optical pyrometer was used to monitor the substrate temperature. The deposition of AlN films was carried out in nitrogen plasma ambient for 2 - 5 hours. The pressure of the induced N$_2$ was performed at 40 mTorr. The dc glow discharge nitrogen plasma was generated by using a ring-shaped electrode biased negatively to about 600 V placed between the target and the substrate as shown in Fig. 1. In this case, the substrate and the target were grounded. The discharge current was adjusted to be less than 1 mA. A faint violet glow could be observed near the ring.

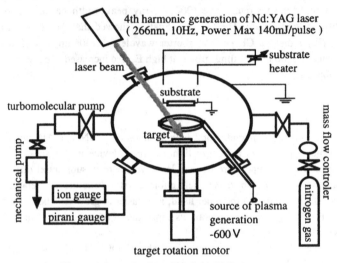

Fig. 1 Schematic diagram of the PLD apparatus.

We have grown AlN films at various substrate temperature (Ts) and various laser power density (Ed), and so investigated the effect of growth parameters on the film quality by scanning electron microscopy (SEM), the X-ray diffraction (XRD) techniques, cathodoluminescence spectroscopy (CL), and X-ray Photoelectron Spectroscopy (XPS).

RESULTS AND DISCUSSIONS

The SEM micrographs of the laser deposited AlN film are shown in Figs. 2. Although splashing of microscopic particles from the target is common problem in PLD[14], the AlN film surface is very smooth and almost free particulates. The film thickness was measured, from the image in Fig. 2 (b), to be ~1 nm; this corresponds to a deposition rate of ~0.006 nm/pulse or ~200 nm/h.

Fig. 2 SEM micrograph of an AlN film grown on Si(111) at 950 °C and 7.0 J/cm² of Ed.

Figure 3 shows XRD patterns of the AlN films grown on Si(100) and Si(111) substrate at substrate temperature (T_s) of 800 °C. The patterns show that the AlN films are highly oriented along the c-axis (AlN (0002)) normal to the Si substrate. The (0002) X-ray diffraction peaks were obtained for the films deposited at 600, 700, 800, 900, and 950 °C. Figures 4 show the full width at half maximum (FWHM) of the AlN (0002) peaks of AlN films at various substrate temperature (T_s), on Si(111) or Si(100). It indicates that Si(111) is better substrate for AlN films than Si(100). The reason for this is a better matching of AlN(0001) plane with a six-fold symmetry {111} plane than four-fold symmetry {100} plane. The crystalline quality of AlN films became better with increasing deposition substrate temperature. The FWHM for the film deposited at 950 °C on Si(111) was 0.282° in θ/2θ mode and 2.05° in ω mode. We wanted to investigate higher T_s, but the Max temperature of our substrate heater was 950 °C. Therefore, we decided that the T_s was 950 °C.

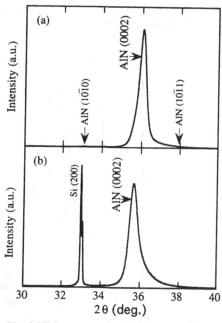

Fig. 3 XRD pattern of AlN film on (a) Si(111) and (b) Si(100).

89

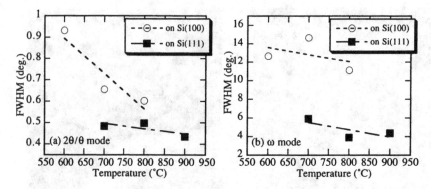

Fig. 4 The effect of substrate temperature (T_s) on the FWHM of the AlN (0002)
X-ray diffraction peak for the AlN in (a) $2\theta/\theta$ mode and (b) ω mode.

The effect of laser energy density (E_d) on the property of deposited AlN films was investigated. Figure 5 shows CL spectra for AlN films at two different E_d. Both films were grown at Ts of 950 °C on Si(111) substrate. It reveals that AlN films at high laser energy density (E_d) indicated CL peak at shorter wavelength (306nm) than that grown at low E_d (394nm). The reason for this shift is not clear yet. In case of the high Ed, evaporated species are dissolved into atomic form and given high kinetic energy. We think that this leads to change of crystallinity of AlN films.

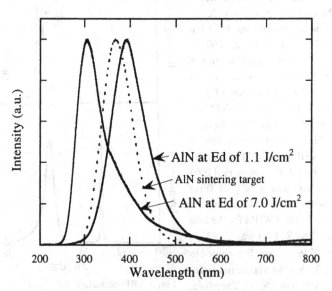

Fig. 5 CL spectra for AlN films at different E_d.

The composition ratio of the AlN films was measured by XPS. Table 1 shows the composition ratio of the same AlN films. N/Al ratio increases by high E_d. The composition of oxygen atom is as small as impossible to detect, and so comparison can't be made between AlN films at high E_d and at low.

XPS depth profile for the AlN film at Ed of 7.0 J/cm² is shown in Fig. 6. The composition ratio is uniform in the depth direction.

Table. 1 Film composition data from XPS.

	AlN sintering target	AlN films at low laser fluence (1.1J/cm^2)	AlN films at high laser fluence (7.0J/cm^2)
Al	47 at.%	46 at.%	40 at.%
N	49 at.%	54 at.%	60 at.%
O	4 at.%	~0 at.%	~0 at.%

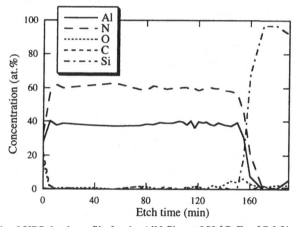

Fig. 6 XPS depth profile for the AlN film at 950 ˚C, E_d of 7.0 J/cm².

CONCLUSION

Oriented AlN films were obtained on Si substrates by pulsed laser deposition. XRD investigation indicated that the orientational relationships of AlN films on Si substrates: AlN (0002) / Si (111) and AlN (0002) / Si (100). According to the FWHM of XRD patterns, the crystalline quality of AlN films became better with increasing deposition substrate temperature. The CL investigation revealed that AlN films at high laser energy density (E_d) indicated shorter wavelength luminescence line (306nm) than at low E_d (394nm). N/Al atomic ratio in AlN films grown at high E_d also increased as comparison with at low E_d.

REFERENCES

1. S. Strite and H. Morkoç, J. Vac. Sci. Technol. B **10**, 1237 (1992).

2. M. A. Khan, J. N. Kuznia, R. A. Skogman,D. T. Olson, M. M. Millan, and W. J. Choyke, Appl. Phys. Lett. **61**, 2539 (1992).

3. P. Kung, A. Saxler, X. Zhang, D. Walker, T. C. Wang, I. Furguson, and M. Razeghi, Appl. Phys. Lett. **66**, 2958 (1995).

4. J. Chaudhuri, R.Thokala, J. H. Edgar, and B. S. Sywe, J. Appl. Phys. **77**, 6263 (1995).

5. S. Kumar and T. L. Tansley, Jpn. J. Appl. Phys. **34**, 4154 (1995).

6. K. S. Stevens, A. Ohtani, M. Kinniburgh, andR. Beresford, Appl. Phys. Lett. **65**, 321 (1994).

7. M. G. Norton, P. G. Kotula, and C.B. Carter, J. Appl. Phys. **70**, 2871 (1991).

8. K. Seki, X. Xu, H. Okabe, J. M. Frye, and J. B. Halpern, Appl. Phys. Lett. **60**, 2234 (1992).

9. R. D. Vispute, J. Narayan, H. Wu, and K. Jagannadham, J. Appl. Phys. **77**, 4724 (1995).

10. W. T. Lin, L. C. Meng, G. J. Chen, and H. S. Liu, Appl. Phys. Lett. **66**, 2066 (1995).

11. R. D. Vispute, H. Wu, and J. Narayan, Appl. Phys. Lett. **67**, 1549 (1995).

12. T. Ogawa, M. Okamoto, Y. Mori, and T. Sasaki in III-Nitride, SiC and Diamond Materials for Electronic Devices, edited by D. K. Gaskill, C. D. Brandt, R. J. Nemanich (Mater. Res. Soc. Proc. **423**, San Francisco, CA, 1996) pp. 391-396.

13. T. Ogawa, M. Okamoto, Y. Mori, A. Hatta, T. Ito, T. Sasaki, and A. Hiraki, Diamond Films and Technology **6**, 87-94 (1996).

14. J. T. Cheung and H. Sankur, CRC Crit. Rev. in Solid State and Mater. Sciences **15** (1), 63-109 (1988).

PYROLYTIC PREPARATION OF GALLIUM NITRIDE FROM [Ga(NEt$_2$)$_3$]$_2$ AND ITS AMMONOLYSIS COMPOUND

Seiichi Koyama*, Yoshiyuki Sugahara*, and Kazuyuki Kuroda*,**
*Department of Applied Chemistry, School of Science and Engineering, Waseda University, Ohkubo-3, Shinjuku-ku, Tokyo 169 JAPAN
**Kagami Memorial Laboratory for Materials Science and Technology, Waseda University, Nishiwaseda-2, Shinjuku-ku, Tokyo 169 JAPAN

Abstract

Gallium nitride (GaN) was prepared by the pyrolytic conversion of both [Ga(NEt$_2$)$_3$]$_2$ and its ammonolysis product at 600 °C for 4 h under Ar. The pyrolyzed residues were analyzed by X-ray powder diffraction and scanning electron microscopy, and the pyrolysis processes of the precursors under He were investigated by thermogravimetry-mass spectrometry. The XRD pattern of the pyrolyzed residue of [Ga(NEt$_2$)$_3$]$_2$ showed well-resolved peaks due to a mixture of cubic and hexagonal close-packed layers of GaN. The broad XRD pattern of the pyrolyzed residue of the ammonolysis product was also attributed to the mixture of cubic and hexagonal close-packed layers of GaN. For the pyrolysis of [Ga(NEt$_2$)$_3$]$_2$, the evolution of hydrocarbons was extensively observed at relatively high temperature, but a large amount of carbon (11 mass%) was still detected in the pyrolyzed residue. On the other hand, the amount of carbon was only 1.1 mass% in the pyrolyzed residue of the ammonolysis product. The pyrolysis results of the ammonolysis product under Ar were very similar to those of [Ga(NEt$_2$)$_3$]$_2$ under NH$_3$.

Introduction

Gallium nitride (GaN) is a promising material as blue light-emitting diodes and lasers because of its wide band gap, and the preparation of thin films has been extensively investigated mainly by chemical vapor deposition (CVD) and molecular beam epitaxy (MBE) [1,2].

The pyrolysis technique of precursors offers attractive routes for the preparation of non-oxide ceramics with desirable shapes, such as coatings and fibers [3,4], if the precursors are soluble in organic solvents or fusible. Thus, extensive work has been reported on the preparation of SiC, Si$_3$N$_4$, BN and AlN via pyrolysis [3,4], and a few investigations have also been reported for the pyrolytic preparation of GaN [5-8]. Gonsalves et al. have reported that the pyrolysis of tris(dimethylamido)gallium, [Ga(NMe$_2$)$_3$]$_2$, under NH$_3$ leads the formation of nano-crystallite GaN [5]. Janik et al. have investigated the reaction of [Ga(NMe$_2$)$_3$]$_2$ with NH$_3$, and the resulting ammonolysis product was converted into GaN containing only a small amount of carbon with high ceramic yield [6]. Furthermore, precursors have been prepared using Me$_3$NGaH$_3$-NH$_3$ [7] and metallic gallium-NH$_3$ [8] systems, and they were also converted into GaN.

Chemistry of gallium amides, Ga(NR^{1}R^2)$_3$ (R^1, R^2 = Me; R^1 = H, R^2 = Bu^t etc.), has been established, and some of their structures have been clarified (monomer or ring-type dimer) [9-11]. Furthermore, it is possible to displace the NR^{1}R^2 groups in Ga(NR^{1}R^2)$_3$ with the NH$_2$ groups by the ammonolysis reaction [6,12,13]. Since the condensation reaction involving the NH$_2$ groups proceeds even at around room temperature, the resulting ammonolysis product is expected to possess a polymeric structure containing the NH and/or NH$_2$ groups. Although both [Ga(NMe$_2$)$_3$]$_2$ and the ammonolysis product of [Ga(NMe$_2$)$_3$]$_2$ were applied to GaN synthesis, the influence of the structures of precursors on the pyrolysis processes and the resulting pyrolyzed residues have not been clarified yet. Thus, we focus here the pyrolytic preparation of GaN under Ar from two kinds of precursors, [Ga(NEt$_2$)$_3$]$_2$ and its ammonolysis product. The pyrolyzed residues were analyzed by X-ray powder diffraction (XRD) and scanning electron microscopy (SEM). Moreover, their pyrolysis processes are also investigated for understanding the elimination process of the organic groups by thermogravimetry-mass spectrometry (TG-MS), which is a valuable tool for mechanistic studies on the pyrolysis process.

Experiment

All the procedures were performed under a protective nitrogen atmosphere using the standard Schlenk technique or a globe box filled with nitrogen. The solvents, diethylether and hexane, were freshly distilled from sodium/benzophenone prior to use.

Tris(diethylamido)gallium was prepared according to the following reaction, based on the previous report [10].

$$GaCl_3 + 3LiNEt_2 \longrightarrow Ga(NEt_2)_3 + 3LiCl \qquad (1)$$

A 1.63 M n-hexane solution of n-BuLi (45.8 mL, 74.6 mmol) was added to a solution of diethylamine (7.8 mL, 75.2 mmol) in 70 mL ether at -10 °C. The resulting clear solution was allowed to warm to room temperature. This solution was added to a $GaCl_3$ solution (4.36 g, 24.8 mmol) in 40 mL ether at -78 °C, and the mixed solution was then refluxed for 12 h. After the removal of the solvents from the reaction mixture under a reduced pressure, 100 ml of hexane was added to a resulting residue. White precipitate of LiCl was filtered off, and the solvent was removed to form a pale-yellow oily product in almost quantitative yield.

Elemental analysis of this oily product provided an empirical formula of $C_{12.1}H_{30.7}N_{2.7}Ga$. ^{1}H-NMR spectrum (270 MHz, C_6D_6) showed the presence of two sets of triplet (CH_3) at δ 1.08 and 1.21, and overlapped quadruplets (CH_2) at δ 3.09 and 3.16. The integrated intensity ratio of the signal at δ 1.21 to the signal at δ 1.08 and that of the signal at δ 3.16 to the signal at δ 3.09 were almost 2. These two sets of the signals are consistent with the structure possessing two environments of ethyl groups. The ^{13}C-NMR spectrum (67.8 MHz, C_6D_6) showed two CH_3 signals at δ 12 and 17 and two CH_2 signals at δ 42 and 45, which are also consistent with the structure possessing two environments of ethyl groups. In the mass spectrum, no fragments larger than $m/e = 570$, corresponding to the parent fragment of dimer, were observed. These analytical data indicate that the oily product is ring-type dimer, $[Ga(NEt_2)_3]_2$ (the terminal NEt_2 groups : the bridging NEt_2 groups = 2 : 1).

An ammonolysis product was prepared as follows. $[Ga(NEt_2)_3]_2$ was loaded into a three-neck flask equipped with a dry ice/acetone condenser and was dissolved in 100 ml of ether. Anhydrous NH_3 was passed into the solution at 0 °C. White solid was immediately precipitated. After passing NH_3 for 2 h, the solution was warmed to room temperature and stirred for 1 h. The supernatant was then decanted. The residual precipitate was washed with ether and dried under a reduced pressure to lead a yellowish solid.

Pyrolysis of these precursors was performed using a tube furnace under Ar (100 mL/min) or NH_3 (20 mL/min) atmospheres. For both pyrolyses, the precursors were heated to 600 °C at the rate of 5 °C/min and held at 600 °C for 4 h.

The analyses were performed using the following instruments. The ammonolysis product was characterized using IR (Perkin Elmer IR-1640), and the pyrolyzed residues were analyzed using XRD (Mac Science MXP^3 diffractometer) and SEM (HITACHI S-4500S). The pyrolysis processes of both precursors were investigated using TG-MS (Shimadzu TGA-50 thermobalance coupled with a Shimadzu QP1100EX quadrupole mass spectrometer).

Results

If $Ga(NEt_2)_3$ ideally reacts with NH_3, $Ga(NH_2)_3$ should be initially formed with the stoichiometric liberation of Et_2NH. Figure 1 shows the infra-red (IR) spectra of $[Ga(NEt_2)_3]_2$ and the ammonolysis product. The IR spectrum of $[Ga(NEt_2)_3]_2$ showed strong C-H stretching bands at ~2700-3000 cm^{-1}, which are consistent with the presence of ethyl groups in $[Ga(NEt_2)_3]_2$. On the contrary, only weak C-H stretching bands were observed in the IR spectrum of the ammonolysis product, indicating the decrease in the amount of ethyl groups. A new broad band at ~3190 cm^{-1} was attributed to the N-H stretching mode. Thus, the amount of the NEt_2 groups appears to be considerably decreased. Furthermore, $Ga(NH_2)_3$ stepwise condenses with the evolution of NH_3 even at low temperature, leading to polymeric products as follows [6,12,13].

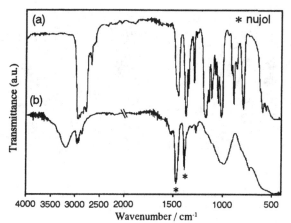

Figure 1. IR spectra of (a) $[Ga(NEt_2)_3]_2$ and (b) the ammonolysis product; the measurements were performed by neat technique for $[Ga(NEt_2)_3]_2$ and by using nujol (2000-400 cm^{-1}) or hexachloro-1,3-butadiene (4000-2000 cm^{-1}) for the ammonolysis product.

$$Ga(NH_2)_3 \longrightarrow Ga(NH_2)(NH) \longrightarrow Ga(NH)_{3/2} \qquad (2)$$

The IR spectrum of the ammonolysis product revealed the presence of the band at 1510 cm^{-1}, ascribable to the N-H bending mode in the NH groups [6,14]. Since the band at 1510 cm^{-1} was broad, a band due to the N-H bending mode in the NH$_2$ groups could be overlapped in this band. Moreover, it is much likely that Ga is tetracoordinated through additional coordination of nitrogen attached to any other Ga atoms [15]. Hence, the resulting ammonolysis product is considered to possess a structure cross-linked mainly with the bridging NH groups, and a small amount of unreacted NEt$_2$ groups still remains in the structure.

Figure 2 shows XRD patterns of the pyrolyzed residues. The pyrolysis of $[Ga(NEt_2)_3]_2$ under Ar produced a black residue. The XRD pattern of the black residue exhibited relatively well-

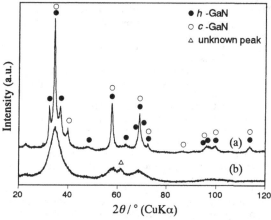

Figure 2. XRD patterns of the pyrolyzed products of (a) $[Ga(NEt_2)_3]_2$ and (b) the ammonolysis product. (pyrolysis; under Ar at 600 °C for 4 h)

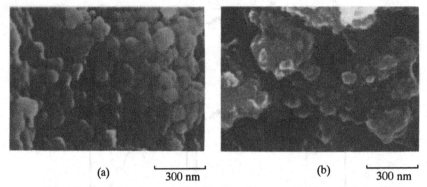

(a) 300 nm   (b) 300 nm

Figure 3. Scanning electron micrographs of the pyrolyzed products of (a) [Ga(NEt$_2$)$_3$]$_2$ and (b) the ammonolysis product. (pyrolysis; under Ar at 600 °C for 4 h)

resolved peaks. A similar XRD pattern has been reported by Hwang *et al.* for the residue prepared by the pyrolysis of (H$_2$GaNH$_2$)$_3$ at 600 °C for 4 h, and was attributed to the mixture of cubic and hexagonal close-packed layers of GaN (*c/h*-GaN) [7]. The XRD pattern for the residue of the ammonolysis product pyrolyzed under Ar is shown in Figure 2b. Only broad peaks were observed at $2\theta = 34$, 58 and 69 °, and the pattern is also assignable to *c/h*-GaN [5,7]. Gonsalves *et al.* have also reported a broad XRD pattern for the residue of [Ga(NMe$_2$)$_3$]$_2$ pyrolyzed under NH$_3$, and the broad peak shapes were attributed to the small crystalline size (~5.5 nm), which was confirmed by TEM image [5]. We preliminary estimated the mean crystalline size to be ~10 nm for the pyrolyzed residue of [Ga(NEt$_2$)$_3$]$_2$ and ~2 nm for the pyrolyzed residue of the ammonolysis product on the basis of Scherrer equation. The formation of the distorted GaN crystallite may also cause broadening of the peaks [6].

Figure 3 shows scanning electron micrographs of the pyrolyzed residues of [Ga(NEt$_2$)$_3$]$_2$ and the ammonolysis product under Ar. Spherical particles were observed on the surface of the pyrolyzed residue of [Ga(NEt$_2$)$_3$]$_2$. The residue possessed a relatively smooth surface as shown in Figure 3a. On the other hand, a rough surface was observed for the pyrolyzed residue of the ammonolysis product (Figure 3b).

Figure 4 shows TG-MS results (TG curves with the behavior of selected fragments, which reflect the behavior of the gas evolution) for the pyrolysis of [Ga(NEt$_2$)$_3$]$_2$ and the ammonolysis product under He at the heating rate of 10 °C/min. A large mass loss occurred in the temperature range from 130 °C to 230 °C during the pyrolysis of [Ga(NEt$_2$)$_3$]$_2$, and Et$_2$NH ($m/e = 30$, a parent fragment) was mainly evolved. A further mass loss was observed up to ~500 °C. The amount of evolved hydrocarbons ($m/e = 41$, a fragment corresponding to intense fragments for the most of hydrocarbons (except CH$_4$, C$_2$H$_4$, C$_2$H$_6$)) gradually increased as pyrolysis temperature increased, and a large amount of hydrocarbons was detected even at ~500 °C. Although the intensity was rather weak, some fragments, which can be attributed to Ga-containing species on the basis of the isotope effect, were also observed at relatively high temperature. The behavior of the most intense fragment ($m/e = 69$) among the fragments attributable to Ga-containing species is shown in Figure 4.

On the other hand, a mass loss occurred even at around room temperature with the evolution of NH$_3$ ($m/e = 17$, parent fragment) during the pyrolysis of the ammonolysis product, as shown in Figure 4b. It should be noted that the evolution of Et$_2$NH was observed at lower temperature compared to the pyrolysis of [Ga(NEt$_2$)$_3$]$_2$. Since the ammonolysis product contains a considerable amount of the NH groups, the condensation reactions involving the NH group and NEt$_2$ group probably occur, leading to the elimination of Et$_2$NH as follows.

$$\equiv Ga - \ddot{N}Et_2 + H - N \equiv \longrightarrow \equiv Ga - N \equiv + Et_2NH \qquad (3)$$

Table I summarizes the pyrolysis results under Ar. A large amount of carbon was detected in the residue of $[Ga(NEt_2)_3]_2$ pyrolyzed under Ar. As described above, the evolution of hydrocarbons was extensively observed at relatively high temperature during the pyrolysis of $[Ga(NEt_2)_3]_2$. The previous report has described that the formation of residual carbon was caused by the thermal decomposition of the organic products at relatively high temperature [16]. Hence, the elimination of hydrocarbons at relatively low temperature is not sufficient enough for the desirable reduction in the amount of residual carbon in the pyrolyzed residue. In contrast, the amount of carbon detected in the pyrolyzed residue was only 1.1 mass% for the pyrolysis of the ammonolysis product. This is mainly attributed to the low contents of the NEt_2 groups in the ammonolysis product, and the elimination of Et_2NH during pyrolysis (shown by the TG-MS results) also contributes to some extent.

Table I also clarified that the ceramic yields for the pyrolysis of the ammonolysis product under Ar is considerably high compared to that for the pyrolysis of $[Ga(NEt_2)_3]_2$ under Ar. Since the elimination of the NEt_2 groups leads to the increase in gallium and nitrogen contents, a higher ceramic yields for the ammonolysis product is expected.

$[Ga(NEt_2)_3]_2$ was also pyrolyzed under NH_3, and the pyrolysis results are summarized in Table I. The analytical result of XRD, SEM observation and the carbon content for the pyrolyzed residue of the ammonolysis product

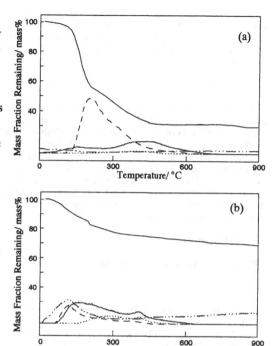

Figure 4. TG-MS analysis of (a) $[Ga(NEt_2)_3]_2$ and (b) its ammonolysis product under He. the fragment at $m/e = 17$ (---) arising from NH_3; at $m/e = 30$ (- -) arising from Et_2NH; at $m/e = 41$ (—) arising from hydrocarbons; at $m/e = 69$ (----) arising from Ga-containing species. [Note that we cannot discuss on the relative intensities of the fragments between Figure 4a and 4b, since the analytical conditions such as sample weight were different.]

Table I. Characteristics of the pyrolysis results.

	Ar		NH_3
	$[Ga(NEt_2)_3]_2$	Ammonolysis product	$[Ga(NEt_2)_3]_2$
XRD pattern	c/h-GaN (well-resolved peaks)	c/h-GaN (broad peaks)	c/h-GaN (broad peaks)
Surface morphology	smooth	rough	rough
carbon content/ mass%	11	1.1	1.1
ceramic yield/ %	32	77	33

under Ar were very similar to those for the residue prepared by the pyrolysis of $[Ga(NEt_2)_3]_2$ under NH_3. A relatively low ceramic yield is ascribed to the low gallium and nitrogen content of the precursor. Thus, it is assumed that the atmospheric NH_3 react with $[Ga(NEt_2)_3]_3$ during pyrolysis at relatively low temperature to form a structure which is close to the ammonolysis product.

Conclusions

We have shown the pyrolytic preparation of GaN from $[Ga(NEt_2)_3]_2$ and its ammonolysis product under Ar. XRD analysis revealed the formation of the mixture of cubic and hexagonal close-packed layers of GaN for both the pyrolysis of $[Ga(NEt_2)_3]_2$ and its ammonolysis product, though the crystalline sizes are different between two systems. Besides GaN, a large amount of carbon was detected in the pyrolyzed residue of $[Ga(NEt_2)_3]_2$. Compared with the pyrolysis of $[Ga(NEt_2)_3]_2$, a substantial decrease in the carbon content was observed for the pyrolyzed residue of the ammonolysis product. The low carbon content in the pyrolyzed residue of the ammonolysis product was mainly due to the low content of the NEt_2 groups in the ammonolysis product, and the elimination of Et_2NH during pyrolysis further reduce the amount of residual carbon.

Acknowledgment

We thank K. Ajima for experimental assistance.

References

1. H. Morkoc, S. Strite, G. B. Gao, M. E. Lin, B. Sverdlov and M. Burns, J. Appl. Phys. **76** (3), 1363 (1994).
2. D. A. Neumayer and G. Ekerdt, Chem. Mater. **8**, 9 (1996).
3. G. Pouskouleli, Ceram. Int. **15**, 213 (1989).
4. R. W. Rice, Am. Ceram. Soc. Bull. **62**, 889 (1983).
5. K. E. Gonsalves, G. Carlson, S. P. Rangarajan, M. Benaissa and M. Jose-Yacaman, J. Mater. Chem. **6** (8), 1451 (1996).
6. J. F. Janik and R. L. Wells, Chem. Mater. **8**, 2708 (1996).
7. J-W. Hwang, J. P. Campbell, J. Kozubowski, S. A. Hanson, J. F. Evans and W. L. Gladfelter, Chem. Mater. **7**, 517 (1995).
8. T. Wade, R. M. Crooks, Mater. Res. Soc. Symp. Proc. **410**, 121 (1996).
9. H. Nöth, P. Konrad, Z. Naturforsch. **30b**, 681 (1975).
10. D. A. Atwood, V. O. Atwood, A. H. Cowley, R. A. Jones, J. L. Atwood and S. G. Bott, Inorg. Chem. **33**, 3251 (1994).
11. K. M. Waggoner, M. M. Olmstead and P. P. Power, Polyhedron **9**, 257 (1990).
12. D. V. Baxter, M. H. Chisholm, G. J. Gama, V. F. Distasi, A. L. Hector and I. P. Parkin, Chem Mater. **8**, 1222 (1996).
13. G. M. Brown and L. Maya, J. Am. Ceram. Soc. **71** (1), 78 (1988).
14. L. Maya, Adv. Ceram. Mater. **1** (2), 150 (1986).
15. M. Veith, Chem. Rev. **90**, 3 (1990).
16. D. M. Narsavage, L. V. Interrante, P. S. Marchetti and G. E. Maciel, Chem. Mater. **3**, 721 (1991).

PULSED LASER DEPOSITION OF GALLIUM NITRIDE ON SAPPHIRE

V. TALYANSKY[*], R. D. VISPUTE[*], R. P. SHARMA[*], S. CHOOPUN[*], M. J. DOWNES[*], T. VENKATESAN[*], Y. X. LI[**], L. G. SALAMANCA-RIBA[**], M. C. WOOD[***], R. T. LAREAU[***], AND K. A. JONES[***]
[*]Center for Superconductivity Research, Department of Physics, University of Maryland, College Park, MD 20742, vitaly@squid.umd.edu
[**]Department of Materials and Nuclear Engineering, University of Maryland, College Park, MD 20742
[***]U.S. Army Research Laboratory, AMSRL-PS-DB, Fort Monmouth, N.J. 07703

ABSTRACT

We have fabricated high quality single crystalline GaN films on sapphire (0001) substrates using pulsed laser deposition. Our best GaN films on sapphire (0001) featured the FWHM of the GaN (002) peak rocking curve of 7 arcmin, the RBS minimum yield of only 3%, and the energy gap width of 3.4 eV. The effect of the deposition temperature on the crystalline quality of the films is discussed.

INTRODUCTION

Due to immense interest on the part of the materials community to a wide range of GaN opto-electronics applications, a tremendous progress has been made in the improvement of the GaN material quality. MOCVD [1-2] and MBE [3-6] film deposition processes have produced epitaxial films suited for fabrication of efficient LED and laser diode devices [7-9]. Being a relatively inexpensive and versatile thin film growth technique, pulsed laser deposition (PLD) [10] has been tried for GaN growth demonstrating the feasibility of the process by forming the right material phase, though the films were polycrystalline [11-12]. Recently, we have shown that the PLD process can be used for the fabrication of single crystalline high quality AlN [13], TiN [14], and GaN [15] films on sapphire. Here we report on detailed studies of the GaN deposition process. To our knowledge, we are first to report on a successful attempt to grow single crystalline GaN in thin film form using the PLD technique.

EXPERIMENT

Fabrication

Thin films of GaN were grown on sapphire (0001) substrates by laser ablating a GaN target pressed out of nominally stoichiometric 99.9% pure powder at room temperature. The growth experiments involving a liquid target of metallic gallium resulted in poor quality films suffering from metallic droplets and nitrogen deficiency which had been reported earlier by our colleagues [11]. The KrF excimer laser (wavelength = 248 nm, pulse duration = 30 ns) was operated at a pulse rate of 10 Hz and a fluence at the target site of around 3-4 J/cm^2 producing an average growth rate of 0.1 nm per pulse. The deposition was done in a vacuum chamber having 10^{-4} Torr of flowing ammonia (NH_3) as a background gas which got partially decomposed by the laser light, ablated species, and a hot heater and supplied nitrogen ions to be incorporated into the growing film. Utilization of molecular nitrogen (N_2) as a background gas resulted in films with

99

severe nitrogen deficiencies. The substrate temperature was varied from 850°C upto 950°C to study the film's crystalline structure dependence on the deposition temperature.

X-ray diffraction measurement

The GaN films 0.1-1.5 μm thick were transparent and shiny. Fig. 1 displays θ–2θ XRD scans for the films grown at three temperatures T_g = 850°C, 900°C, and 950°C. The scan shown in Fig. 1(a) (T_g = 850°C) presents relatively small peaks of various orientation alongside with the dominating c-axis oriented peaks. The other two scans shown in Fig 1(b-c), which correspond to T_g's of 900°C and 950°C, describe a gradual suppression of all the off c-axis peaks and thus the enhancement of the film crystallinity as the growth temperature is raised.

Fig. 2 reflects the effect of the growth temperature on the full width at half maximum (FWHM) of the GaN (002) peak rocking curve. The scans of the rocking curves given in (a), (b), and (c) correspond to T_g's of 850°C, 900°C, and 950°C, respectively. As it is reasonable to expect, the FWHM reduces from 16 arcmin to 10 arcmin to 7 arcmin each time the T_g is raised by 50°C. The rocking curve profile in Fig. 2(c) is so narrow that the $K_{\alpha 2}$ double of the GaN (002) becomes visible indicating a high degree of crystallinity of the film.

The joint Φ-scan of the sapphire (113) and GaN (101) peaks of the finest quality film is presented in Fig. 3. The three peaks of sapphire reflect its trigonal structure. The hexagonal structure of GaN produces six distinct equally spaced peaks. The absence of any other random peaks suggests that the film's grains are predominantly locked in the a-b plane. The 30 degree shift between the GaN peaks and those of sapphire is explained by in-plane rotation of the GaN lattice with respect to the sapphire substrate which has been well documented [4, 6, 15-16].

Rutherford Backscattering study

Rutherford Backscattering (RBS) study was carried out using a well collimated 3 MeV He$^+$ ion beam incident on 1.5 and 0.55 μm thick GaN films. In ion channeling the incident He$^+$ beam was first aligned along a major crystallographic direction in the film [001], and the RBS yield was measured. The beam direction was then changed to a random one, and the spectrum was collected again. The crystalline quality of the film can be evaluated by finding a minimum scattering yield, χ_{min}, which is the ratio of the yield with the ion beam aligned with one of the crystallographic axis of the film to that with the beam incident at a random angle to the film's surface. The RBS spectra for the 1.5 μm film, deposited while the substrate temperature was ramped from 850°C upto 950°C, are plotted in Fig.4(a). In the near surface region up to a depth of 150 nm, the RBS counting rate goes down to 6%, indicating that the crystalline quality is reasonably good in this region. At larger depths the crystalline quality has deteriorated, and, close to the interface, where the drop in the counting rate amounts to only 50%, it is very poor. This is attributed to the lattice mismatch between GaN and sapphire which accumulates defects near the interface; as the film grows thicker and the mismatch strain relaxes, the defect density decreases. The 550 nm thick film was grown at a fixed T_g of 950°C. The aligned RBS spectrum of this film shown in Fig. 4(b) also describes a higher defect density near the interface with a χ_{min} of 18% as well as a gradual improvement in the film's crystallinity in the near surface region with a χ_{min} of as low as 3%. The latter result compares favorably to the χ_{min} = 1-2% obtained for single crystal substrates. Also, the RBS results and the XRD data agree on the high degree of the film's crystallinity. The effect of the growth temperature on the crystalline quality of the films is remarkable: the film, whose temperature of the deposition was fixed at 950°C, had a much

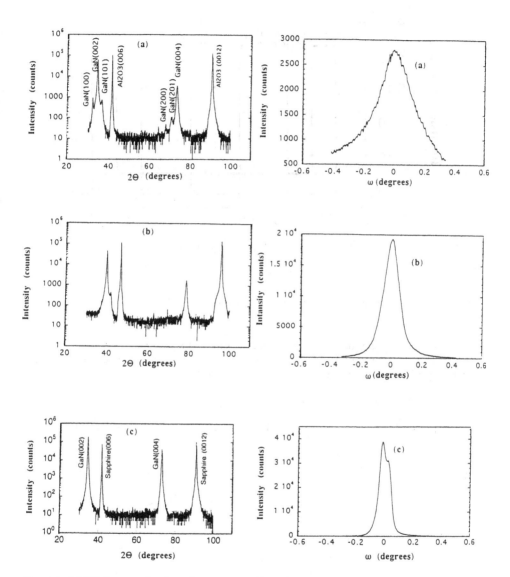

Fig. 1 θ–2θ XRD scans for GaN films grown at various temperatures (a) 850°C, (b) 900°C, and (c) 950°C.

Fig.2 GaN (002) peak rocking curve for films grown at (a) 850°C, (b) 900°C, and (c) 950°C.

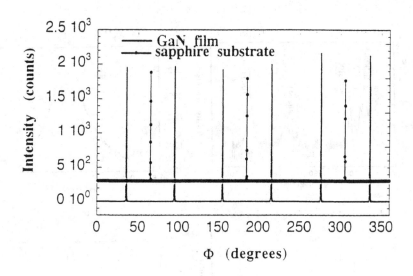

Fig. 3. Φ-scans of the sapphire (113) and GaN (101) peaks.

superior crystal structure compared to that of the one grown at 850°C-950°C as evidenced by the RBS study despite the fact that the thickness of the former was reduced to a half of that of the latter sample.

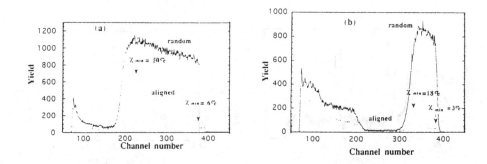

Fig. 4. RBS random and aligned spectra for (a) 1.5 mm thick GaN film grown at a varying temperature 850°C-950°C and for (b) 0.55 mm thick GaN film grown at a fixed temperature of 950°C.

UV-Visible spectroscopy measurement

The UV-Visible transmission spectra of the samples mentioned above are displayed in Fig. 5. The spectra profiles feature interference peaks as well as a step reflecting a strong light absorption. The film becomes opaque for the light of wavelengths below that corresponding to the energy gap of the material. The film grown at 950°C possessed a more homogeneous structure manifested in a steeper absorption drop given by curve (a) compared to that of the film grown at 850°C represented by curve (b). The corresponding energy gaps of the films are 3.4 eV and 3.3 eV, respectively. The higher growth temperature film appears to have good optical properties featuring the energy gap width of 3.4 eV which is close to the accepted value of 3.4-3.5 eV [8].

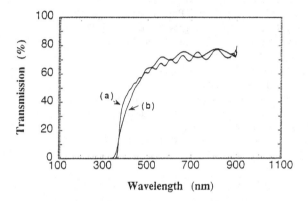

Fig. 5. UV-Visible transmission spectra of (a) 0.55 μm thick film grown at 950°C and (b) 1.5 μm thick film grown at 850°C-950°C

CONCLUSIONS

We have demonstrated the suitability of the PLD technique to growing single crystal epitaxial GaN films. Our best GaN films on sapphire (0001) featured the FWHM of the GaN (002) peak rocking curve of 7 arcmin, the RBS minimum yield of only 3%, and the energy gap width of 3.4 eV. The XRD, RBS, and optical characterization studies revealed the high crystalline quality of the PLD grown GaN thin films. The film crystallinity was found to be enhanced by raising the deposition temperature.

ACKNOWLEDGMENTS

We thank Henry Chen for his help in finalizing the paper. This work was supported with an Army grant DAAL 019523530.

REFERENCES

1. S. Nakamura, Y. Harada, and M. Seno, Appl. Phys. Lett. 58, 2021 (1991).
2. P. Kung, A. Saxler, X. Zhang, D. Walker, T. C. Wang, I. Ferguson, and M. Razeghi, Appl. Phys. Lett. 66, 2958 (1995).
2. W. Kim, O. Aktas, A.E. Botchkarev, A. Salvador, S.N. Mohammad, and H. Morkoc, J. Appl. Phys. Lett. 79, 7657 (1996).
4. R.C. Powell, N.-E. Lee, Y.-W. Kim, and J.E. Greene, J. Appl. Phys. 73, 189 (1993).
5. Z. Yang, L.K. Li, and W.I. Wang, Appl. Phys. Lett. 67, 1686 (1995).
6. M.J. Paisley, and R.F. Devis, J. Crystal growth 162, 537 (1990).
7. H. Morkoc, S. Strite, G. B. Gao, M.E. Lin, B. Sverdlov, and M. Burns, J. Appl. Phys. 76, 1363 (1994), and references therein.
8. "GaN and Related Materials for Device Applications", Materials Research Society Bulletin, 22, 1997, and references therein.
9. R.J. Molnar, R. Singh, and T.D. Moustakas, Appl. Phys. Lett. 66, 268 (1995).
10. "Pulsed laser deposition of thin films", Eds. D. Chrisey and G. Hubler, Wiley-Interscience, New York, 1994, and references therein.
11. R. E. Leuchtner, W. Block, Y. Li, and L. Hristacos, 1995 Fall MRS meeting
12. R.-F. Xiao, X.W. Sun, H.B. Liao, N. Cue, and H.S. Kwok, J. Appl. Phys. 80, 4226 (1996).
13. R.D. Vispute, H. Wu and J. Narayan, Appl. Phys. Lett. 67, 1549 (1995).
14. V. Talyansky, R. P. Sharma, S. Choopun, M. Downes, T. Venkatesan, Y. X. Li, L. G. Salamanca-Riba, M. C. Wood, and K. A. Jones, submitted to J. Appl. Phys.
15. R. D. Vispute, V. Talyansky, R. P. Sharma, S. Choopun, M. Downes, T. Venkatesan, K.A. Jones, A. A. Iliadis, M. A. Khan, and J. W. Yang, accepted to Appl. Phys. Lett.
16. W. Qian, M. Skowronski, M.D. Graef, K. Doverspike, L.B. Rowland, and D.K. Gaskill, Appl. Phys. Lett. 66, 1252 (1995).

CONTROL OF VALENCE STATES BY A CODOPING METHOD
IN P-TYPE GaN MATERIALS

T. YAMAMOTO *, **, H. KATAYAMA-YOSHIDA *,***
*Department of Condensed Matter Physics, Institute of Scientific and Industrial Research,
Osaka University, 8-1 Mihogaoka, Ibaraki 567, Japan, tetsu36@sanken.osaka-u.ac.jp
**Department of Computational Science, Asahi Chemical Industry Co., Ltd., 2-1 Samejima,
Fuji 416, Japan
***PRESTO, Japan Science and Technology Corporation, Kawaguchi, Saitama 332, Japan

ABSTRACT

We propose a new valence control method, the "codoping method (using both n- and p-type dopants at the same time)", for the fabrication of low-resistivity p-type GaN crystals based on the *ab-initio* electronic band structure calculations. We have clarified that while doping of acceptor dopants, Be_{Ga} and Mg_{Ga}, leads to destabilization of the ionic charge distributions in p-type GaN crystals, doping of Si_{Ga} or O_N give rise to n-type doped GaN with high doping levels due to a large decrease in the Madelung energy. The codoping of the n- and p-type dopants (the ratio of their concentrations is 1:2) leads to stabilization of the ionic charge distribution in p-type GaN crystals due to a decrease in the Madelung energy, to result in an increase in the net carrier densities.

INTRODUCTION

In previous works [1-3], we proposed materials design using donor dopants, In_{Cu} with poly valence, to fabricate low-resistivity p-type $CuInS_2$ having almost a band gap of 1.5 eV with chalcopyrite structure for solar cell applications. Very recently, we have applied the codoping method using donor dopants (Si and O) and acceptor ones (Be and Mg) to GaN crystals [4].
. GaN with a wide band gap, 3.39 eV, has recently gained attention because of its possible use in blue small-emitting diodes and UV-emitting laser diodes. Valence controls to realize low-resistivity p-type wide-band-gap semiconductor GaN crystals are one of very important problems to develop large-scale application of GaN and its related semiconductor technology. Several authors reported low-resistivity p-type GaN films [5-8]. Very recently, Brandt *et al.* produced high-conductivity p-type GaN using acceptor dopants, Be, and donor dopants, O [9]. While the properties of the GaN crystals are known, little is known about the relation between the cause of difficulty in fabricating p-type GaN and a change in the electronic band structure or in the nature of chemical bonds of p-type doped GaN crystals. In this work, first, we discuss in more detail the difference in effects of between p-type Be or Mg doping and n-type Si or O doping on the electronic structures employing the *ab-initio* electronic band structure calculations. Then we investigate crystal structures and electronic structures of GaN codoped with n-type dopants (Si, O) and p-type dopants Be in order to clarify the advantage of the codoping method.

CALCULATION METHOD

The results of our band structure calculations for GaN crystals are based on the local-density approximation (LDA) treatment of electronic exchange and correlation [10-12] and on the augmented spherical wave (ASW) formalism [13] for the solution of the effective single-particle equations. For the calculations, the atomic sphere approximation (ASA) with a correction term is adopted. For valence electrons, we employ outermost s and p orbitals for each atom. For doped or codoped GaN crystals, we use the supercell method: (1) The composition of GaN doped with X_{Ga} (X=Be, Mg or Si) is Ga: 46.875, X: 3.125 and N: 50.0 at. %, or $(Ga_{0.9375}X_{0.0625})N$; (2) For GaN:O_N, the composition is Ga: 50.0, N: 46.875, O: 3.125 at. %, or $Ga(N_{0.9375}O_{0.0625})$; (3) For GaN with codoping of Si_{Ga} and two numbers of Be_{Ga} atoms, it is Ga: 40.625, Si: 3.125, Be: 6.25 and N: 50.0 at. %, or $(Ga_{0.8125}Be_{0.125}Si_{0.0625})N$; (4) For GaN crystals:$O_N$ and $2Be_{Ga}$, Ga: 43.75, Mg: 6.25, N: 46.875 and O:3.125 at. %, or $(Ga_{0.875}Be_{0.125})(N_{0.9375}O_{0.0625})$.

RESULTS AND DISCUSSION

Density of States

We show the total and site-decomposed density of states (DOS) of undoped GaN as a standard reference and those of GaN crystals doped with Si (O) atoms in Fig. 1. Energy is measured relative to the Fermi level (E_F). From Figs. 1 (a) to 1 (c) for undoped GaN, some features are important : (1) The letter A in the region between -16.0 and -12.5 eV indicates N 2s - Ga 3s and 3p bonding states with a higher N 2s contribution; (2) The upper valence band located above approximately -7.0 eV consists of two groups of peaks; the letter B indicates N 2s - Ga 3s bonding states and the letter C, N 2p - Ga 3p bonding states with a higher N 2p contribution. From the site-decomposed DOS, we see little indication of sp hybridization at N sites compared with those at Ga sites in the upper valence band. In addition, we find that p states at N sites mainly dominate the DOS of the upper valence band, which indicates that a mechanism that shifts N 2p levels towards lower energy is necessary in fabricating p-type GaN crystals with the stable ionic charge distributions. We noted that for GaN crystals the calculated Madelung energy, E_M, is 4.08 Ry (= 55.5 eV) in the unit cell with 4 atoms. The ionic features at the N sites is due to its large E_M.

For GaN doped with Si$_{Ga}$ or O$_N$, from Figs. 1 (d) and 1 (f), we find the conduction type is n-type and an impurity delocalized band just below the conduction band in a heavily doped n-type GaN causes little change in DOS features of host crystals. We will find shallow donor levels at the very low doping levels for the two crystals.

The O 2s states are included in the calculation as valence states, but they are omitted in Fig. 1 (f) and 1 (g) since these electrons form a narrow band showing little interaction with other states. Figures 1 (f) and 1 (g) show little indication of sp hybridization at O sites for GaN:O$_N$, which is similar to the features at N sites of undoped GaN.

Population analysis and site-decomposed DOS indicate that compared with GaN:O$_N$, we find that great amounts of extra electrons are introduced into the other sites in the vicinity of the Si-sites. Moreover, Fig. 1 (e) shows s band with a wide width at Si sites, resulting in small effective mass of the electrons introduced in the conduction band. From these findings, we can expect low-resistivity n-type GaN:Si$_{Ga}$ crystals.

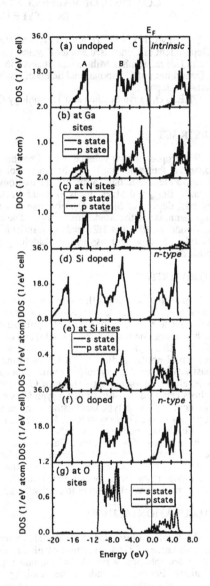

Fig. 1. (a) Total, (b) and (c) site-decomposed DOS of undoped GaN, (d) Total and (e) site-decomposed DOS of GaN:Si and (f) Total and (g) site-decomposed DOS of GaN:O.

We present the total and site-decomposed DOS for GaN:Be$_{Ga}$ and GaN:Mg$_{Ga}$ in Fig. 2, which shows that the conduction type of the two crystals is p-type. As well as the two n-type doped GaN above, we find that electronic structures of the doped GaN crystals with higher doping levels, $2.3*10^{21}$ cm^{-3}, have the same band shape as that of undoped GaN, that is, there is little change in electronic structures except for the energy regions near the band edge. As well as for undoped GaN, their impurity states for the two p-type crystals are in the energy region at the top of the valence band, Γ_6 (= doublet), and are delocalized on sites of N atoms, as determined from site-decomposed DOS and dispersion relations. Based on the discussion above, we predict that for GaN:Be$_{Ga}$ or Mg$_{Ga}$ at very low dopant concentrations, we will find shallow acceptor levels.

From Figs. 2 (b) and 2 (d), we find some of splitting between s and p states in the uppermost valence band for GaN:Be compared with for GaN:Mg. In other words, for GaN:Be, the DOS of impurity Be bands near the Fermi level has a high p states contribution to DOS, which is similar to the DOS features of host crystals. This originates in the difference in *sp splitting* (= the difference in energy between levels, or $\varepsilon_{2(3)p}$ - $\varepsilon_{2(3)s}$) of between a free Be and Mg atom. As a result, at the very low doping levels, we can expect that impurity levels for Be would be shallower than that for Mg.

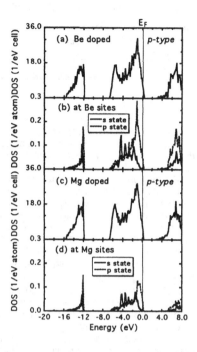

Fig. 2. (a) Total and (b) site-decomposed DOS of GaN:Be and (c) Total and (d) site-decomposed DOS of GaN:Mg.

Madelung Energy

It is of interest to study the effects of p- or n-type doping on stability of the ionic charge distribution. The wurtzite structure is favored by more ionic compounds. Thus, at the low doping levels, 10^{15} -10^{18} cm^{-3}, we can estimate the overall change in the total energy (=ΔE) composed of the electronic and lattice system energies as given by

$$\Delta E = \Delta\mu + \Delta E_M, \qquad (1)$$

where the first and second term refer to a change in chemical potential that is an extrinsic property for the former system and in the electrostatic called the Madelung energy, E_M, that is an intrinsic property for the latter system, respectively. We note that ΔE_M with negative (positive) sign means that compared with undoped crystals under consideration we find a larger attractive (repulsive) force at the anion site resulting in shifts of energy levels of valence orbitals of outermost s and p levels of anion atoms towards lower (higher) energy-regions, leading to destabilization of ionic charge distributions. In fabricating low-resistivity p-type GaN crystals we should prevent the native-defect reactions such as donor-levels generated by anion-vacancies from occurring due to the destabilization above. Thus, we focus on ΔE_M which is a good parameter for stability of the ionic charge distribution, in particular, shifts of p levels of anions, N, towards higher or lower energies. We summarize ΔE_M for GaN doped with Be, Mg, Si or O crystals in Table I.

Table I. The difference in the calculated Madelung energies (units: eV) between undoped and doped GaN crystals.

dopants	undoped	p-type doped		n-type doped	
		Be	Mg	Si	O
ΔE_M	-	+16.3	+9.52	-8.57	-3.26

Table I shows that while p-type doping using acceptor dopants, Be or Mg, leads to an increase in E_M, n-type doping gives rise to a decrease in E_M. In other words, as has already pointed out the importance of E_M in stabilizing the ionic charge distribution, the calculated results for doped GaN suggest the occurrence of doping problems which is called "$uni\text{-}polarity$" that both high conductivity p- and n-type crystals are difficult to fabricate.

For doping of Be, which we propose to be eminently suitable as acceptor dopants for the fabrication of low-resistivity p-type GaN crystals, we find an large increase in E_M. In addition, we have investigated what happens when we increase the concentration of Be and Mg in the supercells whose size is same that of the supercells for GaN:Be or Mg. For p-type GaN whose supercells include two numbers of Be_{Ga} (see Fig. 3 (a)), we find a 1.58 Å increase in distances between the two acceptor dopants due to a large Coulomb repulsive interaction between them compared with that for Mg-doped crystals under the condition that the total energy is minimized.

We give ΔE_M for the two p-type doped GaN crystals in Table II. From these findings, we reach a conclusion that it is very difficult to obtain p-type GaN:Be_{Ga} with high doping levels.

For n-type GaN crystals doped with Si or O, from Table I, we find a decrease in E_M. It follows that valence control using donor dopants, Si and O, is likely to prove extremely useful in fabricating high-conductivity n-type doped GaN crystals with stable ionic charge distributions. For n-type GaN:Si, a decrease in E_M is mainly due to an increase in electric charge at N sites transferred from Si. For n-type GaN doped with O_N, from site-decomposed DOS, we find a strong interaction between the O $2p$ and $4p$ orbitals of Ga close to the O sites. In other words, O doping leads to an increase in charge transfer from the Ga to the O atoms, to decrease E_M for n-type doped GaN crystals.

The codoping method

Band-structure calculations on n-type GaN doped with Si_{Ga} or O_N have shown the effects of the donor dopants on electronic band structure and the ionic charge distribution. In considering the important role of the donor dopants on a decrease in the Madelung energy, that is, the stabilization of the ionic charge distribution in doped GaN crystals, we propose a new type doping method, the "codoping method", using the acceptor dopants and the donor dopants (the ratio of their concentrations is 1:2) at the same time for the fabrication of high-conductivity p-type GaN crystals with the stable ionic charge distribution.

In this study, we focus on GaN codoped with acceptor dopants, Be, and donor dopants, Si or O atoms. For codoped GaN crystals using acceptor dopants, Mg, and the donor dopants, our recent paper has been already published [4]. We show the crystal structure of GaN codoped with Be and Si in Fig. 3 (b) and Be and O in Fig. 3 (d), respectively. We determined them under the condition that the total energy is minimized. We summarize the difference in the calculated Madelung energy between the doped (see Fig. 3 (a) and Fig. 3 (c)) and the corresponding codoped GaN crystals. Moreover, we show the total DOS of undoped and the two codoped GaN crystals in Fig. 4.

A comparison with figures 3 (c), figure 3 (d) shows that the donor dopants, in particular, O codoping leads to the condensation of acceptor dopants by the fromation of the local cluster Be_{Ga}-O_N-Be_{Ga}. At the same time, it gives rise to a decrease in E_M as shown Table II.

For p-type GaN: Si, 2Be, from site-decomposed DOS and population analysis, we find a large delocalized impurity states near the Fermi level, which results in small effective mass of positive holes introduced. In addition, Table II shows that the codoping leads to the condensation of acceptor dopants (see distances between the two Be atoms in Fig. 3) with a decrease in E_M, which gives rise to an increase in the net carrier densities, to fabticate low-resistivity p-type GaN crystals.

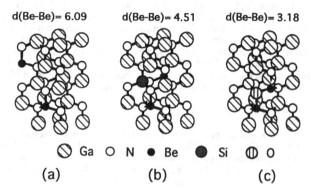

Fig. 3. Crystal structures of (a) GaN: 2Be where a distance between the two Be atoms is 6.09 Å, (b) GaN: Si, 2Be where the distance is 4.51 Å and (c) GaN: O, 2Be where the distance is 3.18 Å.

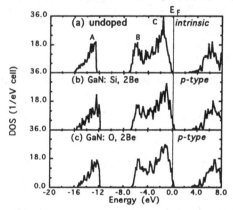

Fig. 4. Total DOS of (a) undoped GaN crystals, GaN crystals codoped with (b) Si and 2Be and (c) O and 2Be. For crystal structures of (b) and (c), see Fig. 3 (b) and 3 (c).

Table II. The calculated Madelung energies (units: eV) doped and codoped GaN crystals. Energy is measured relative to the Madelung energy of undoped GaN crystals.

dopants	undoped	p-type doped 2Be	p-type codoped	
			2Be and Si	2Be and O
ΔE_M	-	+34.02	+26.09	+31.93

CONCLUSIONS

We have studied the electronic structures of undoped, p- or n-type doped and codoped GaN crystals using *ab-initio* electronic band structure calculations. The main results obtained are shown below: (1) p-type doping of Be_{Ga} or Mg_{Ga} leads to the destabilization of the ionic charge distributions due to an increase in the Madelung energy; (2) n-type doping of Si_{Ga} or O_N gives rise to the stabilization of the ionic charge distributions due to a decrease in the Madelung energy; (3) Codoping of n-type dopants, Si and O, and p-type dopants, Be and Mg, being in the ratio of 1 to 2 will cause a decrease in the Madelung energy, resulting in preventation of self-compensation.

From these results and delocalized holes in wide impurity bands of the codoped crystals, we propose a valence control method, the "codoping method" for the fabrication of high-conductivity p-type GaN crystals. This awaits experimental verification.

ACKNOWLEDGMENTS

The authors thank Dr. J. Sticht (Molecular Simulations, Inc.) for his technical support. We use ESOCS code of MSI. Their sincere thanks are also due to Profs. S. Gonda and H. Asahi of ISIR, Osaka university for their fruitful discussion. Thanks are also offered to general manager Dr. T. Yamada and senior scientist Dr. Y. Ueshima (Asahi Chemical Industry Co., Ltd.) who gave support and encouragement during the course of this study.

REFERENCES

1. T. Yamamoto and H. Katayama-Yoshida in Electronic structure of p-type CuInS$_2$, edited by D. Ginley, A. Catalano, H. W. Schock, C. Eberspacher, T. M. Peterson and T. Wada (Mater. Res. Soc. Proc. 426, Pittsburgh, Pennsylvania 1996), p. 201-206.

2. T. Yamamoto and H. Katayama-Yoshida, Proc. of the Int. Conf. Shallow Level in Semiconductors (1996), in press.

3. T. Yamamoto and H. Katayama-Yoshida, Proc. of Photovoltaic Science Engineering Conf. (1996), in press.

4. T. Yamamoto and H. Katayama-Yoshida, Jpn. J. Appl. Phys. **36**, L180 (1997).

5. H. Amano, M. Ito, K. Hiramatsu and I. Akasaki: Jpn. J. Appl. Phys. **28**, L2112 (1989).

6. H. Amano, M. Ito, K. Hiramatsu and I. Akasaki: Inst. Phys. Conf. Ser. **106**, p. 825 (1989).

7. S. Nakamura, T. Mukai, M. Senoh and N. Iwasa: Jpn. J. Appl. Phys. **31**, L139 (1992).

8. S. Nakamura, T. Mukai, M. Senoh and N. Iwasa: Jpn. J. Appl. Phys. **31**, p. 1258 (1992).

9. O. Brandt, H. Yang, H. Kostial, and K. H. Ploog: Appl. Phys. Lett. **69**, p. 2707 (1996).

10. W. Kohn and L. J. Sham: Phys. Rev. **140**, A1133 (1965).

11. L. Hedin and B. I. Lundquist, J. Phys. **C4**, p. 3107 (1971).

12. U. von Barth and L. Hedin, J. Phys. **C5**, p. 1629 (1972).

13. A. R. Williams, J. Kuebler and C. D. Gelatt: Phys. Rev. **B19**, p. 6094 (1979) .

STRUCTURE, ELECTRONIC PROPERTIES, DEFECTS AND DOPING OF AlN USING A SELF-CONSISTENT MOLECULAR DYNAMICS METHOD

PETRA STUMM AND D. A. DRABOLD*
*Department of Physics and Astronomy, Ohio University, Athens, OH 45701
stumm@helios.phy.ohiou.edu

ABSTRACT

Molecular dynamics simulations are employed to study native defects and dopants in AlN. We use local basis density functional theory within the local density approximation where charge transfer between the ions is included in a self-consistent fashion. Employing this code we find reasonable agreement for the band structure compared to other recent calculations, suggesting the suitability of our method to adequately describe AlN.

Wurtzite and zincblende 96 atom AlN cells are used to study the relaxations and electronic properties of common defects in the crystal structure, including vacancies and antisites. We investigate the electronic signatures of these defects. The local topology of column-IV impurities in anion and cation sites is studied. We analyze the lattice relaxations and electronic consequences of these impurities and identify midgap defect, donor and acceptor levels.

INTRODUCTION

III-V nitride semiconducting materials are some of the most promising wide-bandgap materials, because of their suitability for optoelectronic devices [1]. AlN is of particular interest, because it exhibits a combination of desirable physical properties. It has a large band gap (6.2 eV), high thermal conductivity and high temperature stability [2]. AlN commonly crystallizes in the wurtzite structure, with lattice parameters of $a_0 = 3.11$ Å and $C_0 = 4.98$ Å. There are indications that a metastable zincblende phase of AlN with a lattice constant of 4.38 Å has been synthesized recently [3]. Zincblende AlN is expected to show improved electronic properties over the wurtzite phase due to the higher lattice symmetry.

The electronic properties of AlN are not as well known as those for GaN. Unintentionally doped AlN films show high resistivities in the range of 10^{11} to 10^{13} Ωcm. In comparison to GaN this indicates that if intrinsic defects are present in AlN they are not shallow, but exhibit levels deep enough in the band gap to not be ionized at room temperature.

Several calculations on the band structure of wurtzite and zincblende AlN have been reported [4, 5, 6, 7]. Only a few studies of intrinsic defects of wurtzite AlN have been performed with tight binding models [4] and first-principles methods [8]. Experimental measurements of the optical-absorption spectrum in wurtzite AlN exhibits band-edge tails and midgap absorption features which are either due to intrinsic crystal defects or unintentional doping [9].

Si can act as an amphoteric impurity in AlN, since it can either substitute as a donor in an anion (N) site or as an acceptor in a cation site (Al). Theoretical studies of substitutional Si doping of AlN have been performed using a tight-binding method [4]. However, our results described in this paper indicate that significant lattice relaxations occur upon Si substitution in most sites (wurtzite: Si_N, Si_{Al}; zincblende: Si_N). These network relaxations result in significantly different electronic properties for the relaxed configuration than for the substitutional dopant atom incorporation.

Device applications for Aln require an understanding of the nature of intrinsic defects in the material, and in particular the energies of defect states generated by vacancies and antisites. The main aim of this paper is to describe the structure and relaxation of intrinsic defects in zincblende and wurtzite AlN in large supercells. We have determined the role of native defects in zincblende and wurtzite AlN and identified their structural and electronic consequences. Further, relaxations of the network and changes in the electronic structure upon Si doping have been studied.

METHOD

Our method is based on an approximate first-principles electronic-structure approach, first introduced by Sankey and coworkers in 1989 [10]. Demkov, Sankey, Ortega and Grumbach [11] generalized this non-self-consistent local basis Harris functional LDA scheme to an approximate self-consistent form, "Fireball96".

In this approach, Demkov and coworkers [11] exploited the original idea of the Harris functional which allowed *non neutral* input charge densities. Spherical *charged* atom densities are used as Harris input charge, and these charges are then self-consistently determined. The method is very efficient, combining the advantages of charge transfer with a fixed atom-centered basis (and therefore efficient look-ups for matrix elements). Basis functions of four pseudoatomic orbitals per site are used, with a confinement [10] radius of $r_C = 3.8a_B$ and $r_C = 5.4a_B$ for nitrogen and aluminum, respectively.

APPLICATION TO WURTZITE AND ZINCBLENDE AlN

The electronic band structures for wurtzite and zincblende AlN were computed at the theoretical equilibrium lattice constant, which is 7% smaller than the experimental value. Too small lattice constants in GaN have typically been associated with treating the Ga 3d states as core states, which leads to a 3% smaller lattice constant [12]. Since Al does not contain 3d states the incompleteness of the basis set might be the reason for the smaller lattice constant. The energy dispersion along the high-symmetry lines in the first Brillouin zone are shown in Fig. 1 for the zincblende phase and in Fig. 2 for wurtzite AlN.

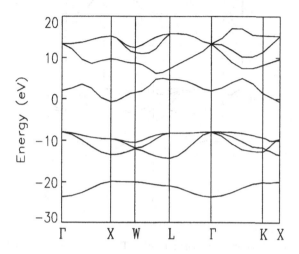

Figure 1: Energy band structure of zincblende AlN. Note the indirect bandgap from Γ to X.

The slightly excited four orbital basis set used for these calculations typically overestimates the bandgap by a factor of 1.6. Wurtzite AlN is a direct band gap semiconductor at the Γ point, while zincblende AlN shows an indirect band gap from Γ to X. Comparing these band structures to other *ab initio* calculations [6, 5, 7], we find good agreement between the bands. Even the lower conduction band states are represented rather well.

N Vacancy and N antisite in Wurtzite AlN

Calculations for Al and N vacancies were carried out in a supercell that would contain 96 atoms for the defect free crystal. The cell was relaxed for 100 steps in each case using the Γ point for BZ zone sampling.

The N vacancy shows only small structural relaxations, equivalent nearest neighbors show no relaxation, while the inequivalent Al atom moves away from V_N by about 0.02 Å. A singlet and a doublet state are formed at the conduction band edge. Scaling our calculated bandgap of the perfect crystal to the experimental value, we find that the singlet state is located 0.44 eV below the conduction band. It is singly occupied, therefore the N vacancy acts as a single donor. This result compares well to that of Reference [8], who use a plane wave basis set and investigated the N vacancy in a 72 atom cell.

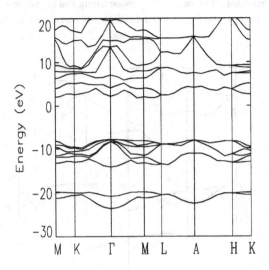

Figure 2: Energy band structure of wurtzite AlN. The wurtzite phase is a direct semiconductor with the minimum bandgap at the Γ point.

For the N antisite we find a large and asymmetrical relaxation of the N nearest neighbor atoms. Two of the in-plane nearest neighbors increase their distance from the antisite atom to 2.05 Å. The third in-plane N_{Al} neighbor forms a short N-N bond of only 1.66 Å with the antisite atom. The inequivalent N atom along the z-direction increases its bond lengths to 2.24 Å. This asymmetrical relaxation can be attributed to the different symmetry for in plane versus out of plane atoms.

The N antisite introduces three new unoccupied midgap states. A split doublet is located about 1 eV above the valence band edge. These states are strongly localized on the antisite atom and the three nearest neighbor atoms forming long bonds (greater than 2.0 Å). The third defect state is positioned 2 eV below the conduction band edge and is localized on N_{Al} as well as the neighboring N atom forming the short bond of 1.66 Å.

N Vacancy and N antisite in Zincblende AlN

Relaxation effects of the lattice around the N vacancy in zincblende AlN are small (less than 0.1 Å) and limited to the Al nearest neighbor atoms. A singly occupied state is formed 0.3 eV below the conduction band edge. Therefore, the N vacancy in zincblende AlN acts as a single donor.

Distortion of the network surrounding the N antisite are more pronounced than for the N vacancy. The N vacancy nearest neighbor N atoms move outward by 0.25 Å to a distance of 2.02 Å from N_{Al}. Three new unoccupied defect states are created 2 eV above the valence band edge. These states are localized on the antisite atom and the surrounding N nearest neighbors.

GROUP IV DOPANTS IN ANION AND CATION SITES

Si was substituted in Al and N sites in the wurtzite as well as the zincblende phase. The doped 96 atom supercell was thoroughly relaxed, which was essential to achieve the stable lattice configurations, as large network relaxations are present. We analyzed the structural and electronic changes upon Si incorporation in the cell.

Si in Wurtzite AlN

For Si incorporation in a N site in the wurtzite phase we find a large spherical outward relaxation of the Al nearest neighbors surrounding the Si atom. The Si - Al nearest neighbor distance increases to 2.1 Å. A split triplet is formed slightly below midgap, with wave functions that are very strongly localized on the Si atom.

Upon substitution of a Si atom in an Al site we find a strong relaxation of the lattice. The Si atom is 3-fold coordinated, with equal in plane bond lengths of 1.78 Å to the three N neighbors. A new midgap state, pinning the Fermi level was created, that is localized on the Si atom. Neither of these two Si incorporations in the AlN network is a dopant configuration, since lattice relaxations lead to midgap defect states, rather than shallow dopant states.

Si in Zincblende AlN

Substituting a Si for a N atom in zincblende AlN leads to a large symmetric outward relaxation of 0.3 Å for all Si nearest neighbor Al atoms. Three midgap gap states are formed that are highly localized on the Si atom.

Replacing an Al atom by a Si atom in zincblende AlN is the only configuration for which substitutional doping occurs. There is no relaxation present in the network, and a singly occupied shallow donor state is formed 0.2 eV below the conduction band edge.

CONCLUSIONS

In summary we have reported the first large scale, detailed *ab initio* treatment of vacancies and native defects in zincblende and wurtzite AlN. We find that the electronic consequences of defects in wurtzite and zincblende AlN are similar for the N vacancy, which is a single donor. For the N antisite the lower symmetry of the wurtzite structure leads to a different bonding and electronic environment than for zincblende N_{Al}.

Substitutional Si doping is only possible for the Al site in zincblende AlN. All other Si incorporations (wurtzite: Si_N, Si_{Al}; zincblende: Si_N) show large lattice relaxations around the Si atom, which leads to midgap defect states, rather than shallow dopant states.

ACKNOWLEDGEMENTS

This work was supported in part by the Ballistic Missile Defense Organization through the Office of Naval Research Grant No.: N00014-96-1-1183. We thank Dr. Alex Demkov and Prof. Otto Sankey for the use of Fireball96 and for helpful discussions.

References

[1] S. Strite and H. Morkoc, J. Vac. Sci. Technol., B 10(4), 1237 (1992).

[2] J. R. Edgar, J. Mater. Res. **7**, 235 (1992).

[3] I. Petrov, E. Mojab et al. Appl. Phys. Lett **60**, 2491 (1992).

[4] D. W. Jenkins and J. D. Dow, Phys. Rev. B **39**, 3317 (1989).

[5] M. Suzuki, T. Uenoyama and A. Yanase, Phys. Rev. B **52**, 8132 (1995).

[6] A. Rubio et al., Phys. Rev. B **48**, 11810 (1993).

[7] W. R. L. Lambrecht and B. Segall, in Properties of III-Nitrides, edited by J. H. Edgar, Ed., INSPEC, London, 1994.

[8] P. Boguslawski, E. L. Briggs, T. A. White, M. G. Wensell and J. Bernholc, Mat. Res. Soc. Symp. Proc. **339**, 693 (1994).

[9] T. L. Tansley and R. J. Egan Phys. Rev. B **45**, 10942 (1992).

[10] O.F. Sankey and D.J. Niklewski, Phys. Rev. B **40**, 3979 (1989); O. F. Sankey, D. A. Drabold, G. B. Adams, Bull. Am. Phys. Soc. **36**, 924 (1991).

[11] A. A. Demkov, J. Ortega, O. F. Sankey and M. Grumbach, Phys. Rev. B **52**, 1618 (1995).

[12] J. Neugebauer and C. G. Van de Walle, Mat. Res. Soc. Symp. Proc. **339**, 687 (1994).

SPECTROSCOPIC IDENTIFICATION OF THE ACCEPTOR-HYDROGEN COMPLEX IN Mg-DOPED GaN GROWN BY MOCVD

W. GÖTZ,
Hewlett-Packard Company, San Jose, California 95131, USA
M.D. MCCLUSKEY, N.M. JOHNSON, AND D.P. BOUR
Xerox Palo Alto Research Center, Palo Alto, California 94304, USA
E.E. HALLER
Lawrence Berkeley Laboratory and the University of California, Berkeley, California 94720, USA

ABSTRACT

Mg-doped GaN films grown by metalorganic chemical vapor deposition were characterized by variable-temperature Hall-effect measurements and Fourier-transform infrared absorption spectroscopy. As-grown, thermally activated, and deuterated Mg-doped GaN samples were investigated. The existence of Mg-H complexes in GaN is demonstrated with the observation of a local vibrational mode (LVM) at 3125 cm^{-1} (8 K). At 300 K this absorption line shifts to 3122 cm^{-1}. The intensity of the LVM line is strongest in absorption spectra of as-grown GaN:Mg which is semi-insulating. Upon thermal activation, the intensity of the LVM line significantly decreases and an acceptor concentration of 2×10^{19}cm^{-3} is derived from the Hall-effect data. After deuteration at 600°C the resistivity of the Mg-doped GaN increased by four orders of magnitude. A LVM line at 2321 cm^{-1} (8 K) appears in the absorption spectra which is consistent with the isotopic shift of the vibrational frequency when D is substituted for H.

INTRODUCTION

The accomplishment of p-type doping in III-V nitrides was a major step towards the realization of bright light emitting and laser diodes with these materials [1,2]. Originally, the incorporation of potential acceptors such as Mg or Zn resulted in semi-insulating GaN films for growth methods which include hydrogen in the growth ambient such as metalorganic chemical vapor deposition (MOCVD) or hydride vapor phase epitaxy [3]. In 1989, Amano and coworkers [4] demonstrated that exposure of Mg-doped GaN to low energy electron beam irradiation activates the acceptors. It was subsequently demonstrated by Nakamura and coworkers [5] that thermal annealing near 600°C in a hydrogen-free ambient also produces p-type conductivity, similar to what has been observed for other semiconductors (e.g., GaAs:Zn [6]).

It has been established by both theory and experiments that hydrogen forms stable complexes with shallow acceptor dopants in a host of elemental and compound semiconductors [7]. For GaN, Neugebauer and Van de Walle [8] concluded from first principles total energy calculations that hydrogen, which is likely to act as a donor under MOCVD growth conditions for GaN:Mg, compensates Mg acceptors and thereby suppresses the formation of impurity-related or native donors. However, during the cool-down, Mg-H complexes form which are electrically inactive.

To date, experimental evidence for the existence of Mg-H complexes has been only indirect. Nakamura and coworkers [5] showed that exposure of p-type, Mg-doped GaN to an NH$_3$ atmosphere at temperatures >600°C reduces the conductivity of the sample by six orders of magnitude whereas the conductivity remains unchanged when the anneal is performed in an N$_2$ ambient. Subsequently, variable-temperature Hall-effect measurements were conducted for p-type, Mg-doped GaN before and after exposure to monatomic deuterium [9]. The incorporation of deuterium which was established by secondary ion mass spectrometry (SIMS) led to a

Mat. Res. Soc. Symp. Proc. Vol. 468 © 1997 Materials Research Society

reduction of the acceptor concentration but to an increase of the hole mobility. This observation is consistent with removal of acceptors implying the formation of Mg-H complexes.

In this paper, spectroscopic evidence for the existence of Mg-H complexes is reported. A local vibrational mode (LVM) of the Mg-H complex is observed by Fourier-transform infrared spectroscopy (FTIR) at sample temperatures between 8 and 300 K. The involvement of H is established by the observation of the isotopic shift of the vibrational frequency when H is replaced by deuterium. The involvement of Mg acceptors is established by the correlation of the intensity of the LVM signal with acceptor concentrations deduced from variable-temperature Hall-effect measurements. The intensity of the LVM line was observed to decrease as the Mg acceptors were activated.

EXPERIMENTAL

The GaN epitaxial layers utilized in this study were grown by MOCVD and doped with Mg during growth. The films were 4 μm thick and deposited on double-polished sapphire substrates. To achieve p-type conductivity, the Mg-doped GaN samples received a standard post-growth thermal treatment in a rapid-thermal-anneal (RTA) system. Deuteration was performed with a remote plasma hydrogenation system [10] at 600°C for two hours. Such treatment introduces deuterium into p-type, Mg-doped GaN at concentrations above 10^{19} cm^{-3} and significantly increases the resistivity of the films [9, 11].

The Hall-effect measurements were conducted in the temperature range from 80 to 500 K. The magnetic field was 17.4 kG. For the Hall-effect measurements, samples of 5x5 mm^2 size were cut from the wafers and metal dots were vacuum evaporated in the four corners to obtain electrical contacts in the Van der Pauw geometry. The contacts were annealed at 230°C for 20 min in vacuum and were Ohmic over the entire temperature range of the measurement.

Infrared absorption measurements were performed with a FTIR spectrometer (Bomem DA8) equipped with a MCT detector and a glowbar light source. The instrumental resolution was set at 2 cm^{-1}. We used a CaF$_2$ and KBr beam splitter for the H- and D-related peaks, respectively. For the absorption measurement, samples of 12x12 mm^2 were cut from the wafers and placed in a Janis liquid-He dewar. The sample temperature was varied between 8 and 300 K. The Mg-doped GaN samples and the reference were placed on the same mounting plate to ensure accurate and reliable subtractions of the background. The absorption measurements were performed in transmission at normal incidence.

SIMS depth profiles for Mg were measured using an O$^-$ primary ion beam in a CAMECA IMS 4F system with GaN implantation standards.

RESULTS AND DISCUSSION

Results from variable-temperature Hall-effect measurements for the Mg-doped GaN after post-growth activation are displayed in Fig. 1. Figure 1a shows the experimental hole concentration (solid squares) as a function of the reciprocal temperature. The hole concentrations (p) are obtained from the experimental Hall constants (R_H) with $p(T)=r_H/q/R_H(T)$ (q = electronic charge). The Hall scattering factor (r_H) was assumed to be temperature-independent, isotropic, and of unity value. After the post-growth treatment our p-type GaN exhibits a room temperature hole concentration of ~10^{17} cm^{-3}. The hole mobilities are displayed in Fig. 1b as a function of the temperature. At 300 K, a hole mobility of 16 cm^2/Vs was measured. The mobility peaks at 190 K with a value of 27cm^2/Vs.

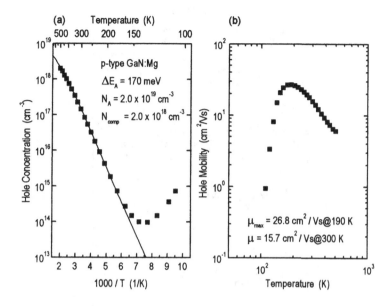

Fig. 1: (a) Hole concentration vs reciprocal temperature and (b) hole mobility vs temperature as determined from Hall-effect measurement (solid squares). The solid line (Fig. 1a) results from a fit of the charge neutrality equation to the experimental data assuming a single acceptor and donor compensation. Fit parameters are the activation energy for acceptor ionization (ΔE_A), the concentration of the acceptors (N_A), and the concentration of compensating donors (N_{comp}).

The temperature dependence of the hole concentration was analyzed in terms of a single acceptor model and the presence of donor compensation [12]. Fit parameters were the acceptor concentration (N_A), the thermal activation energy for acceptor ionization (ΔE_A) and the total concentration of donors (N_{comp}). The calculated hole concentration is represented by the solid line in Fig. 1a. The fit parameters are depicted in Fig. 1a.

The atomic Mg concentration in our GaN material was measured to be $\sim 3 \times 10^{19}$ cm^{-3} by SIMS. The Mg atoms were found to be uniformly distributed throughout the sample thickness. The acceptor concentration as estimated from the Hall-effect measurements (Fig. 1a) agrees approximately with the total Mg concentration indicating that $\sim 2/3$ of the Mg atoms act as acceptors (Mg_{Ga}). However, the SIMS data are only ~ 50 % accurate. Due to this uncertainty it is impossible to decide whether all or only 50 % of the Mg atoms are acceptors in our Mg-doped GaN material. The hydrogen concentration profile was also recorded with SIMS. A concentration of $\sim 2 \times 10^{19}$ cm^{-3} was measured, well above the H background in the SIMS chamber which accounts for $\sim 5 \times 10^{18}$ cm^{-3} of the H signal. In SIMS measurements for Mg-doped GaN films, the H concentration is typically found to be of the order of the Mg concentration.

Results from FTIR absorption spectroscopy for a sample temperature of ~ 8 K are summarized in Fig. 2. The relative absorbance ($\equiv -\ln(T/T_0)$, where $T/T_0 =$ transmittance with/without sample) is shown in two different wavenumber ranges. For the as-grown material (1), the spectrum exhibits an absorption line at 3125 cm^{-1}. After thermal activation of the p-type conductivity (2)

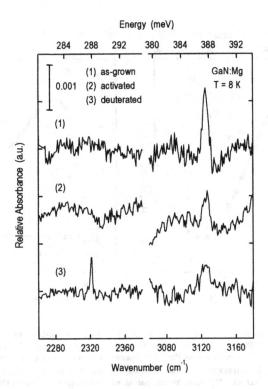

Fig. 2: Infrared absorption spectra for Mg-doped GaN grown by MOCVD. Spectra are shown for (1) as-grown material, (2) after RTA activation of the p-type conductivity , and (3) after deuteration. The vertical bar indicates the magnitude of the absorption scale.

the absorption line is still present in the spectrum, however, the intensity of the line is reduced in the logarithmic absorbance scale which is proportional to the areal density of the vibrating complexes by about a factor of two. At this stage, ~2×10^{19} cm^{-3} acceptors are present in the Mg-doped GaN samples (Fig. 1). After deuteration (3), a new absorption line appears at 2321 cm^{-1}. This line disappears from the spectrum after the thermal activation treatment. Neither of these lines was observed for unintentionally-doped or Si-doped n-type GaN.

The temperature dependence of the LVM line at 3125 cm^{-1} (8 K) is demonstrated in Fig. 3. With increasing temperature its position slightly shifts to smaller wavenumbers and arrives at 3122 cm^{-1} at 300 K. A shift of the frequency to lower wavenumbers with increasing sample temperature is expected for LVMs and due to the vibration of neighboring atoms as well as the expansion of the GaN lattice [6,13].

The experimental evidence presented in this study establishes the absorption line at 3125 cm^{-1} (8 K) as a vibrational mode of the Mg$_{Ga}$-H complex in GaN. First, this line appears only in Mg-doped material. For a given Mg concentration the intensity of this line is strongest for as-grown material where most of the Mg atoms form complexes with hydrogen. At this stage, the Mg-doped GaN is highly resistive (~10^{10} Ωcm, 400 K) [11]. Second, upon thermal activation, the

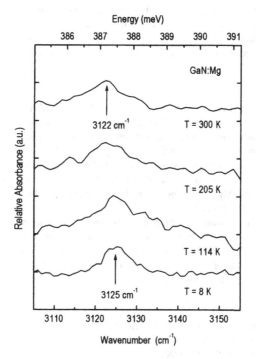

Fig. 3: Infrared absorption spectra for Mg-doped GaN grown by MOCVD. The spectra are shown for four different sample temperatures. The position of the Mg-H LVM related line is indicated by arrows for a sample temperature of 300 and 8 K.

intensity of the line decreases significantly. Now the sample is conductive (~2 Ωcm, 300 K) with an acceptor concentration of ~2×10^{19} cm^{-3} (Fig. 1a). Consequently, a significant portion of the Mg-H complexes was dissociated due to the thermal activation treatment. Third, the involvement of hydrogen in the vibrating complex is clearly established by the isotopic shift of the vibrational frequency to 2321 cm^{-1} (8 K). The isotopic ratio of the two absorption line frequencies (r-factor) is ~1.346. Assuming a one-dimensional harmonic oscillator, a r-factor of 1.369 and 1.387 is expected for a N-H and Mg-H oscillator, respectively [6,14].

Recent computational studies by Neugebauer and coworkers [8] suggest a novel atomic structure for Mg-H complexes in (zincblende) GaN which is qualitatively different from the bond-centered hydrogen configuration that has been well established for acceptor-hydrogen complexes in other semiconductors. In GaN, the hydrogen atoms prefer the antibonding site of one of the N neighbors of the substitutional Mg. Consequently, the H-stretch mode frequency was found to be 3360 cm^{-1} which is close to that of H in NH$_3$ (3444 cm^{-1}). Our experimental value for the LVM frequency is consistent with the N-H stretch frequency of the suggested model due to a the uncertainty in the theoretical value.

In wurtzite GaN, one might expect two distinct stretch vibrational modes of the Mg-H complex due to two inequivalent possible orientations of the complexes. They can be aligned either parallel or perpendicular to the c axis of the GaN crystal. However, for illumination normal to the surface only complexes which possess a dipole moment projected into the c-plane can be observed. However, the deviation of the GaN wurtzite structure from the perfect diamond structure is rather small. Therefore, the expected differences in the N-H stretch vibrational frequencies are expected to be small, perhaps too small to be resolved.

CONCLUSION

The existence of Mg-H complexes in Mg-doped GaN was established by the observation of an absorption line at 3125 cm^{-1} (8 K) which is related to a stretch mode of the complexes. The intensity of the absorption line reaches a maximum for as-grown GaN:Mg. At this stage, most of the Mg$_{Ga}$ acceptors are structurally associated with hydrogen to form Mg-H complexes which are electrically inactive. The observed frequency of the LVM is consistent with a theoretical model that positions the H atom in an anti-bonding configuration opposite a host N atom which has a substitutional Mg atom (Mg$_{Ga}$) as a neighbor. Only a portion of the Mg atoms are active as acceptors to compensate residual donors. It is now clear, that the dominant mechanism that renders Mg-doped GaN semi-insulating is acceptor neutralization (passivation). Thermal annealing at elevated temperatures (>600°C) in a hydrogen-free ambient or minority carrier injection (LEEBI treatment) leads to the dissociation of Mg-H complexes and, therefore, to p-type conductivity. It is unlikely that hydrogen leaves the sample during activation processes. H atoms may form H molecules or agglomerate at native point defects or extended defects.

ACKNOWLEDGMENT

The authors are pleased to acknowledge fruitful discussions with Jörg Neugebauer and Chris van de Walle and to thank J. Walker for technical support. The work at Xerox was supported by DARPA, agreement # MDA972-95-3-008.

REFERENCES

1 I. Akasaki, H. Amano, N. Koide, M. Kotaki, and K. Manabe, Physica B **185**, 428 (1993)
2 S. Nakamura, M. Senoh, S. Nagahama, N. Iwasa, T. Yamada, T. Matsushita, H. Kyoku, and Y. Sugimoto, Jpn. J. Appl. Phys **35**, L74 (1996)
3 S. Strite and H. Morkoç, J. Vac. Sci. Technol. B **10**, 1237 (1992)
4 H. Amano, M. Kito, K. Hiramatsu, and I. Akasaki, Jpn. J. Appl. Phys. **28**, L2112 (1989)
5 S. Nakamura, T. Mukai, M. Senoh, and N. Iwasa, Jpn. J. Appl. Phys. **31**, L139 (1992)
6 J. Chevallier, B. Clerjaud, and B. Pajot, Hydrogen in Semiconductors, edited by J.L. Pankove and N.M. Johnson (Academic San Diego, 1991), Chap. 13.
7 *Hydrogen in Semiconductors*, eds. J.I. Pankove and N.M. Johnson, Semicondcutors and Semimetals Vol. 34 (Academic Press, San Diego, 1991)
8 J. Neugebauer and C.G. Van de Walle, Phys. Rev. Lett. **75**, 4452 (1995)
9 W. Götz, N.M. Johnson, R.A. Street, H. Amano, and I. Akasaki, Appl. Phys. Lett. **66**, 1340 (1995)
10 N.M. Johnson, in *Hydrogen in Semiconductors*, edited by J.L. Pankove and N.M. Johnson (Academic, San Diego, CA, 1991), p. 118
11 W. Götz, N.M. Johnson, D.P. Bour, M.D. McCluskey, and E.E. Haller, Appl. Phys. Lett. **69**, 3725 (1996)
12 R. Schaub, G. Pensl, M. Schulz, and C. Holm, Appl. Phys. A **34**, 215 (1984)
13 R.C. Newman, in *Infrared Studies of Crystal Defects*, (Taylor and Francis Ltd., London, (1973)
14 M. Stavola and S.J. Pearton, Hydrogen in Semiconductors, edited by J.L. Pankove and N.M. Johnson (Academic San Diego, 1991), Chap. 8.

INCORPORATION AND OPTICAL ACTIVATION OF Er IN GROUP III-N MATERIALS GROWN BY METALORGANIC MOLECULAR BEAM EPITAXY

J. D. MACKENZIE,* C. R. ABERNATHY,* S. J. PEARTON,* S. M. DONOVAN,* U. HÖMMERICH,** M. THAIK,** X. WU,** F. REN,*** R. G. WILSON,† J. M. ZAVADA‡
*Department of Materials Science and Engineering, University of Florida, Gainesville, FL 32611.
**Department of Physics, Research Center for Optical Physics, Hampton University, Hampton, VA 23668
***Lucent Technologies, Murray Hill, New Jersey 07974
†Hughes Research Laboratory, Malibu, California 90265
‡U.S. Army Research Office, Research Triangle Park, NC 27709

ABSTRACT

Metalorganic molecular beam epitaxy has been utilized to incorporate Er into AlGaN materials during growth utilizing elemental and metalorganic sources. Room temperature 1.54 μm photoluminescence was observed from AlN:Er and GaN:Er. Photoluminescence from AlN:Er doped during growth using the elemental source was several times more intense than that observed from implanted material. For the first time, strong room temperature 1.54 μm PL was observed in GaN:Er grown on Si. Temperature-dependent photoluminescence experiments indicated the 1.54 μm intensities were reduced to 60% and 40% for AlN:Er and GaN:Er, respectively, between 15 K and 300 K. The low volatility of Er(III) tris (2,2,6,6 - tetramethyl heptanedionate) and temperature limitations imposed by transport considerations limited maximum doping levels to $\sim 10^{17}$ cm^{-3} indicating that this precursor is unsuitable for UHV.

INTRODUCTION

The scope of this work includes an investigation of the incorporation and optical activation of Er introduced during growth of III-N materials by plasma-assisted metalorganic molecular beam epitaxy. Radiative intra-4f shell transitions in Er matching the attenuation loss minima in silica fibers have stimulated research interest in rare earth-doped optoelectronics. Considerable research work has been devoted to studying the incorporation and optical activity of Er in Si[1, 2] and III-V materials.[3, 4, 5, 6] Si:Er research uncovered two significant aspects of Er optical activation. The luminescence efficiency of Si:Er decreases dramatically with increasing temperature such that the Er luminescence at room temperature was only a small fraction of the low temperature emission. Also, the symmetry and local electronic environment of the Er dramatically impacts the luminescence efficiency.[7, 8, 9] Influencing the local environment with the introduction of electronegative species such as O, N, and F resulted in significant increases in Er^{3+}-related emission. Favennec et al.[10] demonstrated a reduction in thermal quenching with increasing bandgap for a range of Group IV, III-V, and II-VI semiconductors doped by ion-implantation. It has also been suggested that the ionic nature of a particular host matrix can enhance the radiative efficiency of Er^{3+}.[11] Neuhalfen and Wessels[12] confirmed this trend in In$_x$Ga$_{1-x}$P:Er. Since the III-N materials can access an energy gap range spanning the visible spectrum and into the near UV (1.9 -6.3 eV), they are attractive hosts for Er.

Virtually every compound semiconductor growth technique has been adapted to produce Er-doped III-V materials. These have included ex-situ methods such as ion implantation and diffusion. Ion implantation is a highly adaptable processing technique used widely throughout III-V FET production technologies making it an attractive choice for integrated Er-based III-V optoelectronics. A number of groups have demonstrated Er^{3+}-related 1.54 μm luminescence in GaAs, InP and recently, GaN[13, 14] and AlN[13] with implantation. However, the large Er mass and the high energies required to implant Er hinder the application of implantation to device structures. The depth and distribution of the Er implants are limited. Also, at higher doses, the implantation process can cause significant damage in the host matrix. This can be particularly troublesome in III-V materials that have a limited ability to recover damage through post-implant annealing. Silkowski *et al.*[15] observed an intense band of optically-active damage-induced mid-gap states for III-N samples implanted with Nd and Er. Annealing at 1000 °C for 90 minutes did not eliminate these peaks. In-situ doping during III-V:Er growth has been performed by MOCVD,[12, 16, 17] MBE,[18, 19] and MOMBE.[20] Er has been introduced in a variety of forms including evaporation of elemental In and by the transport of Er-metalorganic molecules. Cyclopentadienyl-Er compounds[17] have been used in GaAs MOCVD but lead to high C levels and degraded film quality. Heptanedionate and amide-based Er sources have been used in vapor phase epitaxy of Si[21] and GaAs[22] as well as wider band gap GaP grown heteroepitaxially on Si.[23]

MOMBE allows for the growth of III-N:Er utilizing elemental and metalorganic sources. In this work we have looked at metallic Er and Er(III) tris (2,2,6,6 - tetramethyl heptanedionate) {Er(tmhd)$_3$}. It is likely that the incorporation of Er from this source is accompanied by oxygen which has been shown to enhance luminescence from rare-earths in Si and III-V hosts. Unfortunately, the volatility of Er(tmhd)$_3$ requires high source temperatures and heated lines making it difficult to work with in UHV.

EXPERIMENTAL PROCEDURES

Films were grown by MOMBE in an INTEVAC Gas Source Gen II on In-mounted (0001) Al$_2$O$_3$ and semi-insulating GaAs. GaN and AlN films were preceded by a low temperature AlN buffer (T_g 425 °C). Dimethylethylamine alane (DMEAA), triethyl gallium (TEGa), and trimethyl indium (TMIn) provided the group III fluxes. Reactive nitrogen species were provided by a SVT radio frequency (RF) plasma source and a Wavemat MPDR 610 electron cyclotron resonance (ECR) plasma source. Further details of the plasma-assisted growth technique employed here have been presented previously.[24, 25] A shuttered effusion oven charged with 4N Er was used for solid source doping. A special mounting and transport setup was employed to introduce Er(tmhd)$_3$ into the growth chamber. The heated source bubbler (90-110°C) was mounted directly to the growth chamber source flange. The short stainless steel line connecting the bubbler outlet to the injector and the injector itself were heated in order to prevent condensation of the precursors before entry into the growth chamber.

Er incorporation was profiled by secondary ion mass spectrometry (SIMS). Al, Ga, N, O, and C levels were also monitored. Room and low temperature photoluminescence was used to evaluate 1.54 μm luminescence. For the AlN:Er thermal quenching measurements Er^{3+} PL was excited using the 488 nm line of an Ar ion laser and measured with a liquid-nitrogen cooled Ge detector. An optical parametric oscillator pumped by a Q-switched Nd:YAG laser was the excitation source used for the GaN:Er PL experiments. A He refrigerator capable of reaching 12K was used to cool the sample. The surface morphology of the nitride films was characterized by scanning electron microscopy (SEM) using a JEOL 35CF. Crystal phase composition was determined by X-ray powder diffraction.

RESULTS AND DISCUSION

PL studies detected 1.54 μm luminescence from MOMBE-derived III-N:Er materials deposited on different substrates and using elemental and metalorganic precursors. Er^{3+}-related signal from AlN:Er doped during growth was orders of magnitude higher than that observed for implanted material. Figure 1 shows the thermal quenching behavior of several semiconductors implanted with Er compared to MOMBE-derived AlN:Er and GaN:Er doping during growth on Al_2O_3 with the elemental source.[6, 7] Strong Er^{3+}-related luminescence persisted to room

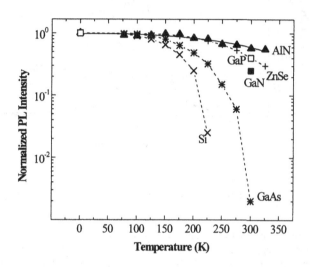

Figure 1. Thermal quenching of Er^{3+} photoluminescence from implanted Si, GaAs, and ZnSe and GaP:Er, GaN:Er, and AlN:Er doped during growth.

temperature for both materials. A quenching factor of only ~60% was observed for AlN:Er/Al_2O_3 while the GaN:Er/Al_2O_3 1.54 μm signal quenched to ~30% of its low temperature level. These promising results agree with the general trend of decreasing thermal quenching with increasing band gap. Doping with the Er(tmhd)$_3$ source resulted in Er concentrations near 10^{17} cm^{-3} as determined by SIMS analysis. However, room temperature, 1.54 μm luminescence was detectable only in AlN:Er. This signal is several orders of magnitude weaker than that observed in AlN films doped more heavily with the solid Er source. During the growth of the Er(tmhd)$_3$ samples, the bubbler and transport lines were heated to nearly 100 °C. The higher temperatures required to produce an Er(tmhd)$_3$ flux equivalent to the level supplied by the effusion oven, are impractical. Temperatures greater than 100 °C could affect the operation of the flow control valves.

Using the elemental Er source, 1.54 μm emission from GaN:Er grown on Si was observed. GaN:Er deposited at 800 °C on Si with an Er concentration of ~5 × 10^{18} cm^{-3} showed substantially stronger luminescence than material deposited on Al_2O_3 under identical conditions. Figure 2. shows the PL spectra (488 nm) taken from GaN:Er grown on Si and Al_2O_3. The absence of fine structure in the spectra suggests that Er is optically active at a number of different

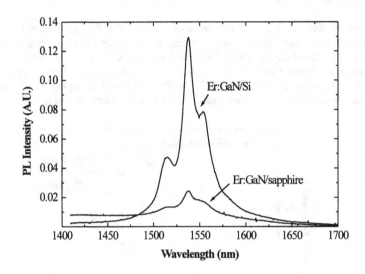

Figure 2. 300 K photoluminescence spectra of AlN:Er doped during growth with Er(tmhd)$_3$ ([Er] ~ 10^{17} cm^{-3}).

sites in the host matrix. X-ray diffraction indicates that GaN on Si is polycrystalline while the material grown on sapphire is columnar-epitaxial. This structural difference is also indicated by the surface morphology observed by SEM. Figure 3 shows the surfaces of the GaN:Er on sapphire and on Si. The film deposited on Al$_2$O$_3$ is relatively smooth with features similar to those observed for AlN:Er previously.[20] The GaN:Er on sapphire has a more pronounced surface texture suggestive of polycrystalline growth with grain features of ~0.2 µm. The increased optical activity from the Er in GaN grown on Si may be attributed to the varied environments for the Er ion caused by the presence of internal surfaces within the polycrystalline film. Temperature-dependent PL experiments indicated a quenching ratio of 40% between 15 K and room temperature. This quenching behavior is represented in Figure 4 through PL spectra collected at incremental temperature steps from 15K to 300 K. This is the first demonstration of Er^{3+}-related luminescence from III-N:Er on Si doped during growth.

CONCLUSIONS

Incorporation and optical activation of Er in AlN and GaN has been demonstrated using elemental and organometallic precursors during growth by plasma-assisted metalorganic molecular beam epitaxy. AlN:Er and GaN:Er films were doped with an elemental source resulting in strong, room-temperature 1.54 µm photoluminescence. The low volatility of Er(tmhd)$_3$ renders it ineffective for UHV application For the first time, room temperature Er^{3+} photoluminescence has been observed for GaN:Er grown on Si. The persistence of strong 1.54 µm luminescence at room temperature from GaN:Er suggests the possibility of integrated GaN:Er/Si optoelectronics.

Figure 3. Plan-view SEM micrographs of GaN:Er {[Er] ~ 5 × 10^{18} cm^{-3}} grown on Al$_2$O$_3$ (top) and Si (bottom).

ACKNOWLEDGEMENTS

The authors would like to acknowledge the support of the National Science Foundation (ECS-952288), the Army Research Office (DAAH04-95-1-0196 and DAAH 04-96-1-0089), the National Aeronautics and Space Administration (NCC-1-251). Growth was conducted at the MICROFABRITECH facility at the University of Florida.

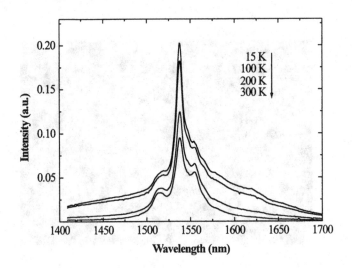

Figure 4. Temperature-dependent PL from GaN:Er $\{[Er] \sim 5 \times 10^{18}$ cm$^{-3}\}$ grown on Si.

REFERENCES

[1] H. Ennen, J. Schneider, G. Pomrenke, and A. Axman, *Appl. Phys. Lett.*, **43**, 943 (1983).

[2] D. J. Eaglesham, J. Michel, E. A. Fitzgerald, D. C. Jacobson, J. M. Poate, J. L. Benton, A. Polman, Y. -H. Xie, and L. C. Kimmerling, *Appl. Phys. Lett.*, **58**, 2797 (1991).

[3] A. Kozanecki, M. Chan, C. Jeynes, B. Sealy, and K. Homewood, *Solid State Commun.*, **78**, 763 (1991).

[4] Y. S. Tang, J. Zhang, K. C. Heasman, and B. J. Sealy, *Solid State Commun.*, **72**, 991 (1989).

[5] J. Nakata, M. Taniguchi, and K. Takahei, *Appl. Phys. Lett.*, **61**, 2665 (1992).

[6] B. W. Wessels, *Mat. Res. Soc. Symp. Proc.*, **422**, 247 (1996).

[7] P. N. Favennec, H. L. L'Haridon, D. Moutonet, M. Salvi, and M. Gauneau, *Jap. J. Appl. Phys.*, **29**, L524 (1990).

[8] D. L. Adler, D. C. Jacobson, D. J. Eaglesham, M. A. Marcus, J. L. Benton, J. M. Poate and P. H. Citrin, Appl. Phys. Lett., 61, 2181 (1992).

[9] F. Priolo, G. Franzo, S. Coffa, A. Polman, S. Libertino, R. Barklie, and D. Carey, *J. Appl. Phys.*, **78**, 3874 (1995).

[10] P. N. Favennec, H. L. Harridon, M. Salvi, D. Moutonnet and Y. L. Cuillo, *Electron. Lett.* **25** 718 (1989).

[11] J. M. Zavada and D. Zhang, *Solid State Electron.*, **38**, 1285 (1995).

[12] A. J. Neuhalfen and B. W. Wessels, *Appl. Phys. Lett.* **59** 2317 (1991).

[13] R. G. Wilson, R. N. Schwartz, C. R. Abernathy, S. J. Pearton, N. Newman, M. Rubin, T. Fu and J. M. Zavada, *Appl. Phys. Lett.* **65**, 992 (1994).

[14] C. H. Qiu, M. W. Leksono, J. L. Pankove, J. T. Torvik, R. J. Feuerstein, and F. Namavar, *Appl. Phys. Lett.* **66**, 562 (1995).

[15] E. Silkowsky, Y. K. Yeo, R. L. Hengehold, B. Goldenberg, and G. S. Pomrenke, *Mat. Res. Soc. Symp. Proc.*, **422**, 69 (1996).

[16] S. J. Chang and K. Takahei, Appl. Phys. Lett., 65, 433 (1994).

[17] J. M. Redwing, T. F. Kuech, D. C. Gordon, B. A. Vaarstra, S. S. Lau, *J. Appl. Phys.*, **97**, 1585 (1994).

[18] A. Rolland, A. Le Corre, P. N. Favennec, M. Gauneau, B. Lambert, D. Lecrosnier, H. L'Haridon, D. Moutonnet, C. Rochaix, *Electron. Lett.*, **24** (1988).

[19] P. S. Whitney, K. Uwai, H. Nakagome, and K. Takahei, *Electron. Lett.*, **24**, 740 (1988).

[20] J. D. MacKenzie, C. R. Abernathy, S. J. Pearton, U. Hommerich, X. Wu, R. N. Schwartz, R. G. Wilson, and J. M. Zavada, *Appl. Phys. Lett.*, **69**, 2083 (1996).

[21] P. S. Andry, W. J. Varhue, E. Adams, M. Lavoie, P. B. klein, R. Hengehold, and J. Hunter, *Mat. Res. Soc. Symp. Proc.*, **422**, 57 (1996).

[22] A. C. Greenwald, K. J. Linden, W. S. Rees, O. Just, N. M. Haegel, and S. Donder, *Mat. Res. Soc. Symp. Proc.*, **422**, 63 (1996).

[23] X. Z. Wang and B. W. Wessels, Appl. Phys. Lett., 67 (1995) 518.

[24] J. D. MacKenzie, L. Abbaschian, C. R. Abernathy, S. M. Donovan, S. J. Pearton P. C. Chow and J. Van Hove, Journal of Electronic Materials, to be published.

[25] J. D. MacKenzie, C. R. Abernathy, S. J. Pearton, V. Krishnamoorthy, S. Bharatan, K. S. Jones, and R. G. Wilson, Appl. Phys. Lett., 67 (1995) 253.

SITE-SELECTIVE PHOTOLUMINESCENCE EXCITATION AND PHOTOLUMINESCENCE SPECTROSCOPY OF Er-IMPLANTED WURTZITE GaN

S. KIM, S.J. RHEE, D.A. TURNBULL, X. LI, J.J. COLEMAN, AND S.G. BISHOP

Microelectronics Laboratory, University of Illinois at Urbana-Champaign, Urbana, IL 61801

ABSTRACT

Site-selective photoluminescence (PL) and photoluminescence excitation (PLE) spectroscopy have been carried out at 6K on the ~1540 nm $^4I_{13/2} \rightarrow {}^4I_{15/2}$ emissions of Er^{3+} in Er-implanted GaN. The PLE spectra exhibit several broad, below-gap, defect- or impurity-related absorption bands which excite three distinct site-selective Er^{3+} PL spectra. The near-band edge spectral position and lineshape of the PLE spectrum of one of the site-selective PL bands suggest that this Er site forms a trap level within the band gap and an exciton bound at this trap is involved in the excitation mechanism. In addition, high resolution PLE spectra obtained with a tunable laser in the 810 nm spectral range reveal a set of sharp PLE peaks due to the $^4I_{15/2} \rightarrow {}^4I_{9/2}$ internal Er^{3+} f-band absorption superimposed on the broad defect PLE band. The site-selective PL spectrum excited by the sharp line ~810 nm Er^{3+} intra-f shell PLE bands is characteristic of a fourth distinct Er^{3+} site. The simple structure of the site-selective PL and PLE spectra associated with direct intra-f shell absorption suggests that the optically active Er site responsible for these spectra is of high symmetry in wurtzite GaN and that it could be attributed to a single Er atom on a Ga site.

INTRODUCTION

Er-doped semiconductors have been shown to exhibit emission from the ~1540 nm $^4I_{13/2} \rightarrow {}^4I_{15/2}$ transitions of Er^{3+}. These emissions coincide with the minimum loss region of silica glass fibers, so Er-doped semiconductors would be ideal source and amplifier materials for optical fiber communication systems. However, the 1540 nm emission from most semiconductor hosts has been limited by severe thermal quenching, hampering the fabrication of optical communication devices operating at room temperature. Previous studies have indicated that the thermal quenching of intra-4f shell emissions from Er^{3+} in semiconductor hosts decreases with increasing band gap [1,2], which has motivated recent research on Er-implanted GaN because of its wide band gap [3,4].

In Er-implanted wurtzite GaN, a recent study on annealing temperatures has demonstrated the existence of multiple Er^{3+} luminescence centers [5] and our previous spectroscopy work [6] has also demonstrated the existence of at least three different Er^{3+} sites. Nevertheless, in wurtzite GaN, neither the microstructure of the optically active Er sites nor the excitation mechanism of the Er^{3+} intra-4f shell emissions at multiple Er sites have been researched in detail yet.

In our previous work [6], photoluminescence excitation (PLE) spectroscopy of the 1540 nm $^4I_{13/2} \rightarrow {}^4I_{15/2}$ emissions of Er^{3+} in Er-implanted GaN detected three broad, below-gap absorption bands, each of which excites a distinct, site-selective Er^{3+} photoluminescence (PL) spectrum. The excitation of two of the site-selective Er PL bands involves optical absorption by defects rather than direct intra-f shell absorption. For the third site-selective PL band, excitons bound at an Er-related trap may be involved in the excitation mechanism. In the current study, a fourth site-selective Er^{3+} PL spectrum has been observed from Er-implanted GaN by high

131

resolution excitation of specific sharp absorption lines in the crystal field split $^4I_{15/2} \rightarrow {}^4I_{9/2}$ manifold of Er^{3+} transitions, suggesting that the excitation mechanism of the fourth optically active Er site involves direct intra-f shell absorption. Both the possible symmetry and origin of this active Er site are discussed in this paper.

EXPERIMENTAL PROCEDURE

GaN films (3 μm thick) were grown on sapphire substrates by atmospheric pressure metalorganic chemical vapor deposition (MOCVD) [7]. For preparation of Er-doped samples, Er ions were implanted with an energy of 280 keV and with implantation dosage of 4 x 10^{12} cm^{-2} into as-grown films at room temperature. The implanted samples were annealed at 900 °C for 30 minutes under a continuous flow of nitrogen gas. PLE and PL spectroscopy were carried out on the Er-implanted GaN at 9 K. Both PLE and PL spectra were excited with light from a Xenon lamp dispersed by a double grating monochromator, a tunable titanium-doped sapphire laser, or selected lines from Ar, He-Cd, and He-Ne lasers. All of the PLE spectra were corrected for the spectral response of the tunable excitation systems. The luminescence was analyzed by a 1-m single grating monochromator and detected by a cooled Ge PIN detector. Samples were cooled to liquid helium temperature in a continuous flow Janis Supervaritemp Cryostat.

RESULTS AND DISCUSSION

Fig. 1 shows PLE spectra detected at the four different wavelengths indicated by arrows in Fig. 2. The PLE spectra plot the PL intensity in the Er-implanted GaN as a function of excitation wavelength. In the PLE spectra (a) and (b), only a broad PLE band is seen. In contrast, both PLE spectra (c) and (d) exhibit five sharp PLE peak pairs superposed on the low energy tail of the broad, mid-gap, defect PLE band discussed below[6]. These sharp peaks are assigned to direct absorption by the $^4I_{15/2} \rightarrow {}^4I_{9/2}$ Er transition. Fig. 2 exhibits the PL spectra excited at three of the sharp PLE peaks as indicated by the arrows in Fig. 1 (henceforward referred to as the PL-f spectra). Fig. 3 shows the site-selective PL (PL-BR) spectra pumped at three different energies in the broad PLE band, as shown in the inset of Fig. 1. Both the PL-BR and PL-f spectra exhibit the sharply structured 1540 nm band characteristic of the $^4I_{13/2} \rightarrow {}^4I_{15/2}$ transitions of Er^{3+}. All the PL-f spectra shown in Fig. 2 have an identical set of PL peaks, but their relative intensities are slightly different from spectrum to spectrum. In contrast, the three PL-BR spectra of Fig. 3 are identical to one another.

Since the broad band PLE underlies the sharp Er PLE peaks, some component of the PL-BR is also excited at 809.3 nm and at the other sharp PLE peaks. This component can be subtracted out, leaving the PL spectrum excited specifically by the sharp intra-f band absorption's of the Er. Fig. 4 illustrates this procedure. The PL-f spectrum [Fig. 2(c)] excited by a 809.3 nm laser line and the PL-BR spectrum [Fig. 3(b)] excited at 810.9 nm were selected as representative PL spectra for the comparison between the PL-f and PL-BR spectra. In Fig. 4, the PL-f spectrum is plotted as a dotted line, the PL-BR spectrum is shown as a dashed line, and the difference between the two spectra is plotted as a solid line. The PL spectrum drawn by the solid line in Fig. 4 isolates the PL peaks excited by the sharp PLE peaks shown in Fig. 1: there are six prominent PL peaks (indicated by arrows in Fig. 4). All the PL-f spectra excited by each of the ten PLE peaks exhibit these six peaks superimposed on the PL-BR spectrum, implying that those ten PLE peaks originate from the same Er sites.

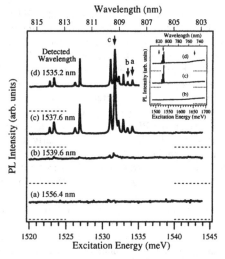

Fig. 1. PLE spectra detected at the four different wavelengths indicated by the arrows in Fig. 2.

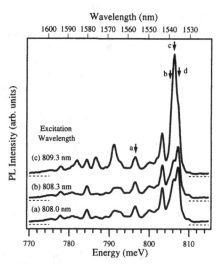

Fig. 2. PL-f spectra excited at three of the sharp PLE peaks as indicated by the arrows in Fig. 1.

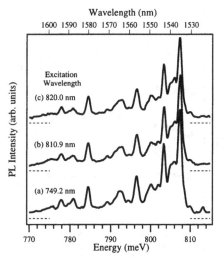

Fig. 3. Site-selective PL-BR spectra pumped at three different energies in the broad PLE band in the inset of Fig. 1.

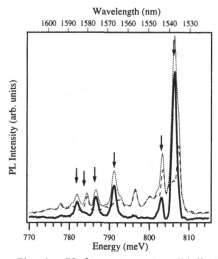

Fig. 4. PL-fs spectrum (a solid line) isolating the PL peaks excited by the sharp PLE peaks shown in Fig. 1. The PL-f spectrum is plotted as a dotted line and the PL-BR spectrum is shown as a dashed line.

In Fig. 5, a comparison is made between the PL-fs spectrum (a) and the three PL spectra [(b), (c), and (d)] associated with the three different Er sites excited by the three broad PLE bands observed in our previous work. This comparison shows that the PL-fs spectrum has distinctively different peaks from the other PL spectra. The details of the PL-fs spectrum (a), the PL spectra pumped by red [632.3 nm, (b)] and blue [457.9 nm, (c)] laser wavelengths and 404 nm light (d) from the xenon lamp are shown in Fig. 5. The "red", "blue", and "xenon" (referred to as PL-R, PL-B, and PL-X, respectively) PL spectra each contain groups or sets of prominent peaks that are dominant in that spectrum. In our previous work [6], the distinct difference among the site-selective PL-R, PL-B, and PL-X spectra was shown, revealing the existence of three different Er^{3+} sites in the GaN crystal. The comparison between the PL-fs spectrum and these three spectra shown in Fig. 5 demonstrates the uniqueness of the PL-fs spectrum. It is clearly distinct from the PL-X and PL-B spectra, with prominent peaks in those spectra absent from the PL-fs spectrum. This illustrates that the PL-fs luminescence originates from a fourth distinct optically active site in the GaN crystal. However, the PL-BR [Fig. 3] and PL-R [Fig. 5(b)] are nearly identical, so both arise from the same Er centers.

Fig. 6 exhibits PLE spectra (a), (b), (c), and (d) detected at the wavelength of the most prominent peak in each of the PL-fs, PL-R, PL-B, and PL-X spectra shown in Fig. 5, respectively: the PLE spectrum (a) was excited with a tunable titanium-doped sapphire laser in this current work while the PLE spectra (b), (c), and (d) with a xenon lamp in the previous work [6]. The PLE spectra (b) and (c) in Fig. 6 demonstrate that the PL-B bands in Fig. 5 are

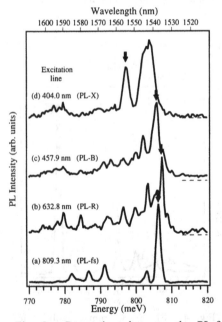

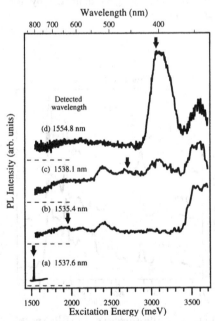

Fig. 5. Comparison between the PL-fs spectrum and the PL-B, PL-R, and PL-X spectra. The arrow in each spectrum indicates the detected wavelength for the corresponding PLE spectrum in Fig 6.

Fig. 6. PLE spectra (a), (b), (c), and (d) taken from the Er-implanted GaN detected at the wavelength of the most prominent peak in each of the PL-fs, PL-R, PL-B, and PL-X spectra in Fig. 5, respectively.

preferentially pumped by exciting light in the 2.6-2.8 eV (blue) range, while the PL-R bands are excited by lower energy <2.0 eV (red) light. The broad, overlapping PLE bands spanning the ~1.6 to 2.9 eV spectral range are defect-related absorptions that are characteristic of the implanted films, and they are not attributable specifically to the presence of Er. In our previous work, the excitation of PL-B and PL-R bands was shown to involve optical absorption by defects rather than direct intra-f shell absorption. The strong PLE band peaking at about 3.1 eV dominates the PLE spectrum of Fig. 6(d). This PLE band is specific to a trap level formed by the Er atoms within the band gap. For the PL-X band, excitons bound at an Er-related trap may be involved in the excitation mechanism, as our previous work suggested. While the PLE (b), (c), and (d) rely upon excitation by broad absorption bands, the PLE (a) was excited by specific sharp absorption lines in the crystal field split $^4I_{15/2} \rightarrow {}^4I_{9/2}$ manifold of Er^{3+} internal transitions. Hence the PL-fs band associated with the fourth Er^{3+} site existing in our sample involves direct intra-f shell absorption.

The PL-fs spectrum has a much simpler structure than other site-selective PL spectra, displaying only six prominent peaks. The PLE spectrum in Fig. 1(c) exhibits five PLE peak pairs. These small numbers imply that the fourth Er^{3+} site observed in this study is one of the high symmetry sites existing in wurtzite GaN. The number of PL and PLE peaks may be interpreted by examining the number of Stark split levels of the f-band multiplets of the Er atom. For Er^{3+} ions in non-cubic (such as triclinic, monoclinic, orthorhombic, hexagonal, trigonal, and tetragonal) crystal fields, the number of Stark levels in a given multiplet is always the same (for $^4I_{15/2}$, $^4I_{13/2}$, $^4I_{11/2}$, and $^4I_{9/2}$ the number of Stark levels is 8, 7, 6, and 5, respectively), while in the cubic case it is 5, 5, 4, and 3, respectively [8].

For an Er^{3+} center with noncubic symmetry the PL spectrum at low temperatures should exhibit eight zero-phonon lines resulting from transitions connecting the ground level of the $^4I_{13/2}$ manifold to the $^4I_{15/2}$ manifold. The excitation spectrum should exhibit five zero-phonon lines resulting from transitions from the ground level of $^4I_{15/2}$ manifold to the $^4I_{9/2}$ manifold. There are five pairs of sharp PLE peaks observed in the PLE spectra in Fig. 1, twice the number expected for an Er^{3+} center with noncubic symmetry. This doubling is caused by the fact that the first excited state of the $^4I_{15/2}$ manifold will still be partially populated, even at liquid helium temperatures. The peak positions are (1522.7 meV, 1523.4 meV), (1526.2 meV, 1526.8 meV), (1531.1 meV, 1531.6 meV), (1532.2 meV, 1532.8 meV), and (1533.5 meV, 1534.1 meV). In each pair, the higher (lower) energy is the transition energy from the ground (first) level of the $^4I_{15/2}$ manifold to the $^4I_{9/2}$ manifold. The average value of the energy difference between the higher and lower energy in each pair is 0.6 meV, which is close to the Boltzmann thermal energy (0.8 meV) of the sample temperature (9 K), verifying that the first level of the $I_{15/2}$ manifold may be populated thermally.

Six prominent PL peaks are observed in the PL-fs spectrum in Fig. 4. As mentioned above, in the $^4I_{15/2}$ manifold, the energy position of the first level is very close to that of the ground level: the energy difference between them is only 0.6 meV in the PLE spectrum in Fig. 1. This difference is smaller than the spectral resolution in the PL-f spectrum. The seven zero-phonon lines resulting from transition connecting the ground level of the $^4I_{13/2}$ manifold with the $^4I_{15/2}$ manifold may be observed. This number is smaller than the value (8) predicted by group theory. However, the missing peak might be not distinguished from other prominent PL peaks or it could have very weak intensity.

The simple structure of the site-selective PL-fs spectrum and the sharp five peak pairs in the PLE spectrum suggests the possibility that the optically active Er site responsible for these spectra is a non-cubic high symmetry site. To check the possibility, the high point symmetry in the wurtzite lattice structure semiconductor GaN is discussed briefly here. Wurtzite GaN belongs to the space group C_{6v}^4 with two formula units in the primitive cell, and all the atoms in this semiconductor occupy sites of symmetry C_{3v} (which is one of the trigonal point groups)[9]. There are two high-symmetry interstitial positions in the wurtzite structure [10]. The point symmetry is also C_{3v} for these interstitial positions. So a single Er atom either at a Ga site or at an interstitial position will have the C_{3v} point symmetry. However, these interstitial positions can not be the equilibrium position of an interstitial defect [10], ruling out them as possible Er sites. Therefore the single Er atom at Ga site could be responsible for the simple structured PL and PLE spectra observed here.

CONCLUSIONS

Site-selective PLE spectroscopy of the 1540 nm $^4I_{13/2} \rightarrow {}^4I_{15/2}$ emission of Er^{3+} in Er-implanted wurtzite GaN shows that the sharp intra-f shell PLE and previously observed broad PLE bands excite different site-selective Er^{3+} PL spectra. The simple structure of the site-selective PL and PLE spectra excited by direct intra-f shell absorption suggests that the optically active Er site responsible for these spectra may be of high symmetry, and that it could be attributed to single Er atoms at Ga sites.

ACKNOWLEDGMENTS

The authors would like to acknowledge P.B. Klein and G. Pomrenke for several helpful discussions. This work was supported by NSF under the Engineering Research Centers Program (ECD 89-43166), DARPA (MDA972-94-1-004), and the JSEP (0014-90-J-1270).

REFERENCES

1. P.N. Favennec, H.L'Haridon, M. Salvi, D. Moutonnet, and Y. Le Guillou, Electron. Lett. **25**, 718 (1989).
2. A.J. Neuhalfen and B.W. Wessels, Appl. Phys. Lett. **60**, 2657 (1992).
3. R.G. Wilson, R.N. Schwartz, C.R. Abernathy, S.J. Pearton, N. Newman, M. Rubin, T. Fu, and J. M. Zavada, Appl. Phys. Lett. **65**, 992 (1994).
4. C.H. Qiu, M.W. Leksono, J.I. Pankove, J.T. Torvik, R.J. Feuerstein, and F. Namavar, Appl. Phys. Lett. **66**, 562 (1995).
5. E. Silkowski, Y.K. Yeo, R.L. Hengehold, B. Goldenberg, and G.S. Pomrenke, Mater. Res. Soc. Symp. Proc. **422**, 69 (1996).
6. S. Kim, S.J. Rhee, D.A. Turnbull, E.E. Reuter, X. Li, J.J. Coleman, and S.G. Bishop, to be published.
7. X. Li, D.V. Forbes, S.Q. Gu, D.A. Turnbull, S.G. Bishop, and J.J. Coleman, J. Electron. Mater. **24**, 1711 (1995).
8. A. A. Kaminskii, Laser Crystals: Their Physics and Properties, Springer-Verlag, Berlin Heidelberg, New York, 1981, pp 120.
9. A. Cingolani, M. Ferrara, M. Lugarà and G. Scamarcio, Solid State Commun. **58**, 823 (1986).
10. P. Boguslawski, E.L. Briggs, and J. Bernholc, Phys. Rev. B **51**, 17255 (1995).

GALLIUM NITRIDE DOPED WITH ZINC AND OXYGEN - THE CRYSTAL
FOR THE BLUE POLARIZED LIGHT-EMITTING DIODES

V.G. Sidorov *, A.G. Drezhuk **, M.V. Zaitsev **, D.V. Sidorov ***
*St.Petersburg State Technical University, 29 Politechnicheskaya Str., St.Petersburg, 195251
Russia, rykov@phsc3.stu.neva.ru
**Politechnical Institute, Vologda, Russia
***A.F. Ioffe Physico-Technical Institute, Russian Academy of Sciences, St.Petersburg, Russia

ABSTRACT

Epitaxial layers and light-emitting i-n-structures with the active region of GaN doped simultaneously with zinc and oxygen have been grown. Effective up to 60% blue polarized luminescence has been observed. Investigated properties of grown layers and structures are presented. A discussion of issues related to GaN utilization based on its found properties is also included.

INTRODUCTION

It is known that LED's emit unpolarized light. The goal of the presented work is to report about creation of LED's (based on M-i-n-GaN-structures with active regions of GaN doped simultaneously with zinc abd oxygen) with blue polarized emission. High-quality (1120) GaN(Zn,O) layers and emitting i-n-structures were grown on (1012) sapphire substrate by chloride-hydride vapor phase epitaxy. The structural, electrical, optical and luminescence properties of GaN(Zn,O) layers and i-n-structures as function of Zn and O concentrations were studied. Spectral maximum of blue LED's lying at 2.55 eV (300 K) had linear polarization up to 60%. To study nature of luminescence the polarized diagram method was used. It is shown that polarization of luminescence is due to intracenter electron transitions in centers formed by Zn and O primary oriented in GaN lattice. The optimum growth conditions for fabrication of LED's with maximum polarization have been found. These LED's can serve as sources in polarization instruments having electronic recording of optical signals.

EXPERIMENT

Emitting i-n-structures were grown by gas-phase technique in chloride-hydride system using helium (99.999%) as the carrier gas [1]. The substrates were sapphire crystals oriented in the (1012) plane with deviation less then $3°$. A layer of undoped n-GaN was grown at a temperature of ~$1050°$ C with an average growth rate of ~20 μm/h. This produced a layer of fairly good structural quality and high conductivity with an electron density of $(6-10) \cdot 10^{19}$ cm^{-3} and electron mobility of 60-110 cm$^2 \cdot$ V^{-1}/ s^{-1}. The thickness of the n-GaN layers was 10-30 μm. The i-layer was grown by the same process but at a low temperature and growth rate. This layer was doped simultaneously with zinc and oxygen. The reactor and the process conditions were designed so that Zn and O interacted in the substrate zone so that not only free Zn but also coupled Zn-O pairs were incorporated into the GaN lattice. With increasing oxygen concentration, the resistivity of i-GaN(Zn,O) layers decreased with that of i-GaN(Zn) layers doped to the same concentrations of Zn. The solubility of Zn in GaN was increased to 8-10% in the presence of oxygen and was (10^{20}-10^{21}) cm^{-3}, where only ~10% of Zn was electrically active. The distributions of Zn and O over a cleaved surface are fairly similar. These observations indicate that dissolution of Zn in GaN may take place in the form of the oxide ZnO. However since ZnO and GaN have the same crystal structure with similar lattice parameters and band gaps, it is fairly difficult to identify ZnO inclusions in GaN as a separate phase. The formation of bound Zn-O pairs is directly indicated by a band (λ=20-40 mm) in the infrared reflection spectrum of GaN(Zn,O) which is similar to the band of ZnO reflection (Fig.1). The luminescence spectra of GaN(Zn,O) layers revealed a previously unknown band with maximum hv$_m$~2.55 eV at 300 K and a half-width Δhv~0.3 eV,

137

having up to 60% linear polarization [2-4].

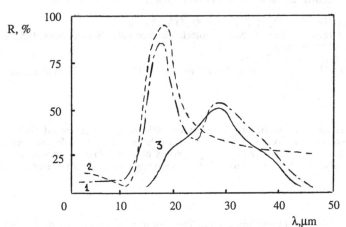

Fig.1. Infrared reflection spectra: 1 - GaN(Zn,O); 2 - n-GaN; 3 - ZnO.

This is the narrowest of all the known impurity luminescence band in GaN. This band has never been observed in GaN(Zn) and in GaN(O). It may be caused by defects, including Zn-O pairs.

An i-GaN(Zn) layer doped only with Zn was grown on the surface of the i-GaN(Zn,O) layer by the same process. The grown i-n-GaN structures were oriented in the (1120) plane. X-ray measurements showed that the C axis in sapphire lies in the same plane as C axis in GaN layer,perpendicular to the sample surface, and angle between them is 17°.

Light-emitting diodes were fabricated from structures (Fig.2). Indium, silver or nickel were used as ohmic contacts to n-GaN. Contacts to the i-layer were made by deposition of aluminum in vacuum. In structures having the highest electroluminescence efficiency, Zn concentration and the average resistivity of the i-layer were $N_{zn} \sim 10^{20}$ cm^{-3} and $\rho=(1-3)$ 10^4 Ω . cm. In this case one polarized band was usually observed (Fig.3). The degree of polarization of the electroluminescence at 300 K was 50-60%, it did not change within the emission band, and did not depend on the voltage applied to the structure. With increasing temperature in the range of 2.4-300 K, the degree of polarization increased by 5-15%, whereas at higher temperatures it decreased and at 60 K, almost vanished. The azimuth of the polarization of radiation emerging through the substrate was always perpendicular to the C axis of the sapphire [3]. This prevents the radiation being influenced by the birefringent properties of sapphire. As temperature decreases, the maximum of the electroluminescence was always shifted toward shorter wavelengths at a higher rate than the increase in the GaN band gap, but the half-width of the band always increased, which indicates that the broadening is of a nonphonon nature. A comparison of the results with published data showed that none of the known luminescence band of GaN doped with various impurities can be compared with the polarized band in the electroluminescence spectra of M-i-n-GaN(Zn,O) structures.

i-GaN(Zn)
i-GaN(Zn,O)
n-GaN
sapphire

+

Emission

Fig.2. Emitting M-I-n-GaN(Zn,O) structure.

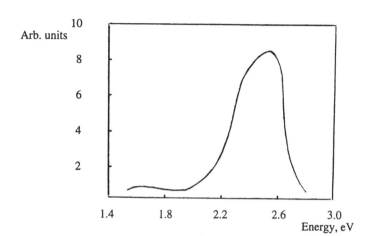

Fig.3. GaN - LED with polarized emission.

RESULTS AND CONCLUSIONS

The narrow half-width of this band, its high degree of polarization, and also the lack of a dependance of the polarization characteristics of the photoluminescence on the polarization of the exciting light suggest that the luminescence has intracenter mechanism. Intracenter nature of the polarized luminescence is confirmed by the character of polarization diagrams at impurity excitation. The main contribution to emission and absorption in GaN(Zn,O) is that of linear oscillators, oriented in plane (0001) of GaN. The radiation polarization is caused by anisotropy of (Zn-O)-centers and their primary orientation in GaN lattice.

The parameters of LED's at 300 K with the negative voltage at the contact with the i-layer had the following typical values: operating voltage 4-40 V, optical power 5-50 μW, brightness 400-600 cd/m^2, response time 20-100 ns. The efficiency was 0.1-1.3%.

The performance of the diodes was maintained over 4.2-900 K. The change of the parameters did not exceed 10% over more than 10,000 h of testing under operating conditions.

On the basis of structures with alternating layers of GaN(Zn) and GaN(Zn,O) the LED's with 1.5-2.5 higher light-efficiency can be made.These LED's can serve as sources in polarization instruments having electronic recording of optical signals and a number of other optoelectronical devices.

ACKNOWLEDGMENTS

The work was partly supported by University of Arizona, USA.

REFERENCES

1. V.M. Andreev and V.L. Oplesnin, Abstracts of Papers Presented at Fifth Intern. Symp. on Growth of Semicond. Film Crystals, (1978) [Novosibirsk, Russia].

2. M.D. Shagalov and A.G. Drezhuk, Pis'ma Fiz. Tekh. Poluprovod. 30, 11-14 (1979).

3. V.V. Rossin and V.G. Sidorov, Fiz. Tekh. Poluprovod. 13, 2411-13 (1979).

4. V.V. Rossin, V.G. Sidorov and A.D. Shagalov, Fiz. Tekh. Poluprovod. 15, 1021-23 (1981).

Part II
Substrates and Substrate Effects

GaN CRYSTALS GROWN FROM A LIQUID PHASE AT REDUCED PRESSURE

V.A. IVANTSOV[1], V.A. SUKHOVEEV[1], and V.A. DMITRIEV[1,2]
[1] PhysTech WBG Research Group, Ioffe Institute, 26 Politechnicheskaya Street,
St. Petersburg 194021, Russia, iva@shuttle.ioffe.rssi.ru;
[2] Material Science Research Center of Excellence, Howard University,
2300 6th Street, Washington, DC 20059, USA, vlad@msrce.howard.edu

ABSTRACT

Gallium nitride crystals were grown from a Ga-based melt at an ambient gas pressure not exceeding 2 atm. Growth temperature was about 1000°C. The crystals were 2H-GaN and had a (0001) plane orientation. Crystal size varied from 0.05x0.05x0.01 mm^3 to 2x2x0.05 mm^3. Lateral growth rate of the crystals ranged from 0.05 to 1 mm/hr. Normal growth rate was about 0.01 mm/hr. Depending on growth conditions, the crystals have a platelet or dendrite shape. Crystals were characterized by optical and electron microscopy, Auger electron spectroscopy, x-ray differential diffraction, photoluminescence and optical absorption. Lattice parameters and band gap value of the grown crystals were determined at 300 K.

INTRODUCTION

Group-III nitride semiconductors, mainly GaN, are a major focus of attention for various applications in blue and green light emitting diodes, ultraviolet and blue laser diodes, ultraviolet photo detectors, and high-frequency electronic devices. Based on their physical properties, novel devices with operating characteristics superior to those made on other wide band gap semiconductors are achievable.

Despite outstanding results in the fabrication of various GaN-based optoelectronic devices, the enormous density of structural defects in III-V nitride device structures cause a serious concern about their reliability. The main reason for the defect formation is the luck of GaN substrates for homoepitaxy.

Bulk crystals of other III-V semiconductor materials such as GaP and GaAs are routinely grown from the liquid phase, namely, from melts and melt solutions. Growth of bulk GaN by liquid phase techniques is hindered due to low temperature of GaN dissociation, high nitrogen pressure over GaN at growth temperatures, and relatively small nitrogen solubility in Ga melts [1-4].

To date, the largest (10x10 mm^2) bulk GaN crystals were grown from nitrogen saturated gallium melts using a high pressure technique [5, 6]. These crystals are grown at a nitrogen pressure of 12-20 kbar and a growth temperature of ~1500°C. Because of the extreme growth conditions, the scaling and commercialization of this technique is doubtful. Recently, polycrystalline gallium nitride has been synthesized at a low pressure [7]. We report on properties of bulk GaN crystals grown from a liquid phase at reduced pressure and temperature [8].

EXPERIMENTS AND RESULTS

Growth processes were conducted in a water-cooled stainless steel chamber supplied by 5 MHz rf generator which provided heating for a graphite susceptor.

143

To make accurate temperature measurements, a W-Re thermocouple connected with a noise suppression unit was mounted near the susceptor. The growth procedure began with the evacuation of the chamber down to 10^{-5} Torr and keeping the Ga-based melt at 500°C to produce initial cleaning of the melt, then the growth chamber was filled with a gas atmosphere containing nitrogen. The gas pressure in the growth chamber was less than 2 atm. Meanwhile, the melt temperature was rapidly rising to 1000-1050°C. This temperature remained constant or slightly decreased during the growth run. Nucleation and growth of GaN crystals occurred spontaneously on the melt surface and in the melt volume. GaN single crystals with average size ranging from 0.05x0.05x0.01 mm^3 to 1x1x0.05 mm^3, depending on growth parameters, could be easily obtained in a 1 hour growth run with a lateral growth rate of ~0.05 mm/hr. Platelet crystals having preferential hexagonal shapes were usually formed (Fig. 1a). Growth spirals are visible on the crystal surface spreading from the center to the periphery. The largest crystals having a size of 2x2x0.1 mm^3 were grown in 2 hours with a lateral growth rate of approximately 1 mm/hr. In all growth runs, the normal growth rate was significantly less than the lateral growth rate and varied from 0.005 mm/hr to 0.05 mm/hr depending on growth conditions. When the growth rate became too high, GaN dendrite formation was observed on the melt surface or at the growth front of single crystals (Fig. 1b). At high growth rates, GaN polycrystalline conglomerates with a maximum lateral size of approximately 10x10 mm^2 were grown in 30 minutes. An Auger electron spectrum measured for the grown GaN crystals looks typical for GaN and contains only nitrogen and gallium peaks (Fig. 2).

X-ray differential diffractometry (XRD) was used for crystal structure investigation. XRD showed that grown platelets were 2H-GaN single crystals with (0001) basal plane. The minimum full width at half maximum (FWHM) of rocking curve measured for grown GaN crystals using $\omega-2\Theta$ scanning geometry for the (0002) reflections was about 29 arcsec (Fig. 3). The lattice spacing along c and a crystallographic axises were determined using a triple-crystal x-ray spectrometer. To determine the lattice parameter a, x-ray rocking curves for the asymmetric (11$\underline{2}$4) reflection were measured. Lattice parameters for the grown crystals were measured to be a = 3.186±0.002 Å and c = 5.1852±0.0001 Å (300 K).

Photoluminescence (PL) of bulk GaN crystals was measured at 80 K and 300 K (Fig. 4). The edge peaks of PL spectra were situated at wavelengths of λ_{max} ~358 nm (hv ~3.5 eV) and λ_{max} ~365 nm (hv ~3.4 eV) for 80 K and 300 K, respectively. The yellow luminescent band at ~550 nm is also present in the PL spectrum.

The study of the optical transmission spectrum of GaN samples was done employing a differential spectrophotometer, Specord UV VIS. Using optical transmission data, a band gap of the grown crystals was determined to be ~3.4 eV (300 K). The temperature dependence of the band gap value for grown crystals is under investigation.

CONCLUSIONS

GaN single crystals were grown from Ga-based melts at 1000°C and ambient gas pressure less than 2 atm. Crystals up to 2x2x0.1 mm^3 in size have been obtained. The maximum lateral growth rate for GaN single crystals was about 1 mm/hr. It was shown by x-ray diffraction that the crystals are 2H-GaN. Lattice parameters for a

0.5 mm

1 mm

Fig. 1. Optical micrographs of GaN crystals grown from a melt at reduced pressure.

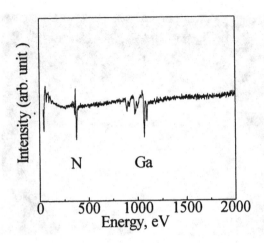

Fig. 2. Auger electron spectrum for a GaN single crystal after 20 sec sputtering by Ar beam.

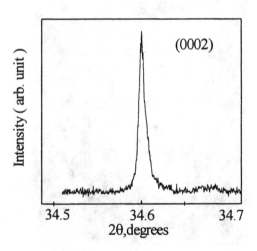

Fig. 3. X-ray diffraction ω–2Θ rocking curve for a GaN crystal.

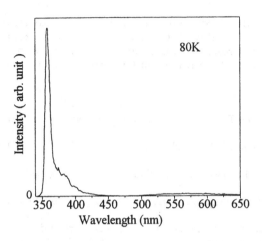

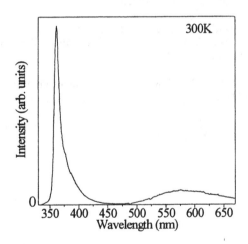

Fig. 4. Photoluminescence spectra measured for GaN crystals.

grown crystal were measured to be a = 3.186±0.002 Å and c = 5.1852±0.0001 Å (300 K). The band gap value determined at room temperature by optical transmission measurements was ~3.4 eV.

ACKNOWLEDGMENTS

The authors would like to thank V.I. Nikolaev, A.P. Kovarskiy, I.P. Nikitina, D.V. Tsvetkov, A.S. Zubrilov, V.E. Bougrov, and A.I. Babanin for material characterization. This research was supported in part by the Crystal Growth Research Center, St. Petersburg, Russia, and the Arizona State University.

REFERENCES

1. R.A. Logan and C.D. Thurmond, J. Electrochem. Soc. **119**, p. 1727 (1972).

2. J. Karpinski, J. Jun and S. Porowski, J. Crystal Growth **66**, p. 1 (1984).

3. J. Karpinski and S. Porowski, J. Crystal Growth **66**, p. 11 (1984).

4. L.A. Marasina, I.G. Pichugin, Izvestiya LETI **433**, p. 3 (1991) (in Russian).

5. S. Porowski, J. Jun, P. Perlin, I. Grzegory, H. Teisseyre and T. Suski, in: Silicon Carbide and Related Materials, Inst. Phys. Conf. Ser. N 137, eds. M.G. Spencer, R.P. Devaty, J.A. Edmond, M.A. Khan, R. Kaplan and M. Rahman (IOP Publishing Ltd, Bristol 1993) pp. 369-372.

6. S. Porowski, M. Bockowski, B. Lucznik, M. Wroblewski, S. Krukowski, I. Grzegory, M. Leszczynski, G. Nowak, K. Pakula and J. Baranowski, Mat. Res. Soc. Symp. Proc., **449**, p. 35 (1997).

7. A. Argoitia, C.C. Hayman, J.C. Angus, L. Wang, J.S. Dyck and K. Kash, Appl. Phys. Lett. **70**, p. 179 (1997).

8. V.I. Ivantsov, V.A. Sukhoveev, V.I. Nikolaev, I.P. Nikitina, V.A. Dmitriev, to be published in Phys. Solid State.

GROWTH OF BULK, POLYCRYSTALLINE GALLIUM AND INDIUM NITRIDE AT SUB-ATMOSPHERIC PRESSURES

John C. Angus[*], Alberto Argoitia[*], Cliff C. Hayman[*], Long Wang[**],
Jeffrey S. Dyck[***] and Kathleen Kash[***]
[*]Chemical Engineering Dept., [**]Materials Science and Engineering Dept., [***]Physics Dept.
Case Western Reserve University, Cleveland, OH 44106

ABSTRACT

Bulk, polycrystalline gallium nitride and indium nitride were crystallized at sub-atmospheric pressures by saturating the pure metals with nitrogen from a microwave electron cyclotron resonance source. Saturation of Ga/In melts with nitrogen led only to the crystallization of gallium nitride. The polycrystalline samples were wurtzitic. The gallium nitride was well faceted, with narrow Raman lineshapes, and showed near-band-edge and yellow band photo-luminescence at both 4K and 300K. The indium nitride was formed in smaller amounts, was less well faceted, and showed no photoluminescence.

INTRODUCTION

Synthesis of bulk GaN has primarily been at high pressures because of the high equilibrium pressures of N_2 [1-4]. To our knowledge, there is no practical method for synthesis of bulk InN. We have recently shown that GaN can be synthesized at low pressures by use of atomic nitrogen, N, rather than molecular nitrogen, N_2, to saturate gallium with nitrogen [5,6].

In this paper we describe the crystallization of polycrystalline gallium nitride and indium nitride from the respective metals that have been saturated with nitrogen obtained from a microwave electron cyclotron resonance (ECR) source. We also report preliminary experiments on crystallization from Ga/In melts.

THERMODYNAMICS OF GROUP III NITRIDE SYNTHESIS

Because of its extreme stability, when N_2 is the source of nitrogen, very high pressures are required to form GaN and InN. Activated forms of nitrogen, $e.g.$, N, N^+, N_2^+, and excited states of N_2 from a plasma can also be used as the nitrogen source [7]. Because these species have a much higher chemical potential than N_2, much lower pressures are required to saturate the melt with nitrogen. For simplicity, we consider only atomic nitrogen, N, in the following discussion.

Nitrogen from the plasma source is dissolved in the liquid metal and subsequently forms the solid nitride.

$$N_{(gas)} = N_{(dissolved)} \tag{1}$$

$$Ga_{(liquid)} + N_{(dissolved)} = GaN_{(solid)} \tag{2}$$

The overall process is the sum of reactions (1) and (2).

$$Ga_{(liquid)} + N_{(gas)} = GaN_{(solid)} \tag{3}$$

Similarly, from liquid indium one has the overall reaction

$$In_{(liquid)} + N_{(gas)} = InN_{(solid)} \tag{4}$$

The solid nitrides can decompose to the elemental metals and N_2.

$$2GaN_{(solid)} = 2Ga_{(liquid)} + N_{2(gas)} \qquad (5)$$

$$2InN_{(solid)} = 2In_{(liquid)} + N_{2(gas)} \qquad (6)$$

Reactions (5) and (6) are in direct competition with the desired reactions (3) and (4). The decomposition of InN, reaction (6), is thermodynamically more favored than the decomposition of GaN by reaction (5). Consequently, one would expect it to be more difficult to form InN than GaN by the present process. This is borne out by the experimental results.

Furthermore, recombination of N to form N_2 is strongly favored thermodynamically at all temperatures.

$$2N_{(dissolved)} = N_{2(gas)} \qquad (7)$$

If reaction (7) is too rapid, it will be difficult to reach dissolved N concentrations high enough to precipitate the solid nitrides. It is not known what form the nitrogen takes when it is dissolved in liquid Ga or In. However, our results indicate that, in the temperature range studied, recombination of N, reaction (7), is sufficiently slow so that formation of GaN and small amounts of InN by reactions (3) and (4) does take place.

EXPERIMENT

Synthesis

Synthesis of GaN and InN was achieved by saturating the appropriate metal with nitrogen from a microwave ECR source. The solid nitrides crystallized from the melt. Details of the experimental procedure have been given earlier [5,6].

At 700C the Ga/N melts did not wet the pyrolytic boron nitride crucibles. At the conclusion of a run, the melt was covered with a crust (skull) of polycrystalline nitride. For GaN, this skull was typically 0.1 mm thick, coherent, and could easily be handled with tweezers. One of the GaN skulls is shown in Fig. 1. At temperatures above 900C, the melt wets the crucible forming a concave surface. A coherent skull was not formed; however, the crystallite size was larger, up to 200 μm, and crystalline quality as measured by Raman spectroscopy was better. Some preliminary experiments were performed using Ga/In alloys. The Raman spectra of the crystals obtained from these melts showed only GaN. Elemental analysis and photoluminescence indicate that less than one atomic percent In may have been incorporated in the crystals. The GaN crystallized from Ga/In melts had a strong tendency to form dendritic crystals; a typical example is shown in Fig. 2.

Synthesis of InN was achieved by using an initial charge of pure In; however, only small quantities of solid nitride were obtained and no coherent skull that could be removed from the indium was formed. Crystals of InN were not observed when the melt was held at 900C, but appeared upon cooling to 650C. At the higher temperatures, decomposition of InN, reaction (6), may be rapid enough to preclude net formation of InN. Although faceted InN crystals were obtained, the morphology was not as well defined as with GaN. See Fig. 3. During runs with In, bubbles were observed coming from the melt. This gas was not identified; however, we believe it is N_2. Increased decomposition of solid nitride and increased recombination of N to N_2 may be expected with indium compared to gallium.

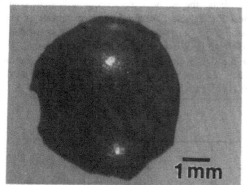

Figure 1. Concave surface of polycrystalline GaN dome formed at approximately 700C. Bright spots are specular reflection from photographic lights.

Figure 2. Scanning electron micrograph of gallium nitride crystals grown from Ga In alloys at approximately 900C. A dendritic structure is evident.

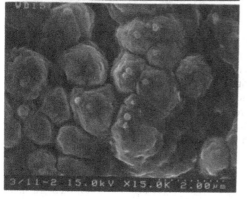

Figure 3. Scanning electron micrograph of indium nitride crystals crystallized from indium metal at about 650C.

Structural Characterization

The gallium nitride crystals were confirmed to be wurtzitic GaN by electron and x-ray diffraction, Raman spectroscopy, and elemental analysis [5,6]. Transmission electron microscopy showed the presence of numerous planar defects; the Seebeck coefficient indicated n-type conductivity [5,6]. Here we show additional Raman and photoluminescence studies of well faceted crystals with the morphology shown in Fig. 2. Raman spectra clearly showed the presence of the low E_2 line at 144 cm^{-1} and the high E_2 line at 568 cm^{-1} [8]. See Figure 4. The FWHM of these peaks are approximately 3 cm^{-1} and 4 cm^{-1} respectively, uncorrected for the instrumental broadening of 2 cm^{-1}. The small shoulder at 534 cm^{-1} is tentatively identified as the A1T line. The feature at 735 cm^{-1} may be the TO phonon line of the zinc blende structure.

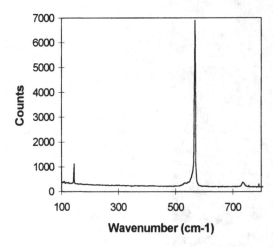

Figure 4. Raman spectrum of GaN. Lines at 144 cm^{-1} and 560 cm^{-1} are the low E_2 and high E_2 respectively.

The indium nitride samples were identified as wurtzitic InN by electron diffraction. An electron diffraction pattern taken along the [$11\bar{2}3$] zone axis is shown in Fig. 5. Energy dispersive x-ray analysis showed an In/N ratio of 1 within the expected experimental error. Only weak Raman spectra were obtained from the InN samples, consistent with the more poorly developed crystallinity shown in Fig. 3. A narrow Raman line observed at 127 cm^{-1} may be the low E_2 line, although this assignment is tentative.

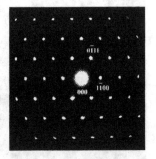

Figure 5. Electron diffraction pattern from InN along the [$11\bar{2}3$] zone axis

Photoluminescence

A representative photoluminescence spectrum from a well-faceted, dendritic GaN sample is shown in Fig. 6. No luminescence was observed from the InN. The near-band-edge peak in Fig. 6 is at 3560 Å (3.480 eV) with a FWHM of 5 Å (5 meV). A shift to longer wavelengths of about 10 Å was observed with GaN grown from In/GaN alloys; however, it is not known whether this shift arises from dissolved In or from other effects. If it is from dissolved In, the concentration of In would be no more than 0.5%.

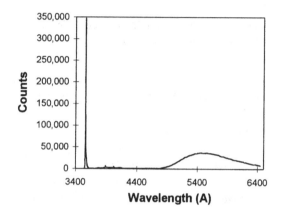

Figure 6 A photo-luminescence spectrum of GaN taken at 4K. Both the near-band-edge and the broad "yellow band" luminescence are present.

DISCUSSION

Dendritic growth forms (See Fig. 2) arise from uncontrolled freezing into a supersaturated melt. This is consistent with our visual observation that sometimes the solid crust forms almost instantaneously after the nitrogen plasma has been on for several hours.

Only very small amounts of In are incorporated into the nitride phase precipitated from Ga/In melts at our experimental conditions. This can mean that there is limited solid solubility of GaN and InN. This behavior can also arise from other reasons well. For example, even if there is solid solubility, there may be very strong segregation during crystallization that suppresses In incorporation. Also, kinetic factors may operate to favor crystallization of GaN over that of InN.

The critical feature that permits the process to work is the favorable competition between the formation reactions (3) and (4) compared to the competitive decomposition reactions (5) and (6). Furthermore, there must be a relatively slow recombination rate of N to N_2 by reaction (7) in the liquid metals, which permits the parallel formation of the nitrides. The slow recombination rate may arise from "solvation" of dissolved N by a coordination sphere of gallium and indium atoms that inhibits the close approach of dissolved N atoms in the melt and thus reduces the recombination rate. The Ga-N bond energy has been reported to be 2.2 eV per bond [9], which would indicate significant attraction between Ga and N atoms in the melt at the temperatures employed in our experiments. The In-N bond energy is 1.93 eV per bond, which would mean a marginally smaller tendency for solvation of N by In in the melt.

CONCLUSIONS

Polycrystalline GaN and InN were crystallized from Ga and In melts saturated with nitrogen derived from a microwave ECR source. GaN was also crystallized from Ga/In melts. The use of atomic nitrogen circumvents the high equilibrium pressures required for the synthesis of bulk group III nitrides from molecular nitrogen and the group III metals. The process depends on the favorable competition between formation and decomposition reactions and a low recombination rate of N to N_2 in the nitrogen saturated melt.

ACKNOWLEDGMENTS

The support of a Materials Research Group grant from the National Science Foundation and an AASERT grant from the Office of Naval Research is gratefully acknowledged.

REFERENCES

1. J. H. Edgar, Editor, Properties of Group III Nitrides, EMIS Datareviews Series, No. 11, INSPEC, (Institution of Electrical Engineers, London, UK, 1994).

2. J. Karpinski and S. Porowski, J. Crystal Growth **66**, 11 (1984)

3. S. Porowski, J. Jun, P. Perlin, I. Grzegory, H. Teisseyre and T. Suski, Inst. Phys. Conf. Series, No. 137, Chapter 4, (5th Conf. on SiC and Related Materials, Washington, DC, 1993) p. 369.

4. P. Perlin, I. Gorszyca, N. E. Christensen, I. Grzegory, H. Teisseyre, and T. Suski, Phys. Rev. B **45**, p. 13,307 (1992).

5. A. Argoitia, C.C. Hayman, J.C. Angus, L. Wang, J.S. Dyck, and K. Kash, Appl. Phys. Lett., **70**, 179 (1997)

6. A. Argoitia, C.C. Hayman, J.C. Angus, L. Wang, J.S. Dyck, and K. Kash, Proceedings III-V Nitride Symposium, Materials Research Society Meeting, Boston, MA, December 1996.

7. N. Newman in III-V Nitride Materials and Processes, edited by T.D. Moustakas, J.P. Dismukes, and S.J. Pearton, Proceedings Volume 96-11, (Electrochemical Society, Pennington, NJ, 1996) p. 1-19

8. L. E. McNeil, Properties of Group III Nitrides, edited by J.H. Edgar, EMIS Datareviews Series, No. 11, INSPEC, (Institution of Electrical Engineers, London, UK, 1994) p. 254

9. S. Porowski and I. Grzegory, Properties of Group III Nitrides, edited by J.H. Edgar, EMIS Datareviews Series, No. 11, INSPEC, (Institution of Electrical Engineers, London, UK, 1994) p. 71-75.

POLAR-TWINNED DEFECTS IN LiGaO$_2$ SUBSTRATES LATTICE MATCHED WITH GaN

TAKAO ISHII, YASUO TAZOH AND SHINTARO MIYAZAWA
NTT System Electronics Laboratories, 3-1 Morinosato Wakamiya, Atsugi-shi, Kanagawa, 243-01, Japan

ABSTRACT

We investigated the Czochralski growth of LiGaO$_2$ single crystal for use as a substrates for the epitaxial growth of hexagonal GaN.   Crossed lines were observed in mechano-chemically polished {001} wafers sliced from a (001) axis boule.   Chemical etching revealed that there exists a difference in chemical stability between two domains separated by a crossed line.   Since LiGaO$_2$ single crystal is polar along the c-axis, the formation of multi-domain structure is due to the polarity inversion of the c-axis, that is, polar twinned defects.   We found that the GaN thin film on the (001) substrate with a multi-domain structure peeled off and that this is closely related with multi-polarity of substrate.

INTRODUCTION

Gallium nitride (GaN) is one of the most promising semiconductors not only for devices that emit light in blue, violet, and ultraviolet spectral regions but also for a high-power, high-temperature devices because of its wide band gap (3.39eV) and high melting point (~1700°C). However, it is very difficult to obtain bulk single crystals due to the extremely high equilibrium pressure of the nitrogen in GaN at growth temperature.   Therefore, heteroepitaxial thin film growth on various substrates has been investigated extensively.   One of the most commonly used substrates is sapphire (Al$_2$O$_3$), but it is fairly difficult to grow high-quality epitaxial GaN film because of the large lattice mismatch (14%).   Large mismatch produces a large dislocation density (10^8-10^{10}/cm^2), which may be a crucial problem in electron devices.   Amano et al. reported that high-quality GaN films can be grown on sapphire substrates using AlN buffer layers[1].   It is necessary to grow a thick epilayer to obtain high quality films.   Therefore it is very important to find a lattice-matched substrate suitable for the epitaxial growth of thin films with high crystallinity.

LiGaO$_2$ is orthorhombic with cell dimensions a = 5.402 Å, b = 6.372 Å and c = 5.007 Å[2]. From these cell dimensions, the lattice mismatch to hexagonal GaN is estimated to be about 0.9%. Therefore, LiGaO$_2$ is expected to be a promising substrate material for the epitaxial growth of hexagonal GaN thin film.   Chai et al. proposed first LiGaO$_2$ as a substrate for GaN films[3]. The crystal structure of LiGaO$_2$ consists of an infinite three-dimensional array of oxygen tetrahedra centered on Ga and Li atoms.   The space group of LiGaO$_2$ belongs to Pna2$_1$ and this means that LiGaO$_2$ has polarity along the c-axis.   LiGaO$_2$ crystal shows piezoelectric properties, which was first pointed out by Remeika and Ballman[4].

Recently, there have been some reports of GaN thin film growth on LiGaO$_2$ substrate by MBE and MOCVD [3,5].   We also investigated MBE growth of hexagonal GaN film on the substrate, but we encountered a film-peel-off phenomenon. There was little information about the qualification of LiGaO$_2$ as a substrate material in their reports.

In the present study, we report Czochralski (Cz) growth of LiGaO$_2$ and show polar twinned defects in the {001} wafers cut from a boule pulled along the [001] axis for the first time.   We found that the peel-off of the film correlated strongly with polar-twinned structure of the substrate. We investigated the relation between the growth morphology of hexagonal GaN thin films and the polarity of the substrate.

EXPERIMENT

Starting LiGaO$_2$ materials for Cz pulling were prepared from a stoichiometric mixture of Li$_2$CO$_3$ and Ga$_2$O$_3$ (5N purity). The mixture was calcinated at 1300°C for 20 h in air. The X-ray powder diffraction pattern of the calcined specimen was completely indexed as a β-LiGaO$_2$ structure. Single crystals of LiGaO$_2$ were grown by the Cz pulling technique. An Ir crucible was used and it was heated inductively by a 450 kHz radio-frequency generator. The calcined specimen was charged into the crucible and the growth chamber was filled with N$_2$ gas before melting the starting material. Seed crystals oriented [100], [010] and [001] in the orthorhombic system were used. The pulling rate was 3-5mm/h and the crystal rotation rate was 30-60rpm. The steep temperature gradient above the melt was reduced by the use of a cone shaped after-heater made of Ir. The axial temperature gradient was optimized experimentally by adjusting the RF coil position and the distance between the crucible and the after-heater. The measured temperature gradient just above the melt was about 30°C/cm.

GaN thin films were grown using MBE in a radical nitrogen atmosphere. Radical nitrogen was generated by an electrical discharge from inductively-coupled RF excitation at 13.56 MHz and supplied to the heated substrate. Ga metal was evaporated from a K-cell and deposited on the substrate. The substrates temperature was fixed to about 700°C. The film thickness was 50nm~400nm. The in-plane orientation of the films was examined by the reflection high energy electron diffraction (RHEED). The crystallinity of as-grown films was examined by double-crystal X-ray diffractometry.

RESULTS AND DISCUSSIONS

(A) Crystal growth of LiGaO$_2$

Figure 1 shows examples of as-grown crystals pulled along [001] and [010] axes. The crystals were colorless and transparent. The surface of the as-grown crystals is very smooth. Many cracks were often produced in the as-grown crystals with [100] and [010] pulling when they were taken out of the growth chamber, but no cracks were produced in the crystal with [001] pulling. The shape of the solid-liquid interface of all the as-grown crystals was convex toward the melt and did not change with increases in the crystal rotation rate from 30 rpm to 60 rpm. In the molten states, thermal convection due to the temperature gradient in the melt was clearly observed. Therefore, one possible explanation of the convex interface of the as-grown crystals is that the thermal convection exceeds the forced convection associated with crystal rotation because of the very low viscosity of the melt.

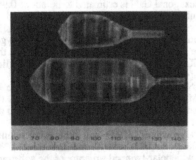

Figure 1 As-grown LiGaO$_2$ single crystals pulled along different axes.
Upper crystal : [001] pulling.
Lower crystal : [010] pulling.

Figures 2 shows a mechano-chemically polished (001)-oriented wafers sliced from the [001] axis-grown boule. Straight lines (We call them "crossed lines".) were clearly observed along the [110] direction in all the wafers from one boule.

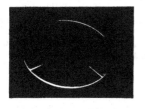

(100)

(010)

Figure 2 Mechano-chmically polished
{001}-oriented wafer sliced from
the [001] axis-grown boule.
White lines run along two [110]
directions.

Microscope observation revealed that there were long and narrow domains along the [110] direction. We examined the surface topography of a mechano-chemically polished {001} wafer with a mechanical profilometer (Solan DEKTAK 3030). Figure 3(a) and(b) show profilometer traces for the two sides of an as-polished surface[6]. The scanning direction was perpendicular to the crossed lines. As shown in these figures (on the left in Fig. 3), the surface topography inverts on either side of the area containing the same crossed line. A schematic cross-sectional view of the polished wafer is illustrated on the right in Fig. 3.

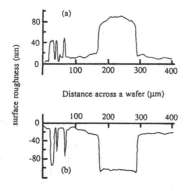

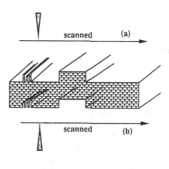

Figure 3 Surface profiles measured with a DEKTAK profilometer (left): (a) one side
(b) the other side of an as-polished wafer. Schematic cross-sectional view
of the polished wafer (right).

We believe that these topographies resulted from the difference in chemical etching rates during the mechano-chemical polishing[7]. In fact, such lines were not observed only after mechanical polishing. This suggests that there are two different domains having different chemical stability: a face that is hardly etched and a face that is easily etched. Considering that LiGaO$_2$ is a polar crystal, we believe that this difference is due to the polarity inversion. Therefore, the wafer cut from the [001] axis-grown boule has a multi-domain structure. We noticed that this multi-domain structure is a crucial problem in using this material as a substrate. This is because, as discussed later, the as-grown films often peeled off when we used {001}-oriented wafers sliced from the [001] axis-grown boule as a substrate.

We investigated the domain structure of the crystals with [010] and [100] pulling. Microscopic observation revealed that a mechano-chemically polished {001}-oriented wafer sliced from the [100] and [010] axes-grown boules did not contain crossed lines. Moreover, we found that these wafers have a uniform etching pattern, which means a single domain structure. Figures 4(a) and (b) show the etched pattern of the two sides of an as-polished {001}-oriented wafer sliced from the [010] axis-grown boule. We used a mixture of H$_2$O:HNO$_3$=1:1 as an etchant. Not only the etched pattern but also etching rate were different between the two sides of a substrate with a single domain. The etched pattern in Fig 4(a) is lozenge-shaped and the

157

etching rate is high, while that in Fig 4(b) is elliptical and the etching rate is low. These etching patterns are the same as the ones we observed in a multi-domain wafer cut from [001] axis-grown crystals [6]. Hereafter, we refer to the easily etched surface as "A-surface" and the hardly etched one as "B-surface". The reason for the single domain formation in [010] or [100] axis-grown crystals is probably associated with a "self-poling" effect due to internal stress, but whether this is actually the case or not is not clear because of the lack of exact information about the lattice distortion at elevated temperature.

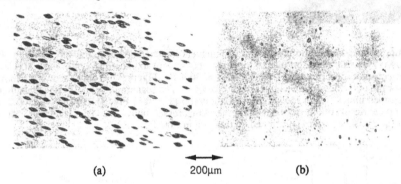

(a)   200μm   (b)

Figure 4 Etched patterns of the two sides of {001}polished wafer sliced from the [010] axis-grown boule. (a) easily etched A-surface (b) hardly etched B-surface

Figure 5 shows the differential thermal analysis of a LiGaO$_2$ powder sample. Just below the endothermic peak corresponding to melting (1585°C), a small but clear peak was observed at 1504°C. This suggests a phase transition in LiGaO$_2$ at this temperature. The formation of the polarity inversion may be attributed to this transition, which is probably to noncentrosymmetry from centrosymmetry. This means that LiGaO$_2$ exhibits a structural phase transition with a displacement of Ga and Li ions at 1504°C. Therefore, strain due to the structural phase transition produces an electric field in the piezoelectric crystal. Then it is expected that this electric fields becomes large enough to align the multi-domains of the crystal in case of [010] or [100]-axis pulling if its magnitude depends on the pulling axes of the orthorhombic LiGaO$_2$ crystal.

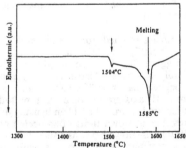

Figure 5 Differential thermal analysis curve of a LiGaO$_2$ powder sample[6].

(B) Thin film growth of GaN

To examine the suitability of LiGaO$_2$ as a substrate for GaN films, we grew GaN thin films on (001) LiGaO$_2$ substrates using MBE in a radical nitrogen atmosphere. When we took out an as-grown film from the growth chamber, we clearly saw with naked eyes that only the film on A-surface peeled off.   Figures 6(a) and (b) respectively show the microscope photograghs of films 50nm thick grown on A-surface and B-surface of (001) LiGaO$_2$ substrate with a single domain.   The stripes in Fig. 6(a) are due to interferrence fringes produced between the peeled-off film and substrate surface.   When we used (001) LiGaO$_2$ substrate sliced from the [001] as-grown boule, which contains a multi-domain structure, we found that both the grown part and peeled-off part coexist on the same substrate surface.   These experimental results lead us to conclude that the peeling-off of the film is due to the multi-domain structure (coexistence of A-surface and B-surface) of the {001}-oriented substrates sliced from the [001] axis-grown boule.

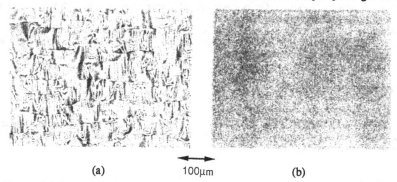

(a)   100μm   (b)

Figure 6 Microscope photograghs of films 50-nm thick grown on A-surface (a) and B-surface (b) of (001) LiGaO$_2$ substrate.

The θ-2θ scan X-ray diffraction (XRD) pattern of an as-grown film only 0.4μm thick had a strong preferred c-axis orientation of hexagonal GaN perpendicular to (001) LiGaO$_2$ substrate surface.   From the RHEED patterns for the initial growth stage, it was found that the epitaxial relation is [010]LiGaO$_2$//[11$\bar{2}$0]GaN.   Figure 7(a) and (b) respectively show ω-scan XRD rocking curves of the (0002) peak of GaN thin film grown on the B-surface of (001)LiGaO$_2$ substrate and on the (0001) Al$_2$O$_3$ substrate.   The film thickness were the same: 0.4μm.   The full width at half maximum (FWHM) value of the thin films grown on the B-surface LiGaO$_2$ substrate was 260arc sec, about seven times narrower than that of GaN thin film on the (0001) Al$_2$O$_3$ substrate.   This indicates that hexagonal GaN thin film with good crystallinity grew on (001) LiGaO$_2$ substrate and LiGaO$_2$ is a promising substrate for epitaxial growth of hexagonal GaN thin films.   The details of the GaN thin film growth on LiGaO$_2$ substrates will be reported in separated paper.

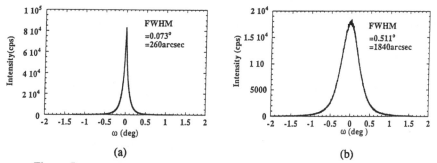

(a)   (b)

Figure 7 ω-scan XRD rocking curves of the (0002) peak of GaN films grown on (a) LiGaO$_2$ substrate and (b) Al$_2$O$_3$ substrate.

CONCLUSION

We examined the suitability of $LiGaO_2$ as a substrates for epitaxial growth of hexagonal GaN.   We found the multi-domain structure in mechano-chemically polished {001} wafers sliced from a (001) axis boule for the first time.   Chemical etching revealed that there exists a difference in chemical stability between two domains separated by a crossed line.   The formation of the multi-domain structure is associated with the polarity inversion of the c-axis, that is, polar twinned defects, because $LiGaO_2$ single crystal is polar along c-axis.   It was found that the GaN thin films peeled off when they were deposited on the $LiGaO_2$ substrate with multi-domain structure.   We succeeded in growing crystal with a single-domain by [100] or [010] pulling.   We found that the GaN thin film grew only on the hardly etched surface of (001) $LiGaO_2$ substrates with a single-domain.

ACKNOWLEDGMENTS

The authors would like to express their thanks to Masahiro Sasaura for his useful comments on crystal growth and to Dr Takashi Matsuoka of NTT Basic Research Laboratories for his helpful comment and discussion on GaN thin film growth.

REFERENCES

[1] H. Amano, N. Sawaki, I. Akazaki and Y. Toyoda: Appl. Phys Lett. **4 8** 353 (1986)
[2] M. Marezio: Acta Cryst. **1 8** 481 (1965)
[3] J. H. M. Nicholls, H. Gallagher, B. Henderson, C. Trager-Cowan, P. G. Middleton, K. P. O'Donnell, T. S. Cheng, C. T. Foxon and B. H. T. Chai: Mat. Res. Soc. Symp. Proc. **3 9 5** 535 (1996)
[4] J. P. Remeika and A. A. Ballman: Appl. Phys. Lett **5** 180 (1964)
[5] P. Kung, A. Saxler, X. Zhang, D. Walker, R. Lavado and M. Razeghi: Appl. Phys. Lett. **6 9** 2116 (1996)
[6] T. Ishii, Y. Tazoh and S. Miyazawa: Jpn. J. Appl. Phys. **3 6** L139 (1997)
[7] The polishing powder was "Baikalox"(Baikowski international corporation).

SUBSTRATE EFFECTS ON THE GROWTH OF InN

S. M. Donovan, J. D. MacKenzie, C. R. Abernathy, S. J. Pearton, P. Holloway, F. Ren[+],
J. M. Zavada**, B. Chai[++]*
*University of Florida, Dept. of Materials Science and Engineering, Gainesville, FL,
32611
[+]Bell Laboratories, Lucent Technologies, Murray Hill, NJ 07974
**U. S. Army Research Office, Raleigh, NC
[++]University of Central Florida, Orlando, FL

ABSTRACT

Auger electron spectroscopy was used to examine the nitridation behavior of
GaAs, sapphire and lithium aluminate (LAO) substrates exposed to an RF nitrogen
plasma. No evidence of nitridation was found for the sapphire and LAO substrates.
GaAs substrates did show evidence of nitridation which led to smooth InN surface
morphology without the need for a low temperature buffer. Comparable InN films were
obtained on sapphire and LAO substrates when a low temperature AlN buffer was used.
Hall measurements indicate background carrier concentrations are relatively insensitive to
substrate type, though mobilities decreased as surface morphology was improved.

INTRODUCTION

III-V nitrides are becoming increasingly important for the development of optical
and high temperature electronic devices due to their wide bandgaps and high saturation
velocities[1]. To maximize the performance of devices such as metal enhanced
semiconductor field effect transistors (MESFETs) or heterojunction bipolar transistors
(HBTs), several key processing issues must be addressed. The ability to fabricate low
resistance, thermally stable Ohmic contacts to n and p-type material is among the most
important of these issues and is a critical technology for III-V nitrides.

In electronics applications, narrow bandgaps and/or high doping levels are usually
needed in order to minimize the resistances arising from Ohmic contacts. In GaAs-based
devices, for example, heavily n-doped InGaAs layers on n-GaAs can be used to reduce
parasitic resistances as much as one order of magnitude relative to contacts directly on
GaAs[2]. The higher doping and smaller bandgap which can be achieved in InGaAs result
in contact resistances of $\sim 2 - 6 \times 10^{-7}$ Ω-cm^2 vs. $\sim 1 \times 10^{-6}$ Ω-cm^2. For the potential of
nitride-based devices to be realized, similar contact schemes must be developed. The high
background carrier concentration and relatively small band gap of InN ($\sim$2eV versus $\sim$3.4
eV for GaN and $\sim$6.2 eV for AlN) make it a promising contact for n-type material.
Thermally stable, low resistance ($\sim 10^{-6}$ Ohm-cm^2) contacts have recently been
demonstrated using InN contact layers graded from InAlN[3]. To further advance the
potential of III-Nitride devices, the development of p-type InN contact layers must be
achieved. Difficulty stems from the high background electron concentration (10^{17} to $5 \times
10^{20}$ /cm^3), the origin of which has not been clearly determined. By comparing the
properties of InN grown on different substrates, including closely lattice matched lithium
aluminate (LAO) and examining the nitridation behavior of the substrates, we hope to

gain a greater understanding of the relationship of this n-type autodoping to defect structure. This may lead to an effective growth strategy for obtaining p-type films. In this article we discuss the effects of pre-growth RF nitrogen plasma exposure and substrate choice on the electrical and morphological properties of InN grown by radio frequency metalorganic molecular beam epitaxy (RF MOMBE)[13].

EXPERIMENTAL

Four substrates were compared for this study, epi-ready (100) GaAs, c-plane sapphire, LAO (100), and LAO (100) miscut 18°. The miscut LAO is believed to posses matching symmetry as well as lattice dimension for nitride epitaxy. Lithium aluminate has an improved lattice match to III-nitrides with LAO (a= 3.13Å) having a near perfect match to AlN (a=3.11 Å).[14] This is an important consideration when growing low temperature AlN buffers. The substrates were indium mounted to a Mo block without any *ex-situ* cleaning steps and loaded into a Varian Gen II MOMBE system. In order to investigate their nitridation behavior, the substrates were exposed to an RF nitrogen plasma. The plasma source was provided by SVT Associates and operated at 400 W forward power with 5 sccm N_2 flow. Substrates were exposed for five minutes at temperatures of 725°, 775°, 825° and 875°C. Samples were exposed to the plasma during heat-up and cool-down as well. These samples were then removed from the system and analyzed using depth-profile auger electron spectroscopy (AES) and atomic force microscopy (AFM).

InN films were grown on the various substrates using trimethylindium (TMI) transported by a He carrier. Reactive nitrogen was provided by the RF plasma source operated under the conditions mentioned above. Substrates were maintained at ~525° C giving a growth rate of ~250 nm/hr. Three cases were examined for this study: 1) InN films grown on as received substrates with no prior plasma treatment, 2) InN films grown on substrates that received a plasma exposure of 875° C for five minutes and 3) InN films grown on substrates that received a plasma exposure of 875° C for five minutes and an AlN buffer. The 30 nm AlN buffers were grown at 425° C using dimethylethylamine alane and RF nitrogen plasma operated as previously described. Electrical transport properties were obtained from Van der Pauw geometry Hall measurements at 300K using alloyed (400° C, 2 min) HgIn Ohmic contacts. Surface morphology was examined by scanning electron microscopy (SEM) and contact mode AFM.

RESULTS AND DISCUSSION

SEMs of InN grown on sapphire and GaAs are shown in Figure 1. The top images show InN films grown on as received substrates while the bottom images are of films deposited on RF nitrogen plasma exposed substrates. The surface morphologies of the "no anneal" films are quite rough. The InN surface morphology shows significant improvement in the case of the GaAs substrate that received plasma exposure prior to growth. However, for InN grown on plasma exposed sapphire there is no morphological improvement. The differences in morphology between the two substrates can be traced to differences in nitridation behavior as measured by AES. GaAs samples that received a nitrogen plasma exposure at 875° C for five minutes yield a strong nitrogen signal at 385 eV, indicating nitridation of the surface. No such signal is discernible for sapphire substrates given the same treatment. This would suggest that no nitride template is being formed on sapphire under the given nitrogen plasma exposure, resulting in no improvement in morphology.

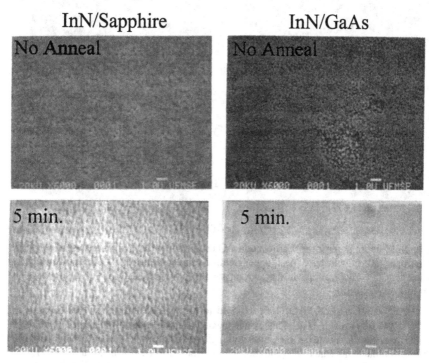

Figure 1. SEMs of InN films grown on sapphire (left) and GaAs (right). The top samples were grown on as received substrates while the bottom samples were grown on nitrogen plasma exposed substrates.

Next examined were InN films grown on LAO (100). Figure 2 shows SEMs of films grown using the three conditions discussed above. The film grown on the as received substrate is seen to be discontinuous and has poor morphology. Notable morphological improvement is seen when the substrate is exposed to nitrogen plasma with even further improvement obtained from the use of an AlN buffer. Figure 3 shows SEMs of InN films grown on LAO (100) miscut 18°. Again, the as received substrates yield discontinuous, poorly formed films. Improvements in film continuity are gained by nitrogen plasma exposure prior to growth with best results obtained using an AlN buffer.

Fig 2. SEM micrographs of InN grown on LAO (100) for the three cases discussed in the text. Upper left: no plasma anneal, upper right: 5 min. anneal at 875°C, and lower left: plasma anneal + AlN buffer. InN grown on sapphire using an AlN buffer is shown at right for comparison.

AES conducted on LAO (100) exposed to nitrogen plasma for five minutes at 875° C shows no nitrogen peak. This would suggest that although the plasma may not be forming a nitride template as measured by AES, it is affecting the surface in some fashion. Similarly, miscut LAO (100) substrates exposed to an RF nitrogen plasma for five minutes at 875° C do not show evidence of nitridation as measured by AES. The observed differences in morphology between InN grown on as received substrates and plasma exposed substrates may be due to cleaning of adventitious contaminants, though this remains to be verified. The dramatic improvement in morphology produced by the AlN buffers is due to the planar surfaces induced by these layers. InN by contrast tends to island severely when grown on highly mismatched or non-III-V surfaces.

Room temperature Hall measurements yield carrier concentrations on the order of 10^{20} cm^{-3} as seen in Figure 4. The background electron concentration was found to be relatively insensitive to substrate type and to pre-growth surface treatment. By contrast, the mobilities dropped substantially as the surface morphology was improved through the various pre-growth treatments. This effect has also been observed in GaN grown by MOCVD[15], and suggests some form of gettering. Further work is in progress to confirm this.

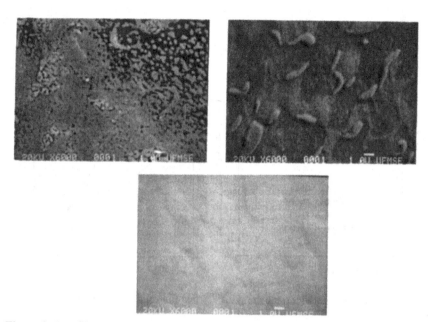

Figure 3. InN films grown on miscut LAO (100) for the three cases discussed in the text. Upper left: no plasma anneal, upper right: 5 min. plasma anneal, and bottom: plasma anneal + AlN buffer.

CONCLUSIONS

The results of AES on LAO (100), LAO (100) miscut and sapphire substrates exposed to an RF nitrogen plasma for five minutes at 875° C indicate no significant nitridation occurs. Since differences in morphology are observed for InN films grown on exposed versus unexposed substrates, there may be some cleaning effect taking place during exposure. By contrast, GaAs substrates show evidence of nitridation with exposure to the nitrogen plasma. Nitrided GaAs leads to smooth InN growth without the use of AlN buffers due to the presence of a smooth GaN template. Comparable InN morphology is obtained when AlN buffers are used on LAO and sapphire substrates. Hall measurements indicate background carrier concentrations are relatively insensitive to substrate type, suggesting that the background electron concentration is probably not related to misfit dislocations since it is expected that these defects would vary significantly for the variety of substrates used in this study.

ACKNOWLEDGMENTS

This work was supported by NSF Grant #ECS-952287 and ARO AASERT DAAHO4-95-1-0196.

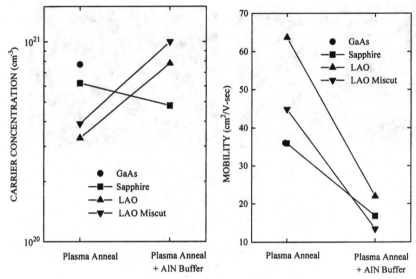

Figure 4. Carrier concentrations and mobilities of InN films given by room temperature Hall measurements.

REFERENCES

1. S. Strite and H. Morkoc, J. Vac. Sci. Technol. **B10** 1237 (1992) and references therein.
2. F. Ren, S. N. G. Chu, C. R. Abernathy, T. R. Fullowan, J. Lothian and S. J. Pearton, Semicond. Sci Tech. **7** 793 (1992).
3. S. M. Donovan, J. D. MacKenzie, C. R. Abernathy, S. J. Pearton, F. Ren, K. Jones and M. Cole, submitted to Appl. Phys. Lett.
4. H. J. Hovel and J. J. Cuomo, Appl. Phys. Lett. **20**, 71 (1972).
5. K. Kubota, Y. Kobayashi, and K. Fujimoto, J. Appl. Phys. **66**, 2984 (1989).
6. T. L. Tansley and R. J. Egan, Mater. Res. Soc. Symp. Proc. **242**, 395 (1992).
7. B. T. Sullivan, R. R. Parsons, K. Westra and M. Brett, J. Appl. Phys. **64** 414 (1988).
8. W. A. Bryden, S. A. Ecelberger, J. S. Morgan, T. O. Poehler, and T. J. Kistenmacher, Mat. Res. Soc. Symp. Proc. **242**, 409 (1992).
9. T. J. Kistenmacher, S. A. Ecelberger, and W. A. Bryden, Mater. Res. Soc. Symp. Proc. **242**, 441 (1992).
10. K. Osamura, S. Naka and Y. Mukakami, J. Appl. Phys. **46**, 3432 (1975).
11. T. Matsuoka, J. Cryst. Growth **124**, 433 (1992).
12. R. J. Molnar, R. Singh and T. D. Moustakas, J. Elec. Matls, **24**, 275 (1995).
13. C. R. Abernathy, S. J. Pearton, F. Ren, P. Wisk, J. Vac. Sci. Technol. **B11** 179 (1993).
14. R. G. Wilson, B. L. H. Chai, S. J. Pearton, C. R. Abernathy, F. Ren, and J. M. Zavada, Appl. Phys. Lett., **69** 3848 (1996).
15. S. DenBaars, presented at the Electrochemical Society Meeting, Los Angeles, 1996.

MICROSTRUCTURES OF GaN FILMS GROWN ON A LiGaO₂ NEW SUBSTRATE BY METALORGANIC CHEMICAL VAPOR DEPOSITION

Jing-Hong Li, Olga M. Kryliouk*, Paul H. Holloway, Timothy J. Anderson* and Kevin S. Jones
Dept. of Materials Science and Engineering, *Dept. of Chemical Engineering
University of Florida, Gainesville, FL 32611

ABSTRACT

Microstructures of GaN films grown on the LiGaO₂ by metalorganic chemical vapor deposition (MOCVD) have been characterized by transmission electron microscopy (TEM) and high resolution transmission electron microscopy (HRTEM). TEM and HRTEM results show that high quality single-crystal wurtzite GaN films have been deposited on the LiGaO₂ and that the GaN film and the LiGaO₂ have the following orientation relationship: $[2\bar{1}\bar{1}0]_{GaN} \parallel [010]_{LiGaO2}$; $(0002)_{GaN} \wedge (002)_{LiGaO2} < 5\text{-}8°$. A higher density of threading dislocations and stacking faults have been observed near the GaN/LiGaO₂ interface, even though the lattice mismatch of GaN to LiGaO₂ is only ~1%. Threading dislocations with burgers vector $b=<0001>$ and $b=a/3<11\bar{2}0>$ are predominant in the GaN films. Also the GaN films contain some columnar inversion domain boundaries (IDBs). Both TEM and HRTEM results reveal that there is an unexpected amorphous or nano-crystalline inter-layer between the GaN and the LiGaO2 with a thickness of 50-100 nm.

INTRODUCTION

III/V nitrides especially GaN are being received extensive scientific and technological attention at present because of their potential applications in short wavelength opto-electronic devices such as high performance transistors, light emitting diodes (LEDs) and laser diodes, etc. One of the main problems in developing these kinds of materials is the crystalline quality of grown films which usually contain a high density of defects, including dislocations on the order of $10^{10}/cm^2$ and stacking faults, etc. Opto-electronic properties, such as spatial luminescence are correlated well to the distribution of dislocation. Defects may decrease the efficiency of a device by acting as nonradiative recombination sites.[1-3] Surprisingly, LEDs operate quite well despite these defects. However, for blue emitting laser diodes, these defects may ultimately limit the device life-times.[1]

One of major obstacles in improving crystalline quality of films is the lack of suitable substrates which are lattice and thermally matched with GaN. So far, the most commonly used substrate is sapphire (Al_2O_3). Due to large lattice-mismatch (~13%) to the GaN, a high density of dislocations in the GaN film is normally observed.[4] In addition to the threading dislocations and stacking faults, other structural defects including stacking mismatch boundaries (SMBs), inversion domain boundaries (IDMs) and double positioning boundaries (DPBs) have been found by other groups.[5,6] Recently, it was reported that LiGaO₂ is a promising substrate for growth of GaN film because of its close lattice mismatch of ~1% to the GaN.[7-9] LiGaO₂ has a orthorhombic structure with lattice parameters of a=5.402 Å, b=6.372 Å and c=5.007 Å. To optimize the GaN film growth and its opto-electronic properties, it is necessary to understand extended structural defects and other types of defects. There are a few reports on microstructures of GaN/LiGaO₂ so far.[9] We present here our microstructural study on the MOCVD GaN films on the new LiGaO₂ substrate using TEM and HRTEM.

167

Mat. Res. Soc. Symp. Proc. Vol. 468 ©1997 Materials Research Society

EXPERIMENTAL PROCEDURE

The GaN films were grown on the (001) LiGaO$_2$ which is grown by the Czochralski technique and subsequently mechanically and chemically polished on both sides. The GaN/LiGaO$_2$ samples were deposited in a low pressure, horizontal cold-wall MOCVD reactor on the (001) LiGaO$_2$ substrate with triethygallium (TEGa) and ammonia NH$_3$ as precursors and N$_2$ as carrier gas using a V/III ratio of 3324 and substrate temperature of 850 °C under a reactor pressure of 130 Torr for 1 hour. The thickness of the GaN films was typically 0.35-0.5 μm. Cross-sectional TEM (XTEM) analysis and HRTEM have been carried out on the GaN thin films. XTEM samples were prepared by firstly sticking two pieces of film-face-to-film, cutting by low speed saw, then grinding to ~100 μm, dimpling down to 20 μm, and finally ion-beam milling from both sides. Both XTEM and HRTEM were carried in a JEOL 200C and a JEOL FX4000, respectively.

RESULTS AND DISCUSSION

Threading Dislocations and Stacking Faults

Fig. 1 shows a XTEM micrograph of: (a) GaN/LiGaO$_2$ interface and the corresponding selected area diffraction patterns (SADPs) from a GaN/LiGaO$_2$ interface, (b) with a big aperture and (c) with a smaller aperture covered with half GaN and half LiGaO$_2$. The corresponding SADP in Fig.1(b) shows superposition of the GaN [2$\bar{1}$10] and the LiGaO$_2$ [010] diffraction patterns, suggesting that the GaN film is single crystal and grew epitaxially on the LiGaO$_2$ substrate. Also the corresponding SADP in Fig.1 (b) illustrates that the GaN film and the LiGaO$_2$ have the following orientation relationship: [2$\bar{1}$10]$_{GaN}$ ‖ [010]$_{LiGaO2}$; (0002)$_{GaN}$ ∧(002)$_{LiGaO2}$<5-8°. It was found that the GaN (0002) plane tilted 5-8° from the LiGaO$_2$ (002) plane, as shown in Fig. 1(b). The observed 5-8° mis-orientation between the grown GaN and the LiGaO$_2$ resulted from mis-cutting of the LiGaO$_2$ substrate. The XTEM micrograph in Fig. 1 (a) revealed the presence of a high defect density, as indicated by "d" in Fig. 1(a). The threading dislocation density was high near the interface, but it decreased with increasing distance from the interface. Dislocation analysis has been conducted by conventional XTEM using the $\vec{g}.\vec{b}$ =0 criteria. Fig. 2 shows two weak-beam XTEM micrographs taken with different reflections of [2$\bar{1}$10] zone axis. The Burger's vectors for some of the threading dislocations were determined to be [0001] and $\frac{1}{3}$[11$\bar{2}$0]. These dislocations are predominantly edge-type in nature, assuming that the growth direction is the same as the translation vector of the dislocations. In addition to the diffraction spots, streaking along the [0002] was observed. The streaks are attributed to a basal plane stacking faults which were parallel to the interface or the LiGaO2 {001} planes. Near the GaN/LiGaO$_2$ interface, the GaN film has a higher density of stacking faults, compared with a lower density with increasing of distance from the interface. Large numbers of dislocations present in heteroepitaxy GaN can act as nucleation sites for stacking faults.[10] This high density of stacking faults near the GaN/LiGaO$_2$ interface was associated with the high density of threading dislocation near the interface, as shown in Fig. 1(a).

GaN/LiGaO$_2$ Interface

Contrast analysis shows that there was residual strain at the interface between the GaN and the LiGaO$_2$ substrate, even through the GaN and the LiGaO$_2$ have a very small mismatch of 1%, indicating that the lattice mismatch was not completely relieved by misfit dislocations. The interface of GaN/LiGaO$_2$ was quite rough. The XTEM micrograph in Fig.1(a) revealed that there

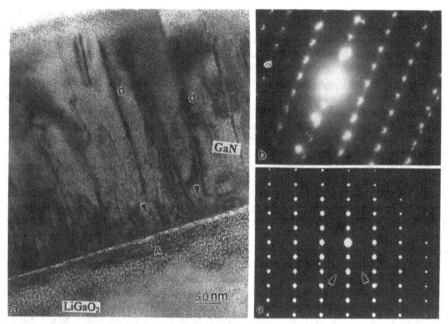

Fig. 1 (a) XTEM micrograph of the GaN film; (b) a SADP with a bigger aperture and (c) a SADP with a smaller aperture

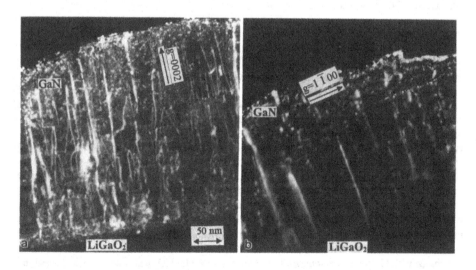

Fig. 2 Weak-beam XTEM micrographs of GaN film imaged with different diffraction reflections: (a) g=0002 and (b) g=1$\bar{1}$00

was an inter-layer between the GaN film and the LiGaO$_2$ substrate, marked by "A" in Fig. 1 (a). Correspondingly, the weak spot-ring SADP with a smaller aperture from the GaN/LiGaO$_2$ interfacial region, as indicated by arrows in Fig. 1(c), show that the inter-layer is amorphous or nano-crystalline with a thickness of 50-100nm. HRTEM was carried out on the GaN/LiGaO$_2$. Fig. 3 shows a HRTEM micrograph of the GaN/LiGaO$_2$ interface. The HRTEM micrograph shows two dimensional lattice images, both along the GaN [2$\overline{1}$$\overline{1}$0] direction. The HRTEM micrograph also confirmed the XTEM observations that there was a nano-crystalline or amorphous inter-layer between the GaN film and the LiGaO$_2$ substrate. Stacking faults were also observed near the GaN/LiGaO$_2$ interface by HRTEM, as indicated by arrows in Fig. 3. The presence of the amorphous or nano-crystalline inter-layer indicates that there probably is a interfacial chemical reaction during the GaN film growth. This is critical to the crystalline quality of the GaN film since the amorphous or nano-crystalline may change lattice mismatch, leading to a high density of threading dislocation, even through the GaN and the LiGaO$_2$ substrate have a very small lattice mismatch of 1%. A smaller lattice mismatch should result in lower misfit strain and is therefore favorable to minimize the density of dislocations which are inevitably formed by the epitaxial GaN film to alleviate the lattice-mismatch and the strain of post-growth cooling. However, since there is a small lattice mismatch of 1% between the GaN and the LiGaO$_2$, the mechanism driving formation of this layer must be something other than mismatch. This inter-layer might be formed during the film growth at high temperature because of inter-diffusion of N and Li or O, or reaction of the film and the substrate. This inter-layer may play a important role in formation of defects in the GaN film. In our previous work, we have found evidence of Li diffusion into GaN. This may lead to the formation a new inter-phase such as Li$_5$GaO$_2$. This reaction appears to occur after the nucleation of the GaN since single crystal GaN is still grown which would not be possible on an amorphous or polycrystalline substrate. As the GaN film is cooled down from high temperature, the thermal-mismatch between the GaN and the Li$_5$GaO$_2$ may cause the observed high density of dislocations.

<u>Inversion Domain Boundaries</u>

In addition to the threading dislocations and stacking faults, columnar inversion domains have been found in GaN films. Fig. 4 (a) and (b) show two-beam dark-field TEM micrographs imaged with g=0002 and g=000$\overline{2}$, respectively, inserted with convergent beam electron diffraction (CBED) patterns taken from two adjacent inversion domains. From the CBED patterns, we are able to identify the columnar defects are inversion domain boundaries (IDBs) where different domains show opposite contrasts when imaged with opposite diffraction reflections. These results are very similar to observations of IDBs in GaN grown molecular beam epitaxy (MBE), MOCVD, and hydride vapor phase epitaxy (HVPE) on Al$_2$O$_3$ substrate.[6, 10-13] At an inversion domain boundary in a wurtzite structure, the identity of the atoms occupying the two sublattices is interchanged.[6] The Ga and N atoms are interchanged as one cross the (10$\overline{1}$0) plane, leading to both Ga-Ga and N-N bonds at the boundary. Within the columnar IDBs, the GaN polarity is reversed. IDBs are formed due to the atomic steps at the surface of a substrate. The IDB is nucleated at a step on a substrate. On the left side of the IDB, the orientation of the GaN is the [0001] while on the right side of the IDB the orientation is the [000$\overline{1}$] direction. In other words, on one side of the IDB, the GaN is terminated with Ga while on the other side, the GaN is terminated with N. In our previous work, the surface of the LiGaO$_2$ has some disordered region,[14] producing some steps at surface. This is believed to contribute the formation of IDBs.

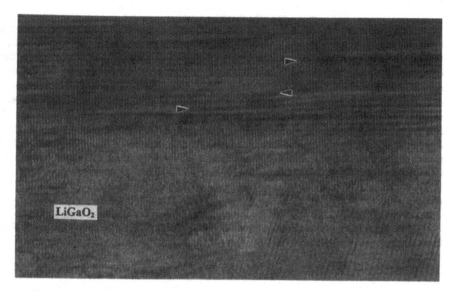

Fig. 3   HRTEM micrograph of the GaN/LiGaO₂ interface

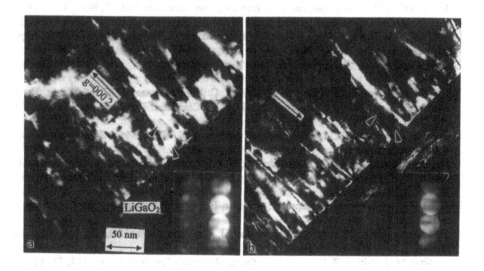

Fig. 4   Two-beam dark-field XTEM micrographs of the GaN film imaged with opposite diffraction reflections: (a) g=0002 and (b) g=000 $\overline{2}$

CONCLUSIONS

Microstructures of GaN films grown on the $LiGaO_2$ by MOCVD have been characterized by transmission electron microscopy (TEM) and high resolution transmission electron microscopy (HRTEM). TEM and HRTEM results show that high quality single-crystal wurtzite GaN films have been deposited on the $LiGaO_2$ and that the GaN film and the $LiGaO_2$ have the following orientation relationship: $[2\bar{1}\bar{1}0]_{GaN} \parallel [010]_{LiGaO2}$; $(0002)_{GaN} \wedge (002)_{LiGaO2} < 5\text{-}8°$. A higher density of threading dislocations and stacking faults have been observed near the GaN/$LiGaO_2$ interface, even though the lattice mismatch of GaN to $LiGaO_2$ is only 1%. Threading dislocations with burgers vector $b=<0001>$ and $b=a/3<11\bar{2}0>$ are the predominant types formed in the GaN films. Also the GaN films contain some columnar inversion domain boundaries. Both TEM and HRTEM results reveal that there is an unexpected amorphous or nano-crystalline inter-layer between the GaN and the LiGaO2 with a thickness of 50-100 nm.

ACKNOWLEDGMENT

We acknowledge the financial sponsorship from DARPA and the Office of Naval Research under contract N00014-92-J-1895.

REFERENCES:

1. S. Nakamura, M. Senoh, N. Iwasa, and S. Nagahama, Jpn. Appl. Phys. **34**, L979 (1995).
2. S. Nakamura, M. Senoh, S. Nagahama N. Iwasa, T. Yamda, T. Matsushita, H. Kiyoku, and Y. Sugimoto, Jpn. Appl. Phys. **35**, L74 (1996).
3. J. E. Northrup, J. Neugebauer, and L. T. Romano, Phys. Rev. Lett. Vol. **77**, No.1, 103, 1996.
4. F. A. Ponce, D. Cherns, W. T. Young, and J. W. Steeds, Appl. Phys. Lett., **69**(6), 770 (1996)
5. D. J. Smith, D. Chandraskhar, B. Sverdlov, A. Botchkarev, A. Salvador and H. Morkoc, Appl. Phys. Lett., **67** (13), 1830 (1995)
6. L. T. Romano, J. E. Northrup and M. A. O'Keefe, Appl. Phys. Lett., **69** (16), 2394, October 1996.
7. J. F. H. Nicholls, H. Gallagher, B. Henderson, C. Trager-Cowan, P. G. Middleton, K. P. O'Donnell, T. S. Cheng, C. T. Foxon, and B. H. T. Chai, Mat. Res. Soc. Symp. Proc., Vol **395**, 1996, 535
8. S. Limpijumnong, W. R. L. Lambrecht, B. Segall and K. Kim, 1996 MRS Meeting Proc., Boston, 1996 (accepted)
9. O. M. Kryliouk, T. W. Dann, T. J. Anderson, H. P. Maruska, L. D. Zhu, J. T. Daly, M. Lin, P. Norris, B, H. T. Chai, D. W. Kisker, J. H. Li, and K. S. Jones, 1996 MRS Meeting Proc., Boston, 1996 (accepted)
10. X. H. Wu, L. M. Brown, D. Kapolnek, S. Keller, B. Keller, S. P. Denbaars and J. S. Speck, J. Appl. Phys. **80**, 3228 (1996)
11. Z. Liliental-Weber, H. Sohn, N. Newman, and J. Washburn, J. Vac. Sci. Technol., **13**, 1578 (1995)
12. Z. Sitar, M. J. Paisley, B. Yan, and R. F. Davis, Mater. Res Soc. Symp. Proc. **162**, 537 (1990)
13. B. N. Sverdlov, G. A. Martin, H. Morkoc, and D. J. Smith, Appl. Phys. Lett., **67**, 2063 (1995)
14. J. H. Li, O. M. Kryliouk, B. H. T. Chai, T. J. Anderson and K. S. Jones (to be published)

DEPENDENCE OF THE RESIDUAL STRAIN IN GaN ON THE AlN BUFFER LAYER ANNEALING PARAMETERS

Y.-M. Le Vaillant, S. Clur, A. Andenet, O. Briot, B. Gil and R.L. Aulombard
GES-CNRS, CC074 Université Montpellier II, Place E.Bataillon, 34095 Montpellier Cedex 5
France
R. Bisaro, J. Olivier, O. Durand, J.-Y. Duboz
THOMSON-CSF-LCR, Domaine de Corbeville, 91404 Orsay Cedex, France

ABSTRACT

The problem of residual strain in GaN epilayers is currently the attention of many studies, since it affects the optical and electrical properties of the epilayers. In order to discuss the origin of this residual strain, we have grown a series of GaN epilayers onto AlN buffer layers, sapphire (0001) being used as substrate. The buffer layer is usually deposited in an amorphous state and is recrystallized by a thermal annealing . Here we have made a systematic study of the buffer recrystallization by changing the annealing temperature and the annealing time. The surface morphology is probed using Atomic Force Microscopy (AFM). The lattice parameter c is carried out from accurate x-ray diffraction measurements. The GaN layers were studied by low temperature photoluminescence and reflectivity. The amount of residual strain is calibrated from the position of the A exciton and the optical quality of the layers is assessed from the photoluminescence linewidths. The longer the annealing time the better the strain relaxation in AlN buffer layers and the higher the lattice mismatch with GaN overlayers.

INTRODUCTION

Gallium nitride has recently been the subject of numerous studies, since optoelectronic devices of stunning quality have been produced from these materials. It has been observed by many researchers that important residual strain is present in GaN heteroepitaxial layers. These strains greatly affect the optoelectronic properties [1,2], but their origin remains unclear. In Metal Organic Vapor Phase Epitaxy (MOVPE), high quality epilayers are obtained using a double-step growth process involving the pre-deposition of a low temperature AlN [3] or GaN [4] buffer layer. This buffer layer is heat-treated prior to the growth of GaN. As a result, numerous factors can be invoked to explain the residual strain in GaN, some of them have already been pointed out: the epilayer thickness [5, 6], the III/V molecular ratio [7]. However, we expect that other growth parameters could also play a role.

The buffer layer annealing transforms the textured, as-deposited buffer layer into a highly defective monocrystalline layer. In this paper, we will show that the lattice mismatch between GaN and AlN depends on the crystalline quality and on the mean lattice parameter of the AlN buffer layer. This results in a strain state varying with the recrystallization treatment of the buffer layer.

Such a dependence of the buffer layer lattice parameter on heat treatment has already been reported by Wickenden et al. [8] for GaN buffer layers. To our knowledge, no similar results have been reported concerning AlN buffer layers. Moreover, we establish here a clear correlation between the GaN layer strain state and the buffer layer heat treatment.

EXPERIMENTAL

The layers were grown using low pressure MOVPE (76 Torr), the precursors being triethylgallium and ammonia. The buffer layers were deposited onto C-plane sapphire substrates at 800°C, and the heat treatment was performed at 980°C (the GaN layer growth temperature), changing the annealing time from 30 seconds to 15 minutes. In a first series of samples, the buffer layers were grown and annealed, but no GaN overlayer was deposited. In a second set of samples, the first series of buffer layers was reproduced and one micrometer thick GaN layers were grown on top of it, with a V/III molecular ratio of 10,000. The layers were characterized by X-ray diffraction. A high performance x-ray goniometer was used in Bragg-Brentano geometry. AlN (0002) and (0004) peak positions were determined from experimental measurements using a computer-fit procedure. The c- lattice parameters was then obtained with a regression method to correct the experimental errors. The surface morphology was determined by AFM (contact mode in a liquid cell or dynamic mode in order to avoid charge problems). The strain state of the epilayers was determined by reflectivity at low temperature (2 K).

RESULTS AND DISCUSSION

In figure 1, we show the morphologies of AlN buffer layers, observed by AFM. In a), the buffer is as-grown, without any heat treatment. The morphology is mainly constituted of two types of grains: small and large ones. In b), the buffer layer has been annealed for 2 minutes at the growth temperature (980°C).

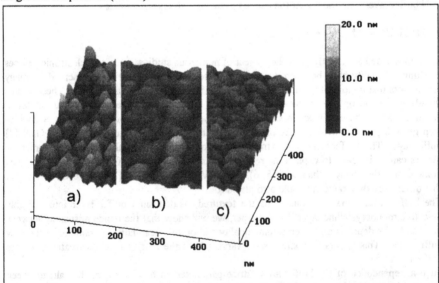

Figure 1: AFM pictures of AlN buffers: non- annealed a), annealed at 980°C for: 2 minutes b), 15 minutes c).

The morphology appears flattened, some degree of coalescence is clearly visible. In c), the coalescence is even more pronounced with a lower roughness after 15 minutes annealing at the same temperature.

In figure 2, we report the lattice parameter c of the AlN buffer layer, measured by x-ray diffraction. Its value increases sharply with increasing heat treatment and for longer annealing times tends towards a value close to the bulk lattice parameter c as reported by Chuang et al. [9]. We can interpret this result by a relaxation of the tensile stress of the as-grown buffer layer due to the annealing treatment.

From these experimental measurements, we may suppose that the in-plane parameter of the AlN buffer layer will change according to the c-parameter. In figure 3, the a parameter is calculated, assuming an elastic behavior of the AlN buffer. This hypothesis is likely, since the maximum strain in the annealed buffer layers and their thicknesses are about 0.27% and 500 Å respectively.

The in-plane lattice parameter was calculated for each annealed buffer layer using the following expressions :

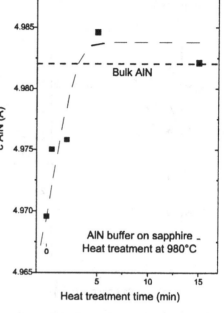

Figure 2: Measured lattice parameter of AlN buffer layers versus annealing time.

$$\varepsilon_{xx} = -\frac{C_{33}}{2C_{13}}\varepsilon_{zz} \qquad \varepsilon_{zz} = \frac{c_{measured} - c_0}{c_0} \qquad \varepsilon_{xx} = \frac{a_{AlN} - a^0_{AlN}}{a^0_{AlN}}$$

where $c_0 = 4.982$ Å is the value of the bulk lattice parameter c of AlN [9], $a^0_{AlN} = 3.112$ Å is the value of the lattice parameter a of bulk AlN. The elastic coefficients for AlN have been taken from ref. 10. The calculated values, displayed in figure 3, decrease from 3.120 Å to 3.112 Å for the AlN layers and the corresponding mismatches with GaN, also given in figure 3, change from 1.9% to 2.5% with increasing buffer annealing time.

From these results, we can understand that part of the residual strain in the GaN layer may originate from the buffer/layer lattice mismatch, which changes with the buffer annealing treatment.

To assess whether these effects effectively induce residual strains in the GaN epilayers, we have grown GaN layers onto buffer layers annealed in the conditions previously described. The GaN layers were characterized by reflectance, in order to determine the position of the transition corresponding to the A exciton. As it has been previously demonstrated [1], the energy position of the A exciton is linearly correlated to the amount of residual biaxial stress.

As it can be seen in the reflectance spectra (figure 4), the excitonic structures are shifted towards higher energies when the annealing time increases, with an apparent saturation for annealing times equal to or above 5 minutes. Since all the growth parameters have been held constant, this demonstrates the influence of the buffer recrystallization treatment on the strain state of the layer.

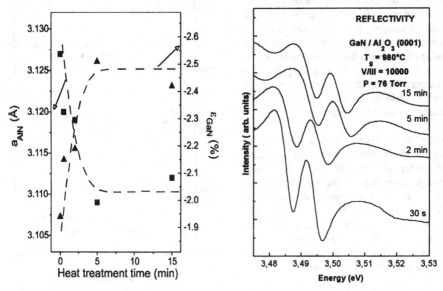

Figure 3: Calculated a-lattice parameter of AlN buffer layers (squares) and lattice mismatch with GaN (triangles) versus buffer annealing time.

Figure 4: 2K reflectance of 1 μm GaN epilayers grown on AlN layers annealed for 30 seconds to 15 minutes.

CONCLUSION

We have demonstrated that the thermal annealing affects the morphology and the crystalline properties of AlN buffer layers. The as-grown buffer layer exhibits tensile stress, which relax during the annealing, full relaxation being reached for treatment time equal to or above 5 minutes, at an annealing temperature of 980°C. The residual strain in GaN epilayers grown on buffers annealed for 30 seconds to 15 minutes is observed to increase with the annealing time. Correlatively, the lattice mismatch between the AlN buffer and the GaN overlayer increases. This demonstrates that among other parameters (thickness, V/III molar ratio, ...), the buffer annealing plays a significant role in the control of the residual strain value in GaN epilayers.

ACKNOWLEDGMENT

This work has been partly supported by the DRET/DGA.

REFERENCES

[1] B. Gil, O. Briot and R.L. Aulombard, Phys. Rev. B **52** (1996) R17028

[2] M. Tchounkeu, O. Briot, B. Gil, J.P. Alexis and R.L. Aulombard, J. Appl. Phys. **80** (1996) 5352

[3] H. Amano, N. Sawaki, I. Akasaki and Y. Toyoda, Appl. Phys. Lett. **48** (1986) 353

[4] S. Nakamura, Jpn. J.Appl. Phys. **30** (1991) L1705

[5] W. Rieger,T. Metzger, H. Angerer, R. Dimitrov,O. Ambacher and M. Stutzmann, Appl. Phys. Lett. **68**, 7 (1996) pp 970-972

[6] O. Gfrörer, T. Schlüsener, V. Härle, F. Scholz and A. Hangleiter, Symposium-C E-MRS 1996 Spring Meeting UV, Blue and Green Light Emission from Semiconductor Materials *(symp. C-x.4 : Relaxation of thermal strain in GaN epitaxial layers grown on sapphire)*

[7] O. Briot, J.P. Alexis, B. Gil and R.L. Aulombard, Mat. Res. Soc. Symp. Proc. Vol **395** (1996) 411

[8] A. Wickenden, D. Wickenden and T. Kistenmacher, J. Appl. Phys. **75** (1994) 5367

[9] S. L. Chuang and C. S. Chang, Phys. Rev. B **54,** 4 (1996) pp 2491-2504

[10] J. Chaudhuri, R. Thokala, J.H. Edgar and B.S. Sywe, Thin Solid Films **274** (1996) 23

EXCITONS BOUND TO STACKING FAULTS IN WURTZITE GaN

Y.T.REBANE, Y.G.SHRETER AND M.ALBRECHT*

A.F.Ioffe Physico-Technical Institute, Russian Academy of Sciences, 26 Politechnicheskaya, St. Petersburg 94021, Russia
*Universität Erlangen-Nürnberg, Institut für Werkstoffwissenschaften, Mikrocharakteri-sierung, Cauerstr.6, 91058 Erlangen, F.R.G.

ABSTRACT

A model of the exciton bound to stacking faults (SF) in GaN is suggested. It is shown that SFs are potential wells (depth ~ 120 meV) for electrons and potential barriers (~ 60 meV) for holes. The binding energy of the exciton at stacking faults is estimated as 30 - 60 meV. The 364 nm line in GaN photoluminescence is attributed to excitons at stacking faults.

INTRODUCTION

Recent investigations of the 364 nm line in the photoluminescence spectra of GaN-epilayers have shown that it can be attributed to extended structural defects. The reasons for this are the small value of the Huang-Rhys factor, the spatial anticorrelation of the line intensity with respect to the bound exciton (BE) line, an increase of the line intensity near the buffer layer and the disappearance of the line in thick films [1-5]. Screw dislocations [2-4] and isolated cubic phase crystallites in the hexagonal matrix [5] were suggested as possible candidates for such defects. However, a detailed microscopical investigation of GaN epilayers grown on vicinal surfaces of the SiC substrates where 364 nm line is particularly intense with respect to BE line have shown very high concentration of stacking faults [6]. This gives us reasons to suggest that 364 nm line is related to excitons bound to stacking faults (SFE). In this paper we develop a theoretical model that allows to calculate the wave functions and the binding energies of SFE.

STACKING FAULT AS A POTENTIAL WELL FOR ELECTRONS

The three most common stacking faults (SF) in wurtzite GaN with an $\alpha\beta\alpha\beta$ structure have the following stacking sequence $\alpha\beta\underline{a\alpha b\beta c\gamma}b\beta c\gamma$ (I1), $\alpha\beta\underline{a\alpha b\beta c\gamma a\alpha}c\gamma a\alpha$ (I2) and $ab\underline{\beta a\alpha b\beta c\gamma a\alpha b\beta}c\gamma b\beta$ (E). The corresponding Burgers vectors of the surrounding dislocations are $b(I_1) = 1/6[2,-2,0,3]$, $b(I_2) = 1/3[1,-1,0,0]$ and $b(E) = 1/2[0,0,0,1]$ [7]. It can be easily seen that they include layers of cubic phase with widths $w(I_1) = 1.5c_0$, $w(I_2) = 2.0c_0$ and $w(E) = 2.5c_0$ respectively, inserted into the host wurtzite crystal, where c_0 is the lattice constant. The wurtzite crystal can be considered as a cubic crystal that is uniaxially deformed along the [111]-axis of the cubic crystal with an corresponding strain $\varepsilon_{zz} = 0.612c_0/a_0 - 1$. For wurtzite GaN with lattice constants $c_0 = 0.5185$ nm and $a_0 = 0.3189$ nm [8] the strain is $\varepsilon_{zz} = -0.005$. This strain shifts the edges of conduction and valence bands by values $\Delta Ec = \Xi\varepsilon_{zz}$ and $\Delta E_v = (a-b)\,\varepsilon_{zz}$ respectively, where Ξ, a, and b are the deformation potential constants for cubic GaN. The experimental values of Ξ, a and b for cubic GaN are not available. However, the combination

$$\Xi - a = dE_g / \ln V \tag{1}$$

179

can be found from the experimental value of dEg/lnV = -9.8 eV [9]. The constants a and b for cubic GaN are related to the deformation potential constants D_2, D_3 and D_5 for wurtzite GaN in the cubic approximation by the expressions

$$a = D_2 - D_3/3, \quad b = 2D_5 + 2D_3/3 \qquad (2)$$

This gives a = -14.6 eV and b = -2.1 eV for D_2 = -0.91Ry, D_3 = 0.22Ry, D_5 = -0.15Ry found from calculations of the band structure of strained wurtzite GaN [10]. Then from Eqs.(1),(2) we can find Ξ = -24.4 eV. We would like to notice that the value of Ξ is much higher compared to other semiconductors, where $\Xi \sim - (4 - 8)$ eV [11]. Since the SFs are layers of cubic phase surrounded by wurtzite phase that can be considered as uniaxially compressed cubic phase they form potential wells for electrons. The depth of the wells is $\Delta E_c = \Xi \varepsilon_{zz}$ = 122 meV. The widths of the potential wells L depend on the interface positions between cubic and hexagonal phases that are in the range $L(I_1) = (1.0 \pm 0.5)$ c_0 = 0.26nm - 0.78 nm, $L(I_2) = (1.5 \pm 0.5)$ c_0 = 0.52 - 1.04 nm, and $L(E) = (2.0 \pm 0.5)$ c_0 = 0.78 - 1.30 nm. Thus, we can consider the SFs as quantum wells for electrons.

In the valence band the deformation potential a is negative and, therefore, SFs are potential barriers for holes. The widths of the barriers are the same as the widths of potential wells for electrons and the height of the barriers is $\Delta E_v = (a-b)\varepsilon_{zz}$ = 62 meV. Thus, electrons are attracted to the stacking faults but the holes are repelled from them and the interface between wurtzite and cubic GaN is similar to the type II heterojunction.

It should be noted that this model is a semiphenomenological one. To check its accuracy we can calculate the difference in band gaps for wurtzite and cubic phases $\Delta E_g = \Delta E_c - \Delta E_v$ = 60 meV. The experimental values of ΔE_g are in the range of 90 - 190 meV [9]. Thus, the theoretical estimate based on this model is lower but in a reasonable agreement with the experimental value.

EXCITONS AT STACKING FAULTS

Since the stacking faults contain a few atomic layers the quantum effects are significant for the calculation of the electron binding energy E_e at the stacking fault. Thus, the binding energy can be found from a well known solution of the one-dimensional Schrödinger equation for a square quantum well and E_e is given by a solution of the equation

$$\sqrt{\Delta E_c - E_e} tg\sqrt{m_e L^2 (\Delta E_c - E_e)/\hbar} = \sqrt{E_e} \qquad (3)$$

For the case $E_e \ll \Delta E_c$, E_e can be found in the δ-potential well approximation as

$$E_e = m_e (\Delta E_c L)^2 / 2\hbar^2 \qquad (4)$$

For the stacking fault of I_2-type with L ~ 1 nm and m_e = 0.2 m_0 [13], the binding energy found from Eq.(4) is E_e ~ 25 meV. Thus, the SFs can bound electrons even at room temperature.

The holes can be attracted to the electrons bound to the stacking faults via Coulomb force, forming excitons bound to stacking faults (SFE). The binding energy of the stacking fault exciton E_{SFE} can be estimated in an approximation that the electron at SF is immobile. An account of the electron motion in the SF plane should reduce this energy to some extent and, therefore, this approximation gives an upper estimate for E_{SFE}. Since the SF strongly repels the

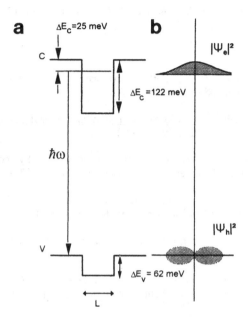

Fig.1: Schematic showing (a) the potential well of a stacking fault embedded in a hexagonal wurtzite lattice; (b) the scheme for wave functions of electrons and holes bound to the stacking fault.

hole its wave function should go to zero at the SF plane. Therefore the binding energy of the hole is given by its lowest p-state in the Coulomb potential. The energy of this level is $E_0^{J=3/2}, - = 0.23$ Ry* , where Ry* = $m_{hh} e^4/2\, \varepsilon^2\, \hbar^2$ and m_{hh} is the mass of the heavy hole, ε is dielectric constant [14]. For GaN with m_{hh} = 1.76 m_0 [10] and ε = 9 [15], Ry* = 295 meV and the upper estimate for SFE binding energy is ~ 60 meV. From another hand the lower estimate for E_{SFE} is the binding energy of the free exciton ~ 30 meV. Thus, the binding energy of SFE is in the range E_{SFE} = 30 - 60 meV.

CONCLUSIONS

On the base of this model we can attribute the line observed at ~ 3.4 eV in the h-GaN samples with high density of stacking faults to the excitons bound to stacking faults. To check the model we can compare the shift of the line with respect to the h-GaN band gap $E_g - \hbar\omega$ = 3.5eV - 3.4eV = 0.1 eV with the sum of the binding energies of carriers to the stacking fault E_{SFE} + E_c ~ 0.08 eV. Thus, a reasonable agreement exists between the model and the experimental data.

ACKNOWLEDGEMENTS

Y T R thanks the Russian Fund of Fundamental Studies (grant no.96-02-17825-a)

REFERENCES

1. C.Wetzel, S.Fisher, J.Krüger, E.Haller, R.J. Molnar, T.D. Moustakas, E.N.Mokhov, and P.G.Baranov, Appl.Phys.Lett. **68**, 2556 (1996).

2. Y.G.Shreter and Y.T.Rebane, Proceedings of the 23d International Conference on the Physics of Semiconductors Berlin, V.1.D.22, p.2937 (1996).

3. Y.T.Rebane and Y.G.Shreter, Proceedings of 23d International Symposium on Compound Semiconductors, Astoria Hotel, St. Petersburg, Russia, September 23-27, 1996 (in press).

4. Y.G.Shreter, Y.T.Rebane, T.J.Davis, J.Barnard, M.Darbyshire, J.W.Steeds, W.G.Perry, M.Bremser and R.F.Davis, Mat.Res.Soc.Symp.Proc. **449**, 683 (1997).

5. W.Rieger, R.Dimitrov, D.Brunner, E.Rohrer, O.Ambacher and M.Stutzmann, Phys.Rev.B, to be published.

6. M.Albrecht, S.Christiansen, G.Salviati, C.Zanotti-Fregonara, Y.T.Rebane, Y.G.Shreter, M.Mayer, A.Pelzmann, M.Kamp, K.J.Ebeling, M.D.Bremser, R.F.Davis, H.P.Strunk (paper at this conference).

7. P.Vermaut, P.Ruterana, and G.Nouet, Phil. Mag. a **75**, 239 (1997).

8. H.P.Maruska and J.J.Tietjen, Appl. Phys. Lett. **15** 327 (1969).

9. P.Perlin, I.Gorczyca, N.E.Christensen, I.Grzegory, H.Teisseyre, and T.Suski, Phys.Rev. B **45** 13307 (1992).

10. M. Suzuki and T.Uenoyama, J.Appl.Phys. **80** 6868 (1996).

11. C.G.Van de Walle, Phys.Rev. B **39** 1871 (1989).

12. S.Strite and H.Morkoç, J.Vac.Sci. Technol. B **10** 1237 (1992).

13. A.S.Baker and M.Ilegems, Phys.Rev. B **7** 743 (1973).

14. Y.T.Rebane, Phys.Rev. B **48** 11772 (1993).

15. H.Morkoc, S.Strite, G.B.Gao, M.E.Lin, B.Sverdlov, and M.Burns, J.Appl.Phys. **76**, 1363 (1994).

CHARACTERIZATION OF THE SUBSTRATE/FILM INTERFACE IN GaN FILMS BY IMAGE DEPTH PROFILING SECONDARY ION MASS SPECTROMETRY (SIMS)

Salman Mitha, Robert Clark-Phelps, Jon W. Erickson and Y. Gao
Charles Evans and Associates, Redwood City, CA 94063, USA
Wook Kim and Hadis Morkoç
University of Illinois at Urbana-Champaign, Coordinated Science Laboratory, 1101 West
Springfield Avenue, Urbana, IL 61801

Epitaxial GaN films are normally grown on substrates, such as sapphire, that are not an exact lattice match for GaN. Thus during the early stages of film growth, defects may be introduced in the film. These defects can lead to islanding and form voids or other defects just above the interface. These defects produce nonuniformity in the films and affect the quality of the final film. SIMS depth profiling is a widely used to characterize GaN films. The normal SIMS depth profiles provide chemical and depth information but do not provide any lateral information. We show that image depth profiling with SIMS is a technique that can be used to identify the defects and also chemically identify other interface features with lateral dimensions down to 1 µm.

Introduction.

One major difference between GaN and all other widely used commercial semiconductors is that no bulk material exists to manufacture substrates. Therefore all of the material has to be grown hetero-epitaxially on foreign substrates. The choice of substrate material is based on several criteria such good lattice match, availability of single crystal material, and inertness at the high temperatures, about 1000 °C[1], that are needed to grow GaN. For these reasons and economic considerations sapphire and SiC have been the most important substrates for GaN growth.

The lattice mismatch between GaN and sapphire is about 14%. Between SiC and GaN the mismatch is 3.5% [2]. In either case the lattice mismatch has a substantial effect on the crystal quality. Most commercially produced GaN is of very poor crystal quality. Typical dislocation densities are on the order of 10^{10} per cm^2 [3]. At such high dislocation densities no other semiconductor material would maintain any of its useful properties. GaN is unique and remarkable in that even such poor quality crystal can be used to build high powered optoelectronic devices. GaN is expected to completely dominate the market for blue/green and white light emitting diodes and lasers.

As the commercial market for GaN expands there are efforts to control and improve the crystal quality of the material. Bulk GaN substrates are not commercially available and not expected to be commercially available in the foreseeable future. Therefore hetro-epitaxy will continue to be used to manufacture GaN and the GaN/substrate interface will continue to play a very important role in the final material quality. Good post-mortem techniques would greatly enhance the chances of understanding the microstructure during the early stages of growth. This is especially important for the commercial R&D environment where it may be difficult to rapidly develop and construct in-situ monitoring tools.

During the initial stages the film can experience several types of problems. The first types of problems arise from the lattice mismatch. The hetro-epitaxi and the resulting stacking faults seriously degrade crystal quality. These stacking faults, other types of dislocations and other atomic scale interface irregularities are best studied with Transmission Electron Microscopy

Mat. Res. Soc. Symp. Proc. Vol. 468 ©1997 Materials Research Society

(TEM). However there are other potential problems with film growth. During the initial stages of growth the film may not deposit uniformly resulting in defects that are at a larger scale than the atomic scale defects. These defects would then result in highly localized regions of differing quality material. Therefore the tools used to identify and charecterize these defects would need to measure areas on the order of several hundred microns.

SIMS is a very important analysis tool in semiconductor industry. It has also has been used extensively to characterize GaN. However the analysis of GaN presents a unique challenge during SIMS analysis. Since GaN is the only commercial semiconductor where extremely poor crystal is economically useful, artifacts from the defects in the material can affect the data. In this paper we show how imaging SIMS analysis can be used to study the GaN film quality and identify large microstructural defects that originate at the interface.

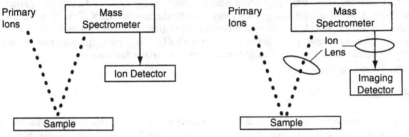

Figure 1.   Schematic of a SIMS instrument   Figure 2.   Schematic of an imaging SIMS instrument

Experimental

In the SIMS technique, a beam of energetic primary ions sputters material from the surface and near surface region of the sample. The sputtered material comes off as a mixture of neutral and charged molecular and atomic species. These secondary ions are electrostatically collected and directed into a mass spectrometer. The mass spectrometer selects the desired ion species and sends them through to an ion detector. A schematic of a SIMS instrument is shown in Fig. 1. A depth profile is acquired by setting the mass spectrometer to the desired ion species and counting those ions as the primary beam sputters into the sample, digging a crater. If the crater maintains a flat bottom, the instantaneous ion count rate is related to the concentration at the crater depth at that instant. The raw data that comes out as ion counts versus time can be converted into concentration versus depth profiles.

In the typical way of acquiring a depth profile, all lateral information is lost. However it is possible to maintain lateral information by allowing secondary ions to pass through ion imaging optics as well as a mass filter. The ions are then imaged on a two dimensionally position sensitive detector. As the sample is sputtered the detector then creates an elemental map layer by layer through the sample. Therefore it becomes possible to construct a 3-D map of a particular element in the sample. These 3-D maps then can be used to identify and characterize the defects in the film and the interface. It should be noted that normal depth profiling is usually much less time consuming and provides better detection limits than image depth profiles.

The SIMS data for this experiment was acquired on a CAMECA ims-4f double focussing ion microscope that was equipped with a Resistive Anode Encoder (RAE) position sensitive detector [4]. A 14.5 keV Cs+ primary beam was used for sputtering and negative atomic secondary ions from the sample were collected. Similar systems have been used for 3-D mapping of oxygen, carbon and fluorine defects in silicon[5]. The GaN materal used in this experiment was grown on sappire using Reactive Molecular Beam Epitaxy (RMBE) [6].

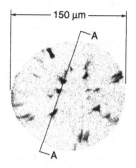

Figure 3a.
Oxygen image at depth of 0.5 µm

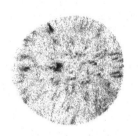

Figure 3b.
Oxygen image at depth of 1.5 µm

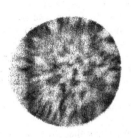

Figure 3c.
Oxygen image at depth of 2.0 µm

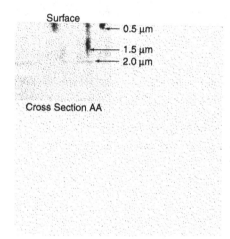

Figure 4a. Oxygen cross section

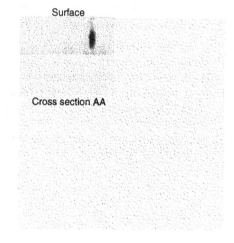

Figure 4b. Aluminum cross section

Results and Discussion

The data for this paper were aquired by profiling a 2-µm thick GaN film on sapphire substrate using imaging SIMS to acquired oxygen and aluminum image depth profiles. The data are presented as images from selected depths and as cross-section of the film through a defect. Figs. 3a–3c are oxygen images at 0.5 µm, 1.5 µm and 2.0 µm depths respectively. Figs. 4a and 4b show O and Al cross sections through the line marked in Fig 3a.

In Fig. 3a most of localized oxygen features are particles on or near the surface of the semiconductor. Note that all but one has disappeared in Fig. 3b. This localized oxygen defect persists until Fig. 3c, down to the interface. Figs. 4a and 4b show a cross section of the film through this defect and two other surface defects. Fig. 4a shows that the surface features end very abruptly close to the surface. Therefore these detects are particles probably resulting from contamination from the surface. Note that there are no counterparts to these shallow oxygen features in the aluminum cross section of the film. The deep oxygen feature has an aluminum counterpart at the same location. Therefore this feature represents a defect in the film. The feature starts at the interface and therefore may represent a pinhole in the film or diffusion pipe for the aluminum and oxygen from the substrate.

Figs. 3b and 3c also show the lateral nonuniformity oxygen contamination in the film. The pockets of oxygen probably reflect the original microstructure of the film as it formed. They may also represent particulate contamination during growth.

Concluding remarks.

We show that SIMS can be used to determine large scale defects in hetro-epitaxially grown GaN films. We also show that latereral non-uniformity of the contamination can be measured with SIMS.

References

1. S. N. Mohammad, W. Kim, A. Salvador and H. Morkoc, Materials Bulletin, **22** (1997), p22

2. S. Nakamura, Materials Bulletin, **22** (1997), p29

3. F. A. Ponce, Materials Bulletin, **22** (1997), p51

4. R. W. Odom, D. H. Wayne and C. A. Evans, Jr. in Secondary Ion Mass Spectrometry SIMS IV, edited by. A. Benninghoven, et. al., Springer-Verlag, New York 1984, pp. 186-188

5. R. S. Hockett, P. B. Fraundorf, D. A. Reed, D. H. Wayne, and G. K. Fraundorf, in Oxygen, Carbon, Hydrogen and Nitrogen in Crystalline Silicon, edited by J. C. Mikkelsen, Jr., S. J. Pearton, J. W. Corbett, and S. J. Pennycook. (Mater. Res. Soc. Proc. 59, 1985), p. 433

6. W. Kim, O. Aktas, A. E. Botchkarev, A. Salvador, S. N. Mohammad and H. Morkoc, J. Applied Phys. **79** (1996) p. 7657

INITIAL STAGES OF MOCVD GROWTH OF GALLIUM NITRIDE USING A MULTI-STEP GROWTH APPROACH

J. T. Kobayashi, N. P. Kobayashi, and P. D. Dapkus
Compound Semiconductor Laboratory,
Departments of Materials Science and Electrical Engineering/Electrophysics,
University of Southern California, Los Angeles, CA 90089-0483

X. Zhang and D. H. Rich
Photonic Materials and Devices Laboratory,
Department of Materials Science and Engineering
University of Southern California, Los Angeles, CA 90089-0241

ABSTRACT

A multilayer buffer layer approach to GaN growth has been developed in which the thermal desorption and mass transport of low temperature buffer layer are minimized by deposition of successive layers at increased temperatures. High quality GaN with featureless surface morphology has been grown on (0001) sapphire substrate by metalorganic chemical vapor deposition using this multilayer buffer layer approach. The lateral growth and coalescence of truncated 3D islands (TTIs) nucleated on low temperature buffer layers at the initial stage of overlayer growth is affected by the thickness of the final buffer layer on which nucleation of TTIs takes place. The effect of the thickness of this buffer layer on the quality of GaN is studied by using scanning electron microscopy, van der Pauw geometry Hall measurements and cathodoluminescence and an optimum value of 400Å is obtained.

INTRODUCTION

The group III nitrides are attractive materials for application to optoelectronic and electronic devices. [1],[2]. A lattice matched substrate for these materials is still not commercially available and they are generally grown on lattice mismatched substrates like sapphire or SiC by a two step growth method. In this approach a buffer layer is grown at low temperature (~450°C) and the second layer is grown at high temperature (950°C - 1050°C) [3],[4]. Because of the lattice mismatch between the sapphire (0001) substrate and GaN, the growth mechanism has been found to proceed in three distinct evolution stages : (1) formation of three dimensional truncated islands (TTIs) on the low temperature buffer layer, (2) coalescence of the TTIs by lateral growth, (3) two dimensional growth after complete coalescence of the TTIs. [5],[6]. To enhance the formation of TTIs on the whole buffer layer and to increase the coverage of them, thermal desorption and / or mass transport of the buffer layer must be suppressed. Recently, we have found that using a multilayer buffer layer approach consisting of layers of GaN grown at different temperature suppresses the thermal desorption and mass transport of the buffer layer and enhances the lateral growth of the TTIs. In this paper, we report the lateral growth behavior of TTIs formed during the initial growth on the low temperature buffer layer and the surface roughness and quality of GaN overlayers grown by this multi-step buffer layer approach.

Mat. Res. Soc. Symp. Proc. Vol. 468 © 1997 Materials Research Society

EXPERIMENTAL

GaN epitaxial films were grown by atmospheric pressure MOCVD using a closed space showerhead reactor. This reactor design utilizes a water-cooled multi-inlet gas distribution showerhead in which the group III and V sources are separately inlet directly (~1cm) above the substrate. Trimethylgallium (TMGa) and NH_3 were used with H_2 carrier gas. C-plane sapphire was used as the substrate. After degreasing the substrate and annealing it in the growth chamber for 30 minutes at 1080°C under H_2 flow, a thin (~100Å) buffer layer of GaN was grown at 450°C and the temperature was raised to 800°C to grow a second buffer layer. The second buffer layer thickness was varied from 0Å to 1600Å to observe the effect of the multi-buffer layer approach and the influence of the buffer layer thickness on the quality of the final overlayer. The overlayer was grown at 950°C with a V/III ratio of 933 and the growth time was varied from 10 minutes to 1 hour to observe the effect of the second buffer layer thickness on the macroscopic evolution of the overlayer. About 2.5μm of GaN was grown at 1000°C on the TTIs formed on the multi-step buffer layers to increase the relative lateral growth rate. This results in filling the pits on the surface that arise when the TTIs begin to coalesce and eventually in 2D growth. [7] To obtain a certain growth rate at 1000°C, the V/III ratio was decreased to 373 when the growth temperature was increased to 1000°C. Growth rates for overlayers were 4.4Å/sec at 950°C and 2.2Å/sec at 1000°C. The evolution stages and surface morphology were examined by SEM, the optical and electrical quality of the overlayer were examined by CL and van der Pauw method Hall effect, respectively. CL measurements were performed at 87K with the acceleration voltage of 15KeV.

RESULTS and DISCUSSION

Figs. 1 (a) - (f) show the cross-sectional and planview SEM pictures of samples grown at 950°C for 30 minutes on buffer layers with three different thicknesses of an 800°C buffer layer. In the planviews one observes that for all buffer layers, the dominant surface features are pyramidal shaped islands with flat tops - TTIs. Fig 1(a) shows the cross-sectional image and (b) shows the planview image for a sample without an 800°C buffer layer. The height of the TTIs is not uniform and the TTIs do not cover the whole area of the buffer layer. The black areas in the planview image are areas which are not covered by TTIs because of thermal desorption and/or mass transport of the low temperature buffer layers. This desorption of buffer layers and nonuniformity of the vertical size of the TTIs cause surface roughness and voids in the grown film even after growing a 4.0μm thick GaN film. Fig. 1(c) shows the cross-sectional image and (d) the planview

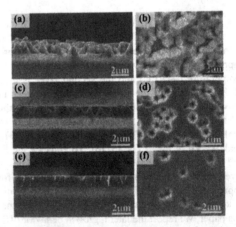

Fig. 1 Cross-sectional and planview SEM images of samples grown 10 minutes at 950°C on buffer layers. The thickness of the second buffer layer grown at 800°C is (a),(b) 0Å, (c),(d) 400Å, (e), (f) 1600Å

image for a sample with a 400Å thick 800°C buffer layer. The height of the TTIs are uniform, cover the whole area, and begin to merge with each other. This sample shows a smooth surface morphology after growing a 4.0μm thick GaN overlayer as shown in Fig. 2.(a). Fig. 1 (e) shows the cross-sectional and (f) the planview image for a sample with a 1600Å thick 800°C buffer layer. The TTIs have merged and a nearly continuous film covers the whole surface. However, after growing a 4.0μm thick GaN overlayer, this sample exhibits a rough surface morphology as shown in Fig. 2.(b). The height of the TTIs for each sample is the same except for the sample grown without an 800°C buffer layer.

Comparison of these figure shows that the lateral growth rate of TTIs depends on the 800°C buffer layer thickness. As the thickness of 800°C layer increases, lateral growth of the TTIs and merging of the TTIs is enhanced, but the vertical growth rate does not change. However, beyond 400Å, as the thickness increases, surface roughness of the final film also increases.

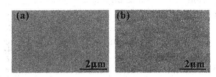

Fig. 2 Planview SEM image of surface of 4μm thick film grown on multi-buffer layer. The thickness of the second buffer layer grown at 800°C is (a) 400Å, (b) 1600Å

Fig. 3 shows the spatially integrated CL spectra of sample grown 4μm at 950°C/1000°C on the buffer layers, 800°C layer thickness of which is varied from 0Å to 1600Å. (a), (b), (c), (d) show the spectrum of sample, 800°C layer thickness of which is 0Å, 400Å, 800Å and 1600Å respectively. Only spectra for samples with 400Å and 800Å thick 800°C buffer layers show near band edge emission at 357nm, and spectrum for sample with the 400Å thick 800°C buffer layer shows higher intensity for the near band edge emission. Decreasing the V/III ratio to obtain higher growth rate during growth of the thick overlayer increases the intensity of the yellowband. By optimizing the growth conditions for the overlayer, the intensity ratio of band edge emission to yellow band emission can be increased to 2.3 in the CL spectrum for the sample with the 400Å thick 800°C buffer under the same conditions for measurement.

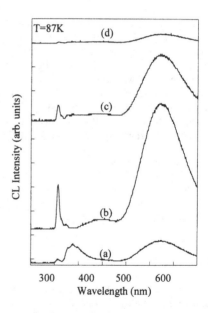

Fig. 3 CL spectra of 4μm thick GaN films grown with four different second buffer layer thickness. The thickness of the second buffer layer grown at 800°C is (a) 0Å, (b) 400Å, (c) 800Å, (f) 1600Å. The spectra were taken at 87K with acceleration voltage of 15KeV

Fig. 4 shows the electron Hall mobility as a function of the thickness of the 800°C buffer layer. The mobility increases as the thickness of the 800°C layer increases to 400Å, reaches a maximum and decreases as the thickness increases from 400Å to 1600Å. The absolute values of the mobility for these samples are low because we selected relatively high growth rate conditions for growth of the thick film. However by growing at 950°C with a 400Å thick 800°C buffer layer, we obtained an electron mobilities as high as 531 cm^2/V sec.

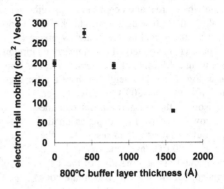

Fig.4. Electron Hall mobility as a function of 800°C buffer layer thickness

From the results of SEM, CL and Hall measurements, it is clear that 400Å is the optimum thickness for the 800°C buffer layer to achieve smooth surface morphology and good crystallinity. If no 800°C buffer layer is used, the desorption of the 450°C buffer layer affects the initial nucleation of TTIs and prevents their coalescence. As a result, defects or grain boundaries are generated between the TTIs and these defects or boundaries lower the mobility and decrease the CL band edge emission. On the other hand, as the thickness of 800°C layer increases, the lateral growth rate of the TTIs increases. Because of this increased lateral growth rate misoriented TTIs become entrapped in the film after coalescence. As a result, the crystallinity of the films after TTI coalescence is reduced which affects the carrier mobility and the intensity of band edge emission. Further study is underway to verify this hypothesis. In summary, we show that a multi-step buffer layer is effective to avoid thermal desorption and/or mass transport of lower temperature buffer layers, to enhance lateral growth of TTIs and to generate TTIs which have uniform height and uniformly cover the buffer layers. The lateral growth rate of TTIs can be controlled by the thickness of the buffer layer grown at higher temperature. There is an optimum lateral growth rate to achieve the maximum mobility , smooth surface morphology and good crystallinity.

ACKNOWLEDGMENTS

The authors gratefully acknowledge the support of the Office of Naval Research and DARPA through the National Center for Integrated Photonic Technology (NCIPT) and the GaN University Consortium at UCSB.

REFERENCES

[1] S. Nakamura, M. Senoh, S. Nagahama, N.Iwasa, T. Yamada, T. Matsushita, H. Kiyoku, and Y. Sugimoto, Appl. Phys. Lett. 68, 3269 (1996)
[2] H. Morkoc, S. Strite, G. B. Gao, M. E. Lin, B. Sverdlov and M. Burns, J. Appl.Phys. 76, 1363 (1994)
[3] H. Amano, N. Sawaki, I. Akasaki, and Y. Toyoda, Appl. Phys. Lett. 48, 353 (1986)
[4] S. Nakamura, Jpn. J Appl. Phys. 30, L1705 (1991)
[5] H. Amano, I. Akasaki, K. Hiramatsu, N. Koide, and N. Sawaki, Thin Solid Films 163, 415 (1988)

[6] K. Hiramatsu, S. Itoh, H. Amano, I. akasaki, N Kuwano, T. Shiraishi, and K. Oki, J. Cryst. Growth 115, 633 (1991)

[7] J. T. Kobayashi, N. P. Kobayashi, X. Zhang, D. H. Rich, P. D. Dapkus (will be presented in "eighth biennial workshop on Organometallic Vapor Phase Epitaxy, and to be published in the Journal of Electronic Materials)

PLASMA CLEANING AND NITRIDATION OF SAPPHIRE SUBSTRATES FOR $Al_xGa_{1-x}N$ EPITAXY AS STUDIED BY ARXPS AND XPD

M. SEELMANN-EGGEBERT, H. ZIMMERMANN, H. OBLOH,
Fraunhofer Institut Für Angewandte Festkörperphysik, Tullastr 72, D-79108 Freiburg
R. NIEBUHR , B. WACHTENDORF,
Aixtron Gmbh, Kackertstr. 15-17, D-52072 Aachen, Germany

ABSTRACT

The influence of plasma and thermal treatments on the structure and composition of sapphire (00·1) surfaces have been studied by hemispherically recorded x-ray photoelectron spectroscopy in view of substrate preparation for the epitaxy of GaN. Producing well-ordered surfaces, O_2 plasma based treatments are found to efficiently remove surface contamination. AlN films with good short-range order are obtained by a simple high temperature nitridation step in the MOCVD reactor.

INTRODUCTION

Presently, much effort is directed towards the technological developments regarding the direct wide band gap material system $Al_xGa_{1-x}N$ and $In_xGa_{1-x}N$[1,2]. Despite the relatively large lattice mismatch of 16% to GaN, sapphire (α-Al_2O_3) of (00·1) orientation is commonly used as substrate for MOCVD epitaxy of $Al_xGa_{1-x}N$.

High quality growth of GaN requires the formation of a buffer layer which is typically grown at low temperature and subsequently annealed at high temperature[3]. Surface conditioning of the sapphire substrate prior to buffer layer growth is found to have an essential influence on the crystalline quality of the final GaN epilayer grown on top of the buffer[4]. Surface conditioning is typically performed in an initial high temperature step and may involve a simple initial heat treatment at 1100°C to desorb surface contaminants[3] or a short exposure to a N_2 plasma[5] or an NH_3[4,6,7] environment. The treatment with nitrogen has been reported to result in a change of the surface structure tentatively ascribed to the transformation of a surface layer into AlN[5]. A nitridation by conversion of the substrate potentially bears the advantage to form a smooth initial layer[7] of improved lattice match by a quasi-epitaxial chemical reaction. Starting from such a surface epilayer growth is facilitated to proceed in a two-dimensional mode, since film nucleation can start from lattice matched crystal regions of well defined orientation.

Since substrate preparation is known to have an important influence on the quality of the resulting epitaxial layers, we performed a systematic investigation of sapphire surfaces. In view of MOCVD of GaN epilayers the objective of this investigation was to set up processing recipes regarding 1.) the preparation of a clean and well ordered substrate surface and 2.) the "in situ" nitridation of Al_2O_3 surfaces.

EXPERIMENTAL

The single crystalline (00.1) oriented sapphire wafers were, as received, polished to an epitaxial grade finish. The samples could be cleaned in a plasma reactor chamber equipped with an ASTEX compact ECR source and transferred under UHV condition ($< 1 \times 10^{-7}$ Pa) to an attached VG ESCALAB MK II surface analysis facility equipped with angle-resolved x-ray photoelectron spectroscopy (ARXPS) capabilities. The measured ARXPS raw data were evaluated in view of x-ray photoelectron diffraction (XPD) effects ("forward-focusing" enhancement and higher order interference) as well as in view of the surface and subsurface composition of the examined samples. The depth profile analysis, in principle, was based on the common concept of a linear relation between the escape depth and the cosine of the polar escape angle of the recorded

Mat. Res. Soc. Symp. Proc. Vol. 468 ©1997 Materials Research Society

photoelectrons. Details on the more involved evaluation techniques used for the quantitative assessment of the depth profiles have been published elsewhere[8]. XPD was used to probe short-range order in a subsurface volume extending from the surface to a depth of a few 10 Å. For the detailed structural analysis a novel direct crystallographic method was employed called CHRISDA[8] (Combined Holographic Real space Imaging by Superimposed Dimer functions Algorithm).

RESULTS AND DISCUSSION

Surface cleaning of sapphire

Polished sapphire surfaces were typically found to be covered with a contamination layer which had a thickness of 7-10 Å and consisted predominantly of carbonaceous species. Frequently, in addition impurities of Ca and F were present on polished sapphire surfaces, whereas oxygen was generally absent in these contamination layers.

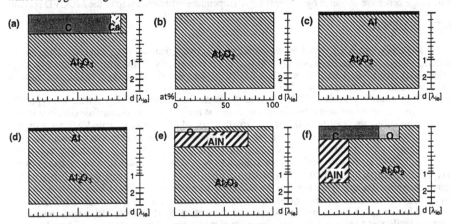

Fig.1 : Depth profiles as obtained on the basis of a simple one-step or two-step model from the polar angle dependence of the O1s-, Al2p-, N1s-, C1s- und Ca2p-photoelectron intensities. To remove the XPD modulation the hemispherically recorded XPS data were processed by an angle-averaging algorithm. (a) typical sample as received, (b) after 5 min exposure to an O_2 plasma, (c) after 5 min exposure to an H_2 plasma, (d) after 30 min at 900°C, (e) after 10 min exposure to an N_2/H_2 plasma, (f) after 3 min exposure to NH_3 at 1100°C

The effects of various surface pretreatments on the surface composition of sapphire are summarized in Fig. 1 which shows compositional depth profiles obtained by analysis of the polar angle dependent ARXPS signal contributions[8]. On the given exponential scale compositional ARXPS profiles are quantitative, though (except for special cases) the respective depth resolution is relatively poor[9]. In Fig.1, the concentration profiles of Al (,N) and O have been expressed as Al_2O_3 (and AlN) mole fractions. The depth unit in the profiles is the escape length λ_{ie} ($\approx$15-25Å).

Fig. 1 a shows a typical depth profile for a polished sapphire (00·1) surface. Within the experimental error the contamination forms a closed layer on the Al_2O_3 substrate. The presence of surface roughness can not be unambiguously detected by ARXPS, since in the framework of our simple model the prevailing effect of surface roughness is to change the result in the ARXPS profile in a sense that two actually distinct compositional zones separated by an abrupt interface will erroneously show considerable intermixing. Hence, the profile of Fig. 1a is a demonstration of the remarkable smoothness reported[10,11] for sapphire surfaces (00·1). Evidence that polishing does only little damage to the near surface region of sapphire (00·1) was found by the observation of XPD patterns with high anisotropy contrast which indicated a good crystalline order in the uppermost 20 Å of the substrate.

The efficiency of cleaning and the detrimental effects on the crystalline surface structure upon plasma exposure of the sapphire samples were investigated for an O_2 plasma as well as for an H_2/Ar plasma. Both plasma treatments reduced the hydrocarbon and flourine contamination below the detection limit (≈ 0.1 ML). Calcium could not be removed by either one of the two plasma treatments.

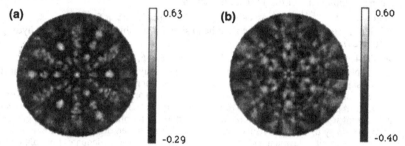

Fig. 3 : Hemispherical diffraction diagrams of (a) the Al 2p and (b) the O1s photoelectrons obtained for a clean sapphire substrate with MgKα excitation. To reflect the anisotropy contrast the patterns are normalized with respect to the averaged polar angle dependent signal.

Differences between the oxygen and the hydrogen plasma based cleaning procedures affecting the resulting surface were observed in two respects, namely, the crystalline order and the surface termination. After O_2 exposure (Fig. 1b) the composition was found to be uniform over the entire probed depth region and to correspond to the nominal composition. The XPD pattern of the Al2p and the O1s photoelectrons (Fig.2) showed a good anisotropy contrast indicating that the near-range order in the surface near region was not impaired by the plasma exposure. In contrast, with an H_2/Ar plasma the XPD anisotropy contrast was reduced by 40 % upon plasma exposure and Ar2p photoelctrons could be detected. Hence, the observed crystal damage has to be attributed to the implantation of Ar in the sapphire matrix implying that ions accelerated by the plasma potential of the ECR source (≤ 20 eV) carry sufficient energy to displace aluminum or oxygen atoms from their regular lattice sites. Upon sputtering with 1 keV Ar ions a complete loss of short-range order was indicated by XPD even for low dose exposure ($\leq 10^{15}$ cm^{-2}) . An amorphous surface region[12] extending over more than 50 Å forms under these conditions and surface cleaning was found to be very unefficient owing to considerable intermixing effects in this region.

H_2/Ar and O_2 plasma treatments result in a different termination of the sapphire surfaces. By interaction with hydrogen, the sapphire surfaces depleted slightly of oxygen and became terminated by Al atoms (Fig. 1c). Very similar depth profiles were found when oxygen cleaned surfaces (Fig. 1b shows an equal O and Al occupation of surface sites) were annealed at 900°C for a period of 30 min (Fig.1d). A surface depletion of oxygen upon heat treatments of sapphire has also been reported by Gautier et.al.[12]. At 1400°C further oxygen depletion leads to the formation of an ordered double layer of metallic aluminum[10,12]. The (non-conducting) Al terminated surface of sapphire (00·1) has been also predicted by theory to be the stable one[13]. The amount of Al found for the depth profile of Fig. 1c and d in the surface layer corresponds only two half a monolayer and possibly indicates the presence of surface vacancies. Harrison[14] has pointed out that polar sur faces are likely to be stabilized by surface vacancies to avoid dielectric breakdown by an internal field. To eliminate the internal field, the surface terminating Al layer terminating the (00.1) surface of sapphire has to be filled half by surface vacancies, in agreement with the profiles Fig. 1c and d.

Nitridation of sapphire surfaces

On a first view, nitridation of the sapphire surface to obtain AlN nuclei appears to be impossible. On the contrary, AlN decomposes in contact with oxygen according to

(1) $\quad Al_2O_3 + N_2 \quad\quad \Rightarrow 2\ AlN + 3/2\ O_2$; $\quad \Delta G_{298} = +10.4$ eV per formula unit

The energetic situation is improved in favor of AlN formation if atomic rather than molecular nitrogen can be employed. For the nitridation reaction

(2) $\quad Al_2O_3 + 2\ N \quad\quad \Rightarrow 2\ AlN + 1/2\ O_2$; $\quad \Delta G_{298} = +1.0$ eV p.f.u.

the equilibrium is strongly shifted to the right hand side if a suitable reactant, such as hydrogen, is provided for the freed oxygen

(3) $\quad Al_2O_3 + 2\ N + 3\ H_2 \Rightarrow 2\ AlN + 3\ H_2O$; $\quad \Delta G_{298} = -6.1$ eV p.f.u.

Hence, an H_2/N_2 plasma which provides atomic nitrogen may serve as a suitable ambient for the nitridation of sapphire at room temperature. In our experiments, AlN layers of about 6 to 12 Å thickness were formed on sapphire samples upon exposure to an N_2/H_2 plasma. These layers were relatively uniform but extremely prone to reoxidation (Fig. 1e). By the absence of XPD anisotropy for the N1s signal it was evident that nitride layers obtained by plasma exposure are amorphous. The poor crystallinity of the AlN films and their sensitivity to reoxidation upon exposure to air calls in question if ex situ plasma nitridation is a substrate preparation adequate for GaN epitaxy.

A simple possibility for surface nitridation inside the MOCVD reactor is offered by the exposure of the sapphire surface to NH_3 in a high temperature step preceding layer growth. At room temperature no AlN can be formed by the reaction

(4) $\quad Al_2O_3 + 2\ NH_3 \quad\quad \Rightarrow 2\ AlN + 3\ H_2O$; $\quad \Delta G_{298} = 3.68$ eV p.f.u.

At room temperature a formation of AlN via this reaction is impossible. However, owing to an increase in entropy , at high temperatures the free energy ΔG of formation is reduced to about 2.0 eV p.f.u. at 1100°C (kT≈120 meV). If we assume for the MOCVD reactor under NH_3 flow that the water pressure does not rise above 10^{-6} bar at the surface of the sapphire substrate then AlN formation is favored by reaction (5) downto a NH_3 partial pressure as low as 10^{-5} bar.

The result of the ARXPS analysis demonstrates that nitridation of sapphire by exposure to NH_3 is feasible at high temperature. The sapphire sample analysed in Fig. 2f had been exposed in the MOCVD reactor to NH_3 for 3 min (T=1100°C, partial pressure>40 mbar, carrier gas N_2) and subsequently transferred into the UHV system within 10 minutes. An AlN layer of thickness 20-30 Å was found to be formed upon this treatment (at least on a large part of the surface). The respective ARXPS depth profile analysis (Fig. 2f) shows an AlN layer which is not closed and is consistent with the presence of AlN islands which cover about one third of the sample.The formation of AlN islands as a result of a high temperature NH_3 exposure has also been reported in the literature[7].

The AlN layer formed upon high temperature nitridation has a well established (short-range) order as the XPD pattern of the N1s electrons showed pronounced diffraction features (Fig.3). For the O1s and the Al2p photoemission the similarity between the diffraction patterns of Fig. 3 and Fig.2 implies that these patterns arise predominantly from the sapphire substrate. A comparison of the N1s pattern with the O1s and Al2p pattern shows that the nitride layer is in registry with the oxygen sublattice. To identify the crystal structure and polarity of the AlN layer, the N1s pattern of Fig. 3 c was subjected to a CHRISDA analysis. Fig.4 a shows the occupation probability profiles (OPP) obtained for the three possible emitter positions (see insert) within the unit mesh of the two-dimensional lattice. With these OPPs the experimental data of Fig. 3c are well

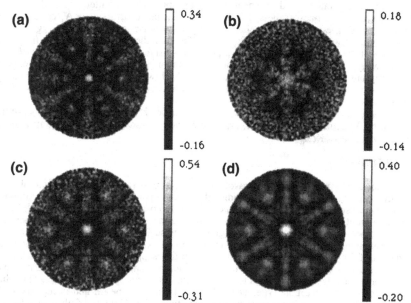

Fig.3: Anisotropy contrast patterns of the (a) the Al 2p and (b) the O1s and (c) the N1s photoelectrons recorded after nitridation of a clean sapphire surface. (d) is the pattern reproduction obtained upon CHRISDA analysis of (c).

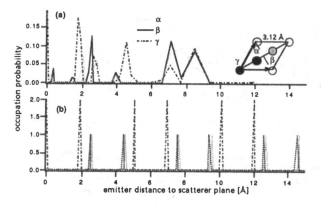

Fig.4: Occupation probability profiles obtained by CHRISDA analysis of the N1s pattern of Fig. 3 c. (a) analysis based on the shown emitter configurations α, β, γ. (b) Nominal joint occupation density of an hexagonal AlN (00·1) surface for emission from nitrogen atoms. The surface is assumed to be terminated by nitrogen atoms.

reproduced (Fig. 3d). The OPPs of Fig. 4a are compared with the (joint) occupation density profiles of a hexagonal AlN crystal with (00.1) orientation and nitrogen termination (Fig. 4b). Since hexagonal AlN contains two nitrogen atoms per primitive unit cell, in Fig. 4b the peaks at 2.5 Å and 4.4 Å are smaller and have no unique configuration assignment.

The coincidence of the oxygen sublattice and the AlN layer lattice is confirmed in Fig 4a by the peak for configuration γ at d=0. The nearest plane above the N emitters is found for scatterers in top sites at a distance of about 1.9 Å. Up to a distance of about 6 Å (beyond which a determination of layers becomes unreliable owing to the noise in the pattern of Fig. 3c) there is good agreement between the OPPs of Fig. 4a and Fig. 4b, indicating that the AlN is fully relaxed. In particular, the absence of a significant response at a distance of 0.6 Å confirms that the AlN is (predominantly) of nitrogen termination. For d≤5 Å similar layer images would also be expected if twinned regions of cubic AlN (111) were analyzed. However, with the OPPs of Fig. 4a the presence of the hexagonal phase becomes unambiguous by a response at about 5 Å for configuration γ.

Hence, the epitaxial relationship is confimed to be $[21.0]_{AlN} \parallel [11.0]_{Al2O3}$ as previously suggested by Yamamoto et.al.[6]. However, the observed termination of the AlN layer disagrees with the prediction of Kung[15].

CONCLUSIONS

Using ARXPS as a technique providing depth compositional as well as structural information of the surface near region we have studied the effect of plasma treatments on the composition, termination and structure of sapphire surfaces with (00·1) orientation. O_2 as well as H_2/Ar based plasma treatments are shown to adequately remove surface contamination. Short-range order is preserved in both cases, however, a higher grade of the surface crystallinity is obtained with the oxygen treatment.Thermal or H_2 plasma treatments of sapphire are found to result in an Al termination of the sapphire surface.We have shown that sapphire substrates can be nitrided by either a plasma process or by NH_3 exposure at high temperatures in the MOCVD reactor, however, only the latter process leads to the formation of a crystalline AlN phase. This phase is identified to be of the wurtzite type. The AlN layer is found to be formed in registry with the oxygen sublattice and to be terminated by nitrogen atoms. The AlN has a fully relaxed structure and tends to conglomerate in extended crystallites rather than forming a closed thin film.

REFERENCES

1. S. Nakamura, M. Senoh, S.Nagahama, N. Iwasa, T. Ymada, T. Matsushita, H. Kiyoku, Y. Sugimoto, Appl. Phys. Lett. **68**, 2105 (1996)
2. B.W.Lim,Q.C.Chen,J.Y.Yang, M.A.Khan, Appl. Phys. Lett. **68**, 3761 (1996)
3. C.F. Lin, G.C. Chi, M.S. Feng, J.D.Guo, J.S.Tsang, J.Minghuang Hong, Appl. Phys. Lett. **68**, 3758 (1996)
4. S. Keller, B.P. Kemmer, Y.-F. Wu, B. Heyring, D. Kapolneck, J. S. Speck, U.K. Mishra, S.P. Denbaars, Appl. Phys. Lett. **68**, 1525 (1996)
5. R.J. Molnar, T.D. Moustakas, J. Appl. Phys. **76**, 4587 (1994)
6. A. Yamamoto, M. Tsujino, M. Ohkubo, A. Hashimoto, J. Cryst. Growth **137**, 415 (1994)
7. K. Uchida, A. Watanabe, F. Yano, M. Kouguchi, T. Tanaka, S. Minagawa, J. Appl. Phys. **79**, 3487 (1996)
8. M. Seelmann-Eggebert, G.P. Carey, R. Klauser, H.J. Richter, Surf. Sci. **287/288**, 495 (1993); M. Seelmann-Eggebert, Surf. Sci.(in press)
9. M. Seelmann-Eggebert, R.C.Keller, Surf. and Interf. Anal. **23**, 589 (1995)
10. G. Renaud, B. Vilette, I. Vilfan, A. Bourret, Phys. Rev. Letters **73**, 1825 (1994)
11. Yan Yu, R.J. Lad, Mat. Res. Soc. Symp. Proc. **317**, 583 (1994)
12. M. Gautier, J.P. Durand, L. Pham Van , M. J. Guittet, Surf. Sci. **250**, 71 (1991)
13. J. Guo, D.E. Ellis, D.J. Lam, Phys. Rev. B **45**, 13647 (1992)
14. W.A. Harrison, J. Vac. Sci. Technol. **16**, 1492 (1979)
15. P. Kung, C.J. Sun, A. Saxler, H. Ohsato, M. Razeghi, J. Appl. Phys. **75**, 4515 (1994)

Part III
Characterization

MECHANICAL PROPERTIES OF GALLIUM NITRIDE
AND RELATED MATERIALS

M. D. DRORY
Crystallume, 3506 Bassett Street, Santa Clara, CA 95054, mddrory@batnet.com

ABSTRACT

Basic mechanical properties, such as hardness and fracture toughness are examined for GaN and related materials. The mechanical properties are explored by indentation with a diamond (Vickers) pyramid under less than 20N load. Testing on bulk single crystal GaN indicated hardness and fracture toughness similar to GaP, but with much greater values than GaAs. Data for compound semiconductors are compared with values of a number of substrate materials. The following materials are examined in bulk form: AlN, Al_2O_3 (sapphire), Al_2O_3-TiC (ALTIC), GaAs, GaN, GaP, Ge, Si, SiC, ZnS, and ZnSe.

INTRODUCTION

Basic mechanical properties can determine the practical use of materials where fracture and deformation lead to failure of electronic devices, interconnects, and packaging. There are several sources of catastrophic failure including the thermal cycling of devices, delamination of metal interconnects, and fracture of dielectric layers. The mechanical loads on materials may be the result of processing to produce "intrinsic" or growth stresses, and during use to cause thermally-induced stresses. Other mechanical loads may arise from shock loading in mobile systems. The failures may be present as macroscopic features, such as cracks, or through defects generated as a result of permanent deformation (e.g. dislocations). Other mechanical failure modes can result from the application of electromagnetic fields[1], or environmentally-assisted slow crack growth. Permanent deformation and crack formation are explored here by measuring the hardness and fracture toughness of a number of bulk device and substrate materials (Table 1).

Permanent deformation of bulk materials is readily measured by indentation of a relatively smooth surface with a hard penetrator which is typically diamond. The permanent deformation profile depends on deformation resistance of the material along with the indenter shape and applied load. The material hardness, H, is a parameter describing the resistance to permanent deformation and follows from the penetration profile, however also depends on details of the test, such as loading cycle and indenter material and shape. The Vickers Hardness Test is commonly used for ceramic materials since the pyramidal diamond penetrator can produce easily measured impressions without failure of the indenter tip. A rectangular impression is visible from the in-plane view (Fig. 1) with the cross pattern matching the pyramid shape. Measurements are made with an optical microscope under

Mat. Res. Soc. Symp. Proc. Vol. 468 ° 1997 Materials Research Society

≤1000X magnification. Other hardness tests have been devised for different classes of materials, such as metals, with separate definitions of hardness [2]. In recent years, efforts have focused on measuring the hardness of thin films under very low loads through "nanoindentation" with a Berkovich diamond penetrator [3].

A further utility of using the Vickers Hardness Test, is that the resistance to fracture can be measured from the same impression. Under sufficient load, cracks emanate from the corners of the hardness impression. The fracture toughness, K_c, follows from the crack length, applied load, and materials properties as described in the next section[4, 5]. This method of measuring Kc has considerable uncertainty estimated as twenty percent[5], however more accurate methods require relatively large volumes of materials to produce samples with a precise pre-crack which is propagated under an applied (tensile) load[6]. Considerable costs are associated in testing with the latter methods in specimen fabrication and equipment. In contrast, numerous values of the hardness and fracture toughness are made by indenting a sample over a few mm^2 surface area(Fig. 1).

The mechanical properties data presented here should be used for relative comparison purposes since there are several complicating factors such as the indentation of single crystals and differences between bulk and thin film properties. Fracture toughness data obtained by Vickers indentation testing is most appropriate for homogenous and isotropic materials.

Table 1 – Summary of Material Properties

Material	Structure	Theoretical Density [g/cm³]	Young's Modulus E[GPa]	Vickers Hardness, H[GPa]	Fracture Toughness, K$_c$[MPa√m]
AlN*		3.25	300	9.2	2.2
Al$_2$O$_3$ Sapphire	Corundum	3.99	380	19	2.8
Al$_2$O$_3$-TiC*	(composite)	4.26	420	23	7.4
GaAs	Zincblende	5.32	120	5.8	0.52
GaN	Wurtzite	6.10	287	12	0.79
GaP	Zincblende	3.85	103	7.2	0.84
Ge	Diamond	5.33	100	7.2	0.63
Si [100]	Diamond	2.34	168	9.3	1.0
SiC*	Hexagonal	3.09	380	19	6.7
ZnS	Zincblende	4.08	75	1.7	1.5
ZnSe	Zincblende	5.27	67	1.1	0.73

*polycrystalline

EXPERIMENT

The hardness and fracture toughness of materials in Table 1 were measured by Vickers indentation using a Zwick Hardness Tester(Windsor, CT). The testing machine applies a load at relatively low rates to achieve a specified level within ~20sec. A minimum of six indentations are made at a given load and with four loads over a range determined by the impression size, i.e. material hardness (Fig. 2). An X-Y stage on the hardness tester allows for positioning the sample for successive indentations. After an indentation is made the sample is repositioned to a distance greater than 20X the crack length observed in an optical microscope attached to the testing machine. Adequate separation is needed to avoid interaction with the damaged material which may affect apparent values of the hardness and fracture toughness.

Materials were obtained from commercial sources as wafers or thin sheets polished to a specular surface finish, with the exception of the bulk single crystal GaN which was grown by Unipress(Polish Academy of Sciences) and tested in the as-grown condition. Details of the GaN crystal obtained for this purpose are described elsewhere[7]. A reaction-bonded SiC was obtained from Goodfellow Corp. (Berwyn, PA).

The hardness, H, is determined by measuring the diagonal lengths, 2a, of the impression in an optical microscope with reflected illumination(Fig. 1A), and is related to the applied load, P by

$$H = \frac{P}{2a^2}$$

$$(1)$$

The fracture toughness, K_c, is determined from the crack length, c, measured with transmitted illumination when feasible (Fig. 1B) [5]:

$$K_c = \xi \left(\frac{E}{H} \right)^{1/2} \frac{P}{c^{3/2}}$$

$$(2)$$

where E is the Young's modulus, and $\xi(=0.016\pm0.004)$ is a calibration constant. Literature values of the Young's modulus (Table 1) were used with eqn. (2).

RESULTS AND DISCUSSION

Hardness and fracture toughness data obtained by indentation testing are provided in Fig. 2 with average values listed in Table 1. A minimum of two indentation loads were used, with the exception of the indenting the GaN bulk single crystal where a hardness of 15.1, 12.0, and 9.9GPa was measured at 2N load. The indentation fracture toughness values for GaN were 0.76, 0.91 and

0.71MPa√m. Catastrophic failure of the bulk GaN single crystal occurred at a higher indentation load.

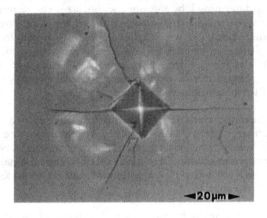

(A)

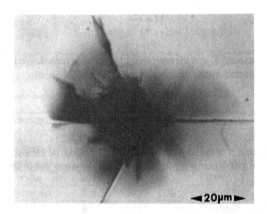

(B)

Fig. 1 Optical micrograph of indented GaP under
1.5N load with (A) reflected, and (B) transmitted
illumination.

(A)

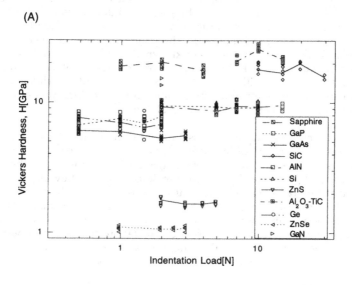

(B)

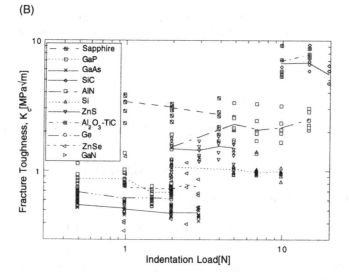

Fig. 2 Vickers indentation (A) hardness and (B) fracture toughness.

Indentation of the single crystals materials produced considerable crack deflection, particularly in anisotropic materials, such as sapphire. In other cases, non-planar cracks were caused by grain boundary deflection (SiC and AlN). The largest values of the hardness and toughness were found with the substrate materials (Al_2O_3-TiC, and SiC). The poorest mechanical behavior was observed with ZnSe and GaAs which are known for low toughness (GaAs) and poor erosion resistance (ZnSe). Nearly all electronic device materials have low hardness and toughness, with the notable exception of SiC. However, the values measured on bulk materials may differ from thin film properties as a result of differences in chemical and structural defects.

The hardness and fracture toughness measurements are consistent with published reports within experimental error of ~5% for the hardness measurements and ~20% for fracture toughness[2]. Causes for the large expected error in K_c are described in the previous section. Nevertheless, it is surprising that the fracture toughness values are consistent with more accurate (and expensive) measurements obtained through other testing methods with double torsion and double cantilever beam specimens. Further work is needed in each single crystal material to determine orientation effects of hardness and fracture toughness. The latter involves identifying the crystal planes associated with the direction of cracks emanating from indentations.

ACKNOWLEDGMENTS

Important discussions with J. W. Ager, A. J. Moll and F. Faili are acknowledged. The author also thanks E. Y. Luh of W. R. Grace & Co. for supplying the aluminum nitride sample.

REFERENCES

1.   H. Okabayashi, Mat. Res. Soc. Symp. Proc., Vol. 337, 503 (1994).

2.   H. W. Boyer, ed. Hardness Testing, ASM International, Metals Park, OH, 1987.

3.   M. F. Doerner and W. D. Nix, J. Mater. Res., Vol. 1, 601 (1986).

4.   A. G. Evans and E. A. Charles, J. Am. Ceram. Soc., Vol. 59, 371 (1976).

5.   G. R. Anstis, P. Chantikul, B. R. Lawn, and D. B. Marshall, J. Am. Ceram. Soc., Vol. 64, 533 (1981).

6.   M. D. Drory, R. H. Dauskardt, A. Kant, and R. O. Ritchie, J. Appl. Phys., Vol. 78, 3083 (1995).

7.   M. D. Drory, J. W. Ager III, T. Siski, I. Grzegory, and S. Porowski, Appl. Phys. Lett., Vol. 69, 4044 (1996).

MOSAIC STRUCTURE AND CATHODOLUMINESCENCE OF GaN EPILAYER GROWN BY LP-MOVPE

Shukun Duan*, Xuegong Teng*, Yenran Li+, Yutian Wang+, Peide Han*** and Dacheng Lu**

* National Integrated Optoelectronics Laboratory, Institute of Semiconductors, Chinese Academy of Sciences, P.O. Box 912, Beijing 100083, CHINA, E-mail: skduan@red.semi.ac.cn
+ Institute of Semiconductors, Chinese Academy of Sciences, Beijing CHINA 100083
** Laboratory of Semiconductor Materials of Sciences, Institute of Semiconductors, The Chinese Academy of Sciences, P.O. Box 912,Beijing 100083, CHINA, E-mail: dclu@red.semi.ac.cn
*** Beijing Laboratory of Electron Microscopy Center of Condensed Matter Physics, Chinese Academy of Sciences, P.O.Box 2724, Beijing 100080, CHINA, E-mail: pdhan@image.blun.ac.cn

ABSTRACT

We have studied the growth of GaN on (0001) sapphire and (111) spinel substrates by LP-MOVPE and compared the mosaic structure and cathodoluminescence for the heteroepitaxial films of GaN grown on these substrates.

INTRODUCTION

Epitaxial GaN films have recently attracted much interest due to their optoelectronic applications in the ultraviolet through green wavelength region and for their high temperature stability [1,2]. Since bulk GaN substrates are not currently available, the films are generally grown on basal plane sapphire (α-Al$_2$O$_3$) substrates. Sapphire is the substrate of choice of the easy of cleaning and its stability at the high temperature required for GaN growth. The high lattice mismatch results an extremely high density of structural defects in the GaN layer. The dislocation density in GaN films is generally on the order of 10^{10} cm^{-2} [3], which is much higher than conventional compound semiconductor materials. Additionally, X-ray analysis of these films reveals a substantial degree of "mosaic" structure wherein domains of single crystal GaN are slightly misoriented with respect to neighboring domains, with a small amount of lattice parameter variation. We have also studied (111) MgAl$_2$O$_4$ as another substrate for comparison. It has smaller lattice mismatch and thermal expansion coefficient mismatch with GaN than sapphire. It is easy to cleavage, which requires for fabricating the cavity mirror for a current-injection laser[4]. In this study, we compare the mosaic structure and cathodoluminescence for the heteroepitaxial films of GaN grown on α-Al$_2$O$_3$ and MgAl$_2$O$_4$ substrates with different morphologies.

EXPERIMENT

The GaN thin films used for this study were grown in a horizontal MOVPE reactor at 50 mbar[5]. Trimethylgallium (TMG) and ammonia (NH$_3$) were used as Ga and N precursors. The TMG was utilized by bubbling hydrogen through the liquid; then diluted by another flow of H$_2$ (3000 standard cm^3 min^{-1}). The typical TMG flow was 25 μmole/min. The input V/III ratio was 4500. The sources were mixed at the entrance of the reactor in order to suppress the parasitic reaction. The substrates used in the study were (0001) α-Al$_2$O$_3$ and (111) MgAl$_2$O$_4$. Undoped epilayers were deposited at 1050^{0}C with 20nm GaN buffer layer grown at low temperature. The GaN films discussed in this paper have average thickness of 3 μm.

The morphology was observed by Normarski interference contrast microscopy, scanning electron microscopy (SEM) and atomic force microscopy (AFM). The crystalline quality was measured by X-ray diffraction (XRD) and two or three-crystal X-ray diffraction (DXRD or TXRD). In TXRD, the X-ray diffractometer had a third analyzer crystal in front of the detector to limit the acceptance angle, which allowed separation of lattice parameter and misorientation contributions to diffraction peak broadening. Cathodoluminescence (CL) spectra and images were performed in a refitted electron probe microanalyzer.

RESULTS AND DISCUSSION

The samples grown on α-Al$_2$O$_3$ substrates have two kinds of morphologies: Sample A has mirror-like surface. Sample B has large hexagonal crystalline structure. Sample C is GaN grown on a MgAl$_2$O$_4$ substrate. The MgAl$_2$O$_4$ substrate has mirror-like surface prepared by mechanical polishing. After chemical etching, however, many scratches can be seen on the surface by microscope. Although sample C has mirror-like GaN surface to the naked eye, scratch features can still be observed on a AFM image and CL image.

The XRD from (0002), (0004) and (0006) diffraction of GaN film grown on the α-Al$_2$O$_3$ substrates and MgAl$_2$O$_4$ substrate are shown in Fig.1(a) and (b), respectively. All other peaks in Fig.1 are attributed to the substrates. It is well known that GaN films grown on (0001) α-Al$_2$O$_3$ and (111) MgAl$_2$O$_4$ substrate have mosaic structure and variation of lattice spacing. If an epitaxial film has a mosaic structure and a variation of lattice spacing, the magnitude of the full width at half maximum (FWHM) of X-ray diffraction profile depends on the method of measurement. We made further measurements of the samples by TXRD in ω-mode and 2θ/θ-mode. During recording of an ω/2θ rocking curve (RC), both the detector (2θ) and the sample

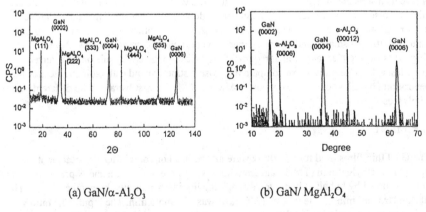

(a) GaN/α-Al$_2$O$_3$ (b) GaN/ MgAl$_2$O$_4$

Fig.1 XRD patterns of GaN for (a): sample A and (b): sample C.

rock through the Bragg angle (θ), with the detector moving at twice the angular velocity of the sample. During an ω scan, only the sample rocks through θ. The $2\theta/\theta$ RC is sensitive to lattice plane spacing [6], and the ω RC is sensitive to lattice plane tilt or mosaic structure. The FWHM $\Delta\theta_1$, which is broadening due to misorientations of the GaN grains, is measured from a rocking curve obtained by the ω-mode. The FWHM $\Delta\theta_2$, which is broadening due to a variation of the lattice spacing, was measured from the diffraction profile by the $2\theta/\theta$-mode. We also measured the FWHM $\Delta\theta$ from a conventional double crystal rocking curve and noticed that the superposition rule, $\Delta\theta = \Delta\theta_1 + \Delta\theta_2$, holds roughly in the samples. All these results are listed in the table 1. We found that the $\Delta\theta_1$ is about one order of magnitude larger than $\Delta\theta_2$ in the samples, which means that the film consists of mosaic crystallites. However, the films with mirror-like surface (sample A and C)have better crystalline quality than that one with rough surface. Although (111) $MgAl_2O_4$ has smaller lattice and thermal mismatch with GaN, the crystalline quality of sample C is the poor than that of sample A. We think the reason maybe due to the scratches on the $MgAl_2O_4$ surface and the un-optimized growth condition. We also found that the value of $\Delta\theta_1/\Delta\theta_2$ of the sample A is larger than that of the sample B and C.

TABLE 1. HWFM of X-ray rocking curve for GaN films ([0002] reflection)

Sample#	$\Delta\theta_1$ (arcmin)	$\Delta\theta_2$ (arcmin)	$\Delta\theta$ (arcmin)	$\Delta\theta_1+\theta_2$ (arcmin)	$\Delta\theta_1/\Delta\theta_2$
A	7.86	0.42	8.76	8.28	18.71
B	11.16	1.08	11.70	12.24	10.33
C	9.18	0.69	9.78	9.87	13.30

The CL spectra and CL images for sample A, B and C are shown in Fig.2. From the Fig.2 we can see that the near band-edge luminescence and yellow luminescence (YL) in the CL spectra. Inhomogeneous light emission has been observed in the sample B and C, respectively. CL allows the local excitation of the material and the observation of the light emitted in the vicinity of the excited region. In the CL image of the sample B, we found two kinds of YL image shape: hexagonal shaped broad emission and fine straight line shaped emission that has triple symmetry. The former is associated with hexagonal grain boundaries. The light emission spreads around the grain boundaries. The latter is associated with cracks. The YL is very confined along the cracks. In the CL image of the sample C, many yellow emission lines can be seen. These lines have scratch type and we believe that this kind yellow emissions are associated with imperfect of crystalline created   by the scratches on the substrate.

The cross-section TEM image of GaN/(111) $MgAl_2O_4$ as shown in Fig. 3. A defect density reduction was observed, within the initial 0.4 μm of GaN film. In the immediate vicinity of GaN/$MgAl_2O_4$ interface the defect density was so high that we were unable resolve the defect individually using conventional TEM.

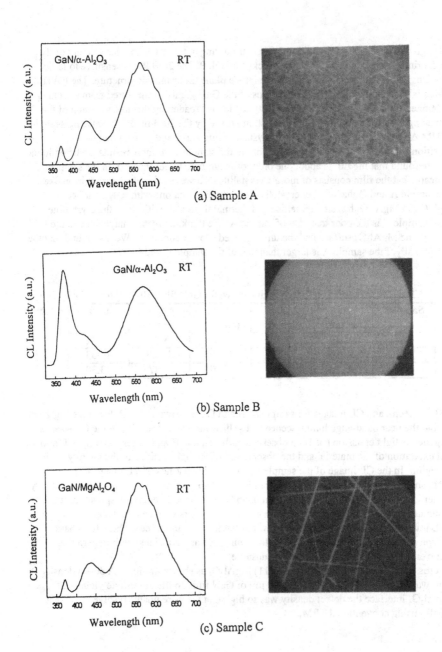

Fig.2 CL spectra and CL image for sample A,B and C.

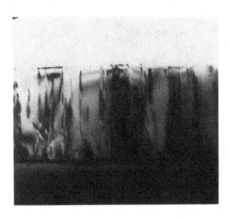

Fig. 3. TEM cross-section image of a GaN/(111) MgAl$_2$O$_4$ sample.

CONCLUSIONS

In conclusion, X-ray rocking curves studied show that mosaic structures exist in all samples. CL studies of GaN films by LP-MOVPE indicated that (a) inhomogeneous yellow light emission has been observed in undoped materials and (b) two kinds of YL image shape: hexagonal shaped broad emission and fine straight line shaped emission that has triple symmetry. The former is associated with hexagonal grain boundaries. The light emission spreads around the grain boundaries. The latter is associated with cracks. The YL is very confined along the cracks. The scratches on the (111) MgAl$_2$O$_4$ substrate can cause inhomogenous yellow emission in the GaN epilayer.

ACKNOWLEDGMENT

The authors would like to thank Prof. Cheng ji Li for discussion on the CL measurements.

REFERENCES

1. S. Nakamura, M. Senoh, N.Iwasa, S.Nagahama, T.Yamada and T.Mukai, Jpn.J. Appl. Phys. Lett. 2, Lett. **34**,L1332 (1995).

2. M.S.Khan, J.N.Kuznia, A.R.Bhattarai and D.T.Olsen, Appl. Phys. Lett. **62**, 1786 (1993).

3. S.D.Lester, F.A.Ponce,M.G.Craford and D.A.Steigerwald, Appl. Phys.Lett., **66**,1249 (1995)

4. S.-K. Duan, X.-G. Teng, W.-B. Gao and Y.-Y. Li, ActaPhotonica Sinica, **24**, p.105 (1995).

5. J. Neugebauer and Chris G. Van de walle, Appl. Phys. Lett. **69,** 503 (1996)

6. S. J. Rosner, E.C. Carr, M. J. Ludowise, G.Girolami, and H.I. Erikson, Appl. Phys. Lett. **70**, p.420 (1997).

Fig. [illegible] TEM cross-section image of [illegible] $MgAl_2O_4$ sample

CONCLUSIONS

[illegible]

ACKNOWLEDGMENT

[illegible]

REFERENCES

1. [illegible]

2. [illegible]

3. [illegible]

4. [illegible]

5. [illegible] Neugeb[illegible] and C. [illegible] Van de Walle, Appl. Phys. Lett. [illegible] (1996)

6. [illegible]

RESONANT RAMAN SCATTERING IN GaN/Al$_{0.15}$Ga$_{0.85}$N AND In$_y$Ga$_{1-y}$N/GaN/Al$_x$Ga$_{1-x}$N HETEROSTRUCTURES

D. Behr, R. Niebuhr, H. Obloh, J. Wagner, K.H. Bachem, and U. Kaufmann
Fraunhofer-Institut für Angewandte Festkörperphysik, Tullastrasse 72, D-79108 Freiburg, Germany, behr@iaf.fhg.de

ABSTRACT

We report on resonant Raman scattering in Al$_{0.15}$Ga$_{0.85}$N/GaN single quantum wells (QWs) and Al$_x$Ga$_{1-x}$N/GaN/In$_y$Ga$_{1-y}$N heterostructures. By choosing appropriate excitation conditions we could probe selectively the GaN quantum well or the Al$_{0.15}$Ga$_{0.85}$N barrier of Al$_{0.15}$Ga$_{0.85}$N/GaN single quantum wells. For the In$_x$Ga$_{1-x}$N material system a linear frequency shift of the E$_2$- and A$_1$(LO) phonon mode to lower frequencies was found with increasing In content. The shift was determined to -0.79cm^{-1} per % In content for the A$_1$(LO) phonon frequency. Resonant excitation of Al$_x$Ga$_{1-x}$N/GaN/In$_y$Ga$_{1-y}$N heterostructures enabled us to detect phonon signals from the In$_x$Ga$_{1-x}$N layer in the heterostructure and to determine its In content.

INTRODUCTION

Al$_x$Ga$_{1-x}$N/GaN/In$_y$Ga$_{1-y}$N quantum wells and heterostructures form the basis of high efficiency light emitting diodes (LED) [1, 2], injection lasers [3] covering the UV, blue, and green spectral range as well as for luminescence converting white LEDs [4]. In spite of the commercial availability of such devices there are up to now only scarce reports on the characterization of these quantum structures. A few photoluminescence [5,6] and reflection [7] studies have been reported on Al$_x$Ga$_{1-x}$N/GaN and Al$_x$Ga$_{1-x}$N/In$_y$Ga$_{1-y}$N QWs and a resonant-Raman study [8] was published recently.

Resonant Raman scattering has found widespread use for the characterization of more conventional semiconductors, such as GaAs/(AlGa)As, with fundamental gap energies in the visible spectral region. Recently we applied this technique to the Al$_x$Ga$_{1-x}$N material system, where we studied multi-phonon scattering by longitudinal optical (LO) phonons [9], acoustic phonon interbranch exciton-polariton scattering followed by LO-phonon emission [10], and the vibrational properties of single QWs [8]. In this paper we report on further resonant Raman experiments on a Al$_{0.15}$Ga$_{0.85}$N/GaN single QW and on Raman- and resonant Raman scattering in Al$_x$Ga$_{1-x}$N/GaN/In$_y$Ga$_{1-y}$N heterostructures.

EXPERIMENT

All samples investigated in this study were grown by metal-organic chemical vapor deposition (MOCVD) on c-plane 2" sapphire substrates, using TMGa, TMAl, TMIn, and ammonia as precursors. The GaN/Al$_{0.15}$Ga$_{0.85}$N single QW consisted of a 3nm GaN well, embedded between a 1μm thick Al$_{0.15}$Ga$_{0.85}$N bottom barrier layer and a 0.1μm thick Al$_{0.15}$Ga$_{0.85}$N top barrier layer. The In$_x$Ga$_{1-x}$N containing heterostructure grown on a 1μm thick GaN buffer was composed of a 150nm thick Al$_{0.25}$Ga$_{0.25}$N barrier, a 50nm thick

213

In$_x$Ga$_{1-x}$N layer with a nominal In content of 15%, another 50nm thick Al$_{0.25}$Ga$_{0.75}$N barrier, a 50nm thick In$_x$Ga$_{1-x}$N layer with a nominal In content of 30% and a 150nm thick Al$_{0.25}$Ga$_{0.75}$N top barrier. For reference purposes 0.75µm thick In$_y$Ga$_{1-y}$N layers grown on a 0.75µm thick GaN buffer with an In content varying between 1.9% and 11.3% as determined by energy dispersive X-ray analysis (EDX) were used. Details of the growth process will be published elsewhere [11].

Raman spectra were taken in backscattering configuration with the light propagating parallel to the c-axis of the crystal. The scattered light was not analyzed for its polarization. For the scattering configuration used, deformation-potential scattering by E$_2$- and A$_1$(LO) phonons as well as Fröhlich induced scattering by A$_1$(LO) phonons can be observed [12]. For optical excitation, either an Ar-ion and a Kr-ion laser was used emitting photons with different energies in the visible and near UV spectral region for resonant- or off-resonant optical excitation. The excitation power at the sample surface was around 50mW, resulting in a power density of about 750W/cm^2.

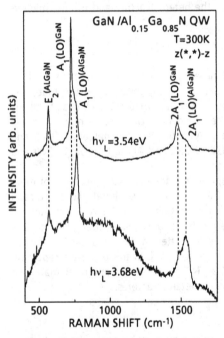

Fig. 1: Raman spectra of a 3nm GaN/Al$_{0.15}$Ga$_{0.85}$N quantum well, taken with optical excitation in resonance with either the fundamental interband transition of the GaN well (top) or the Al$_{0.15}$Ga$_{0.85}$N barrier (bottom).

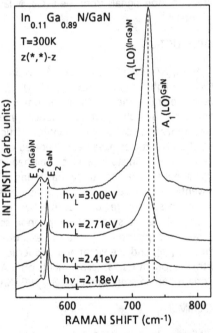

Fig. 2: Raman spectra of a In$_{0.11}$Ga$_{0.89}$N/GaN heterostructure, taken with different optical excitation energies. For excitation at 2.71eV and 3.00eV scattering by the In$_{0.11}$Ga$_{0.89}$N A$_1$(LO) phonon dominates the spectrum.

RESULTS AND DISCUSSION

Fig.1 shows room-temperature Raman spectra of the 3nm $Al_{0.15}Ga_{0.85}N$/GaN single QW. The upper spectrum was taken with optical excitation at 3.54eV, which is close to the lowest interband transition energy of the GaN QW [8]. The lower spectrum, in contrast, was taken with an optical excitation energy of 3.68eV, which is close to the fundamental bandgap energy of the $Al_{0.15}Ga_{0.85}N$ barrier. Both spectra show the E_2 phonon line which appears in the $Al_xGa_{1-x}N$ material system at 569cm^{-1} independent of the Al content [8]. The upper spectrum which was taken in resonance with the GaN well, shows in addition strong lines at 733cm^{-1} and 1475cm^{-1}. These lines can be assigned to A_1(LO)- and 2 A_1(LO) phonon scattering processes originating from the GaN QW [8]. The A_1(LO) and 2 A_1(LO) phonon lines of the $Al_{0.15}Ga_{0.85}N$ barrier at around 770cm^{-1} and 1540cm^{-1}, respectively, appear only as weak shoulders on the high energy side of the GaN phonon lines. In the lower spectrum, taken with optical excitation energy close to the fundamental bandgap energy in the $Al_{0.15}Ga_{0.85}N$ barrier, the relative intensities of the phonon lines of GaN and $Al_{0.15}Ga_{0.85}N$ are reversed. This spectrum shows strong $Al_{0.15}Ga_{0.85}N$ A_1(LO)- and 2 A_1(LO) phonon lines at 770cm^{-1} and 1534cm^{-1}, respectively, whereas the GaN phonon lines are just resolved at the low energy side of the $Al_{0.15}Ga_{0.85}N$ phonon lines. These findings demonstrate, that choosing suitable excitation conditions, i.e. resonant excitation of the

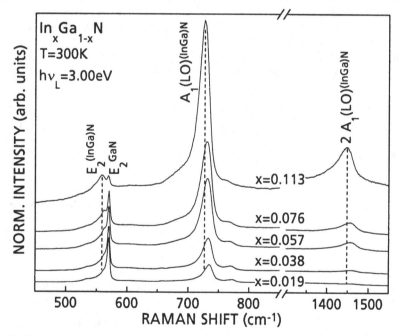

Fig. 3: Room-temperature Raman spectra of $In_xGa_{1-x}N$ samples with varying In content x. All spectra were optically excited with 3.00eV. Besides an increasing resonant enhancement for increasing In content, a frequency shift of the $In_xGa_{1-x}N$ phonon modes is observed.

barrier or the well in the QW heterostructure, enables us to probe selectively the vibrational properties of the different layers in $Al_{0.15}Ga_{0.85}N/GaN$ single QWs.

In Fig.2 room-temperature Raman spectra of an $In_{0.11}Ga_{0.89}N/GaN$ heterostructures are displayed. The spectra were recorded with different optical excitation energies of 2.18eV, 2.41eV, 2.71eV and 3.00eV. All spectra are normalized to the intensity of the E_2 phonon line at 568cm^{-1} originating from the GaN buffer layer. Beside this E_2 phonon line, the spectrum taken with an excitation energy of 2.18eV (bottom) shows the GaN A_1(LO) phonon line at 735cm^{-1}. Additionally a weak shoulder at the low energy side of the E_2 phonon line is just resolved. For increasing excitation energy, the shoulder at the low energy side of the E_2 phonon line gains intensity and an intense phonon line at 726cm^{-1} appears. Keeping in mind that the fundamental gap energy of $In_{0.11}Ga_{0.89}N$ is around 3.16 eV [7], one can explain the observed behavior in terms of resonant enhancement of the scattering efficiency in the $In_{0.11}Ga_{0.89}N$. This resonant enhancement manifests itself most clearly in the gain in scattering strength of the $In_{0.11}Ga_{0.89}N$ A_1(LO) phonon mode [13] for optical excitation at 2.71eV and in particular at 3.00eV.

In Fig.3 room-temperature Raman spectra of samples with varying In content are displayed. Optical excitation was performed at 3.00eV for all spectra. For increasing In content an increase in intensity of the $In_xGa_{1-x}N$ A_1(LO)- and 2 A_1(LO) phonon lines can be observed, indicating an increasing resonant enhancement [14]. This behavior is expected because for increasing In content up to 11.3% in the samples the fundamental bandgap energy decreases from 3.44eV down to 3.16eV [7] and thus providing increasing resonant enhancement for the given fixed exciting photon energy.

Further a decrease of the frequencies of all $In_xGa_{1-x}N$-related phonon lines, and in particular of the A_1(LO) phonon line, was found for increasing In content. This shift is

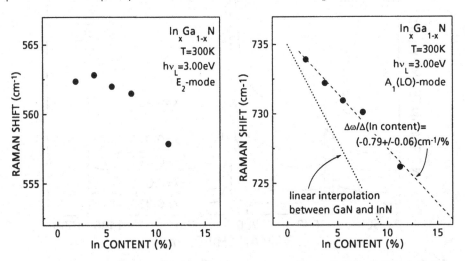

Fig. 4: Frequency dependence of the E_2- (left) and A_1(LO) phonon mode (right) on the In content. For the A_1(LO) phonon a change in the mode frequency of -0.79cm^{-1} per % In content (dashed line) was found.

displayed in Fig.4, where the frequency of the E$_2$ phonon (left) and the A$_1$(LO) phonon (right) derived from the spectra shown in Fig. 3 are plotted versus the In content. Both plots indicate a linear dependence of the phonon frequencies on the In content. But only for the A$_1$(LO) phonon the frequencies could be determined with sufficient accuracy to extract a change in the mode frequency of -0.79cm^{-1} per % In content (dashed line). This value is smaller than that expected from a linear interpolation of the A$_1$(LO) phonon frequencies in GaN and InN (dotted line) [14], indicating a bowing behavior of the dependence of the A$_1$(LO) phonon frequency on the In content.

Finally, also an Al$_x$Ga$_{1-x}$N/GaN/In$_y$Ga$_{1-y}$N heterostructure with two In$_y$Ga$_{1-y}$N layers of different composition was studied. Room-temperature Raman spectra of this sample, recorded with optical excitation at different photon energies, are shown in Fig. 5. The lower spectrum, recorded with optical excitation at 2.41eV, e.g. off-resonant excitation for all layers in the heterostructure, are dominated by the vibrational modes (E$_2$- and A$_1$(LO) phonon) of the GaN buffer. Only a

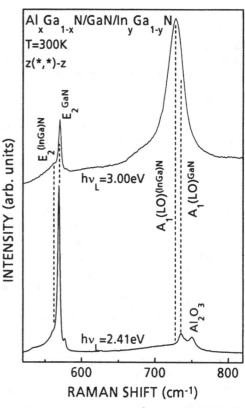

Fig.5: Raman spectra of an Al$_x$Ga$_{1-x}$N/GaN/In$_y$Ga$_{1-y}$N heterostructure, containing two In$_y$Ga$_{1-y}$N layers of different composition, taken with off- resonant excitation (bottom) and with resonant excitation of the In$_y$Ga$_{1-y}$N layer (top).

small shoulder due to the In$_y$Ga$_{1-y}$N E$_2$ phonon at around 555cm^{-1} is detected. Changing the optical excitation energy to 3.00eV, as shown in the upper spectrum, the Raman efficiency for scattering in the In$_y$Ga$_{1-y}$N layer is resonantly enhanced, and the spectrum becomes dominated by the In$_y$Ga$_{1-y}$N A$_1$(LO) phonon mode. From its frequency of 726.3cm^{-1} the average In content in the In$_y$Ga$_{1-y}$N layer can be estimated to 11%. This value is lower than the nominally In content of 15% or even 30% expected from the growth parameter.

CONCLUSION

GaN/Al$_x$Ga$_{1-x}$N and In$_y$Ga$_{1-y}$N/GaN/Al$_x$Ga$_{1-x}$N heterostructures were investigated by resonant Raman scattering. This experimental technique allowed us to probe selectively the vibrational properties of the individual layers in nitride heterostructures and to detect the presence of thin well layers of only a few nm in width via resonantly enhanced LO-phonon scattering. Measurements on $In_xGa_{1-x}N/GaN$ reference samples showed a linear frequency down-shift of the E_2- and $A_1(LO)$ phonon mode with increasing In content, with a slope of $-0.79cm^{-1}$ per % In for the latter phonon mode.

ACKNOWLEDGMENTS

The authors would like to thank C.Hoffmann for performing the EDX analysis and D.Serries and P.Schlotter for valuable assistance.

REFERENCES

1. S. Nakamura, T. Mukai, and M. Senoh, Appl. Phys. Lett. **64**, 1687 (1994).
2. S. Nakamura, M. Senoh, N. Iwasa, and S. Nagahama, Appl. Phys. Lett. **67**, 1868 (1995); Jpn. J. Appl. Phys. **34**, L797 (1995).
3. S. Nakamura, M. Senoh,S. Nagahama, N. Iwasa, T. Yamada, T. Matsushita, H. Kiyoku, and Y. Sugimoto, Jpn. J. Appl. Phys. **35**, L74 (1996); Appl. Phys. Lett. **68**, 2105 (1996).
4. P. Schlotter, R. Schmidt, and J. Schneider, to appear in Appl. Phys. A.
5. E.S. Jeon, V. Kozlov, Y.-K. Song, A. Vertikov, M. Kuball, A.V. Nurmikko, H. Liu, C. Chen, R.S. Kern, and G. Craford, Appl. Phys. Lett. **69** (27), 4194 (1996).
6. W. Li, P. Bergman, B. Monemar, H. Amano, and I. Akasaki, J. Appl. Phys. **81** (2), 1005 (1997).
7. W. Shan, B.D. Little J.J. Song, Z.C. Feng, M. Schurman, and R.A. Stall, Appl. Phys. Lett. **69** (22), 3315 (1996).
8. D. Behr, R. Niebuhr, J. Wagner, K.H. Bachem, and U. Kaufmann, Appl. Phys. Lett. **70**, 363 (1997).
9. D. Behr, J. Wagner, J. Schneider, H. Amano, and I. Akasaki, Appl. Phys. Lett. **68**, 2404 (1996).
10. D. Behr, J. Wagner, R. Niebuhr, C. Merz, K.H. Bachem, H. Amano, I. Akasaki in Proceedings of the 23rd Int. Conf. on the Physics of Semiconductors, edited by M. Scheffler and R. Zimmermann (World Scientific, Singapore, 1996), p.505.
11. R. Niebuhr, K.H Bachem, D. Behr, C. Hoffmann, U. Kaufmann, Y. Lu, B. Santic, J. Wagner, M. Arlery, J.L. Rouviere, H. Jürgensen, to appear in Proceedings of the MRS Fall Meeting 1996; H.Obloh et al., to be published.
12. T. Azuhata, T. Sota, K. Suzuki, and S. Nakamura, J. Phys. Condens. Matter. **7**, 129 (1995).
13. M. Cardona in Light Scattering in Solids II, edited by M. Cardona and G. Güntherodt (Springer Verlag, 1982), pp 19-273.
14. H.J. Kwon, Y.H. Lee, O. Miki, H. Yamano, and A. Yoshida, Appl.Phys.Lett. **69** (7), 937 (1996).

RAMAN SCATTERING AND PHOTOLUMINESCENCE OF Mg DOPED GaN FILMS GROWN BY MOLECULAR BEAM EPITAXY

G. Popovici*, G. Y. Xu*, A. Botchkarev*, W. Kim*, H. Tang*, A. Salvador*, R. Strange**, J. O. White**, and H. Morkoç*+
*University of Illinois at Urbana-Champaign, Coordinated Science Laboratory, 1101 West Springfield, Urbana, IL 61801, e-mail: popovici@uiuc.edu
**University of Illinois at Urbana-Champaign, Materials Science Laboratory, 104 S. Goodwin Avenue, Urbana, IL 61801
+ On leave at Wright Laboratory Wright Patterson AFB under a URRP program funded by AFOSR

ABSTRACT

Raman, photoluminescence, and Hall measurements are reported for Mg doped GaN films grown by molecular beam epitaxy. The compressive and tensile stress determined by the Raman shift of the phonon lines is due to the growth conditions rather than the presence of Mg in the film. The photoluminescence peak of near band-to-band transitions is also shifted to larger (smaller) energies by the compressive (tensile) stress. The study of the longitudinal optical phonon of the A_1 branch shows that its Raman line shape is affected not only by phonon-plasmon interactions but by the crystalline quality of the film, as well.

INTRODUCTION

Raman spectroscopy is a useful tool for assessing crystalline quality. Shifts of Raman peaks indicate macroscopic stresses. In films, where stresses vary randomly from compressive to tensile on a microscopic length scale, Raman peaks are broadened [1,2]. Two- and three-dimensional defects reduce the coherent scattering domain size, adding another contribution to the broadening [3]. Phonon-plasmon interactions permit the determination of doping levels in semiconductors through their influence on Raman spectra [1, 4, 5, 6, 7]. For instance, Ponce et al. [8] mapped the Si spatial distribution in a GaN film grown on sapphire, by measuring the local $A_1(LO)$ peak height. A tacit assumption was made that the quenching of the $A_1(LO)$ peak is determined only by the phonon-plasmon interaction associated with free electrons from the Si dopant, and does not depend on the crystalline quality of the film.

Most GaN samples for Raman experiments have been grown on sapphire, i.e. not under thermal and lattice-matched conditions. Therefore, Raman peaks are usually both widened and shifted, leading to discrepancies among published values for the phonon energies. Tabata et al.[9] provide a table of published phonon energies (in cm^{-1}), showing a large variation: $A_1(TO)$: 531-537, E_1 (TO): 560-563, $A_1(LO)$: 710-737, $E_1(LO)$: 741-745, E_2(high energy): 568-572. The most accurate data were probably obtained by Perlin et al.[2] because their measurements were done on bulk crystals: $A_1(TO)$: 531, E_1 (TO): 560, E_2(low energy): 144, and E_2(high energy): 568 cm^{-1}. The energy of the low branch of the E_2 phonon is found to be the same in all publications (144 cm^{-1}) because it only weakly depends on pressure.

Photoluminescence (PL) spectra provide a characterization complementary to that of Raman spectra. The PL intensity is affected not only by the presence of extended defects but

219

even more strongly by the presence of point defects which are too small to be detected by Raman scattering. Point defects such as impurities, vacancies and interstitials may form deep levels in the bandgap, serving as nonradiative pathways for carrier recombination, and thus affecting the PL intensity. In this paper, we discuss the crystalline quality and stress of Mg-doped GaN films, as determined by PL and Raman measurements.

EXPERIMENTAL

Wurtzite GaN films on sapphire substrates were grown in a Riber 1000 MBE system. The (0001) substrate orientation leads to films with a wurtzite z axis perpendicular to the surface. Nitridation was performed by exposing the sapphire substrate to an ammonia flux of 16 sccm for 1 min. at 850°C. This nitridation procedure was chosen because it provides the smoothest GaN surface morphology [10]. AlN buffer layers 60 - 80 nm thick were then grown. Mg evaporated from a solid source served for p-type doping of the GaN. The thickness of the Mg-doped GaN layers was 1-3 µm. Further details about the growth process are described in a previous report [11].

Electrical contacts were formed from a Ni/Au bilayer alloyed at 600°C for 10 minutes in a nitrogen ambient. Hall measurements for the *as-grown* layers were carried out by the van der Pauw method. Photoluminescence (PL) was excited with the 325 nm line of a He-Cd laser, dispersed with a single monochromator, and detected with a photomultiplier tube. Raman emission was excited with the 514.5 nm line of an Ar ion laser, dispersed with a three-stage monochromator, and detected with a cooled CCD camera. The first two stages of the triple monochromator were used in the zero-dispersion filter mode. The resulting spectral peaks were fit to a Lorenzian to determine the width and center wavelength.

RESULTS AND DISCUSSIONS

Wurtzite has the space group $C6v^4$ (C63mc) with all atoms occupying C3v sites. The following Raman-active phonons were predicted for the wurtzite structure [12, 13] an A_1 branch in which the phonon is polarized in the z direction; an E_1 branch in which the phonon is polarized in the xy plane, and two E_2 branches [9] (the c-axis of the crystal is taken to be the z-axis). The E_1 branch is linearly polarized, but the two E_2 branches do not have a simple polarization behavior. [13] A_1 and E_1 branches are polar with different energies for the longitudinal (LO) and transverse (TO) components. In the backscattering geometry, with the c-axis normal to the surface, the TO branch of the A_1 mode and the TO and LO branches of the E_1 mode are forbidden. The LO branch of the A_1 mode and E_2 modes are allowed.

The hole concentration of the Mg-doped GaN films was determined by Hall measurements to be in the range $(2-7) \times 10^{17}$ cm^{-3}. The density of Mg atoms measured by SIMS is approximately 10 to 100 times greater (10^{18} - 10^{19} cm^{-3}) than the hole concentration, because only a few percent of the Mg atoms are ionized at room temperature [14]. Two samples had even higher Mg concentrations, as measured by SIMS. It was shown [14] that at high Mg concentrations the samples become more resistive, with hole concentrations too low to be measured.

A typical Raman spectrum of a Mg-doped sample and the Lorenzian fit are given in Fig. 1. The E_2 line is observed in all samples. The E_2 peak was in the range 564 - 576 cm^{-1}, depending on the sample. The sapphire line at 750 cm^{-1} is also observed in all samples, because the GaN films are too thin (1-3 µm) to absorb all the incident light.

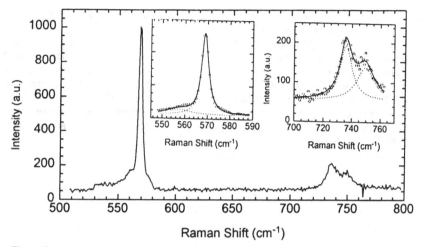

Fig. 1. Raman spectrum of a Mg-doped film. Solid line is the measured spectrum, dashed line is Lorenzian fit. The peaks maxima are at 569 cm^{-1} (E$_2$ phonon) 736 cm^{-1} A$_1$ (LO), 749 cm^{-1} (sapphire) a small peak at 557 cm^{-1} is most likely due to E$_1$ (TO) phonon.

The A$_1$ (LO) line is allowed in the backscattering geometry but observed in only three of the samples. It appears at ~735 cm^{-1} for all three samples, with full widths at half maximum (FWHM) of 4.2, 5.1 and 6.8 cm^{-1}. One of these samples is a good quality p-type sample (p= 7×10^{17} cm^{-3}, mobility μ = 17 cm^2/Vs,), while two other samples were highly resistive. All three samples have a narrow E$_2$ line, with FWHM in the range 1.5-2 cm^{-1}, indicative of good crystalline quality. Other samples on which the A$_1$(LO) was not observed had a smaller hole concentration (p = (2–3)×10^{17} cm^{-3}) and E$_2$ Raman lines with larger FWHM (4 - 8 cm^{-1}).

The LO phonons readily interact with free charged carriers, when the plasma frequency is in the range of the phonon frequency. Phonons interact more strongly with electrons than with holes. This interaction leads to a displacement and widening of the LO line leading to the damping of the Raman line at high charge carrier concentrations.

The plasmon frequency is $\omega_p^2 = (4\pi ne^2)/m^* \, \varepsilon_\infty \varepsilon_0$. The LO phonon-plasmon coupled modes $\omega_p^{\pm}$ are solutions of the equation: [15]

$$(\omega_p^{\pm})^2 = 1/2 \{\omega_L^2 + \omega_p^2 \pm [(\omega_L^2 + \omega_p^2)^2 - \omega_p^2 \, \omega_T^2]^{1/2}\} \qquad (1)$$

where n and m* are the concentration and effective mass of charged carriers,

ε_∞ is the high frequency dielectric constant, and

ω_L and ω_T are the frequencies of LO and TO phonons.

The damping is larger at higher free carrier concentrations. We solved equation (1) for concentrations of 10^{17} and 10^{18} cm^{-3}, taking ω_L = 735 cm^{-1}, ω_T = 532 cm^{-1}, m* = 0.8 m$_0$ and ε_∞ = 5.5. For holes at a concentration of 10^{17} cm^{-3} the plasmon-phonon interaction is negligible. For a concentration of 10^{18} cm^{-3} eq. (1) gives ω^+ = 736 cm^{-1} and ω^- = 37 cm^{-1}.

Thus, within the range of hole concentrations of $(2-7)\cdot 10^{17}$ cm^{-3} in our samples the phonon-plasmon interactions can be neglected.

The preceding arguments suggest that something other than the plasmon-phonon interaction is responsible for the quenching of the $A_1(LO)$ line in our samples. Thus the assumption made in earlier works, [6, 8] that the shape of the $A_1(LO)$ line is a direct measurement of the carrier concentration, may not be entirely correct. In the samples where the $A_1(LO)$ line is absent, the FWHM of the E_2 line is much wider (2.6 - 7.7 cm^{-1}). The E_2 line is known to be broadened by poor crystal quality, and known to be unaffected by the phonon-plasmon interaction. This then suggests that crystalline quality is at least partly responsible for broadening and quenching of the $A_1(LO)$ line.

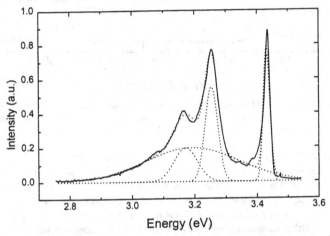

Fig. 2. Low temperature (4 K) PL spectrum. Solid line is the measured spectrum. Dashed lines are due to Gaussian fit. The peak at 3.44 eV is due to transitions involving bound exciton. The peak at 3.26 eV and its phonon A_1 (LO) replica at 3.17 eV is attributed to a transition from a donor level to the Mg acceptor level. The wide peak at 3.19 eV, might be due to Mg complex or native defect level.

The PL spectrum taken at 4 K in Fig. 2 was obtained from the same sample as in Fig. 1. The Gaussian fit yielded four maxima for all the measured samples. The energy of the band-edge line varies for different samples between 3.429 and 3.469 eV with FWHM varying between 21 and 29 meV. This line is due to transitions involving an acceptor-bound exciton. The peaks at 3.36 eV and 3.17 eV are attributed to a donor-acceptor transition and its A_1 (LO) phonon replica. The peak at 3.26 eV has a FWHM = (40 - 44) meV and the peak at 3.17 eV has a FWHM varying between 70 and 80 meV. The 3.26 eV peak is attributed to a transition from a donor level to the Mg acceptor level. The fourth maximum at 3.19 meV is very wide (FWHM = 0.2-0.3 eV). Its nature is not understood, but it may belong to a transition involving a Mg complex or a native defect.

Fig. 3 presents the dependence of the energy of the low- and room-temperature PL band edge emission peak on the E_2 phonon energy. The proportionality of the PL band edge maxima to the phonon energy is not surprising, as both depend on the lattice constant of the material. Compressive stress increases the bandgap and the phonon energy, while tensile stress reduces them. If one takes the E_2 line of the unstressed sample [7] equal to 568 cm^{-1}, it can be seen that some Mg-doped samples have compressive stress while others have tensile stress. It is well known that dopant atoms which are large (small), compared to the host lattice atoms, can introduce tensile (compressive) stress. If the stress were due to the dopant atoms, it would be either compressive or tensile, but not both. The presence of both types of stress shows that the growth conditions, and not the Mg impurity, are the major cause of stress in Mg-doped GaN samples. On the other hand, the samples grown on sapphire should show tensile stress in the plane of growth and compressive stress in the perpendicular direction (direction of measurements). The presence of both compressive and tensile stresses shows that the substrate either does not control the stress of the samples. Once more one can conclude that the growth conditions are responsible the stress in GaN samples.

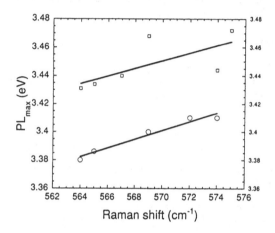

Fig. 3. PL peak energy vs E2 phonon energy as measured by Raman scattering. Circles are for room temperature measurements, squares are for low temperature (4 K) measurements.

The compressive stress value can be estimated using the formula obtained by Perlin et al [2] which gives the Raman shift $(\sigma - \sigma_0)$ as a function of hydrostatic pressure p:

$$\sigma - \sigma_0 = \sigma_1 p + \sigma_2 p^2$$

where p is the pressure in GPa, σ and σ_0 are the positions of the Raman line in the stressed and unstressed sample, and σ_1 and σ_2 are fitting parameters. This formula is only an approximation, because it assumes that the Young's modulus for the film is equal to the bulk value, and independent of the film's crystal orientation. For the maximum shift of 6 cm^{-1} (574 cm^{-1} compared to the unstressed value of 568 cm^{-1}) one can estimate a stress of ~1.4 GPa, using $\sigma_1 = 4.17$ cm^{-1}/GPa and $\sigma_2 = -0.0136$ cm^{-1}/GPa^2 [2]. From the slope of the PL peak dependence on Raman shift (Fig. 3), one can estimate the stress dependence of this PL peak to be 15 meV/GPa.

CONCLUSIONS

The stress of Mg-doped GaN films grown on sapphire is investigated by Raman and PL measurements. It is shown that both compressive and tensile stress exist with only a single dopant, pointing to the growth conditions, rather than dopant atoms, being the major cause of the stress in Mg-doped films. The maximum compressive stress estimated in our films is estimated to be ~1.4 GPa. The shift of PL band-edge line is also proportional to the stress, as determined by the Raman shift.

The LO phonons are strongly interacting with the charge carriers. Therefore, the local intensity of the $A_1(LO)$ phonon line was used in earlier papers as a direct measure of the phonon-plasmon interaction. We have shown here that bad crystalline quality of the film may also be a source of damping. Thus the interpretation of the $A_1(LO)$ line in Raman spectra may be more complicated than previously assumed.

ACKNOWLEDGMENTS

This research is funded by grants from ONR (contract #N00014-95-1-0635, #N00014-89-J-1780), AFOSR (contract #F49620-95-1-0298), under the supervision of Mr. M. Yoder, and Drs. G. L. Witt, Y. S. Park, and C. E. C. Wood.

REFERENCES

1. W. J. Meng and T. A. Perry, J. Appl. Phys. 76, 7824 (1994)
2. P. Perlin, C. Jauberthie-Carillon, J. P. Itie, A. S. Miguel, I. Grezgory, and A. Polian, Phys. Rev. B 45, 83 (199
3. P. M. Faushet, "Light Scattering in Semiconductor Structures and Superlattices", edit. by D. J. Lockwood and J. F. Young (Plenum, NY)1991 pp.229-245.
4. D. Kirilov, H. Lee, J. S. Harris, Jr. Appl. Phys. Lett. 80, 4058 (1996)
5. D. Olego and M. Cardona, Phys. Rev. B 24, 7217 (1981)
6. T. Kozawa, T. Kachi, H. Kano, Y.Taga, M. Hashimoto, N. Koide, K. Manabe, Appl. Phys. Lett. 75,1098 (1994)
7. P. Perlin, J. Camassel, W. Knap, T. Taliercio, J. C. Chervin, T. Suski, I. Grzegory, and S. Porowski, Appl. Phys. Lett. 67, 2524 (1995)
8. F. A. Ponce, J. W. Steeds, C. D. Dyer, and G. D. Pitt, Appl. Phys. Lett. 69, 2650 (1996)
9. A. Tabata, R. Enderlein, J. R. Leite, S. W. da Silva, J. C. Galzerani, D. Schikora, M. Kloidt and K. Lischka, J. Appl. Phys. 79, 4137 (1996)
10..W. Kim, M. Yeadon, A. E. Botchkarev, S. N. Mohhamad, J. M. Gibson, and H. Morkoç, submitted to JVST B.
11. W. Kim, A. Salvador, A. E. Botchkarev, O. Aktas, S. N. Mohhamad, and H. Morkoç, Appl. Phys. Lett. 69, 559 (1996)
12. L. E. McNeil, Raman and IR Reflection Spectra of GaN in "Properties of III-Nitrides", edited by J. H. Edgar, INSPEC, IEE, London, UK, 1994, pp.252-253.
13. C. A. Arguello, D. L. Rousseau, and S. P. S. Porto, Phys. Rev. 181, 1351 (1969)
14. W. Kim, A. E. Botchkarev, A. Salvador, G. Popovici, H. Tang, and H. Morkoç, to be published in J. Appl. Phys.
15. A. Mooradian and and G. B. Wright, Phys Rev. Lett. 16, 999 (1996)

HIGH-PRESSURE RAMAN SCATTERING OF
BIAXIALLY STRAINED GaN ON GaAs

H. SIEGLE*, A. R. GOÑI*, C. THOMSEN*, C. ULRICH**, K. SYASSEN**,
B. SCHÖTTKER***, D. J. AS***, D. SCHIKORA***
*Institut für Festkörperphysik, TU Berlin, Hardenbergstraße 36, 10623 Berlin, Germany
**Max-Planck-Institut für Festkörperforschung, Stuttgart, Germany
***Institut für Optoelektronik, GHS Paderborn, Germany

ABSTRACT

We present results of high-pressure Raman-scattering experiments on bulk GaN and GaN grown on GaAs. We determined the Grüneisen parameters of both the cubic TO and LO phonon modes and the hexagonal A_1, E_1 and E_2 modes. Our measurements reveal that the Grüneisen parameters for the GaAs substrate are about 30% smaller than those of bulk GaAs. This is a consequence of the lower compressibility of GaN compared to GaAs, which results in a pressure-induced biaxial strain on the substrate. From the pressure behavior of the GaAs modes and by comparing with our results for bulk GaN we obtained information about the biaxial strain in the GaN epitaxial layer.

INTRODUCTION

A major problem in growing GaN layers on standard substrates as, e.g., sapphire, 6H-SiC, or GaAs is the large lattice-mismatch and the difference in the thermal expansion coefficients between layer and substrate. This causes a large biaxial stress in the layers, which is either compressive for hexagonal GaN on sapphire or tensile for GaN on 6H-SiC [1]. In order to handle thin-film heterostructures based on GaN a knowledge of the internal stress is necessary. Experimental information about the main strain/pressure parameters in particular for cubic GaN is still lacking. We therefore performed Raman-scattering experiments on predominantly cubic GaN layers grown on GaAs and bulk hexagonal GaN crystals at high hydrostatic pressures up to 6 GPa.

We determined the Grüneisen parameters of the cubic and hexagonal phonon modes. The application of hydrostatic pressure on the GaN/GaAs sample resulted in a tensile stress of the GaAs substrate while the GaN layer became compressively strained. This is a consequence of the different bulk moduli of GaN and GaAs. From the pressure dependence of the different phonon modes we could estimated the stress distribution. While the GaAs substrate was heavily biaxially strained due to the applied hydrostatic pressure, we found only a weak biaxial stress in the GaN layer. This can partly be explained by stress relaxation at the interface to the substrate.

Mat. Res. Soc. Symp. Proc. Vol. 468 © 1997 Materials Research Society

EXPERIMENT

The samples under study were an 1 μm thick GaN layer grown on (001) GaAs using molecular beam epitaxy (MBE) and a 50 μm thick GaN crystal grown on 6H-SiC by hydride vapor phase epitaxy (HVPE) of which the substrate was mechanically removed by polishing. While the latter sample crystallized in the hexagonal modification, the thin film on GaAs was predominantly cubic with about 5 % hexagonal minority phase built-in.

The Raman-scattering experiments were performed in a gasketed diamond-anvil cell (DAC) at low temperatures in case of the cubic layer and at room temperature in case of the hexagonal sample. For low-temperature investigations we used condensed helium as a pressure medium while a 4:1 methanol-ethanol mixture was used for room-temperature measurements. The pressure was measured using the ruby luminescence method. [2] The 458 nm and the 514.5 nm lines of an Ar^{2+}-ion laser were used for excitation. The scattered light was detected in back-scattering geometry, corresponding to z(..)z̄, and analyzed by a triple-grating spectrometer equipped with a liquid-nitrogen cooled CCD detector. This setup allowed the observation of Raman shifts smaller than 0.2 cm^{-1}.

RESULTS AND DISCUSSION

Depending on the growth conditions GaN crystallizes either in the hexagonal wurtzite structure or in the cubic zincblende form. Because of the metastability of the cubic phase these layers often exhibits a small content of hexagonal minority phase. The hexagonal modification of GaN belongs to the point group C_{6v} (6mm) having four atoms per unit cell, while cubic GaN belongs to the point group T_d (43m) with two atoms per primive unit cell. Hence group theory predicts in the case of hexagonal GaN the following zone-center optical modes: one totally symmetric A_1, one doubly degenerate infrared and Raman-active E_1, two doubly degenerate non-polar E_2, which are only Raman-active and two silent B_1 modes. In case of cubic GaN one finds in first-order Raman-scattering experiments a doubly degenerate TO and a single LO phonon mode.

Figure 1 shows low-temperature Raman spectra taken from the GaN layer grown on

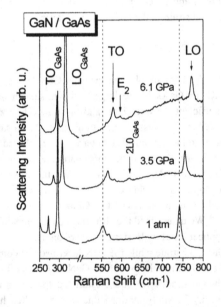

Fig. 1: Low-temperature Raman spectra taken from GaN/GaAs for different hydrostatic pressures. Excitation wavelength was at 514.5 nm.

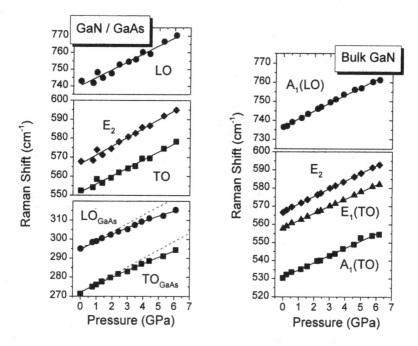

Fig. 2: Frequencies of the observed GaN phonon modes as a function of the applied hydrostatic pressure. Frequencies obtained from GaN / GaAs are shown on the left and from bulk hexagonal GaN on the right. The dashed lines indicate the expected pressure dependence of the GaAs phonon modes for bulk material.

GaAs for different pressures up to 6.1 GPa. Apart from the GaAs phonon modes we observed the cubic GaN TO and LO phonon modes as well as the E_2 mode of the hexagonal minority phase. The increasing background signal with increasing pressure originates from luminescence of the diamond anvils. Although forbidden by symmetry in back-scattering geometry from an (001) face the TO modes are appearant in the spectra due to forward scattering effects [3], a rough interface, and a non strictly back-scattering configuration in the diamond-anvil cell. The structure located on the high-energy side of the E_2 mode, which can be seen for higher pressures, originates from 2 LO scattering of the GaAs substrate.

The results for the measured frequencies of the GaN phonon modes are displayed in Fig. 2 as a function of the applied pressure. All modes exhibit the expected shift to higher frequencies with increasing hydrostatic pressure. On the right-hand side of Fig. 2 the pressure dependences of the phonon modes of bulk GaN are shown for the same pressure range. The solid lines in the left picture of Fig. 2 for the GaAs data represent least-square fits using a

Table I: Fitting parameters of the pressure dependence of phonon modes in cubic and hexagonal GaN and the corresponding Grüneisen parameters. We assumed a bulk-modulus of $B_0 = 200$ GPa for both cubic and hexagonal GaN.

Sample	Mode	ω_0 (cm^{-1})	$\partial\omega/\partial P$ (cm^{-1} / GPa)	γ
GaN/GaAs	TO	553 ± 2	4.0 ± 0.2	1.4 ± 0.1
	LO	743 ± 2	4.5 ± 0.2	1.20 ± 0.05
	E_2	568 ± 2	4.5 ± 0.2	1.5 ± 0.1
GaN	A_1(TO)	530.9 ± 0.3	4.0 ± 0.1	1.51 ± 0.04
(hexagonal	E_1(TO)	558.2 ± 0.1	3.94 ± 0.03	1.41 ± 0.01
bulk material)	E_2	567.0 ± 0.1	4.24 ± 0.03	1.50 ± 0.01
	A_1(LO)	736.4 ± 0.2	4.4 ± 0.1	1.20 ± 0.03

second-order polynomial. To determine the Grüneisen parameters the experimental data were fitted by the linear law

$$\omega(P) = \omega_0 + \frac{\partial\omega(P)}{\partial P} \tag{1}$$

where P is the pressure in GPa. Assuming a bulk modulus of $B_0 = 200$ GPa [4-6] for both cubic and hexagonal GaN we determined the Grüneisen parameters γ for all observed phonon modes. The Grüneisen parameter is defined as

$$\gamma = B_0 \frac{\partial \ln\omega}{\partial P}. \tag{2}$$

The resulting parameters are summarized in Table I. The pressure coefficients and Grüneisen parameters obtained from our measurements agree well with those of Ref. [7]. Differences in the Grüneisen parameters in case of the A_1(TO) and E_2 modes are due to the larger bulk modulus used in [7]. For the E_1(TO) mode, on the contrary, we found a much smaller value. To the best of our knowledge up to now no experimental data were available for the hexagonal A_1(LO) mode and for the phonon modes in cubic GaN. Our measurements show that the cubic modes behave quite similar as the hexagonal E_1(TO) and A_1(LO) modes.

One interesting observation as a result of the application of hydrostatic pressure on the GaN/GaAs sample is the pressure behavior of the GaAs phonons. As can be seen from Fig. 2 their pressure dependences differ strongly from the expected behavior of bulk GaAs as depicted by the dashed lines. We found Grüneisen parameters which were about 30% lower than those obtained from bulk material ($\gamma_{TO} = 1.23$, $\gamma_{LO} = 1.11$ [8]). In addition, a deviation from the linear pressure dependence can clearly be seen as a negative bending of the fitting curve. We interpret this effect as a result of the different compressibilities of GaN and GaAs. Their bulk moduli differ by about a factor of three: $B_{0, GaAs} = 74.7$ GPa [9]; $B_{0, GaN} = 200$ GPa [4-6]. Consequently, the applied hydrostatic pressure induced a biaxial tensile strain in the

GaAs substrate and a compressive strain in the GaN layer. This pressure-induced strain results in a softening of the frequencies of the substrate phonons while the GaN modes shifts to higher frequencies. From the frequency shift we could estimate the biaxial stress in the sample. Following Cerdeira et al. [10] the shift of the cubic LO phonon mode can be calculated by

$$\Delta\omega_{LO} = \frac{p+2q}{3\omega_0}(S_{11} + 2S_{12})X - \frac{p-q}{3\omega_0}(S_{11} - S_{12})X \qquad (3)$$

where p and q are the phonon deformation potentials, S_{ij} the compliances, and X the biaxial stress. For an applied hydrostatic pressure of 6.1 GPa we measured differences with respect to the GaAs bulk values of 6.6 cm^{-1} for the LO and 4.9 cm^{-1} for the TO modes. Using the phonon deformation potentials given in Ref. [8] and the compliances derived from the elastic constants given in Ref. [11] our calculation yields a biaxial tensile stress in the GaAs substrate of 1.6 GPa. This corresponds to a change of 1.3% in lattice constant.

In order to obtain also some information about the stress in the GaN layer we considered the pressure dependence of the E_2 mode which appears in both samples, in the hexagonal bulk sample as well as in the thin layer due to the small content of hexagonal minority phase in the predominantly cubic GaN. From the comparison of the different pressure dependences of this mode in the different samples we can infer the strain in the GaN layer. Using Murnughan equation of state [12] we calculated the change of the GaN and GaAs lattice constants under pure hydrostatic pressure. While at 6.1 GPa the GaAs lattice constant should have changed by 2.3% the one of GaN should have changed by only 1% because of the lower compressibility. In case of the GaN/GaAs sample the difference in the alterations should be built in the sample as biaxial strain. From the shift of the GaAs modes we already determined the tensile stress contribution to the substrate. The residual pressure-induced stress should strain the GaN layer compressively. Using the elastic constants of GaN given in [13] one would expect a biaxial stress in the layer of 1.8 GPa.

Considering the different pressure dependences of the E_2 mode our measurements reveal a strain-induced shift for the E_2 mode of 2.5 cm^{-1}. This shift corresponds to a compressive stress in the range between 0.4 and 0.6 GPa following the relations given in [14] and [4]. This unexpected low biaxial stress in the layer may be caused by a stress relaxation near the GaN/GaAs interface. Considering the E_2 frequencies measured while decreasing the hydrostatic pressure we found that they are slightly lower than those obtained while increasing the pressure. This weak hystereses can also be understood in terms of the stress relaxation.

SUMMARY

In summary, we investigated the influence of stress on the phonon modes of GaN by high-pressure Raman-scattering experiments on cubic and hexagonal GaN samples. We determined the Grüneisen parameters of the hexagonal $A_1(TO)$, $E_1(TO)$, E_2 and $A_1(LO)$ as well as of the cubic TO and LO modes. We found that due to the different compressibilities of GaN and GaAs the application of hydrostatic pressure induced a large biaxial strain in the epitaxial sample, which is tensile for the GaAs substrate and compressive for the GaN film. From the strain-induced phonon shifts we could estimate the stress distribution. We gave a possible explaination of the unexpected weak biaxial strain in the GaN layer found in our measurements.

ACKNOWLEDGMENTS

This work was in parts supported by the Stifterverband für die Deutsche Wissenschaft.

REFERENCES

1. For a review, see for example: H. Morkoç, S. Strite, G. B. Gao, M. E. Lin, B. Sverdlov, and M. Burns, J. Appl. Phys. **76**, 1363 (1994); Properties of Group III Nitrides, edited by J. H. Edgar, Electronic Materials Information Service (EMIS) Datareviews Series (Institution of Electrical Engineers, London, 1994)
2. H. K. Mao, J. Xu, and P. M. Bell, J. Geophys. Res. 91, 4673 (1986). For temperature corrections see R. A. Noack and W. B. Holzapfel, in High Pressure Science and Technology, edited by K. D. Timmerhaus and M. S. Barber (Plenum, New York, 1979), Vol. 1, p. 748
3. H. Siegle, L. Eckey, A. Hoffmann, C. Thomsen, B. K. Meyer, D. Schikora, M. Hankeln, K. Lischka, Solid State Commun. **96**, 943 (1995)
4. C. Kisielowski, J. Krüger, S. Ruvimov, T. Suski, J. W. Ager III, E. Jones, Z. Lilienthal-Weber, M. Rubin, E. R. Weber, M. D. Bremser, R. F. Davis, Phys. Rev. B **54**, 17745 (1996)
5. V. A. Savastenko and A. U. Sheleg, Phys. Status Solidi A **48**, K135 (1978)
6. K. Kim, W. R. L. Lambrecht, and B. Segall, Phys. Rev. B **53**, 16310 (1996)
7. P. Perlin, C. Jauberthie-Carillon, J. P. Itie, A. San Miguel, I. Grzegory, A. Polain, Phys. Rev. B **45**, 83 (1992)
8. P. Wickboldt, E. Anastassakis, R. Sauer, and M. Cardona, Phys. Rev. B **35**, 1362 (1987)
9. Landolt-Börnstein Tables, edited by O. Madelung, M. Schulz, and H. Weiss (Springer, Berlin 1982), Vol. 17a
10. F. Cerdeira, C. J. Buchenauer, F. H. Pollak, and M. Cardona, Phys. Rev. B **5**, 580 (1972)
11. J. S. Blakemore, J. Appl. Phys. **53**, R123 (1982)
12. F. D. Murnaghan, Proc. Natl. Acad. Sci. USA **30**, 244 (1944)
13. A. Polian, M. Grimsdich, and I. Grzegory, J. Appl. Phys. **79**, 3343 (1996)
14. T. Kozawa, T. Kachi, H. Kano, H. Nagase, N. Koide, K. Manabe, J. Appl. Phys. **77**, 4389 (1995)

ELECTRON-PHONON SCATTERING IN VERY HIGH ELECTRIC FIELDS

B.K. Ridley*
School of Electrical Engineering, Phillips Hall, Cornell Unviersity, Ithaca, NY 14853
*(Home address,: Department of Physics, University of Essex, Colchester, UK)

ABSTRACT

Large-bandgap materials can support very high electric fields (>1MV/cm) without breaking down. The possibility then exists for an electron in a conduction band to become quasi-localized in Wannier-Stark states, a possibility that depends on the scattering rate being less than the Bloch oscillation frequency. If this condition is met the scattering rate itself is affected and the description of transport must be changed from the usual model in which the electron is assumed to be virtually free. Here, we examine the feasibility of obtaining this condition in GaN using a simple three-dimensional tight-binding model for the bandstructure and taking the dominant scattering mechanism to be the polar and non-polar interaction with optical phonons and short-wavelength acoustic phonons.

INTRODUCTION

In all materials the interaction between electrons and phonons provides the basis for intrinsic scattering mechanisms. In pure materials at high energies the electron-phonon interaction is usually the only scattering mechanism and is the one that will determine whether Bloch oscillations [1] can occur before the material breaks down. If Bloch oscillations are established the conduction band transforms to a ladder of Wannier-Stark (WS) states [2] and all transport properties including the impact-ionization become profoundly altered. The criterion for a Bloch oscillation to occur is that the electric field F must be large enough to move the electron ballistically through the whole extent of the first Brillouin zone. In a direction where the extent of the zone is $2\pi/a$, were a is the lattice constant, the condition is

$$\omega_F \tau \geq 1, \qquad \omega_F = eFa/\hbar \qquad (1)$$

where τ is the average scattering time-constant and ω_F is the Bloch angular frequency. The problem is to estimate τ. Scattering rates, hitherto, have been described only for states near the low-energy edge of the band. Here we describe an estimate in which we assume a simple form for the band structure and obtain full-band scattering rates for the polar and non-polar interactions. Parameters relating to GaN have been used to provide concrete figures.

BANDSTRUCTURE MODEL

Real bandstructures are highly complex. For an analytic estimate of scattering rates that encompass the whole-band we choose a very simple bandstructure, namely, that for cubic tight-binding:

$$E = E_b \sum_\alpha \sin^2(k_\alpha a/2) \qquad \alpha = x, y, z \qquad (2)$$

where E_b is the bandwidth along a cubic edge, (Fig. 1a). An important feature affecting scattering is the density-of-states function, N(E). For simplicity we ignore the complicated

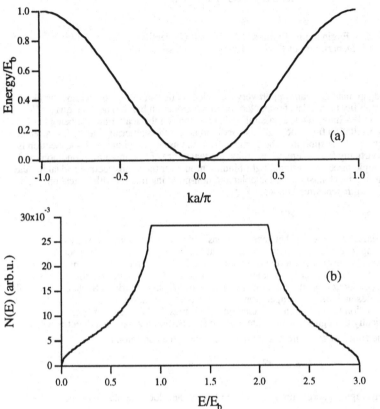

Fig.1. Bandstructure model. (a) Bandstructure along a principal direction.
(b) Density-of-states function.

directional dependence and adopt a quasi-spherical model leading to a density-of-states function of the form shown in Fig. 1b. The total number of states in the spherical mdoel is taken to be the same as in the cubic structure. The lower and upper halves of the band are symmetrical in the dependence of energy on wavevector. Thus

$$\hbar^2 k^2 / 2m^* = \gamma(E), \qquad\qquad E \leq (3/2)\ E_b$$

$$\hbar^2 (k - k_o)^2 / 2m^* = \gamma(3E_b - E) \qquad E \geq (3/2)\ E_b$$

(3)

where k_o is the zone-edge wavevector, $\gamma(E)$ is a function of energy derivable from N(E), and the effective mass is related to E_b as follows:

$$m^* E_b = 2\hbar^2 / a^2$$

(4)

PHONON SCATTERING RATES

Phonon scattering rates in typical direct-gap semiconductors are relatively small for electrons in the Γ-valley conduction band, but they increase dramatically once transfer to upper conduction-band valleys takes place. For our purpose the scattering rate at high-energies is the germain parameter and this implies electron-phonon interactions with phonons with large wavevector. For simplicity we will assume that the polar interaction is solely with LO modes with a scattering rate of the same form as at low energies [3]:

$$
\frac{1}{\tau_p} = W_0 \left(\frac{\hbar \omega_{LO}}{\gamma(E)} \right)^{1/2} \left[\begin{array}{l} \left(\dfrac{d\gamma}{dE} \right)_{E+\hbar\omega} \tanh^{-1}\left[\dfrac{\gamma(E)}{\gamma(e+\hbar\omega)} \right]^{1/2} n(\omega_{LO}) + \\[2mm] \left(\dfrac{d\gamma}{dE} \right)_{E-\hbar\omega} \tanh^{-1}\left[\dfrac{\gamma(E-\hbar\omega)}{\lambda(E)} \right]^{1/2} (n(\omega_{LO})+1) \end{array} \right]
\tag{5}
$$

where

$$
W_0 = \frac{e^2}{4\pi\hbar} \left(\frac{2m^* \omega_{LO}}{\hbar} \right)^{1/2} \left(\frac{1}{\varepsilon_\infty} - \frac{1}{\varepsilon_s} \right) \, ,
\tag{6}
$$

and $n(\omega_{LO})$ is the phonon occupation number.

We take the non-polar interaction to be modelled by intervalley scattering with phonon frequencies for LO, TO, LA and TA modes taken to be zone-edge values. This rate is simply proportional to the single-spin density of states [4]:

$$
\frac{1}{\tau_{np}} = \sum_i \frac{\pi D_i^2}{\rho \omega_i} \left[n(\omega_i) N(E+\hbar\omega_i) + (n(\omega_i+1) N(E-\hbar\omega_i) \right]
\tag{7}
$$

where ρ is the mass density, D_i is the deformation-potential constant, and the sum is over longitudinal and transverse modes.

For both types of scattering we assume that the overlap integral involving cell-periodic wavefunctions is unity. In eq. (5) contact between the function $\gamma(E)$ and $N(E)$ is made via:

$$
N(E) = \frac{(2m^*)^{3/2} \gamma^{1/2}(E)}{4\pi^2 \hbar^3} \frac{d\gamma(E)}{dE}
\tag{8}
$$

RESULTS

One problem in assessing the non-polar rate is that the relevant deformation-potential constants are unknown for large bandgap materials like GaN. However, values of D for many of the III-V compounds have been calculated [5,6] for particular transitions. In these, the orbital symmetry of the initial and final electron states determine the type of phonon that can participate. Making crude averages over the III-V compounds and symmetries we have estimated the non-polar strengths of the various types of phonon summarized in Table 1 along with data relevant to GaN.

Table 1. Parameters used to calculate rates

a_0 (Å)	ρ (g/cm^3)	$\omega_{LO}(0)$ $|\omega_{LO}(\pi/a)$ $|\omega_{LA}$ $|\omega_{TO}$ $|\omega_{TA}$ 10^{14}s^{-1}	$\varepsilon_\infty/\varepsilon_0$	$\varepsilon_s/\varepsilon_0$	D_{LO} D_{LA} D_{TO} D_{TA} 10^{18} eV/cm
2.25	6.1	1.40 1.20 0.53 1.05 0.38	5.35	9.5	3.5 2.5 0.4 0.1

The rates are shown in Fig. 2 with E_b= 2eV. Note that because of the non-parabolicity of the band the polar rate remains the strongest throughout. The average total rate is about 1.5 x 10^{14}s^{-1} leading to a critical field for WS states to form of 4.4 MV cm^{-1}. The breakdown field of GaN is not very well known but it is thought to be around 3MV cm^{-1}. If so then it is unlikely that WS states can form.

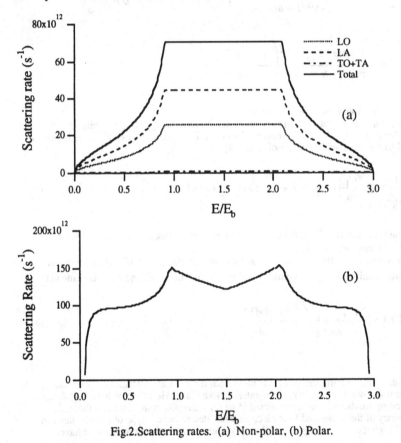

Fig.2.Scattering rates. (a) Non-polar, (b) Polar.

CONCLUSIONS

It is obvious that approaches to the problem of the ontology of WS states need to be more sophisticated than presented here, and should take into account real band structures. This is something that is within the capability of Monte Carlo techniques. Since higher-lying conduction bands may nearly overlap the lowest, the calculation would need to be extended beyond the one-band model. Nevertheless, our estimate of scattering rate is unlikely to be far out because dominated by the polar rate and by the density of states. Uncertainties relating to deformation-potential constants are therefore not fatal, and the number of states is not controversial, being determined by the lattice constant. Our assumption of unity for the cell-periodic overlap integral suggests that our estimate might be an upper limit.

ACKNOWLEDGEMENTS

The author is grateful for support from the ONR and ONR MURI contract # N00014-96-1-1223 monitored by Dr. K. Sleger. He is also grateful to Professor Lester F. Eastman for providing once again congenial surroundings for his visit to Cornell.

REFERENCES

1. F. Bloch, Zeits. f. Physik 52, 555 (1928).
2. G. Wannier, Phys. Rev. 117, 432 (1960).
3. E.M. Conwell and M.O. Vassell, Phys. Rev. 166, 797 (1968).
4. E.M. Conwell, High Field Transport in Semiconductors (Academic Press, NY, 1967).
5. S. Zollner, S. Gopalan and M. Cardona, Appl. Phys. Lett. 54, 614 (1989).
6. S. N. Grinyaev, G.F. Karavaev and V.G. Tyurerev, Sov. Phys. Semicond. 23 905 (1990).

OPTICAL-GAIN MEASUREMENTS
ON GaN AND $Al_x Ga_{1-x} N$ HETEROSTRUCTURES

L. ECKEY, J. HOLST, V. KUTZER, A. HOFFMANN, I. BROSER,
O. AMBACHER*, M. STUTZMANN*, H. AMANO°, I. AKASAKI°

Technische Universität Berlin, Hardenbergstraße 36, 10623 Berlin, Germany

*Walter-Schottky-Institut, Technische Universität München, 85748 Garching, Germany

°Meijo University, Department of Electrical and Electronical Engineering, Nagoya, Japan

ABSTRACT

Optical gain processes in thin GaN and AlGaN are compared by means of gain spectroscopy using the stripe length method and high-excitation photoluminescence, both performed at various densities and temperatures. We find that inelastic excitonic scattering processes and biexciton decay are important at low temperatures and low excitation densities Both materials are similar in that increasing the excitation density results in gain spectra dominated by the electron-hole plasma and phonon-assisted band-to-band recombination. These also prevail at high temperatures.

INTRODUCTION

The physical processes causing stimulated emission in GaN-based laser structures are still subject to discussion. Localized biexcitons [1], the electron-hole plasma [2] and band-to-band recombination [3] have been claimed responsible for the laser mechanism in group-III-nitride heterostructures. Meanwhile, the properties of GaN and especially AlGaN at high excitation levels are not understood in detail. Most investigations have been limited to room temperature not allowing an unambiguous identification of the electronic processes involved. With increasing temperature the contribution of phonon-assisted processes becomes larger resulting typically in a broad spectral region of optical amplification. In our previous work [4] we studied the gain spectra of thick quasi-bulk GaN between 2 K and room temperature and found that excitonic processes add to the gain at high temperatures. The purpose of the present paper is to analyze optical gain spectra obtained from thin GaN epilayers at very high excitation levels and to study the influence of increasing Al content on the gain in AlGaN epilayers. For the identification of the mechanisms providing optical amplification in GaN and AlGaN useful information is obtained from high-density effects on the spontaneous and stimulated photoluminescence (PL).

EXPERIMENTAL

The results presented here were obtained from a MOCVD-grown GaN/SiC epilayer of 3 μm thickness [5] whose low-density free-exciton resonance energy is 3.468 eV at 1.8 K, and from a series $Al_xGa_{1-x}N$ samples grown by MBE on (0001) sapphire with a thickness of about 1 μm [6]. To obtain the high excitation density necessary for our investigations we used a dye laser pumped by an excimer laser, providing pulses with a duration of 15 ns at a rate of 30 Hz and a total energy of up to 20 μJ at 340 nm. AlGaN samples were pumped by the excimer laser at 308 nm using similar pulse energies. The samples were either mounted in a bath cryostat at 1.8 K or in a helium flow cryostat at temperatures varied between 4 K and 300 K. Gain measurements were performed using the stripe length method [e.g., 7].

RESULTS

Dependence of High-Excitation PL from Thin GaN Epilayers on Intensity and Temperature

In a previous paper [8] we demonstrated that the radiative decay of biexcitons (M-band) can be the dominating low-temperature emission and gain process in GaN at excitation densities up to 5 MW/cm². In order to study the properties of the electron-hole plasma (EHP) in GaN we further increased the excitation density. The dependence of the PL of GaN on excitation density up to 50 MW/cm² is displayed in Fig. 1. The most striking observation is the appearance of a new emission band around 3.43 eV at 10 MW/cm². It exhibits a superlinear growth of emission intensity while strongly shifting its peak position to lower energies. This behavior is typical for stimulated emission from an electron-hole plasma. Similar observations in GaN at low temperatures were made before by Cingolani *et al.* [9]. Wiesmann *et al.* recently showed by room temperature measurements that the observation of this strong peak from the surface of the epilayer is due to scattering of in-plane stimulated emission [10]. We ascribe the high-energy maximum around 3.46 eV to a superposition of various spontaneous and stimulated emission processes. With the fast decay of the density of the plasma after the excitation pulse band-to-band recombination (B), inelastic scattering processes between excitons (P), and the formation of biexcitons (M) become possible and all give rise to luminescence in the same energy range. The gain measurements presented below will prove this interpretation.

Fig. 2 compares the temperature dependencies of the observed luminescence peak positions at 3 MW/cm² and at 50 MW/cm². The biexciton luminescence M is quenched at higher lattice temperatures. From the Arrhenius plot of the integrated M-band intensity up to 80 K (shown in the insert) we deduce an activation energy of 3.5 meV which in spite of its high tolerance of 40% confirms the interpretation of this band as well as the previously determined biexciton binding energy of 3.7 meV [8]. Above 60 K the observed peak position of the dominating luminescence at 3 MW/cm² agrees with that expected for inelastic exciton-exciton scattering (P-band) as indicated by the dashed line labeled P. Longitudinal optical (LO) phonon assisted exciton decay appears throughout the measured temperature range up to 250 K at the energy expected. Neither a screening of the exciton

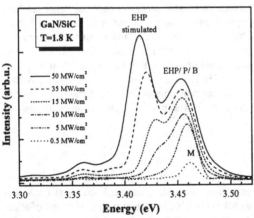

Fig. 1: Low-temperature surface PL from a 3 μm thick GaN/SiC epilayer at excitation densities between 0.5 and 50 MW/cm².

binding nor a band gap renormalization can be seen at this excitation level. In contrast, at 50 MW/cm² no LO replica of free-exciton emission is detected. The stimulated EHP peak exhibits a strikingly strong red shift with temperature. Above 200 K its position is near that of X_A-LO at lower densities. However, up to room temperature the integrated intensity of the EHP peak follows the predicted T^{-3}-rule [11]. This observation indicates that also at room temperature the emission is caused by an EHP. At 270 K the measured value of 3.221 eV is virtually equal to those often reported in the literature for stimulated emission from GaN at room temperature [e.g.,12]. The high-

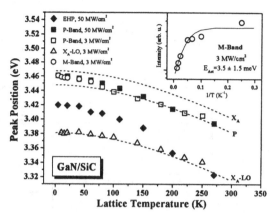

Fig. 2: Peak positions vs. lattice temperature for the PL of the GaN/SiC sample of Fig. 1 at excitation densities of 3 MW/cm² (open symbols) and 50 MW/cm² (full symbols). Dotted lines indicate expected peak positions for the respective emissions. The insert displays the integrated intensity of the M-Band as a function of lattice temperature.

energy band seen in the luminescence spectra at 50 MW/cm² exhibits a red shift with temperature that is weaker than that of the stimulated EHP peak but stronger than the low-density excitonic band gap given by the dashed line labeled X_A. At higher temperatures it appears at the energy position expected for the P-Band. Summarizing the main results of this section, at very high excitation densities the PL of the thin GaN epilayer is dominated by the spontaneous and scattered stimulated emission of the EHP. In addition, a high-energy broad emission is attributed to a superposition of band-to-band recombination and excitonic processes. In the following section we will investigate the processes contributing to the optical gain at different excitation levels and lattice temperatures.

Temperature and Intensity Dependent Gain Properties of Thin GaN Epilayers

Fig. 3 displays two series of gain spectra taken at various excitation densities (a) and temperatures (b), respectively. The spectra presented here were smoothed to enhance the visibility of the observed spectral features. In the density series of Fig. 3 (a) a broadening of the region of optical gain with increasing pump intensity is observed. Additional structures B and EHP appear on both the high- and low-energy shoulders of the main peak M. Their relative strength with respect to M increases with excitation. This main peak is due to biexciton decay below 8 MW/cm². Above, inelastic exciton-exciton scattering P is likely to contribute in the same energy range. The low-energy peak EHP can be identified as due to the electron-hole plasma by its energy position and shift characteristics which are identical to those of the stimulated EHP emission shown in Fig. 1. At 30 MW/cm² the EHP represents the dominating low-temperature gain mechanism with a peak gain value of 250 cm⁻¹. The high-energy gain peak B appears at energies near or above the band gap of the sample. This peak

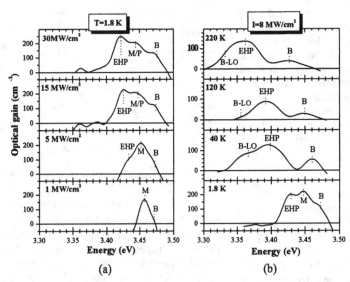

Fig. 3: Optical gain spectra of a 3μm GaN/SiC epilayer (a) at 1.8 K and various excitation densities, (b) at a fixed excitation density of 8 MW/cm^2 and various temperatures. Labels are explained in the text.

does not correlate to a pronounced spontaneous emission peak. The zero-crossing energy due to this process shifts to higher values at increased pump intensities. We ascribe this gain peak to band-to-band recombination. The blue shift with increasing pump intensity is typical for this gain mechanism and is caused by a shift of the respective quasi-Fermi levels of holes and electrons into the valence and conduction bands [13].

Our observation of gain caused by plasma, free carriers, and excitonic processes at highest excitation levels does not imply that they all occur simultaneously. In our time-integrated measurements we detect all processes which produce optical gain after the excitation pulse. Time-resolved measurements of the optical gain in CdS showed that carrier diffusion as well as the formation of biexcitons and excitons contribute to the ultrashort decay of the plasma within 100-200 ps [14]. It is well known that free carriers at high densities as well as biexcitons and inelastic excitonic scattering processes give rise to optical gain themselves. Thus, our observations are in perfect agreement with the typical behavior observed in other direct wide-gap semiconductors. The temperature-dependent gain measurements of Fig. 3 (b) taken at a fixed excitation density of 8 MW/cm^2 show that below the near-gap gain band a second broad low-energy band appears and grows with increasing temperature. It is this structure that is responsible for the stimulated emission at room temperature. In the high-energy band only band-to-band recombination B can be identified without doubt due to its zero-crossing energy above the band gap of the sample at the respective temperatures. Since the shape of this gain band is untypical for pure band-to-band transitions excitonic processes probably also contribute to a small extent to the gain in this region at higher temperatures. Inelastic scattering processes between excitons or between excitons and free carriers are typical gain processes at higher temperatures and were observed in thick GaN epilayers before

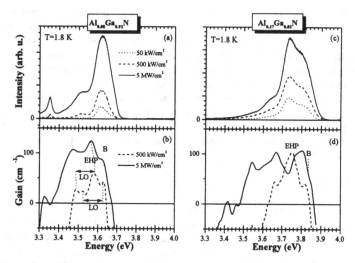

Fig. 4: Comparison of low-temperature high-excitation luminescence (a), (c) and optical gain spectra (b), (d) from two $Al_xGa_{1-x}N$ samples. Legends on the left-hand side also apply to the plots on the right-hand side.

[4]. However, in the thin epilayer investigated here excitonic processes do not play an important role for the gain at high excitation densities and high temperatures. Instead, the electron-hole plasma and LO-assisted band-to-band recombination are the dominating processes causing the low-energy gain band at temperatures above 40 K.

High-Excitation PL and Gain in Thin AlGaN Epilayers

Fig. 4 displays a comparison of high-excitation photoluminescence and gain spectra taken at 1.8 K from two $Al_xGa_{1-x}N$ samples with x equal 0.08 on the left-hand side and 0.17 on the right-hand side. The excitation density was varied between 50 kW/cm² and 5 MW/cm². From low-density PL measurements the excitonic band edge of these samples is found at 3.63 eV (x=0.08) and 3.83 eV (x=0.17) at 1.8 K. Both the luminescence and gain spectra of the two samples extend to energies well above these values, shifting further to higher energies with increasing excitation density. This behavior gives strong evidence for band-to-band recombination. The gain spectrum of $Al_{0.08}Ga_{0.92}N$ in Fig. 4 (b) also exhibits a pronounced peak around 3.57 eV strongly shifting to lower energies with increasing excitation density. This allows an attribution of this band to the electron-hole plasma. No indication of a stimulated peak is observed in the luminescence spectra up to 5 MW/cm². The two weak low-energy gain structures around 3.5 eV are found one LO phonon energy below the dominating lines and strongly gain intensity with increasing pump power. The role of phonon-assisted processes is stronger here than in the experiments on GaN because the excitation energy (4 eV) is higher above the band gap producing a plasma with higher effective temperature whose relaxation towards thermal equilibrium creates a larger number of phonons available for a stimulated emission process. Unlike the results for GaN even at low temperatures excitonic processes do not seem to contribute much to the optical gain at the given excitation densities in this AlGaN sample. We ascribe this to the higher density of excited carriers due to the small thickness of 1 μm which is a factor of three lower than that of the GaN sample studied. Similar considerations apply to the gain spectra of the $Al_{0.17}Ga_{0.83}N$ sample in Fig. 4 (d). It is obvious by the spectrally broad region of optical amplification that phonon-assisted processes play the dominant role at 5 MW/cm^2. From the results obtained to this point we conclude that the AlGaN samples investigated show gain mechanisms very similar to those of high-quality GaN. The electron-hole plasma and band-to-band recombination are the main processes providing optical gain at high excitation densities. While the absolute values are somewhat lower our results show that the fabrication of laser structures with active AlGaN layers operating in the ultraviolet spectral range is possible.

CONCLUSIONS

In conclusion, we compared the luminescence and optical gain properties of thin GaN and AlGaN epilayers at high excitation densities. The GaN and AlGaN samples investigated exhibit a similar behavior. Excitonic processes play an important role only at relatively low densitiy and at low temperatures. With increasing pump intensity the electron-hole plasma and band-to-band recombination exhibit the highest gain reaching values of 250 cm^{-1} in GaN and 150 cm^{-1} in AlGaN. Towards room temperature phonon-assisted band-to-band recombination and electron-hole-plasma create a gain structure roughly 100 meV wide giving rise to the often reported stimulated emission peak.

REFERENCES

1. M. Sugawara, Jpn. J. Appl. Phys. **35**, 124 (1996)

2. W. W. Chow, Appl. Phys. Lett. **66**, 3000 (1995)

3. G. Frankowsky, F. Steuber, V. Härle, F. Scholz, A. Hangleiter, Appl. Phys. Lett. **68**, 3746 (1996)

4. L. Eckey, J.-Chr. Holst, A. Hoffmann, I. Broser, T. Detchprohm, K. Hiramatsu, MRS Internet J. of Nitride Semiconductor Research **2**, 1 (1997)

5. H. Amano, K. Hiramtsu, I. Akasaki, Jpn. J. Appl. Phys. **27**, L1384 (1988)

6. H. Angerer, O. Ambacher, R. Dimitrov, T. Metzger, W. Rieger, M. Stutzmann, MRS Internet J. of Nitride Semiconductor Research **1**, 15 (1996)

7. K. L. Shaklee, R. E. Nahory, R. F. Leheny, J. Lumin. **7**, 284 (1973)

8. L. Eckey, J. Holst, A. Hoffmann, I. Broser, H. Amano, I. Akasaki, T. Detchprohm, K. Hiramatsu, Proc. 23rd Int. Conf. Phys. Semicond., ed. M. Scheffler, R. Zimmermann; World Scientific, Singapore, 1996, p. 2861

9. R. Cingolani, M. Ferrara, M. Lugarà, Solid State Communications 60, 705 (1986)

10. D. Wiesmann, I. Brener, L. Pfeiffer, M. A. Khan, C. J. Sun, Appl. Phys. Lett. **69**, 3384 (1996)

11. Y. Pokrovskii, phys. stat. sol **82**, 385 (1972)

12. H. Amano, T. Asahi, M. Kito I. Akasaki, J. Lumin. **48&49**, 889 (1991)

13. G. Lasher, F. Stern, Phys. Rev. B **133**, A553 (1964)

14. H. Saito, E. Göbel, Phys. Rev. B **31**, 2360 (1985)

OPTICAL ANISOTROPY OF THE OPTICAL RESPONSE FOR STRAINED GaN EPILAYERS GROWN ALONG THE <10-10> DIRECTION

A.ANDENET, B.GIL, Y.-M.LE VAILLANT, S.CLUR, O.BRIOT, AND R.L.AULOMBARD
CNRS-GES, Université de Montpellier II -Case courrier 074 - 34095 Montpellier Cedex 5
France

ABSTRACT

Group theory states that the spectroscopy of excitons in GaN epilayers grown with biaxial stress on M Plane (10-10)-oriented substrates is strongly anisotropic in the growth plane, in contrast to the situation if growth occurs on conventional C plane substrates. In addition, we predict existence of nine radiative transitions instead of five, three for each family of A, B, and C excitons. We calculate the in-plane anisotropy and show that, due to the crossed configuration of the wurtzite crystal field and the built-in strain one, the A exciton is significantly coupled to the electromagnetic field in π polarization (E//c).

INTRODUCTION

The lack of large scale substrates, lattice matched to GaN has long been thought to thwart possibility to realize long life operating devices based on group III nitrides. Alternative substrates such as 6H-SiC, or ZnO, or oxides have been proposed to tentatively by-pass some of the drawbacks of the «classical» epitaxy on C-plane sapphire[1]. Growth on other oriented faces may be performed too, in order to reduce the lattice mismatch between the substrate and the epilayers, or, in heterostructures, to modify the confinement masses, the in-plane dispersion relation relative to the hole motion[2,3]. From this, one expect subsequent implications on laser thresholds. In the present paper we study optical properties of the exciton in GaN layers grown with strain on M-plane substrates. A comparison with similar properties but for conventional (0001)-grown epilayers without in-plane anisotropy is addressed.

GROUP THEORY ANALYSIS

Unstrained GaN preferentially crystallizes in the wurtzite phase belonging to the C_{6v}^4 ($P6_3mc$) space group, with two formula units per primitive cell. The conduction band transforms like Γ_7^c whilst the symmetry of the valence band writes $\Gamma_9^v + \Gamma_7^v + \Gamma_7^v$. The corresponding band to band transitions can be distinguished by appropriate selection rules: the transition between the Γ_9^v valence band and the Γ_7^c conduction band is forbidden in π configuration *i.e.* when the electric field of the interacting photon is z-oriented[4]. Such a simple description does not include the excitonic effect, which we recently studied for excitons in epilayers grown on C-plane[5]. We found it, when included, to produce fine structure splitting between Γ_5 and Γ_1 exciton sublevels, radiatively coupled to the σ- and π- polarized photons respectively, and with hole wave functions mainly built from Γ_7^v valence states. Figure 1 left-hand side illustrates the relationships between bands symmetries and the excitonic sublevels levels in unstrained GaN. If epitaxy of GaN is performed along the <001> direction, on (001) plane substrates the symmetry of the crystal is not altered[6]. This does not remain true in case of growth along a lower symmetry direction, if for instance M-plane substrates are used. The crystal point group reduces from the hexagonal symmetry to an orthorhombic one[3]. We note from the right hand side of figure 1, that the twelve fold exciton splits into three Γ_1 levels which selection rules report to be observed for Z polarized photons, three Γ_2 levels optically active under X polarization, three Γ_4 levels optically active under Y polarization and three dipole forbidden states having Γ_3 symmetry. If the orthorhombic distortion of the lattice is small, we anticipate that the strengths of optical transitions will be a reminiscence of the strengths they have in wurtzite symmetry.

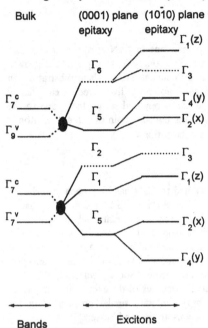

Bands   Excitons

Figure 1: Left hand side: the construction of the exciton states from zone center conduction and valence band Bloch states of unstrained GaN. Middle in the figure are given the symmetry of the exciton states in wurtzite environnement. Γ_1 and Γ_5 levels are dipole allowed for E // c and E $\perp$ c polarizations respectively. Γ_2 and Γ_6 states are not coupled with the electromagnetic field. Right hand part: the exciton states in orthorhombic symmetry (growth on (10-10) plane). The polarizations of the light which can create this excitons are given between parenthesis.

THE RESOLUTION OF THE STRAIN ORBITAL HAMILTONIAN

The exciton hamiltonian Ξ_{exc} is written under the assumption that the strain field can be treated as a perturbation of the wurtzite crystal field[5]:

$$\Xi_{exc} = H_{c=0} + H_{v=0} + H_{cstrain} + H_{vstrain} + H_{exc}$$

where $H_{v=0}$ ($H_{c=0}$) are the strain-free valence (conduction) band hamiltonians and $H_{vstrain}$ ($H_{cstrain}$) are their respective, which account of strain-related effects on the evolution of band extrema.

$$H_{v=0} = <E_{0v}> + H_{cr} + H_{so}$$

$<E_{0v}>$ is the average energy in the valence band, $H_{cr} = \Delta_1 L_z^2$, where L_z is the z projection of the hole (spinless) angular momentum. If we now introduce the spin, two parameters are required to describe the spin-orbit interaction which we write as a function of the components of the angular momentum and spin of the hole as:

$$H_{so} = \Delta_2 L_z \sigma_z + \Delta_3 (L_x \sigma_x + L_y \sigma_y).$$

The last operator H_{exc}, we write:

$$H_{exc} = R^* + \tfrac{1}{2} \gamma \, \sigma_h \cdot \sigma_c$$

where R^* is the customary exciton binding energy and the last term is the crystalline exchange interaction. Operators σ_h and σ_c operate on valence hole and conduction electron spin functions. In the most general way, the strain hamiltonian writes using six *spin-independent* deformation potentials[6]

$$H_{vstrain} = (C_1 + C_3 L_z^2)e_{zz} + (C_2 + C_4 L_z^2)e_\perp + C_5(L^2 \cdot e_+ + L^2_+ e_-) + C_6 \{[L_z L_+] \, e_{+z} + [L_z L_-] e_{-z}\}$$

where

$$L_\pm = 1/\sqrt{2}(L_x \pm iL_y), \ 2[L_i L_j] = L_i L_j + L_j L_i \ , \ e_\perp = e_{xx} + e_{yy} \ , \ e_\pm = e_{xx} \pm 2ie_{xy} - e_{yy} \ , \ e_{\pm z} = e_{xz} \pm ie_{yz}$$

The strain hamiltonian acts as follows on the conduction band:

$$H_{cstrain} = D_1 e_{zz} + D_2 e_\perp$$

Now, in this paper, for which strain fields are restricted to (10-10)-oriented ones, the non vanishing components of the strain tensor are e_{zz}, e_{xx} and e_{yy}. It is thus very convenient for the clarity of the presentation, to introduce the following notations :

$\delta_1 = (D_1 + C_1) e_{zz} + (D_2 + C_2) e_\perp = a_1 e_{zz} + a_2 e_\perp$, $\delta_2 = C_3 e_{zz} + C_4 e_\perp$, and $\delta_3 = C_5(e_{xx}-e_{yy})$. In addition, the exciton basis functions are constructed from the P-like basis functions, which are noted p_x, p_y, and p_z or even simpler: x, y, z, we define eigenvectors $|+>$, $|p_+>$, $|->$ and $|p_->$ as the following linear combinations:

$|+> = |p_+> = -(p_x + ip_y)/\sqrt{2}$, and $|-> = |p_-> = (p_x - ip_y)/\sqrt{2}$.

Spin components of the missing valence electron are ↑ and ↓, and α and β represent the spin components of the conduction electron.

The three Γ_1 and the three Γ_3 excitons levels are easily obtained as the solutions of two 3x 3 block-diagonal matrices:

$\|+\uparrow\beta>$	$\|(p.\uparrow\beta+p_+\downarrow\alpha)/\sqrt{2}>$	$\|p_z(\uparrow\alpha+\downarrow\beta)/\sqrt{2}>$	$\|-\downarrow\alpha>$	$\|(p.\uparrow\beta - p_+\downarrow\alpha)/\sqrt{2}>$	$\|p_z(\uparrow\alpha-\downarrow\beta)/\sqrt{2}>$
$\Delta_1 + \Delta_2 +$ $\delta_1 + \delta_2 +\gamma/2$	$\delta_3/\sqrt{2}$	0	0	0	0
$\delta_3/\sqrt{2}$	$\Delta_1 - \Delta_2 + \delta_1 + \delta_2 +$ $\gamma/2$	$\sqrt{2}\Delta_3$	0	0	0
0	$\sqrt{2}\Delta_3$	$\delta_1 - 3\gamma/2$	0	0	0
0	0	0	$\Delta_1 + \Delta_2 +$ $\delta_1 + \delta_2 +\gamma/2$	$-\delta_3/\sqrt{2}$	0
0	0	0	$-\delta_3/\sqrt{2}$	$\Delta_1 - \Delta_2 +$ $\delta_1 + \delta_2 +\gamma/2$	$\sqrt{2}\Delta_3$
0	0	0	0	$\sqrt{2}\Delta_3$	$\delta_1 + \gamma/2$

Where the orthorhombic excitons having Γ_1 (respectively Γ_3) symmetry are eigenstates of the left hand part (resp. right hand part) block. It is a straightforward elementary quantum mechanics exercise to demonstrate that, for the z-polarization, the oscillator strength of Γ_1 modes is proportional to the square of the module of the contribution of $|p_z(\uparrow\alpha+\downarrow\beta)/\sqrt{2}>$ in the three eigen vectors of the resolved problem (the transition is not spin-flip)[5].

The situation is a little bit more delicate if it is to resolve levels derived from the three doubly degenerate Γ_5 wurtzite excitons. As said above, group theory predicts six different energy levels gathered into two groups of three, respectively having Γ_2 and Γ_4 symmetry in the orthorhombic GaN. We work in a basis obtained by treating the interaction of the two Γ_7 valence band states, setting γ and $\delta_3= 0$. The solutions E_B and E_C are the well known solutions of a two-level system. We thus can derive the corresponding sets of doubly degenerated eigen vectors: $\{ |\Psi_B>, |\Psi_B'> \}$ and $\{ |\Psi_C>, |\Psi_C'> \}$.

$$|\Psi_B> = a |{+}{\downarrow}\beta> + \sqrt{(1-a^2)} |z{\uparrow}\beta> \quad , \quad |\Psi_B'> = a |{-}{\uparrow}\alpha> + \sqrt{(1-a^2)} |z{\downarrow}\alpha>,$$

$$|\Psi_C> = \sqrt{(1-a^2)} |{+}{\downarrow}\beta> - a |z{\uparrow}\beta> \quad , \quad |\Psi_C'> = \sqrt{(1-a^2)} |{-}{\uparrow}\alpha> - a |z{\downarrow}\alpha>$$

where parameter a is obtained from the resolution of the eigenvector problem for the valence band[7].Then additive and substractive linear combinations give Γ_4 and Γ_2 exciton basis

Γ_4 orthorhombic exciton states are :

$$|A^+> = (|{+}{\uparrow}\alpha> + |{-}{\downarrow}\beta>)/\sqrt{2} = |x({-}{\uparrow}\alpha + {\downarrow}\beta)>/\sqrt{2} - i |y({\uparrow}\alpha + {\downarrow}\beta)>/\sqrt{2}$$

$$|B^+> = (|\Psi_B> + |\Psi_B'>)/\sqrt{2} = a |x({\uparrow}\alpha - {\downarrow}\beta)>/\sqrt{2} - i a |y({\uparrow}\alpha + {\downarrow}\beta)>/\sqrt{2} + \sqrt{(1-a^2)}/\sqrt{2} |z({\downarrow}\alpha + {\uparrow}\beta)>$$

$$|C^+> = \sqrt{(1-a^2)} |x({\uparrow}\alpha - {\downarrow}\beta)>/\sqrt{2} - i \sqrt{(1-a^2)} |y({\uparrow}\alpha + {\downarrow}\beta)>/\sqrt{2} - a/\sqrt{2} |z({\downarrow}\alpha + {\uparrow}\beta)>$$

and Γ_2 orthorhombic ones are :

$$|A^-> = - |x({\uparrow}\alpha + {\downarrow}\beta)>/\sqrt{2} + i |y({-}{\uparrow}\alpha + {\downarrow}\beta)>/\sqrt{2}$$

$$|B^-> = -a |x({\uparrow}\alpha + {\downarrow}\beta)>/\sqrt{2} - i a |y({-}{\uparrow}\alpha + {\downarrow}\beta)>/\sqrt{2} - \sqrt{(1-a^2)}/\sqrt{2} |z({\downarrow}\alpha - {\uparrow}\beta)>$$

$$|C^-> = - \sqrt{(1-a^2)} |x({\uparrow}\alpha + {\downarrow}\beta)>/\sqrt{2} - i \sqrt{(1-a^2)} |y({-}{\uparrow}\alpha + {\downarrow}\beta)>/\sqrt{2} + a/\sqrt{2} |z({\downarrow}\alpha - {\uparrow}\beta)>$$

From the spin parts of the wave functions above, it is obvious that Γ_4 and Γ_2 exciton states are radiative for Y and X polarizations respectively.

In the novel basis, we obtain:

$|A^+> (\Gamma_4, Y)$	$|B^+> (\Gamma_4,Y)$	$|C^+> (\Gamma_4,Y)$	$|A> (\Gamma_2,X)$	$|B> (\Gamma_2,X)$	$|C> (\Gamma_2,X)$
$E_A - 1/2\gamma$	$-a(\gamma-\delta_3)$	$-\sqrt{(1-a^2)}(\gamma-\delta_3)$	0	0	0
$-a(\gamma-\delta_3)$	$E_B + \gamma(1-2a^2)/2$	0	0	0	0
$-\sqrt{(1-a^2)}(\gamma-\delta_3)$	0	$E_C - \gamma(1-2a^2)/2$	0	0	0
0	0	0	$E_A - 1/2\gamma$	$-a(\gamma+\delta_3)$	$-\sqrt{(1-a^2)}(\gamma+\delta_3)$
0	0	0	$-a(\gamma+\delta_3)$	$E_B + \gamma(1-2a^2)/2$	0
0	0	0	$-\sqrt{(1-a^2)}(\gamma+\delta_3)$	0	$E_C - \gamma(1-2a^2)/2$

The excitons eigenstates can be expanded along the basis vectors according with the expressions:

$$\Psi_{exc\Gamma2} = \upsilon_2 |A> + \omega_2 |B> + \varpi_2 |C>$$
$$\Psi_{exc\Gamma4} = \upsilon_4 |A^+> + \omega_4 |B^+> + \varpi_4 |C^+>$$

Because the transitions we interest about are not spin-flip between the valence missing electron state and the conduction electron one, the oscillator strengths Φ_i in X and Y polarizations are given by:

$$\Phi_X \sim (\upsilon_2^2 + a^2\omega_2^2 + (1-a^2)\varpi_2^2)/2$$
$$\Phi_Y \sim (\upsilon_4^2 + a^2\omega_4^2 + (1-a^2)\varpi_4^2)/2$$

NUMERICAL RESULTS

For GaN grown on M-Plane substrates, we take the components of the deformation tensors $e_{xx} = (S_{12} + S_{13})\sigma$, $e_{yy} = (S_{11} + S_{13})\sigma$, and $e_{zz} = (S_{33} + S_{13})\sigma$. where σ is a residual biaxial stress. Next, using for the crystal field splitting (Δ_1), spin-orbit (Δ_2,Δ_3) and exchange (γ) interaction, and deformation potentials parameters below: $\Delta_1 = 10 \pm 0.1$ meV $\Delta_2 = 6.2 \pm 0.1$ meV, $\Delta_3 = 5.5 \pm 0.1$ meV, $\gamma = 2$ meV, $a_1 = -5.32$ eV, $a_2 = -10.23$ eV, $C_3 = -2C_4 = -4.91$ eV and $C_5 = -2$ eV[8]. The components of the compliance tensor are obtained from the recent determination of the stiffness coefficients in GaN [9] : $C_{11} = 365$ GPa, $C_{12} = 135$ GPa , $C_{13} = 114$ GPa and $C_{33} = 381$ Gpa.

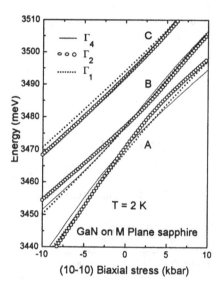

Figure 2: Evolution of the exciton states under (10-10) biaxial stress. The influence of the crossed configurated wurtzitic and strain field is obvious if comparing this data with figure 2 of reference 5. Note that the splittings between in-plane polarized photons Γ_1 and Γ_4, although small are in the range of experimental resolution.

Figure 2 displays the evolution of the exciton levels under biaxial stress in the M Plane. Positive values of the stress correspond to biaxial compression. This will be also the case for all figures hereafter. We note that our calculation predicts exciton mixings stronger than for (0001)-grown epilayers. Careful examination of the matrices for Γ_4 and Γ_2 excitons reveal that extra diagonal terms which couples $|A^+>$, $|B^+>$ or $|A>$, $|B>$, proportional to $\gamma \pm \delta_3$ vanish for opposite values of the stress. As a consequence we expect to have excitonic oscillator strength to reach maximum values (0.5) for two opposite stresses. Last, the anisotropy of the optical response in the (10-10) plane is associated with measurable splittings between Γ_4 and Γ_1 excitons.

Figure 3 illustrates the stress-induced evolution of the oscillator strength of Γ_1 excitons. For the convenience of the discussion we keep the standard A,B,C notations which are only meaningful at zero stress. We note that this anisotropic strain field gives oscillator strength to the Γ_1 (A) exciton, which was not obtained in standard growth conditions. This property is very interesting and this effect merits further detailed examination. Combined with calculations of the joint density of state, we could eventually suggest the design of heterostructures for lasing with differential optical gain[3], or detectors or modulators taking advantage of the gain of oscillator strength produced by the strain fields related to growth on low symmetry M Plane. The latter excitons Γ_1 (B) and Γ_1 (C) exchange their oscillator strengths in a manner comparable to the situation in (0001)-grown epilayer for σ-polarized light, when the electric field of the incident photon is in crossed-configuration with the wurtzite crystal field[5]. If the Γ_1 (C) exciton looses Z-oriented oscillator strength when the biaxial compression increases, it is not surprising to expect an increase in the other polarization situations. This is effectively observed for Γ_4 (C) and Γ_2 (C) excitons in the adapted polarization situations, as shown on figure 4 for Γ_4 excitons. The anisotropy of the in-plane optical response is well evidence by the difference between oscillator strength of Γ_4 and Γ_1 excitons in the Y and Z polarizations respectively.

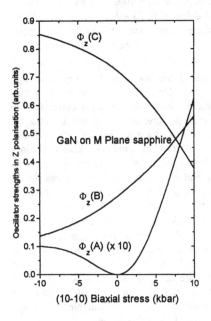

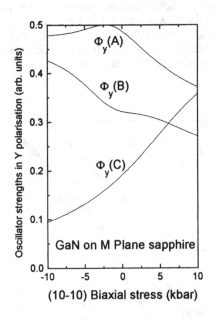

Figure 3: Oscillator strengths we calculate in case of Z polarization for the three A, B, C -like Γ_1 sub levels of the twelve fold exciton in GaN with orthorhombic biaxial lattice distortion.

Figure 4: Oscillator strengths we calculate in case of Y polarization for the three A, B, C -like Γ_4 sub levels of the twelve fold exciton in GaN with orthorhombic biaxial lattice distortion.

CONCLUSION

We have developed the theoretical examination of the optical properties of excitons in GaN epilayers biaxially strain in the (10-10) growth plane. This gives different behaviors under strain comparing to the case of growth on C plane substrates with in particular a strong in-plane anisotropy of the optical response and measurable fine structure splittings. In addition, these effects suggest us to envision that quantum wells grown on M plane oriented sapphire, SiC and ZnO substrates might exhibit anisotropic non linear optical effects, probably more dramatic than for quantum wells grown along the conventional <0001> orientation.

Last there still remains an uncertainty concerning the band gap of unstrained GaN. Previous investigations of epilayers grown along the (0001) direction do not allow to unambiguously adress this determination. By growing GaN on M plane and c plane oriented Al_2O_3 and 6H-SiC one could switch the sign of the residual strain and thus extract the value of unstrained GaN by cancelling the strength of A line in z polarization.

ACKNOWLEDGEMENTS

This work has been supported by the DRET/DGA and by THOMSON-CSF-LCR.

REFERENCES

[1] -see for instance chapter 1 of « Gallium Nitride and Related Materials » in Materials Research Society Symposium Proceedings volume **395** of Edited by F.A.Ponce, R.D.Dupuis, S.Nakamura, and J.A.Edmond (1996)
[2] T Matsuoka, Materials Research Society Symposium Proceedings volume **395** Edited by F.A.Ponce, R.D.Dupuis, S.Nakamura, and J.A.Edmond , 39 (1996)
[3] M.Susuki, and T.Uenoyama, Jpn. J.Appl. Phys. **35**, L953, (1995)
[4] in this paper, we use the notationsof « Semiconductor Optics » by C.F.Klingshirn Springer Verlag- Berlin Heidelberg New York (1995)
[5] B.Gil and O.Briot, Phys. Rev. B. **55**, 2530, (1997)
[6] B.Gil, O.Briot, and R.L.Aulombard, Phys. Rev. B **52**, R 17 028, (1995)
[7] B.Gil, F.Hamdani and H.Morkoç, Phys. Rev. B, **54**, 7680, (1996)
[8] M. Tchounkeu, O.Briot, B.Gil, and R.L.Aulombard J. of Applied Physics **80**,5352,(1996)
[9] A.Shikanai,T.Azuhata, T.Sota, S.Chichibu, A.Kuramata, K.Horino, and S.Nakamura J.Appl.Phys.**81**, 417,(1997)

ELECTRONIC AND OPTICAL PROPERTIES OF BULK GaN AND GaN/AlGaN QUANTUM WELL STRUCTURES

M. SUZUKI and T. UENOYAMA
Central Research Laboratories, Matsushita Electric Industrial Co., Ltd.,
3-4 Hikaridai, Seikacho, Sourakugun, Kyoto 619-02, Japan, suzuki@crl.mei.co.jp

ABSTRACT

Electronic structures and optical gains of bulk GaN and GaN/AlGaN quantum wells (QWs) are theoretically investigated for the wurtzite and the zincblende structures, using the k·p theory. It is found that the lower crystal symmetry, that is the wurtzite, is preferable for the lower threshold carrier density in the bulk. Although the QW structure leads to symmetry lowering only in the zincblende, we can not find a significant benefit of the zincblende QWs. As for the reduction of the threshold carrier density, biaxial strains are more effective in the zincblende. However, the threshold carrier density is still higher than in the wurtzite. It is proposed that the uniaxial strain in the c-plane of the wurtzite is more useful for reducing it.

INTRODUCTION

The III-V nitrides are attractive for optoelectronic devices at the short wavelengths. During the past few years, remarkable progress has been made in development of the optical devices based on wurtzite (WZ) GaN. At present, high-brightness blue light emitting diodes (LEDs), based on WZ InGaN/AlGaN single quantum well (SQW) structure [1], are commercially available. Very recently the lasing action of InGaN multi quantum well (MQW) laser diodes (LDs) by current injection was first demonstrated [2]. Furthermore, the recent development of crystal growth techniques made it possible to obtain high quality zincblende (ZB) crystals as well. However, many fundamental material and device characteristics of the III-V nitrides are less understood than conventional ZB compounds.

We have already studied the electronic and optical properties of WZ GaN/AlGaN QWs, including strain effects, based on the k·p theory [3, 4, 5, 6]. The physical parameters have been derived from the first-principles band calculations [4, 5, 7, 8]. It was found that the threshold carrier density of WZ nitride LDs is higher than conventional ZB LDs [3, 4] and that the biaxial strain is not so effective to reduce it [4, 6]. In addition, it was suggested that the uniaxial strain in the c-plane is very useful for reducing it [5, 6]. On the other hand, it is expected that ZB nitrides might have special advantages to the LD operation. Although there have been a few studies on ZB GaN LDs [9, 10, 11], the comprehensive study, taking the detail electronic structure into account and comparing it with WZ GaN LDs, has not been reported yet. Thus, it is very important to clarify the relation between the LD performance and the crystal symmetry in III-V nitrides.

In this paper, the electronic and optical properties of bulk GaN and GaN/AlGaN QW structures are discussed from the point of view of crystal symmetry. Since the crystal symmetry largely affects the feature of the valence band maximum (VBM), WZ and ZB

251

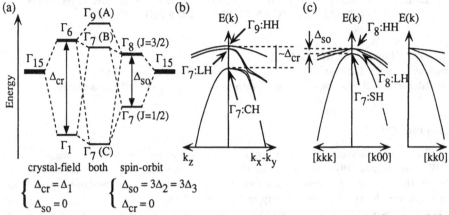

Figure 1: (a) Splitting at the top of valence bands of GaN under the influence of the crystal-field and the spin-orbit coupling. (b) and (c) show the schematic band structures around the top of valence bands of the wurtzite and the zincblende GaN, respectively.

symmetries should be considered in the 6 × 6 k·p and strain Hamiltonians. The detail can be found in the references [6, 12]. The parameters of ZB GaN have been also derived, recently [13]. It is also shown that the lower crystal symmetry is one of the preferable approaches for improving the LD performance but that there is no significant benefit of ZB nitrides [14].

ELECTRONIC BAND STRUCTURES

The electronic band calculations have been performed by using the full-potential linearized augmented plane wave (FLAPW) method [15]. We focus on the features of electronic structures of WZ and ZB GaN near the VBM. They are very different from conventional ZB III-V compounds, even in the ZB structure. Figure 1 shows (a) the splitting at the top of the valence bands of GaN and (b), (c) the schematic band structures around the VBM of WZ and ZB GaN, respectively.

In the WZ structure, the VBM is split into two-fold Γ_6 and non-degenerate Γ_1 states even without the spin-orbit interaction. The order between them depends on the kind of materials and the lattice parameters. In case of GaN (AlN), the Γ_6 (Γ_1) level is higher. Considering the spin-orbit interaction, the Γ_6 state is split into Γ_9 and Γ_7 levels, and the Γ_1 state is labeled Γ_7 as well. As shown in Fig. 1 (b), the hole effective masses have the large k-dependence. We label three hole bands as HH (heavy), LH (light) and CH (crystal-field split-off), based on the feature in the k_x-k_y plane. The HH mass is heavy along any k direction. On the other hand, the LH (CH) mass is light (heavy) in the k_x-k_y plane but is heavy (light) along k_z direction. Note that the anisotropy of the hole masses is not negligible when we carry out the material design or the characteristic analysis of the GaN-based QW devices like LDs.

In the ZB structure, the three-fold degenerate Γ_{15} state is not split into the two-fold Γ_8 state and the non-degenerate Γ_7 state unless the spin-orbit interaction is taken into

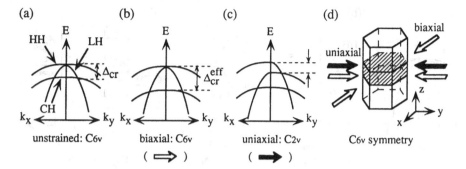

(a)   (b)   (c)   (d)

unstrained: C6v   biaxial: C6v   uniaxial: C2v   C6v symmetry

Figure 2: Schematic band structure in the k_x-k_y plane around the top of valence bands of the wurtzite GaN, (a) without a strain, (b) with a biaxial strain and (c) with a uniaxial strain in the c-plane. (d) shows the direction of each strain.

account. We label three hole bands as HH (heavy), LH (light) and SH (spin-orbit split-off), conventionally. However, the LH mass is as heavy as the HH one along any **k** direction, and it is quantitatively different from ZB GaAs. This difference is very significant, and it originates from very weak spin-orbit coupling. Then, we must equivalently treat the upper six valence bands in the analysis of the GaN-based QW devices.

STRAIN EFFECT ON ELECTRONIC BAND STRUCTURES

Figure 2 shows the strain effect on the valence band of bulk WZ GaN. It is found that each energy dispersion of HH, LH and CH bands is almost unchanged even under stress. Because the weak spin-orbit coupling makes the orbital character such as $|X\rangle$, $|Y\rangle$ and $|Z\rangle$ dominant. Under the compressive (tensile) biaxial stress, only the crystal-field splitting Δ_{cr} becomes effectively large (small). Because the strained crystal has still C_{6v} symmetry, as well as the unstrained one. Then, the HH mass is still heavy, and the density of states (DOS) at the VBM is not so reduced. This is very different from the ZB crystals, where the HH mass becomes light and the DOS decreases considerably since the T_d symmetry is changed to the D_{2d} one and the degeneracy is removed. On the other hand, the uniaxial strain in the c-plane gives the anisotropic energy splitting in the k_x-k_y plane since it causes the symmetry lowering from C_{6v} to C_{2v}. When the compressive uniaxial strain along the y-direction is applied, the HH band along the k_x direction as well as the LH band along the k_y direction moves to higher energy side. Then, the DOS at the VBM is largely reduced, compared with the conditions without a strain or with a biaxial strain.

Figure 3 shows the strain effect in the (001) plane on the valence band of bulk ZB GaN. It is found that the strain effect on HH and LH bands is almost the same as conventional III-V compounds, where the HH (LH) mass becomes light (heavy) due to the removal of the degeneracy. However, the HH mass of the strained ZB GaN is still quite heavy because both HH and LH masses of the unstrained one are very heavy. Here, under the biaxial

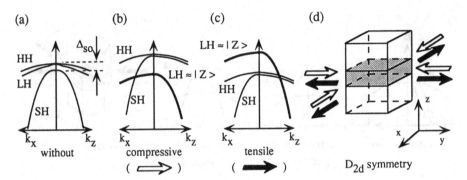

(a)   (b)   (c)   (d)

without   compressive   tensile D_{2d} symmetry

Figure 3: Schematic band structure in the k_x-k_y plane around the top of valence bands of the zincblende GaN, (a) without a strain, (b) with a compressive biaxial strain and (c) with a tensile biaxial strain in the (001) plane. (d) shows the direction of each strain.

stress, Δ_{cr} is effectively induced by the symmetry lowering to D_{2d}. Then, it looks as if compressive biaxial strained ZB GaN is WZ GaN.

VALENCE SUBBAND STRUCTURES

Assuming that the heterojunction is perpendicular to the z-axis, a wave vector k_z becomes an operator in the k·p Hamiltonian. The well-type potential by the heterojunction must be added to the Hamiltonian. Then, we can obtain the subband structure by solving a matrix differential eigenvalue problem. The parameters used in the Hamiltonians are theoretically derived, except for the elastic stiffness constants, as which the observed (transformed) ones for WZ (ZB) GaN are adopted [16]. Using the parameters listed in Tables 1 and 2, the subband structures in WZ and ZB GaN/Al$_{0.2}$Ga$_{0.8}$N SQWs have been calculated.

Subband in Wurtzite Quantum Wells

Figure 4 (a) shows the valence subband structure in the unstrained WZ GaN/Al$_{0.2}$Ga$_{0.8}$N QWs. As long as the Hamiltonian is within the second order of k, the energy dispersions are isotropic in the k_x-k_y plane. This is one of the remarkable features in WZ QWs. The energy bands can be labeled as HH$_i$, LH$_i$ and CH$_i$ as well as in bulk WZ GaN. For any well length, the coupling between HH$_i$ bands and the other ones is weak. On the other hand, the LH$_i$ bands are strongly coupled with the CH$_i$ bands with a different parity. However, the feature, typically shown by the above three labels, is almost unchanged even in the QWs. Because the crystal symmetry is still C_{6v} as well as in the bulk. In conventional ZB QWs, the heavy hole mass becomes light, compared with the bulk. In the WZ nitride QWs, however, the decrease of the hole effective mass by the two dimensional carrier confinement is prevented by the weak spin-orbit coupling.

254

Table 1: Electron effective masses (m_0), spin-orbit and crystal-field splitting energies (meV), Luttinger (-like) parameters ($\hbar^2/2m_0$) and deformation potentials (eV). $\parallel$ and $\perp$ denote the parallel and perpendicular to the c-axis, respectively

wurtzite	GaN	AlN	zincblende	GaN	AlN
$m_e^{\parallel}$	0.20	0.33	m_e^*	0.17	0.30
$m_e^{\perp}$	0.18	0.25			
Δ_{so}	16	20	Δ_{so}	20	20
Δ_{cr}	72	-58			
A_1	-6.56	-3.95	γ_1	2.70	1.50
A_2	-0.91	-0.27	γ_2	0.76	0.39
A_3	5.65	3.68	γ_3	1.07	0.62
A_4	-2.83	-1.84			
A_5	-3.13	-1.95			
A_6	-4.85	-2.92			
A_7	0.00	0.00			
D_1	-15.35	-12.92	a	-13.33	-9.95
D_2	-12.32	-8.46	b	-2.09	-2.17
D_3	3.03	4.46	d	-1.75	-2.57
D_4	-1.52	-2.23			
D_5	-2.05	-2.57			
D_6	-3.66	-4.12			

Table 2: Elastic stiffness constants (10^{11} dyn/cm^2) and band offsets (eV).

	GaN					GaN/Al$_{0.2}$Ga$_{0.8}$N	
	C_{11}	C_{12}	C_{13}	C_{33}	C_{44}	ΔE_v	ΔE_c
wurtzite	29.6	13.0	15.8	26.7	2.41	0.11	0.43
zincblende	25.3	16.5	—	—	6.05	0.07	0.44

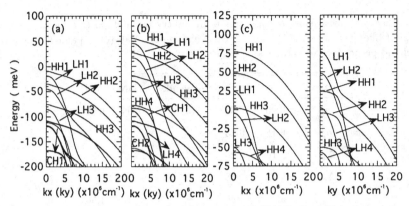

Figure 4: Valence subband structures of the wurtzite GaN/Al$_{0.2}$Ga$_{0.8}$N single quantum wells (a) without a strain, (b) with a compressive biaxial strain and (c) with a tensile uniaxial tensile strain along the x-direction. The well length L_z is 50 Å.

Figure 4 (b) shows the valence subband structure in WZ GaN/Al$_{0.2}$Ga$_{0.8}$N QWs with a compressive biaxial strain. It is induced only into the GaN well layer by 0.5 % due to the lattice mismatch with the barrier layers. The main feature is the same as in the unstrained WZ QWs. The difference can be qualitatively understood by the change of the well's depth and the strain effect on the bulk. Considering compressive strains, the QWs become deep and the number of subbands tends to increase. For $L_z = 50$ Å, CH$_2$ band as well as HH$_4$ and LH$_4$ bands comes into the QWs. Then, the LH$_1$-CH$_2$ coupling causes the non-parabolicity of the LH$_1$ band as well as the LH$_2$ band. Moreover, the energy difference between HH$_1$ and HH$_2$ bands becomes large due to the deeper well's depth, and the energy difference between LH$_i$ and CH$_i$ bands also becomes large due to the biaxial strain effect. The situation for tensile strain is opposite to that for compressive one. This is the reason why the DOS at the VBM is smaller (larger) in the compressive (tensile) biaxial strained QWs than in the unstrained QWs. Namely, the compressive (tensile) biaxial strain plays a decreasing (increasing) role in the DOS at the VBM. However, the change of the DOS is not so large because the symmetry is not changed and the degeneracy is not removed further.

Figure 4 (c) shows the valence subband structure in WZ GaN/Al$_{0.2}$Ga$_{0.8}$N QWs with a uniaxial strain. The uniaxial strain is introduced into both well and barrier layers, and it is 1.0 % tensile strain along the x-direction. This situation can be realized by the epitaxial growth on the uniaxial strained hexagonal substrate, namely orthorhombic (C_{2v}) or monoclinic (C_{2h}) substrates. In the uniaxial strained QWs, each energy dispersion of HH$_i$, LH$_i$ and CH$_i$ bands is similar in the unstrained QWs. However, the energy splittings between HH$_i$ and LH$_i$ bands are not isotropic in the k_x-k_y plane due to the symmetry lowering from C_{6v} to C_{2v}. The tensile (compressive) strain along the x-direction (y-direction) makes not only the LH$_i$ bands along the k_y direction but also the HH$_i$ bands along the k_x direction move higher energy side. Therefore, the DOS at the VBM is more remarkably reduced than in the biaxial strained QWs, and the population inversion would be more easily realized.

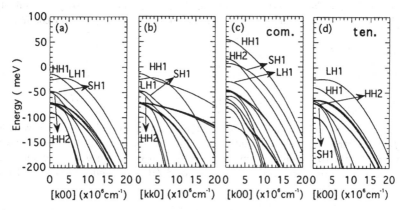

Figure 5: Valence subband structures of the zincblende GaN/Al$_{0.2}$Ga$_{0.8}$N single quantum wells (a), (b) without a strain along [k00] and [kk0] directions, respectively, and (c), (d) with compressive and tensile biaxial strains, respectively. The well length L_z is 40 Å.

Subband in Zincblende Quantum Wells

Figure 5 shows the valence subband structures in the unstrained ZB GaN/Al$_{0.2}$Ga$_{0.8}$N QWs along (a) the [k00] and (b) the [kk0] directions. The energy bands are labeled as HH$_i$, LH$_i$ and SH$_i$, conventionally. The energy dispersions in the k_x-k_y plane are not isotropic in the longer wave lengths. This is different from WZ GaN/AlGaN QWs and ZB GaAs/AlGaAs QWs. For any well length, since the LH$_i$ bands strongly coupled with the SH$_i$ bands even at $k = 0$, the energy dispersions show more complicated behavior. On the other hand, the HH$_i$ bands couple with no LH$_i$ and SH$_i$ bands at $k = 0$ and little at $k \neq 0$. Furthermore, in ZB nitride QWs, the HH$_1$ mass is lighter than the LH$_1$ mass. This is the same as in conventional ZB QWs.

Figures 5 (c) and (d) show the valence subband structures in ZB GaN/Al$_{0.2}$Ga$_{0.8}$N QWs with biaxial strains. The compressive strain is induced into the GaN well layer by 0.5 %. The tensile strain is introduced by 0.2 %, virtually. Here, 0.5 % tensile strain causes almost zero barrier potential, then we have no interest in it. In the biaxial strained ZB QWs, the energy dispersions very close to the Γ point are similar to those in the unstrained QWs. Because there is no further removal of degeneracy due to the same D_{2d} symmetry as in the unstrained QWs. However, the biaxial strain makes the order among the energy levels changed. Under the compressive (tensile) biaxial stress, HH$_i$ and SH$_i$ bands move higher (lower) energy side and LH$_i$ bands move lower (higher) energy side. In the compressive strained QWs, the SH$_1$ band is higher than the LH$_1$. Then, the VBM is still HH$_1$ arising from almost $|X\rangle$ and $|Y\rangle$ states. On the other hand, in the tensile strained QWs, the LH$_1$ band is higher than the HH$_1$ and the VBM. It has much $|Z\rangle$ character. In both cases, the biaxial strain can not cause the decrease of the hole mass at the VBM. However, the HH$_1$ and the LH$_1$ bands are more separated each other. Thus, the DOS around the VBM is considerably reduced. This is the reason why the biaxial strains are more effective on the reduction of the DOS at the VBM than in WZ QWs. Here, the biaxial strain effect on ZB nitride QWs looks quite strange, compared with conventional ZB QWs. But, it is essentially the same, and the difference is only caused by the weak spin-orbit coupling.

OPTICAL GAINS

Generally, optical gain is expresses as

$$g(\omega) = \frac{2\pi\bar{n}}{\hbar c} \sum_{\mathbf{k}} \delta(\omega - E_{\mathbf{k}}^e - E_{\mathbf{k}}^h) |\langle e, \mathbf{k}| \frac{e}{m_0 c} \mathbf{A} \cdot \mathbf{p}|h, \mathbf{k}\rangle|^2 \{f_e(E_{\mathbf{k}}^e) + f_h(E_{\mathbf{k}}^h) - 1\}, \quad (1)$$

where $\mathbf{k}$ is quantum number in a bulk, and c is the light velocity. $\bar{n}$ is the refractive index. f_e and f_h are electron and hole distribution functions, respectively. The calculations were performed under the condition of room temperature, and we assumed that the intraband relaxation time τ_{in} is the same as that of conventional ZB compounds, i.e., $\tau_{in} = 0.1$ psec, though it is not sure for GaN.

Optical Gain of bulk GaN

In WZ GaN, the small spin-orbit coupling makes it possible to approximately express the three hole bands as follows,

$$|\Gamma_9^6(HH); \pm 3/2\rangle = (1/\sqrt{2})|X \pm iY, \pm 1/2\rangle \quad (2)$$

$$|\Gamma_7^6(LH); \pm 1/2\rangle \sim (1/\sqrt{2})|X \pm iY, \mp 1/2\rangle \quad (3)$$

$$|\Gamma_7^1(CH); \pm 1/2\rangle \sim |Z, \pm 1/2\rangle. \quad (4)$$

The basis $|\Gamma_i^j; m_J\rangle$ on the left-hand side indicates the eigenstate at the Γ point and the quantum number of the z-component of total angular momentum operator J. The basis $|L, m_\sigma\rangle$ on the right-hand side indicates the orbital character and the quantum number of the z-component of spin angular momentum operator σ_z. The lowest CH band is almost composed of the $|Z\rangle$ character, which only yields the optical gain for the TM-mode. Thus, the TE-mode optical gain is dominant in bulk WZ GaN.

In ZB GaN, the three hole bands can be expressed by

$$|\Gamma_8^{15}(HH); \pm 3/2\rangle = (1/\sqrt{2})|X \pm iY, \pm 1/2\rangle \quad (5)$$

$$|\Gamma_8^{15}(LH); \pm 1/2\rangle = (1/\sqrt{6})(|X \pm iY, \mp 1/2\rangle \mp 2|Z, \pm 1/2\rangle) \quad (6)$$

$$|\Gamma_7^{15}(SH); \pm 1/2\rangle = (1/\sqrt{3})(|X \pm iY, \mp 1/2\rangle \pm |Z, \pm 1/2\rangle). \quad (7)$$

Since the hole masses of the upper two bands are very heavy along any $\mathbf{k}$ direction, the DOS at the VBM is larger than in WZ GaN, as long as $\Delta_{cr} > \Delta_{so}$. Here, the LH band as well as the SH band includes the $|Z\rangle$ character, whose coefficients are independent of the spin-orbit splitting energy. Thus, the TM-mode optical gain shall start with the TE-mode one at almost the same carrier density. This is very different from WZ GaN. Figure 6 (a) shows the maximum optical gain of bulk GaN. The transparent carrier density of the TE-mode in WZ GaN is lower than that in ZB GaN due to the small DOS at the VBM. However, the transparent carrier density of the TM-mode in WZ GaN is much higher than that in ZB GaN due to $\Delta_{cr} > \Delta_{so}$ and it is not observed in the range of Fig. 6 (a).

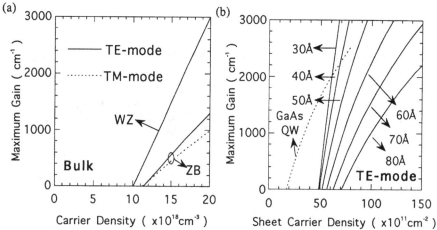

Figure 6: Optical gain of (a) bulk GaN and (b) the wurtzite GaN/Al$_{0.2}$Ga$_{0.8}$N single quantum wells. The solid line stands for the result without a strain. The calculated result for the zincblende GaAs/Al$_{0.3}$Ga$_{0.7}$As single quantum wells, with the well length L_z being 80 Å, is also shown by the dotted line.

Optical Gain of Wurtzite Quantum Wells

In the bulk WZ nitrides, the hybridization of the CH band with the other bands is negligible at the VBM. Thus, the eigenstates along the k_x (k_y) direction can be approximately expressed as $|HH\rangle \sim |Y(X)\rangle$, $|LH\rangle \sim |X(Y)\rangle$, $|CH\rangle \sim |Z(Z)\rangle$. In WZ QWs, the CH$_i$ bands are more split off from the other bands. However, the band mixing is the same as that in the bulk GaN due to no symmetry change. The HH$_1$ mass are still heavy, and the DOS at the VBM is not so reduced. Then, the $|Z\rangle$- character, which yields the optical gain for the TM-mode, is very little at the VBM. This is the reason why the optical gain for the TE-mode is dominant in WZ GaN-based QWs.

Figure 6 (b) shows the well length dependence of the maximum optical gain of WZ GaN/Al$_{0.2}$Ga$_{0.8}$N QWs. In the unstrained QWs, as the well length becomes longer, the DOS around the VBM becomes larger. Thus, it is more difficult to realize the population inversion for longer well length. However, even for $L_z = 30$ Å, the threshold carrier density would be higher than ZB GaAs/AlGaAs QWs. Figures 7 (a) shows the strain effect on the optical gain of WZ GaN/Al$_{0.2}$Ga$_{0.8}$N QWs. In the compressive biaxial strained QWs, the optical gain property is qualitatively improved, and the threshold carrier density would become a little lower than in the unstrained QWs. However, the reduction of the threshold carrier density is quantitatively not so effective. On the other hand, the tensile biaxial strain yields qualitatively negative effect on the threshold carrier density. Thus, no biaxial strain comes up to our expectations in WZ GaN/AlGaN QWs.

Here, on the analogy of the bulk, if it were possible to introduce a uniaxial strain into the c-plane of WZ QWs, it might cause much larger reduction of the DOS at the VBM. According to subband structures, any uniaxial strain in the c-plane reduces the DOS at the VBM. However, considering the optical polarization, the useful uniaxial strain's directions are restricted in the following cases. One is the compressive strain parallel to the optical

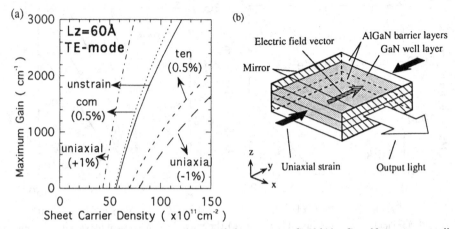

Figure 7: (a) Strain effect on optical gain of the wurtzite GaN/Al$_{0.2}$Ga$_{0.8}$N quantum wells. The well lengths L_z is 60 Å. The solid, dotted, short-dashed, dash-dotted and long-dashed lines stand for the results without a strain, with ± 0.5 % compressive biaxial strains and with ± 1.0 % compressive uniaxial strains along the y-direction, respectively. (b) Schematic structure of useful uniaxial strained wurtzite GaN/AlGaN single quantum well lasers.

polarization for the electric field vector, that is the y-direction. Another is the tensile one perpendicular to it, that is the x-direction. In fact, these are equivalent. Figure 7 (b) shows the schematic structure of the useful uniaxial strained WZ GaN/AlGaN QWs, corresponding to the former case. On the other hand, the useless uniaxial strain's directions are given by changing the position of mirrors to the other lateral sides.

In Fig. 7 (a), the optical gains for uniaxial strained WZ GaN/Al$_{0.2}$Ga$_{0.8}$N QWs are also shown. If we apply the useful uniaxial strain, such as the compressive strain along the y-direction, LH$_i$ bands along the k_y direction and HH$_i$ bands along the k_x direction move to the higher energy side. Then, the orbital component at the VBM becomes almost $|Y\rangle$- character. In other words, such useful uniaxial strains selectively isolate the only $|Y\rangle$-band at the VBM, which can be coupled with the TE-polarized light. Thus, the useful uniaxial strains cause not only the reduced DOS but also give larger differential optical gain. As a result, the threshold carrier density is reduced. On the other hand, the useless uniaxial strains cause the reduced DOS but smaller differential optical gain. As a result, the threshold carrier density is increased. Therefore, if it were possible to introduce a uniaxial strain in the c-plane of WZ QWs, the threshold carrier density would be more efficiently reduced, as long as the relation between the uniaxial strain's direction and the optical polarization is suitable.

Optical Gain of Zincblende Quantum Wells

Figure 8 (a) shows the maximum optical gain of ZB GaN/Al$_{0.2}$Ga$_{0.8}$N QWs. In ZB QWs, since the symmetry is lowered from T_d to D_{2d}, the degeneracy between the HH and the LH bands at the Γ point is removed. Then, the HH$_1$ mass becomes lighter than the LH$_1$ mass, and the DOS at the VBM is reduced. Then, the transparent carrier density of

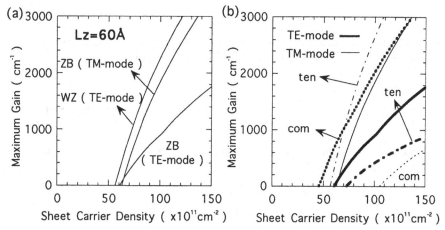

Figure 8: Optical gain of (a) the unstrained and (b) the strained zincblende GaN/Al$_{0.2}$Ga$_{0.8}$N single quantum wells. The well length L_z is 60 Å. The solid, the dotted and the dash-dotted lines stand for the results without a strain, with 0.5 % compressive and 0.2 % tensile biaxial strains, respectively.

the TE-mode becomes low, compared with the bulk. However, since the small spin-orbit splitting enhances the mixing between the LH$_1$ and the SH$_1$ bands, the transparent carrier density is still higher than that in WZ QWs. This enhanced DOS can be observed as the large TM-mode optical gain after the occurrence of the population inversion.

Figures 8 (b) shows the maximum optical gain of biaxial strained ZB GaN/Al$_{0.2}$Ga$_{0.8}$N QWs. The compressive strain strongly depresses the TM-mode optical gain and enhances the TE-mode one. The effect of the tensile strain is reverse to that of the compressive one. These results can be easily understood from the feature of bulk ZB GaN. The compressive biaxial strain lifts the $|X\rangle$ and the $|Y\rangle$ states upper, and the tensile one lifts the $|Z\rangle$ state upper. Thus, the compressive (tensile) biaxial strain is more effective on the reduction of the threshold carrier density for TE- (TM-) mode than that in WZ QWs.

CONCLUSIONS

Electronic and optical properties of bulk GaN and GaN/AlGaN QWs were investigated from the point of view of crystal symmetry. It was found that the lower crystal symmetry would be one of the preferable approaches for reducing the threshold carrier density. In the bulk, the TE-mode optical gain of WZ GaN is obtained at the lower threshold carrier density than that of ZB GaN. In the QWs, the threshold carrier density in the ZB structure is reduced by the symmetry lowering. However, it is still higher than that in WZ structure. In WZ QWs, biaxial strains are not so effective on the reduction of threshold carrier density. In ZB QWs, biaxial strains can more reduce it than in WZ QWs, but the threshold carrier density is almost the same or higher. Therefore, as for the reduction of the threshold carrier density, ZB nitride LDs might have no significant benefit. It was suggested that the threshold carrier density would be considerably reduced if it were possible to introduce a uniaxial strain into the c-plane of WZ QWs.

ACKNOWLEDGMENTS

We are grateful to Emeritus Professor A. Yanase of University of Osaka Prefecture for his helpful discussion and providing us with his FLAPW program.

REFERENCES

1. S. Nakamura, M. Senoh, N. Iwasa and S. Nagahama, Jpn. J. Appl. Phys. **34** (1995) L797.

2. S. Nakamura, M. Senoh, S. Nagahama, N. Iwasa, T. Yamada, T. Matsushita, H. Kiyoku and Y. Sugimoto, Jpn. J. Appl. Phys. **35** (1996) L74.

3. T. Uenoyama and M. Suzuki, Appl. Phys. Lett. **67** (1995) 2527.

4. M. Suzuki and T. Uenoyama, Jpn. J. Appl. Phys. **35** (1996) 1420.

5. M. Suzuki and T. Uenoyama, Jpn. J. Appl. Phys. **35** (1996) L953.

6. M. Suzuki and T. Uenoyama, J. Appl. Phys. **80** (1996) 6868.

7. M. Suzuki, T. Uenoyama and A. Yanase, Phys. Rev. B **52** (1995) 8132.

8. M. Suzuki and T. Uenoyama, Jpn. J. Appl. Phys. **35** (1996) 543.

9. W. Fang and S. L. Chuang, Appl. Phys. Lett. **67**, 751 (1995).

10. W. W. Chow, A. Knorr and S. W. Koch, Appl. Phys. Lett. **67**, 754 (1995).

11. A. T. Meney and E. P. O'Reilly, Appl. Phys. Lett. **67**, 3013 (1995).

12. J. M. Luttinger and W. Kohn, Phys. Rev. **97** (1955) 869.

13. M. Suzuki and T. Uenoyama, Solid State Electron. **41** (1997) 271.

14. M. Suzuki and T. Uenoyama, Appl. Phys. Lett. **69** (1996) 3378.

15. E. Wimmer, H. Krakauer, M. Weinert and A. J. Freeman, Phys. Rev. B **24** (1981) 864.

16. A. U. Sheleg and V. A. Savastenko, Izv. Akad. Nauk SSSR, Neorg. Mat. **15** (1979) 1598.

X-RAY PHOTOELECTRON DIFFRACTION MEASUREMENTS OF HEXAGONAL GaN(0001) THIN FILMS

R. DENECKE [1,2], J. MORAIS[1], C. WETZEL[1], J. LIESEGANG*, E. E. HALLER[1,3], C. S. FADLEY[1,2]
[1]Materials Sciences Division, Lawrence Berkeley National Laboratory, Berkeley, CA 94720, USA
[2]Department of Physics, University of California, Davis, Davis, CA 95616, USA
[3]Department of Materials Science, University of California, Berkeley, Berkeley, CA 94720 USA
Lawrence Berkeley National Laboratory, Materials Science Division, Berkeley, CA 94720, USA

ABSTRACT

We report on the first scanned-angle x-ray photoelectron diffraction measurements on GaN(0001) in the wurtzite structure, as grown on sapphire substrates using MOCVD. These as-grown samples reveal forward scattering peaks in agreement with a theoretical calculation using a single scattering cluster calculation. The surface contamination by O and C does not exhibit any clear structure. From the combination of experiment and theoretical calculation and from a simple intensity ratio argument the surface termination for these samples could be determined to be N. The data also indicate that C is on average closer to the GaN surface than O.

INTRODUCTION

GaN is a promising material for the fabrication of blue light-emitting diodes (LEDs) and lasers due to its large and direct bandgap. The electronic properties depend, however, strongly on the geometric structure and the quality of the samples. There exist two different phases of GaN: a hexagonal wurtzite structure which is the stable structure (called α-GaN), and a zinc-blende structure which can only be achieved by epitaxial growth (the β-GaN phase). Since the normally used technique of X-ray diffraction (XRD) is more a bulk probe and is furthermore not element-specific, it can only determine the overall structure of a given epitaxial sample, and is not able to determine the actual positions of the atoms with respect to the surface. Therefore the use of element-specific x-ray photoelectron diffraction (XPD) promises to give a more detailed view of the near-surface structure, including the nature of the surface termination, which can be via Ga or N, or some mixture of these two growth orientations. This is especially important since surface morphology investigations have revealed a columnar growth for a wide range of growth conditions [1]. So an additional question is whether or not contaminants like O or C are preferentially incorporated in the films in the interstitial regions between the columns.

EXPERIMENT

The samples we used have been grown using metal-organic chemical-vapor deposition (MOCVD). The substrate for the wurtzite structure films was the c-plane (0001) of sapphire, on which layers of about 2-3 μm thickness have been grown without a buffer layer. With XRD the overall quality of the layers has been checked and the presence of a (0001) oriented α-GaN could be confirmed [2]. Looking at the sample surface with a light microscope showed the aforementioned hexagonal columns.

The photoemission measurements have been performed using a standard X-ray source (Al K_α, $h\nu$ =1486.6 eV, Mg K_α, $h\nu$ =1253.6 eV) and a VG ESCALAB electron analyzer, which has been modified so as to permit automated XPD measurements [3]. This system is equipped with a two-axis goniometer, enabling us to rotate the sample on two perpendicular axes so as to cover essentially the full 2π solid angle of emission directions above the surface. The sample for the data shown here was as-grown, meaning that it was not treated in the UHV chamber. Prior to introducing it to the chamber the sample was cleaned using a standard chemical cleaning procedure [4]. XPD measurements were then begun directly after a system bakeout and the attainment of a base pressure of ~1 x 10^{-10} Torr range. We have measured the Ga 3p and N 1s core levels at kinetic energies of 1382 eV and 856 eV, respectively, together with the O 1s and C 1s levels at 722 eV and 1203 eV, respectively, to monitor the surface contamination. The diffraction patterns have been measured with starting angle steps of 3° for both azimuthal (ϕ) and polar (θ) angles. However, the azimuthal angle step was adjusted throughout the measurements to ensure an equal sampling of solid angle. In order to cut the measuring time we used the expected three-fold symmetry of the GaN (0001) surface. Overall measurement times were several days, but the relative intensities of all component (Ga, N, O, and C) were found to be stable from start to finish.

RESULTS

Fig. 1 shows full 2π photoelectron diffraction patterns for all four core levels measured. Shown is the intensity for each core level as a function of emission angle (ϕ,θ), as obtained via a linear background subtraction and an integration over the width of a given peak in an energy distribution curve. Furthermore, an isotropic function I_0 due to unscattered intensity has been subtracted from each diffraction pattern to obtain the so-called χ-function.

One clearly observes strong diffraction peaks for Ga 3p and N 1s emission in a six-fold pattern. Most of these maxima result from scattering along high symmetry directions of the crystal and are mainly brought about by the highly forward peaked nature of the scattering factors at such high kinetic energies. Therefore, these peaks are referred to as forward scattering maxima. Comparing this with the O 1s and C 1s diffraction patterns shows immediately that C and O do not have any ordered scatterers in between them and the detector, as they have basically featureless patterns with no forward scattering maxima. The weak three-fold symmetry in the C 1s pattern is probably artifactual, and arises from the three-fold symmetry operation used to obtain the full diffraction pattern. Just from a simple analysis of the angle of the forward scattering peaks one can thus confirm the overall hexagonal structure of the GaN epilayer.

A more quantitative understanding of the structure can be obtained by comparing these patterns with single-scattering or multiple-scattering diffraction calculations. Using the unit-cell crystal structure as reported in the literature and a cluster with five atomic layers and about 100 atoms we performed preliminary single-scattering calculations based on a Rehr-Albers approach [5]. The results are shown in Fig. 2. We show the diffraction patterns for both Ga 3p and N 1s emission and for a N terminated surface on the cluster.

Both patterns here are three-fold symmetric, which is the symmetry of the GaN(0001) cluster as seen in the surface-sensitive XPD experiment. As the experiment is six-fold symmetric, this suggests the presence of two domain types rotated by 60° with respect to one another in different hexagonal columns. With this in mind, the agreement between experiment and theory is fairly good.

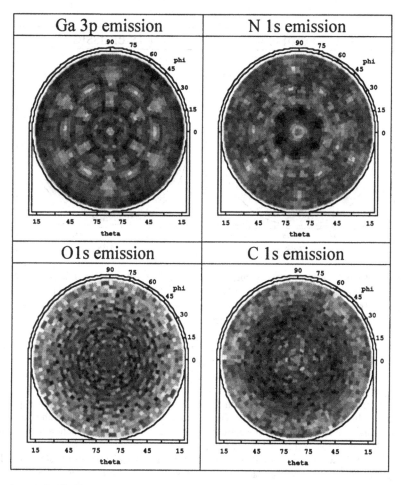

Fig. 1: Experimental photoelectron diffraction patterns (χ-functions) for the four core levels Ga 3p, N 1s, O 1s and C 1s, as excited with Al or Mg K_α radiation, respectively. Light colors correspond to high intensity, dark colors to low intensity. Data obtained over 120° in azimuth have been three-fold symmetrized to yield the full pattern.

There is also an ongoing discussion about whether Ga or N is the terminating layer of this surface [6]. Although not shown here, a comparison of experiment with theory for the Ga-terminated surface yields much less agreement, strongly suggesting a N termination of the sample under study.

In order to further reveal the relative positions of the Ga and N, as well as the contaminant O and C atoms, one can use a rather simple analysis of the diffraction patterns. Since the photoelectron sampling depth varies with polar takeoff angle, θ, due to the finite electron escape depths, Λ_e, according to $\Lambda_e \sin\theta$, plotting azimuthally-averaged intensity ratios of the different core levels as a function of polar angle is a way to get such information. Therefore we have

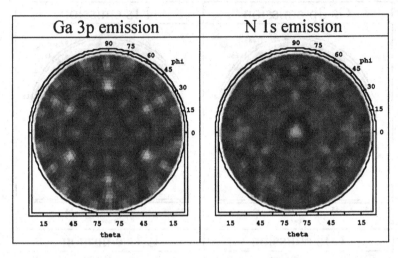

Fig.2: Theoretical photoelectron diffraction patterns for a N-terminated GaN cluster using a single scattering cluster calculation scheme. Shown are χ-functions for Ga 3p and N 1s emission at the same kinetic energies as in Fig. 1. Again light colors mark high intensity and dark colors low intensity.

plotted in Fig. 3 the ratio of these azimuthally-averaged intensities for various combinations of the measured core levels. First of all, the baseline of the ratio Ga 3p/N 1s is very flat, in agreement with the nearly uniform distribution of these atoms in the overall crystal. At first sight, it is a little surprising that there is no enhancement of the N 1s relative intensity at low takeoff angles, in view of our conclusion of N termination based on the XPD patterns. However, this can be explained by the slightly bigger escape depths for the Ga 3p photoelectrons, which have a higher kinetic energy as compared to the N 1s photoelectrons. So this data is not in contradiction to the above finding of N termination of the surface.

Considering now the contaminant peaks, we find that the O 1s/Ga 3p (or O 1s/N 1s) and C 1s/Ga 3p (or C 1s/N 1s) ratios show a dramatic increase for low takeoff angles, indicating that O and C are primarily surface contaminants. The lack of any increase in this ratio for near-normal emission also suggests that not much O or C is present in the interstitial regions between columns, although this might be a reasonable initial conjecture. Finally, the increase in the O 1s/C 1s ratio for low takeoff angles suggests that O is present in the outermost regions of the contaminant layer and thatC is on-average closer to the GaN surface (e.g. as adsorbed CO).

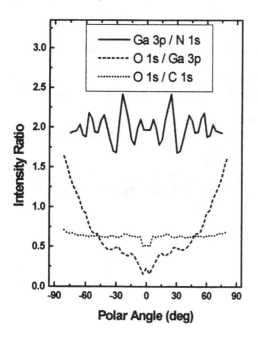

Fig. 3: Ratios of azimuthal-averaged intensities from the diffraction patterns of Fig. 1 plotted versus polar angle. Shown are the ratios Ga3p/N1s (solid line), O1s/Ga3p (dotted line), and O1s/C1s (dashed line). Ratios have not been corrected for the different escape depths of the photoelectrons and the different photoemission cross sections.

CONCLUSION

We have performed the first x-ray photoelectron diffraction (XPD) measurements on an as-grown wurtzite GaN(0001) sample. We obtained diffraction patterns which show the expected crystal structure for the bulk material. A comparison of experiment with single scattering cluster calculations further shows the best overall agreement for a N-terminated surface. An analysis of various azimuthally-averaged peak intensity ratios also permits concluding that both O and C are present as surface impurities, without being significantly incorporated into the interstitial regions between hexagonal columns, and that C is on average closer to the surface than O. These results illustrate the potential of XPD for a more detailed study of the different annealing and cleaning procedures and their effect on the surface structure, and such work is now in progress.

ACKNOWLEDGMENTS

We would like to thank R. X. Ynzunza for helping with the experiments, S. Ruebush for usage of azimuthal-averaging and smoothing routine, and Y. Chen for the use of his program for calculating diffraction patterns. Work has been supported by ONR (Contract N00014-94-1-0162), DOE, BES, Mat. Sci. Div. (Contract DE-AC03-76SF00098), CNPq (Brazil), and DFG (Germany).

REFERENCES

* On leave from La Trobe University, Dept. of Physics, Bundoora 3083, Australia

1.　T. Sasaki, J. Crystal Growth **129**, 81 (1993)
2.　Th. Metzger, H. Angerer, O. Ambacher, M. Stutzmann, E. Born, phys. stat. sol. (b) **193**, 391 (1996)
3.　J. Osterwalder, M. Sagurton, P. J. Orders, C. S. Fadley, B. D. Hermsmeier, D. J. Friedman, J. Electr. Spectr. Relat. Phenom. **48**, 55 (1989); Y.J. Kim, Ph.D. thesis (University of Hawaii, 1995).
4.　V. M. Bermudez, J. Appl. Phys. **80**, 1190 (1996)
5.　J. J. Rehr and R. C. Albers, Phys. Rev. B **41**, 2974 (1993).
6.　M. Asif Khan, J. N. Kuznia, D. T. Olson, R. Kaplan, J. Appl. Phys. **73**, 3108 (1993)

A CHEMICAL AND STRUCTURAL STUDY OF THE AlN-Si INTERFACE

R. Beye, T. George
Center for Space Microelectronics Technology, Jet Propulsion Laboratory, California Institute of Technology, Pasadena, CA 91109

Abstract

Samples of AlN grown on silicon [111] substrates were examined using electron energy loss spectroscopy (EELS) and selected area diffraction (SAD) with high-resolution transmission electron microscopy (TEM) to determine the source of out-of-plane tilts and in-plane rotations of the AlN crystallites at the Si interface. SAD results indicate that the interfacial crystallites are sheared along vertical planes, with random, intercrystalline rotation. The interfacial phenomena are believed to be the result of Si-Al-N interaction. Analytical experiments show no evidence of silicon nitride formation, witnessed by nitrogen-K peak shape, up to the Si interface. No evidence of substrate-epilayer interdiffusion was observed. Chemical interaction within one monolayer of the interface is therefore suspected as the cause of the epilayer tilts and rotations.

Introduction

The growth of low defect density heteroepitaxial AlN has great implications for optoelectronic and high power devices since the AlN can be used either as device material or as a buffer layer for the overgrowth of other group III-Nitrides. Silicon substrates are particularly attractive for the growth of AlN when considering cost, defect density, available sizes, ease of etching and cleaving and finally the possibility of incorporating III-N devices in VLSI circuits. The current limitations of AlN/Si crystalline quality may stem from interactions occurring at the AlN-Si interface during nucleation, observed in this study using transmission electron microscopy, whereby incipient AlN interfacial crystallites are mis-aligned with respect to the substrate.[1] These mis-alignments, witnessed both in and out of the AlN basal plane, are accompanied by a thin amorphous-like region at the AlN/Si interface, and are believed to be a particular feature of III-N/Si growth. Overcoming this errant behavior during subsequent growth, or eliminating these effects entirely would improve epitaxial-layer crystal quality. The focus of the present paper is on the mechanisms giving rise to these features.

Ideally, close-packed planes and directions of the abrupt silicon [111] substrate surface are matched by equivalent planes and directions in the hexagonal AlN epilayer, according to:

$$(111)_{Si} // (0002)_{AlN}$$
$$[\bar{1}10]_{Si} // [11\bar{2}0]_{AlN}.$$

It has been reported that the close-packed planes between the two layers are randomly tilted away from parallel by $\pm 3\text{-}4°$; close-packed directions within these planes between the two layers differ randomly by as much as several times this amount.[1] Chemical interactions in the Si-Al-N system appear to be largely responsible for these phenomena. Such interactions, especially those resulting in amorphous compound formation, may disrupt the ideal epitaxial relationship.

In this work, the results of transmission electron microscopy (TEM) involving both high resolution imaging and electron energy loss spectroscopy (EELS) of AlN/Si layers are reported and the relationship between Si-Al-N interactions and the mis-orientation of AlN nuclei is discussed.

Experimental

Single crystal AlN was deposited over Si[111] substrates using reactive molecular beam epitaxy (MBE) and low-temperature metalorganic chemical vapor deposition (MOCVD). Samples for TEM were prepared as in (1). High-resolution TEM imaging was performed at JPL. Analytical microscopy (chemical) was performed both at Berkeley (see Acknowledgments) and at Tempe. At Berkeley, a JEOL 200cx, operating at 200 keV was used with a Gatan PEELS system, functioning with approximately 2.5 V resolution. Sampling diameters were about 0.1 μm. At Tempe, a VG HB501 scanning TEM was used at 100 keV with approximately 1 V resolution and 0.001 μm probe diameter.

Results and Discussion

Figure 1 shows a typical cross-sectional high-resolution TEM image with corresponding electron diffraction pattern. The image shows the out-of-plane tilt and amorphous-like background at the interface that are the subjects of this paper. Apparent in the diffraction pattern is how the tilting of the planes, in this perspective, is confined to the basal planes, and not to the orthogonal ($1\bar{1}00$)-type planes, as witnessed by the degree of arc for the corresponding reflections. The latter reflections thus remain perpendicular to the AlN/Si interface. These, along with those of ($11\bar{2}0$)-type, show considerable rotation about the basal plane when viewing the epilayer from above (plan-view), however. Simple tilts and rotations of the hexagonal crystallites seem to be the obvious explanation for the observed images and diffraction patterns, but this would render the vertical planes no longer normal to the silicon surface, and no streaking about the central spot is seen for these reflections. (Occasionally, however, arcing about the central spot is observed for the vertical planes, indicating simple tilts and rotations for these occurrences.) Distortion of the unit cell is then necessary, but it must be done such that the basal planes are tilted, while the vertical planes remain normal to the silicon surface. This is possible if the unit cell is sheared along these vertical planes. Plan-view samples show these planes to be significantly and randomly rotated about the silicon surface normal. Distortion of the hexagonal unit cell to realize this rotation would again bring the vertical planes away from parallel to the surface normal. It therefore appears likely that the rotation takes place intergranularly. Both the tilting and the rotation are enough to produce the amorphous-like appearance in high-resolution images.

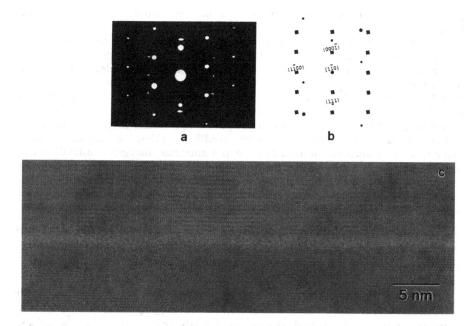

Figure 1: SAD (a), with simulation (b), and corresponding high-resolution TEM cross-sectional image (c), showing interface between silicon (bottom) and AlN (top). Seen, are the diffuse interfacial region and AlN crystallites with mis-orientations.

The arrangement at the silicon interface is such that each crystal is sheared along vertical sets planes and rotation is confined to the interface between the crystallites. This situation suggests the existence of compressive stresses in the nucleating layer, where the tilting of basal planes with larger lattice parameter than that of the substrate yields a projected lattice with more comparable dimensions. Likewise, the rotations may further buckle the epitaxial layer to relieve compressive stress. Compressive stress is not expected for AlN on Si, however. The lattice parameter of AlN is much smaller than that of the silicon substrate, giving a tensile stress in the epitaxial layer. Silicon nitride, on the other hand, can have a hexagonal lattice parameter that is slightly larger than twice that of silicon in the [111] plane. Alpha-Si_3N_4 has a basal-plane lattice parameter of approximately 0.78 nm, and that for the beta phase is around 0.76 nm (about twice the parameter of silicon in the [111] plane).[2] The former Si_3N_4 phase would then yield the necessary compressive stress.

The accompanying distortion would help to preserve the interfacial bonding, while relieving compressive stress. It is therefore likely that, when considering only lattice parameter, the energy expense of distorting the AlN unit cell is lower in magnitude than that for breaking the bond across the AlN/Si interface, and lower than the energy required to initiate an interfacial dislocation. This suggests a quantifiable relationship between the interlayer bonding strength, the energy of distortion, and the dislocation generation energy. However, in addition to lattice parameter-based stresses, silicon nitride-type bonding would introduce different bond angles.

Like silicon and AlN, the fundamental building block for both alpha and beta silicon nitride idealized bulk structures is the tetrahedron, with Si residing at the center and N at the vertices. Each N is common to three tetrahedra, as opposed to the analogous four for the AlN structure. That is, one structure has three Si around each N, and the other has four Al. It is this difference in the angular distribution of tetrahedra that would provide the necessary distorted template for the observed AlN growth. This may not necessarily require more than a monolayer of silicon nitride to produce the interfacial effects observed here. (That the epitaxial structure maintains the directional information of the substrate suggests minimal silicon nitride formation.) It may be that the Si-N bonding across the interface is sufficient for the nitrogen to lower the number of required neighbors at the interface, thus distorting the tetrahedral arrangement and giving bond lengths and directions more toward those of Si_3N_4. Indeed, as this tilting has been observed to a lesser degree for SiC substrates,[3] where the interface is defined by the abrupt termination of the bulk material,[4] the very presence of the single interfacial Si-N bond may be sufficient for this effect. The following examines the extent of Si_3N_4 formation on the present Si substrates.

Figure 2 a-c shows energy-loss spectra obtained with the JEOL 200cx for the nitrogen-K peak for AlN, the AlN/Si interface, and a Si_3N_4 standard, respectively. Examination of the larger peak shapes (particularly the second-largest peak) and intensities for the three spectra shows that no Si_3N_4-type bonding is evident at the Si interface. The spectra show the 2.5 V resolution to be sufficient for distinguishing between the two bonding situations. The 0.1 μm probe size would not be expected to reveal interfacial silicon nitride formation as the relatively small volumes would contribute negligibly to the peak shapes. Figure 3 shows spectra taken using the HB501, utilizing its smaller probe diameter. The similar N-K peak shapes between bulk and interface demonstrate no silicon nitride evidence. Additionally, Si was not found to diffuse into AlN; nor was Al or N found to diffuse into Si. These spectral results show that the chemical interactions responsible for the tilts and rotations of the AlN crystallites are confined to the nearest monolayer to the interface.

Conclusions

It has been shown that the tilts, rotations, and amorphous-like layer reported in (1) consist of shear along vertical planes of the hexagonal unit cell and intergranular rotation of the basal planes. The amorphous-like layer is then the projection of the atomic columns after displacement. Silicon nitride formation was not observed, nor was interdiffusion between the layers at the interface. The influence of the Si-N interfacial bond to alter the nearest-neighbor configuration of the first-layer nitrogen atoms is suspected to be the cause of the out-of-plane tilts and in-plane rotations.

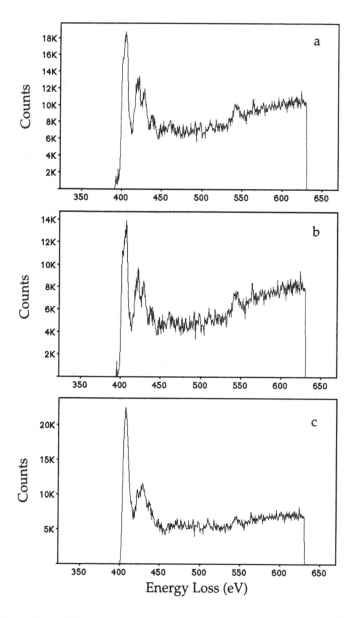

Figure 2:   Electron energy loss spectra for nitrogen K (a) in AlN layer,
             (b) in AlN at Si interface, and (c) in silicon nitride standard.

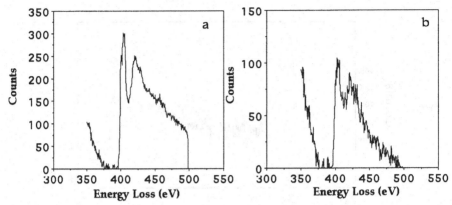

Figure 3: Electron energy loss spectra, showing N-K peaks for AlN layer (a) well away from Si interface and (b) at Si interface, using smaller probe diameter (monolayer scale) than that used in Figure 2. Silicon nitride-type peak is not observed (see Figure 2c).

Acknowledgments

The work described in this paper, performed at the Center for Space Microelectronics Technology, Jet Propulsion Laboratory, California Institute of Technology, and that performed at NCEM (see below) and ASU, was sponsored by the Ballistic Missiles Defense Organization, Innovative Science and Technology Office, through an agreement with the National Aeronautics and Space Administration, and by DOE (see below). The authors are grateful to Amy Chang-Chien for TEM sample preparation, and to Chuck Echer for assistance with the JEOL 200cx analytical microscope at the National Center for Electron Microscopy, a user facility at the Lawrence Berkeley Laboratory, supported by the US department of Energy under Contract No. DE-AC03-76SF00098; and to Peter Crozier for his assistance with the HB501 at the Center for Solid State Science at Arizona State University. Finally, the authors wish to thank J. W. Yang, M. A. Khan at APA Optics for providing the epitaxially-grown material used in this study.

References

1   R. Beye, T. George, J. W. Yang, M. A. Khan , Proceedings of the Fall 1996 MRS meeting, III-V Nitrides symposium.
2   C-M. Wang, X. Pan, M. Rühle, F.L. Riley, M. Mitomo, J. Mat. Sci., 31, 5281(1996).
3   F.A. Ponce, B.S. Krusor, J.S. Major, Jr., W.E. Plano, D.F. Welch, Appl. Phys. Lett., 67(3), 410(1995).
4   F.A. Ponce, M.A. O'Keefe, E.C. Nelson, Transmission electron microscopy of the AlN-SiC interface, Phil. Mag. A, 74(3), 777(1996).

MEASUREMENT OF $In_xGa_{1-x}N$ AND $Al_xGa_{1-x}N$ COMPOSITIONS BY RBS AND SIMS

Y. Gao, J.Kirchhoff, S. Mitha, J. W. Erickson, C. Huang, R. Clark-Phelps
301 Chesapeake Drive, Redwood City, California 94063

ABSTRACT

Secondary ion mass spectrometry (SIMS) and Rutherford Backscattering Spectrometry (RBS) techniques were used to determine $In_xGa_{1-x}N$ and $Al_xGa_{1-x}N$ compositions. While RBS is generally considered a quantitative technique for compositional analysis, SIMS has not been. We have applied a new analytical technique, which reduces the matrix effect in SIMS analysis, to accurately determine stoichiometry. The composition of $In_xGa_{1-x}N$ ($Al_xGa_{1-x}N$) in the multiple layers and quantum well of the LED can be measured by SIMS, but is inaccessible to RBS.

INTRODUCTION

The matrix composition of III-arsenide and III-phosphide ternary alloy can usually be determined very accurately by simple techniques such as X-ray diffraction (XRD) and photoluminescence (PL). However, the application of these techniques to III-nitride material such as $Al_xGa_{1-x}N$ and $In_xGa_{1-x}N$ are hampered by the residual film stress and the difficulty of identifying the band-to-band emission, respectively. Moreover, XRD and PL are not thin film analysis techniques and can not determine the composition for each layer in real multilayer devices. Although Rutherford Back Scattering (RBS) can be used to quantitatively determine the composition, the depth resolution is limited for thin complex structures such as quantum wells.
Secondary Ion Mass Spectroscopy (SIMS) is widely used to profile in-depth the concentration of trace elements because of its high sensitivity and good depth resolution. Since the quantitative analysis of matrix composition under normal analytical conditions is complicated by strong matrix effects, we have employed a new analytical technique (the MCs^+ technique [1]) to reduce matrix effects in SIMS and accurately determine stoichiometry. The results obtained using the two techniques is compared, and the composition of $In_xGa_{1-x}N$ ($Al_xGa_{1-x}N$) in the multiple layers and quantum well of the LED is presented as a SIMS depth profile.

EXPERIMENTAL

SIMS measurements were performed with a magnetic sector instrument (Cameca IMS-4f) and a quadrupole instrument (PHI-6600). The instruments, which are both equipped with Cs^+ and O_2^+ primary ion sources, complement each other in terms of performance. Roughly speaking, the Cameca provides lower detection limits while the PHI has better depth resolution. Depths of sputtered craters were measured by means of a mechanical stylus (Tencor P-10). The RBS spectra were acquired at a backscattering angle of $160°$ with the samples perpendicular to the incident ion beam. The RBS spectra are fit by applying a theoretical model and iteratively adjusting elemental concentration until the theoretical curve agrees with the experimental spectrum. Both the InGaN/GaN and the AlGaN/GaN samples were grown by MOCVD (metallorganic chemical vapor deposition).

RESULTS

The SIMS technique provides depth profiles with high sensitivity and good depth resolution [2], and thus should be well-suited to the characterization of thin GaN films.

Mat. Res. Soc. Symp. Proc. Vol. 468 © 1997 Materials Research Society

However, these advantages are counterbalanced by unpredictable matrix effects which strongly influence the secondary ion yields. This makes quantification or even interpretation of results very difficult, particularly at interfaces. Therefore until recently SIMS has not generally been accepted as a compositional analysis technique such as RBS, ESCA and Auger spectroscopy.

In order to achieve a quantitative SIMS analysis, the matrix effect must to be reduced or corrected. We have made use of the MCs^+ technique, in which molecular secondary ions MCs^+ (where Cs^+ are the primary ions) are monitored rather than M atomic ions. The MCs^+ ions are formed by the recombination of Cs^+ ions with neutral M atoms, a short distance above the sample surface. Hence, the detection of MCs^+ would be proportional to the flux of sputtered atoms M. The matrix dependence of sputter yield is usually much weaker that that of secondary ion yield, which makes MCs^+ ion yields relatively independent of the matrix. This technique has been successfully applied to the III-arsenide system [1]. In the present work, we have tested the validity of the MCs^+ technique in determining the composition of $Al_xGa_{1-x}N$ and $In_xGa_{1-x}N$.

Fig. 1 shows the relationship between $AlCs^+/GaCs^+$ ion ratios versus the composition ratios $x/(1-x)$ for a set of AlGaN samples. The relationship is linear in this range ($0 < x < 0.2$) of composition, indicating that matrix effects are negligible. The samples consisted of an AlGaN layer on top of GaN, and the x values was measured by RBS with an accuracy of ± 0.01.

Fig. 2 shows a similar linear relationship for $InCs^+/GaCs^+$ versus $x/(1-x)$ for $In_xGa_{1-x}N$. Samples of AlGaAs (or AlGaP) and InGaAs (or InGaP), with stoichiometries accurately characterized by XRD and PL, can be used as compositional standards for AlGaN and InGaN respectively. The rationale for this is that the $AlCs^+/GaCs^+$ and $InCs^+/GaCs^+$ ratios should be very similar when the x-values are the same in $Al_xGa_{1-x}V$ or $In_xGa_{1-x}V$ where $V = N, P$, or As. This is suggested by what is known of the physics and chemistry of ion formation mechanisms.

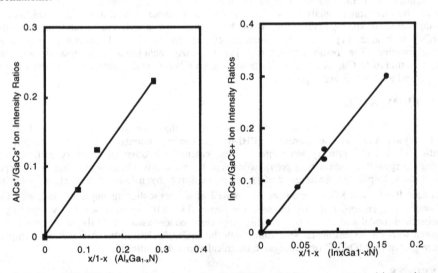

Fig. 1 (left) $AlCs^+/GaCs^+$ ion intensity ratios in SIMS analysis versus composition ratios $x/(1-x)$ for $Al_xGa_{1-x}N$ samples where the x value was determined by RBS.

Fig. 2 (right) $InCs^+/GaCs^+$ ion intensity ratios in SIMS analysis versus composition ratios $x/(1-x)$ for $In_xGa_{1-x}N$ samples where the x value was determined by RBS.

Fig. 3 shows depth profiles of the Al composition determined by SIMS technique using an AlGaAs standard and by the RBS technique. The two measurements for this simple structure agree within 15%, which is excellent.

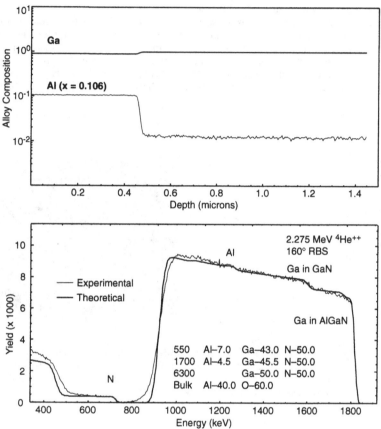

Figure 3. Depth profiles of Al and Ga composition determined by the SIMS technique using an AlGaAs standard (top), along with an RBS spectrum on the same sample (bottom).

The RBS spectrum is a convolution of mass and depth scales. For this reason, RBS analysis can not provide exact information for both composition and thickness, when samples have continuous variations in composition (a popular design feature) or buried layers such as the active layers (double heterojunctions or quantum wells) in LEDs and laser diodes. In contrast with RBS, SIMS has excellent depth resolution quite independent of its mass scale.

The depth resolution in SIMS is determined by several physical processes induced by ion bombardment (cascade atomic mixing, preferential sputtering, etc.) as well as by the initial surface roughness. Generally speaking, one can achieve depth resolution no better than the initial surface roughness of sample. For samples with a very smooth surface such as GaAs and Si, the depth resolution is limited by the ion mixing and by the roughness that is induced by ion bombardment. The ultimate depth resolution in practice is less than 2 nm.

As a practical matter in the case of most present GaN samples, the depth resolution is limited by the initial surface roughness. Although surfaces are locally smooth on an atomic scale, the density of surface pits is great enough that the SIMS analytical area (from 10^2 to 10^3 cm^2) almost always includes some pits. These pits degrade the depth resolution to an extent which depends on their size and nature.

Fig. 4 shows Al depth profiles at the interface of an AlGaN/GaN sample. In one case some pits were visible at the surface, while in the other the pits were far less numerous or invisible. Clearly, the apparent width of interface is made greater by the surface roughness associated with these pits. The loss of depth resolution will make the SIMS composition analysis of quantum well structures much more difficult, because secondary ion intensities characteristic of each thin layer must be measured to allow the composition to be determined accurately.

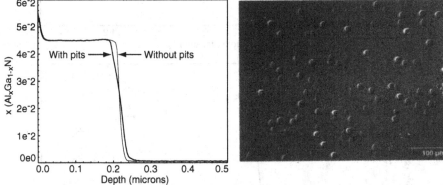

Fig. 4. Al depth profiles at the interface of an AlGaN/GaN sample. One curve was obtained on a sample with a high density of visible surface pits, while the other was obtained on a sample with few if any visible pits. The surface morphology of this sample is illustrated in the optical micrograph shown on the right.

Fig. 5 shows SIMS depth profiles of a GaN/InGaN/GaN quantum well structure. Since the depth resolution (about 15 nm) was not high enough to resolve the well, the In profile does not show a flat top in the InGaN well, necessary to give a single value characteristics of this layer. Consequently, the In concentration value indicated ($x_{max} = 0.084$) in the figure necessarily underestimates the real In concentration. However, a correction can be made: the areal density of In can be calculated by integrating the In profile across the quantum well. An upper limit to the In composition of $x = 0.095$ is obtained by assuming that the quantum well is a perfect delta function with a width of 34 nm (the full-width-at-half-maximum for the In peak). Therefore the real In concentration should be between 0.084 and 0.095.

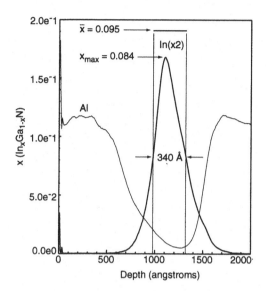

Fig. 5 Composition depth profile by SIMS for a GaN/InGaN/GaN quantum well structure. The quantum well is nominally 40 nm thick.

CONCLUSION

We have applied a quantitative technique recently developed for SIMS [1] to measure alloy compositions of $In_xGa_{1-x}N$ and $Al_xGa_{1-x}N$. The validity of the technique was checked with RBS, which can quantitatively determine the composition of samples with simple structures. For complex structures such as LEDs the SIMS-MCs$^+$ technique provides very accurate measurements of composition with good depth resolution. Thus SIMS analyses now can provide relatively complete information on GaN materials and devices [2]: the concentrations of dopants and impurities, matrix composition, layer thicknesses, and interface quality.

ACKNOWLEDGMENTS

We gratefully acknowledge the contributions of Professor Steven Denbaars of the University of California at Santa Barbara, Dr. Joan Redwing of Advanced Technology Materials Incorporated, Dr. Peter Menz of Philips Research Laboratory, and Professor Ichwara Bhat of Rensselaer Polytechnic Institute.

REFERENCES

(1) Y. Gao, J. Appl. Phys. 64, 3760 (1988).

(2) C. Huang et al. in these proceedings.

REFERENCES

COMPLETE CHARACTERIZATION OF Al$_x$Ga$_{1-x}$N/In$_x$Ga$_{1-x}$N/GaN DEVICES BY SIMS

C. Huang, S. Mitha, J. W. Erickson, R. Clark-Phelps, Jack Sheng, Y. Gao,
Charles Evans & Associates 301 Chesapeake Drive, Redwood City, CA94063

ABSTRACT

SIMS analysis was applied to the characterization of GaN, AlGaN/GaN and InGaN/GaN grown by MOCVD. Such characterization enables the control of purity and doping, and the determination of growth rate and alloy composition. The analysis can be performed on finished optoelectronic and electronic devices and this makes SIMS technique a powerful tool for failure analysis, reverse engineering, and concurrent engineering.

INTRODUCTION

The III-nitrides have recently been the subject of intense research because of their promising applications in blue and ultraviolet optoelectronic devices as well as microwave and electronic devices (1-3). The growth of epitaxial layers, grown by MOCVD and MBE techniques, needs close control of purity, doping, alloy composition, thickness and interface quality. Secondary ion mass spectrometry (SIMS) is a very suitable characterization technique because of its ability to depth profile with high sensitivity and good depth resolution. Figure 1 describes the SIMS application to the characterization of a GaN MOCVD process. We describe depth profiling for the concentration of dopants (Mg, Zn and Si), common impurities (O, C, H and some metals incorporated during the growth), and for determining In and Al composition. Also SIMS measurements on finished LED devices will be described for the purpose of failure analysis, reverse engineering, and concurrent engineering (such as for early identification of failure mechanisms after high-temperature testing). The wealth of information so obtained has proven very useful for solving problems encountered in both research and production.

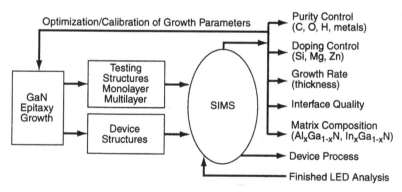

Figure 1. SIMS application to the Characterization of GaN based materials

EXPERIMENTAL

SIMS measurements were performed with a magnetic sector based instrument (Cameca IMS-4f) and a quadruple based instrument (PHI-6600). Both instruments, which are equipped with Cs$^+$ and O$_2^+$ primary ion sources, complement each other in terms of performance. Roughly speaking, the Cameca provides lower detection limits while the PHI has better depth resolution. Depths of sputtered craters were measured by means of a mechanical stylus (Tencor P-10). The reference materials are GaN, AlGaN and InGaN samples implanted with known doses of the elements of interest including Si, Mg, Zn, Cd, Se, B, H, C, O, Cl, Fe, Mo, Ni, Cu and Mn. Fig. 2 shows typical implant profiles for Na, K and Cr in GaN samples. Oxygen ion bombardment was used and positive secondary ions were monitored.

RESULTS

Purity and Doping Control

Control of purity in the epilayer requires very low detection limits because the concentration range of interest is usually below 10^{16} atoms/cm^3. The set-up of the experimental conditions for such analysis is not trivial, depending on the combination of the impurity to be analyzed and the matrix. Although the choice of the primary ions is most important, the appropriate detection techniques for secondary ions, such as energy discrimination and detection of molecular ions, can improve the detection limit further. Table I summarizes the typical detection limits obtained with our instrument for some elements in GaN.

Table I. Typical detection limits obtained with our SIMS instruments for some elements in GaN

Element	Detection Limit (at/cc)
Common Dopants	
Si	1e15
Mg	1e15
Zn	5e15
Cd	1e16
Se	1e14
Common Contaminants	
H	1e17
C	5e15
O	1e16
Cl	1e15
Al	1e16
In	1e16
Metal Contaminants	
Cr	5e14
Fe	5e15
Mo	5e15
Ni	2e16
Cu	2e16
Mn	5e15
Na	1e14
K	5e13

The elements Si, Zn and Mg are currently used as n- and p-dopants for the GaN system. For control of doping, there is a need for an accurate and rapid technique to calibrate the dopant sources. By using ion-implanted standards, SIMS can provide a means of measuring the concentration of impurity with an accuracy of better than 10%. By comparing the atomic concentration of dopant with electrical carriers determined by Hall measurements and C-V profiling, the doping efficiency can be readily obtained. Moreover, with a multilayer structure of different doping levels, a single measurement of the depth profile allows a calibration curve to be

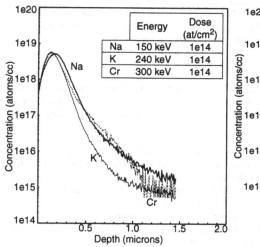

Figure 2. Typical implant profiles for Na, K, and Cr in GaN samples. The implant energy and dose are indicated above

Figure 3. A p-n homojunction in GaN using Mg and Si as p- and n-type dopants. Common contaminants in GaN such as O and C and some transition metals such as Fe, Mo, Cr, and Ni have also been measured.

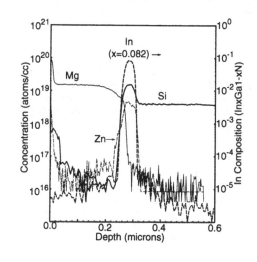

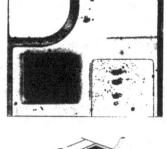

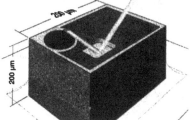

Figure 4. SIMS depth profiles of dopants and compositional profile for a GaN/InGaN/GaN LED structure

Figure 5. SIMS analysis of finished LED chip. A post-SIMS measurement crater on a GaN LED device after de-encapsulation is shown.

determined. Figure 3 shows a p-n homojunction in GaN using Mg and Si as p- and n-type dopants. Common contaminants in GaN such as O and C and some transition metals such as Fe, Mo, Cr and Ni have been also measured. The oxygen and carbon in this sample were above the detection limits while the metals were at or below their detection limits.

Layer Thickness, Matrix Composition, and Interface Quality

Combined with post-bombardment crater depth measurements, SIMS can easily provide the thickness of very thin films with good accuracy. However, for heterostructure samples, a calibration curve for dependence of sputter rates on matrix composition will be needed for an accurate determination of layer thickness.

SIMS quantitation of matrix elements is complicated because a variation in matrix composition changes the secondary ion yield and leads to a non-linear calibration curve. One technique for circumventing the matrix effect is to detect the molecular ion MCs^+ (M is the matrix element to be analyzed) under Cs^+ bombardment (4,5). This technique allows us to use some other well defined III-V compounds such as AlGaAs and InGaAs as the SIMS standards. The x values measured by SIMS agreed well with those obtained by other techniques including RBS (5). Compared with the usual compositional techniques, such as photoluminescence, x-ray diffraction measurements, SIMS can unambiguously provide a chemical composition profile of device structures, in addition to dopant profiles. Fig. 4 shows an example SIMS depth profiles of dopants and compositional profile for a GaN/InGaN/GaN LED structure.

Since the depth resolution of SIMS is excellent, these measurements also provide a straightforward means of identifying diffusion-related degradation mechanisms. Concurrent engineering principles typically prescribe high-temperature testing of device designs, through a series of annealing treatments at temperatures above that of the intended operating environment. The SIMS technique permits the quick analysis of a *batch* of such devices (which have been annealed at various temperatures and times), in order to more swiftly develop robust packaging, contacts, and device designs.

Finished LED Device Analysis

The small size (about 200 μm x 200 μm) of a finished LED device makes SIMS analysis possible but extremely difficult. Fig. 5 shows a post-SIMS measurement crater on a GaN LED device after de-encapsulation. The ability of carrying out the analysis on finished devices makes SIMS a powerful tool for failure analysis, reverse engineering, and concurrent engineering (e.g. early identification of failure mechanisms expected in high-temperature operation).

CONCLUSION

We have demonstrated that SIMS is a very powerful characterization technique for the R&D and production of GaN LED devices. With a single technique, SIMS provides the quantitative analysis for both trace and major elements (concentration ranging from 10^{15} to 10^{22} atoms/cm^3). Information on thickness and interface quality can be obtained with the same measurement. In addition, SIMS analysis of a degraded device (after de-encapsulation) for the diffusion of dopants or metals can help to find possible failure mechanisms.

ACKNOWLEDGMENTS

We gratefully acknowledge the contributions of Professor Steven Denbaars of the University of California at Santa Barbara, Dr. Joan Redwing of Advanced Technology Materials Incorporated, Dr. Peter Menz of Philips Research Laboratory, and Professor Ichwara Bhat of Rensselaer Polytechnic Institute.

REFERENCES

1. S. Nakamura, T. Mukai and M. Senoh, Appl. Phys. Lett., 64(13), 1687(1994)

2. H. Morkoc, S. Strite, G.B. Gao, M.E. Lin, B. Sverdlov, and M. Burns, J. Appl. Phys. 76(3) 1363(1994)

3. M. Asif Khan in Volume 415, Materials Research Society Symposium Proceedings Series, 1995 Symposium BB: Metal-organic Chemical Vapor Deposition of Electronic Ceramics II.

4. Y. Gao, J. Appl. Phys. 64(7), 3762(1988)

5. Y. Gao, J.Kirchoff, S. Mitha, J. W. Erickson, C. Huang, in these proceedings

TEM/HREM ANALYSIS OF DEFECTS IN GaN EPITAXIAL LAYERS GROWN BY MOVPE ON SiC AND SAPPHIRE

S. RUVIMOV, Z. LILIENTAL-WEBER, C. DIEKER and J. WASHBURN,
Lawrence Berkeley National Laboratory, Berkeley, CA 94720;

M. KOIKE,
Toyoda Gosei Co LTD, New Market Technical Division,
Haruhi-cho Nishikasugai-gun, Aichi 452, Japan;

H. AMANO and I. AKASAKI,
Meijo University, Tempakuku-ku, Nagoya 468, Japan

ABSTRACT

High resolution electron microscopy has been applied to study the structure of epitaxial GaN layers grown by MOVPE on SiC and sapphire substrates. Defects in GaN were systematically studied for undoped, and Si- and Mg-doped samples. For both substrates, the Si-doping was found to decrease the dislocation density at the layer surface, while Mg-doping increased it. The density of nanopipes increased with both types of doping. Cracking of GaN layers was observed for SiC substrates. Crack formation was not detected in layers grown on sapphire. Mechanisms of defect generation are discussed in relation to the initial growth stages, the effect of doping, and the type of substrate.

INTRODUCTION

Structure of GaN epitaxial layers has been a focus of many recent studies because GaN is a promising material for optoelectronics applications in a short wavelength range [1-3]. Despite lattice mismatch to GaN, SiC and sapphire are widely used as substrates for epitaxial growth of GaN. High levels of mismatch in lattice parameters and thermal expansion coefficients between epilayer and substrate result in extensive formation of structural defects in the GaN layer during epitaxial growth. A typical dislocation density reported for GaN epitaxial layers is 10^9-10^{10} cm^{-2} [3-8]. Islanding of the GaN during the initial stages of growth and structure of buffer layers have been considered as a primary source of threading dislocations in the layer and the formation of small angle boundaries. The structural quality of the GaN layers has been shown [1,9] to be significantly improved by use of a low temperature AlN buffer layer. In contrast to other semiconducting materials, highly dislocated GaN epilayers grown by different techniques still demonstrate strong photoluminescence (PL). The position of the PL peak was found to be sensitive to in plane strain [8] which varies over a wide range depending on the growth conditions and the type of substrate and buffer layer. Most of the strain is relaxed through formation of misfit dislocations. However, the mechanism of strain relaxation in GaN epilayers is still under discussion and requires a better understanding of defect formation. Point defects and impurities that affect the lattice parameter of the GaN and the dislocation mobility may also play an important role in strain relaxation. Therefore, effects of substrate, buffer layer, and doping of the GaN layer on its structural quality is of special interest.

Here we report the results of an electron microscopy study of defects in GaN epitaxial layers grown on sapphire and SiC and on the effect of doping on the dislocation structure and on strain release.

EXPERIMENTAL

GaN layers were grown by MOVPE on (0001) Al$_2$O$_3$ and SiC substrates with an AlN buffer layer described elsewhere [9]. The AlN buffer layers were grown at 1100 and 400 °C on

SiC and sapphire substrates, respectively. GaN layers were either undoped or doped by Si (up to 4×10^{18} cm^{-3}) and Mg (see Table 1). Mg-doped samples were annealed at 800°C for 30 min.

TEM studies were carried out on a Topcon 002B and JEOL 200CX microscopes operated at 200 kV. Both plan view and cross-sectional specimens were prepared for TEM study by dimpling followed by ion milling. X-ray diffraction experiments were performed using a Siemens D-5000 diffractometer with CuK$_{\alpha 1}$ radiation.

RESULTS AND DISCUSSION

The major types of defects in GaN crystals depend both on growth mode and on the crystallography of the material. Stacking faults are the predominant defects in bulk GaN crystals grown under hydrostatic pressure [10], while dislocations extending parallel to the c-axis usually dominate in epitaxial GaN layers [4-10]. Typical microstructures of the GaN layers grown on sapphire and SiC are shown in Figs. 1 and 2, respectively. Dislocation distribution across the GaN layers grown on sapphire and SiC are similar. Defect density is high near the interface between the GaN layer and the AlN buffer, but drastically decreases over 0.2 μm toward the layer surface.

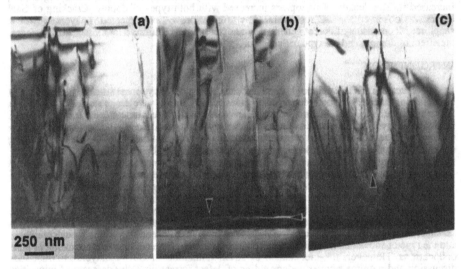

Fig.1, a-c.   Bright-field cross-sectional TEM images of GaN layers grown on SiC using AlN buffer layer: a - undoped GaN, b - Si-doped, c - Mg-doped GaN. Note the micro cracks shown by arrows.

In addition to dislocations, micro cracks were observed in GaN samples grown on SiC (see Fig.1). No such cracks were detected in the layers grown on sapphire. This can be explained by the large tensile thermal stress of GaN layers grown on SiC [3]. Besides the GaN layers grown on SiC are thicker than those on sapphire. Despite the fact that lattice mismatch between GaN and SiC is much less than that between GaN and sapphire, dislocation densities in GaN grown on these two substrates are comparable (Table 1).

Doping affects the density and the distribution of the dislocations in GaN layers similarly for both substrates (compare Figs. 1 and 2, 3 and 4). Si-doping was found to reduce the dislocation density [Figs.3 (b) and 4 (b)] while Mg doping seems to increase it for sapphire substrate [Fig. 4 (c)]. Figs.1 (b) and 2 (b) show a high density of horizontal dislocation segments in Si-doped GaN layers which increases the probability of dislocation interaction and annihilation in agreement with our earlier observation [8, 11]. Formation of such horizontal segments suggests a higher dislocation mobility during the growth for Si-doped GaN compared to that of Mg-doped layers.

Table 1. Structural parameters of GaN layers grown on sapphire and SiC.

Sample N	#6	#16	#5	#15	#18	#17
Substrate	sapphire	SiC	sapphire	SiC	sapphire	SiC
Doping, N, cm^{-3}	undoped	undoped	Si: 2×10^{18}	Si	Mg	Mg
Dislocations, ρ_D, cm^{-2}	1.2×10^9	1.6×10^9	1×10^9	6×10^8	5×10^9	1.2×10^9
Nanopipes, ρ_n, cm^{-2}	$\sim 10^7$	$< 10^6$	$\sim 4 \times 10^7$	$\sim 10^6$	$\sim 10^8$	$\sim 10^6$
FWHM, arcmin	~7	~8	~7	~5	~9	~6

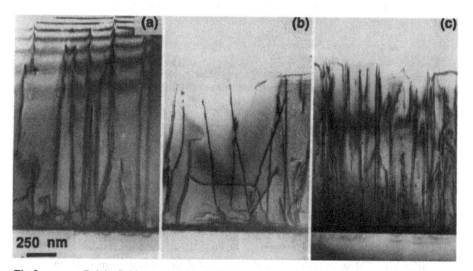

Fig.2, a-c.   Bright-field cross-sectional TEM images of GaN layers grown on sapphire using AlN buffer layer: a - undoped GaN, b - Si-doped, c - Mg-doped GaN.

A high dislocation density leads to the formation of a so called "mosaic" or columnar structure of the GaN layers. Threading dislocations in the GaN layer often are arranged in small angle boundaries which divide the crystal into columnar blocks or domains. Such grain boundaries and domains are seen in Figs. 3 and 4. Domains are slightly misoriented each to the other. Misorientaions between domains have two components, parallel and perpendicular to the layer surface. Individual boundaries were usually complex having both twist and tilt components. This mosaic structure results in an asymmetry of the intensity distribution of x-ray reflections from the layer in reciprocal space.

Three Burgers vectors for perfect dislocations in GaN were observed: 1/3 <11$\bar{2}$0>, <0001> and 1/3<11$\bar{2}$3> [12]. Assuming that most dislocations in the layer lie parallel to the growth, c-axis, those Burgers vectors correspond to edge, screw and mixed dislocations, respectively.

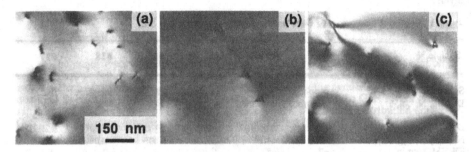

Fig.3, a-c.   Bright-field plan view TEM images of GaN layers grown on SiC using AlN buffer layer: a - undoped GaN, b - Si-doped, c - Mg-doped GaN.

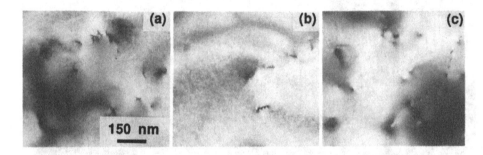

Fig.4, a-c.   Bright-field plan view TEM images of GaN layers grown on sapphire using AlN buffer layer: a - undoped GaN, b - Si-doped, c - Mg-doped GaN.

Diffraction TEM analysis shows the presence of all three dislocation types in both layers, grown on SiC and on sapphire. Dislocations with an edge component dominate in both cases. Pure screw dislocations usually represent less than 10%. The fraction of the pure edge dislocations was found to be different in samples grown on sapphire and SiC being larger for SiC. This difference probably results from the different structure of the GaN-AlN interfaces for the two substrates (Fig.5). Although AlN had a three dimensional growth mode with formation of $\{1\bar{1}01\}$ facets in both cases, the interface between GaN and the AlN buffer for the SiC substrate in Fig. 5 (a) looks more regular than that of the sample grown on sapphire [Fig. 5 (b)]. The latter has a higher density of steps which might increase the density of dislocations with a screw component in the GaN layer. Irregular growth of the AlN layer on sapphire was likely to be related to lower growth temperature and to the deterioration of the substrate during the growth. Deterioration of sapphire associated with vertical defects in the AlN buffer is clearly seen from Fig. 5 (b). This indicates outdiffusion of oxygen toward the GaN layer. This internal source of oxygen in the GaN layer may also lead to the increase of nanopipe density.

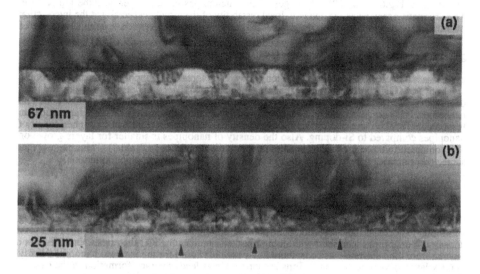

Fig.5, a-b.   Cross-sectional TEM images of AlN/GaN interfaces for samples grown on SiC (a) and sapphire (b). Arrows show the deterioration of sapphire.

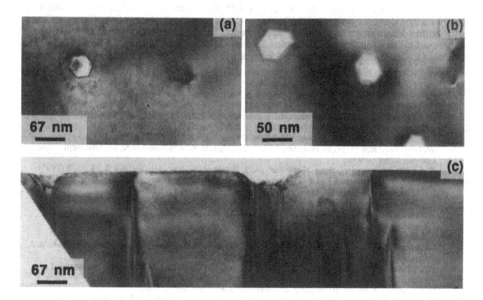

Fig.6   Plan view (a,b) and cross-sectional (c) TEM images of nanopipes in GaN layers grown on sapphire: a- Si-doped and b,c- Mg-doped GaN layer.

Nanopipes are often found in epitaxial GaN to be associated with edge or mixed threading dislocations. Fig.6 shows HREM images of such nanopipes in GaN layer near the top surface. Their formation may be associated with local instability of the growth especially at the intersection of a dislocation or group of dislocations with the growth front. At such a point clustering of impurities would be expected to take place leading to facet formation on $\{1\bar{1}01\}$ planes. A low growth rate of GaN on those planes can lead to nanopipe formation as pointed out earlier [13].

Density of nanopipes was shown to increase with both types of doping (Table 1). This agrees with our previous results [11] which showed an increase of nanopipe density with doping level of Si in GaN layers grown on sapphire. Because the dislocation density in Si-doped layers decreases with doping level, there is no correlation between the densities of dislocations and nanopipes. It is not clear up to now whether the doping atoms themselves or other impurities are responsible for the increased formation of nanopipes. Mg-doping results in a higher density of nanopipes compared to Si-doping. Also the density of nanopipes is smaller for layers grown on SiC compared to those grown on sapphire under similar conditions (Table 1). This suggests that some other impurity such as oxygen may play an important role in nanopipe formation.

CONCLUSIONS

In conclusion, for both substrates, SiC and sapphire, TEM shows comparable dislocation densities in GaN layers that were significantly affected by doping. In both cases, the Si-doping was found to decrease the dislocation density. Mg-doping increased it for sapphire substrates. The density of nanopipes increased with doping but was smaller for SiC substrates. Outdiffusion of oxygen from the substrate was considered as a possible cause for the increased nanopipe density in GaN layers grown on sapphire. Tensile thermal stress leads to crack formation in the case of SiC substrate while cracking was not detected in layers grown on sapphire.

ACKNOWLEDGMENT

This study was supported by the Director, Office of Energy Research, U.S. Department of Energy under Contract No.DE-AC03-76F00098. The use of the facilities of the National Center of Electron Microscopy is gratefully appreciated. The work at Meijo University was partly supported by the Ministry of Education, Science, Sports and Culture of Japan (High-Tech Research Center Project) and the Japan Society for Promotion of Science (Research for the Future Project).

REFERENCES

1. H. Amano, M. Kito, X. Hiramatsu, and I. Akasaki, Jpn. J. Appl. Phys. **28**, L2112 (1989).
2. S. Nakamura, T. Mukai, and M. Senoh, Jpn. J. Appl. Phys. **30**, L1998 (1991).
3. S.N. Mohammad, A. Salvador, and H. Morkoç, Proc. IEEE **83**, 1306 (1995).
4. W. Qian, G.S. Rohrer, M. Skowronski, K. Doverspike, L.B. Rowland, and D.K. Gaskill, Appl. Phys. Lett. **67**, 2284 (1995).
5. Z. Liliental-Weber, H. Sohn, N. Newman, and J. Washburn, J. Vac. Sci. Technol B **13**, 1578 (1995).
6. W. Qian, M. Skowronski, and G.S. Rohrer, Mat. Res. Soc. Symp. Proc., vol. **423**, 475 (1996).
7. F.A. Ponce, D.P. Bour, W. Gotz, and P.J. Wright, Appl. Phys. Lett. **68**, 57, (1996).
8. S. Ruvimov, Z. Liliental-Weber, T. Suski, J.W. Ager, J. Washburn, J. Krueger, C. Kisielowski, E.R. Weber, H. Amano, and I. Akasaki, Appl. Phys. Lett. **69**, 1454 (1996).
9. I.Akasaki, et al. J. Crystal Growth **98** (1989) 209
10. Z. Liliental-Weber, S. Ruvimov, C. Kisielowski, Y. Chen, W. Swider, J. Washburn, N. Newmann, A. Gassmann, X. Liu, L. Schloss, E.R. Weber, I. Grzegory, M. Bockowski, J. Jun, T. Suski, K. Pakula, J. Baranowski, S. Porowski, H. Amano, and I. Akasaki, MRS Proc. v.395, 351 (1996)
11. Z. Liliental-Weber, S. Ruvimov, T. Suski, J.w. Ager III, W. Swider, J. Washburn, H. Amano, I. Akasaki, W. Imler, Mat. Res. Soc. Symp. Proc., vol. **423**, 487 (1996).
12. D. Hull and D.J. Bacon, Introduction to Dislocations, Pegamon Press, 1984
13. Z. Liliental-Weber, S. Ruvimov, Y. Chen, W. Swider and J. Washburn, MRS, v.449, to be published

LUMINESCENCE RELATED TO STACKING FAULTS IN HETEREPITAXIALLY GROWN WURTZITE GaN

M.ALBRECHT,* S.CHRISTIANSEN*, G.SALVIATI** C.ZANOTTI-FREGONARA **, Y.T.REBANE***, Y.G.SHRETER***, M.MAYER****, A.PELZMANN****, M.KAMP****, K.J.EBELING****, M.D.BREMSER*****, R.F.DAVIS*****, and H.P.STRUNK*
*Universität Erlangen-Nürnberg, Institut für Werkstoffwissenschaften, Mikrocharakterisierung, Cauerstr.6, 91058 Erlangen, F.R.G
**CNR-MASPEC, Via Chiavari 18/A, 43100 Parma, Italy
***A.F.Ioffe Institute, Polytechnicheskaya 26, 194021 St.Petersburg, Russia
****Universität Ulm, Abt. Optoelektronik, Albert-Einstein-Allee 45, 89069 Ulm, F.R.G
*****Department of Physics, North Carolina State University, Raleigh, NC 27695-8202, USA

ABSTRACT

We correlate structure analyzed by transmission electron microscopy with photo- and cathodoluminescence studies of GaN/Al$_2$O$_3$(0001) and GaN/SiC(0001) and show that an additional UV line at 364nm/3.4eV can be connected to the occurrence of stacking faults. We explain the occurrence of this line by a model that is based on the concept of excitons bound to stacking faults that form a quantum well of cubic material in the wurtzite lattice of the layer material. The model is in reasonable agreement with the experimental observations.

INTRODUCTION

Photoluminescence and cathodoluminescence spectra of not intentionally doped (hetero)epitaxially grown GaN layers generally show a sharp peak close to the energy gap (at 3.478 eV, attributed to excitons bound to a shallow neutral donor, in the following called bound exciton (BE) line) and in many cases a broad peak centered around 2.2eV, which is well known as "yellow luminescence". Besides these lines a number of additional lines in the energy region between 3.0 eV and 3.40 eV are observed by several authors (e.g. [1-7]). Up to now the structural origin of the yellow and additional UV lines is a controversially debated topic. In principle native defects (Ga or N antisite defects, Ga or N vacancies/interstitials), impurities (e.g. C, O) or extended defects can be involved in luminescence

While the extended defects that form in GaN layers during growth on both SiC and Al$_2$O$_3$(0001) layers are well characterized by transmission electron microscopy [e.g. 8,9] and a large number of papers exist that study the optical properties of these layers by photoluminescence (e.g. [10-14]) and cathodoluminescence (15,16), only few works exist that directly correlate optical and structural defects in the same specimens [6,17,18]. This situation is mainly due to the fact that up to now a very high dislocation density is present in single crystals (10^6 cm^{-2} [4,5]), in homoepitaxial layers (10^7 cm^{-2} [4,5]) and in heteroepitaxial layers (10^8-10^{10} cm^{-2}, e.g. [9,18]) and the spatial resolution of the spectroscopic techniques (CL, PL) is in the range of 1 mm. Since markedly improved GaN layers and improved spatial resolution of luminescence are not available, defects and optical properties can be correlated only at present by comparing the defect populations of two GaN specimens, of which one shows an additional optical line

In this paper we focus on the structural origin of a 'blue' (UV) line at 364nm/3.40 eV that has been reported by several authors [1-7]. We analyze by transmission electron microscopy the defect distribution of two classes of a-GaN (wurtzite lattice) samples grown onto both Al$_2$O$_3$(0001) and SiC[0001] substrates. One of them shows the additional UV line at 364nm/3.40 eV (in the following called sample AB, for additional 'blue' line), while the other does not (in the following called sample "LB", for lacking blue line). We show that the additional UV line at 364nm/3.40 eV can be related to stacking fault bound by Shockley partial dislocations (b=1/3<1-100>) in fact independent of substrate type and growth method.

EXPERIMENTAL

The a-GaN layers grown onto Al$_2$O$_3$(0001) substrates, analyzed here are grown by gas source molecular beam epitaxy (GSMBE) and organo-metallic vapour phase epitaxy (OMVPE). The GSMBE growth took place at 876°C on low temperature (550°C) GaN-buffer layers grown by GSMBE or metalorganic vapour deposition (for details of the GSMBE deposition method see [19]).The OMVPE took place at 1050°C on both SiC[0001]-

293

on-axis ([0001] orientation]) and off-axis substrates (orientation slightly tilted with respect to [0001]) at 1050°C onto AIN buffer layers predeposited at 1100°C (details of the growth process [20]). The electro-optical analysis of our specimens uses cathodoluminescence between 300K and 6K. These measurements are performed in a scanning electron microscope equipped with an Oxford Instruments MonoCL system. The microscope is operated at 20 kV using beam currents between 10 nA and 250nA. Photoluminescenece (PL) was excited by a He-Cd laser. The microstructure of the layer is characterized by conventional transmission electron microscopy in a Philips CM 30 and by high resolution TEM in Philips CM300UT (C_s=0.65 mm, point to point resolution of 0.17 nm), operated at 300 kV. Cross-sectional TEM samples are prepared by standard techniques including mechanical grinding and polishing with diamond paste followed by ion milling to electron transparency.

RESULTS

Electro-optical characterization

Fig. 1a shows typical CL spectra (accelerating voltage: 20 kV, beam current: I=25 nA, T=5K) in the blue spectral region of sample type AB ((GaN/Al$_2$O$_3$(0001): dashed line, "AB$_1$ "; GaN/SiC(0001) off-axis: solid line, "AB$_2$"). Fig.1b shows the corresponding CL spectra that represent sample type LB (GaN/Al$_2$O$_3$(0001): solid line, "LB$_1$"; GaN/SiC(0001)-on-axis: dashed line, "LB$_2$"). Both sample-types show an intense line around 355 nm/3.49 eV, that is usually interpreted as due to an exciton bound to a neutral donor (BE). In the sample grown on SiC this line is shifted to slightly lower energies (357 nm/3.47 eV) compared to that grown on sapphire. Around 382 nm/3.25 eV a line is observed in both types of samples that corresponds to donor acceptor pair recombination. Important for our following considerations is the additional peak at 364 nm/3.40 eV (in the following called SFE) that can be clearly resolved in sample type AB (Fig. 1a), but is only very weakly in sample type LB (Fig.1b). In the AB-sample grown on sapphire a line at 3.32 eV that can be interpreted as phonon replica of the line at 3.40 eV. This line is present only as shoulder of a line at 3.27eV in the AB-sample grown on SiC. The CL spectra taken at different areas of sample type AB (cf. Fig.2a) with a spatial resolution of 1 mm² (20 keV, 25 nA, 5K) show a spatial anti-correlation of BE and SFE, i.e with increasing intensity of L$_1$ BE decreases. With increasing acceleration voltage i.e. with increasing penetration depth of the electrons the relative intensity of SFE with respect to BE increases in the sample grown on sapphire (Fig.2b). Corresponding results are obtained from depth dependent PL measurements on the samples grown on SiC [21]. PL as well as CL measurements show a sublinear dependence of the SFE intensity as a function of excitation and a very rapid reduction of the intensity with temperature [7]. This leads us to conclude that this line can be attributed to an extended defect.

Transmission electron microscopy

A summary of the results of the TEM defect analysis of the samples analyzed in this paper is given in table 1. As an example we show here results of the samples grown on sapphire-substrates. Fig. 2 shows typical cross-sectional TEM micrographs of sample type LB and AB. In both samples a high density of threading dislocations can be seen that extend perpendicularly through the layer from the interface to the surface. The dislocation density in both samples is highest close to the interface. From the micrographs shown in Fig. 3 it becomes obvious that the main structural difference between sample LB and AB consists in the high density of planar defects on (0001) planes in sample AB (Fig. 2b), while no such defects can be observed in sample LB (Fig. 2a). These planar defects are distributed statistically throughout the whole layer volume of sample B, however, occur with increased density near the interface: while down to 1 mm from the surface of the layer 1.5 10^{13}cm^{-3} are present, the density increases to 6.0 10^{13}cm^{-3} close to the interface. From high resolution TEM investigations (e.g. Fig.3) we distinguish four types of planar defects: (i) stacking faults bound by Frank partial dislocations (b=1/2(0001), cf. Fig. 3a), (ii) stacking faults bound by Frank partial dislocations (b=1/6<2-203>, cf. Fig. 3a), (iii) stacking faults bound by Shockley partial dislocations (b=1/3<1-100>, cf. Fig. 3c) and inversion domain boundaries

Comparing the defect distributions in the layer volumes of both samples we can state:
(i)   While in both samples types all types of dislocations are present, planar defects occur only in sample AB.
(ii)   The density of edge-type threading dislocations with Burgers vector b=1/3<11-20> in sample type NB is one order of magnitude higher than in sample AB.

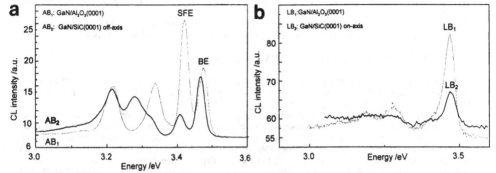

Fig.1. Cathodoluminesence spectra of GaN grown on Al$_2$O$_3$(0001) and SiC. **(a)** Spectra of sample type AB taken at 20K, an electron beam energy of 20 keV and a beam current of 25 nA. Sample AB$_1$ is grown on AL$_2$O$_3$(0001), while sample AB$_2$ is grown on off-axis SiC(0001) substrates. Both samples show a narrow peak at 3.47 eV/358 nm. A pronounced additional UV line at 364nm/3.40 eV is prominent. The additional lines in the lower energy region can be attributed to donor acceptor pair recombination and are different in both samples. **(b)** Spectra of sample type LB. taken at 5 K, 20keV,100 nA. Sample LB$_1$ is grown on Al$_2$O$_3$(0001), while sample AB$_2$ is grown on on-axis SiC(0001) substrates. No additional line is present at 3.4 eV.

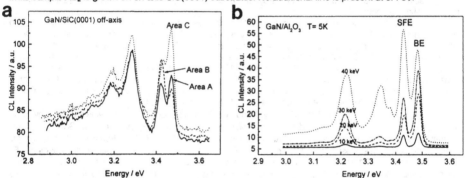

Fig. 2. Spatial distribution of the luminescence. **(a)** CL spectra (5K, 20 keV, 25 nA) of sample AB$_2$ taken at different areas. The intensities of the BE and L$_1$ lines are in spatial anticorrelation. **(b)** CL spectra of sample AB$_1$ taken with different accelerating voltages. With increasing penetration depth (i.e. voltage) the relative intensity of the SFE line increases with respect to the BE line.

Table 1 : Defect distribution of samples investigated in this work as analyzed by TEM. (LB are samples without additional blue line at 364 nm/3.4 eV), AB samples with without additional blue line at 364 nm/3.4 eV). LB$_1$: GaN/Al$_2$O$_3$(0001), LB$_2$: GaN/SiC(0001)-on axis ; AB$_1$: GaN/Al$_2$O$_3$(0001), AB$_2$: GaN/SiC(0001)-off axis.

sample	b=1/3<11-20>	b=1/3<11-23>	b=<0001>	stacking faults	Frank loops	IDB
LB$_1$	$9.9\ 10^9$ cm^{-2}	$< 10^6$ cm^{-2}	$1.3\ 10^6$cm^{-2}	-	-	-
LB$_2$	$1.8\ 10^9$ cm^{-2}	$<10^6$cm^{-2}	$1.3\ 10^9$ cm^{-2}	-	-	-
AB$_1$	$0.4\ 10^9$ cm^{-2}	$0.8\ 10^9$ cm^{-2}	$0.4\ 10^9$ cm^{-2}	$2.3\ 10^{13}$ cm^{-3}	$1.6\ 10^{13}$ cm^{-3}	$1.3\ 10^8$ cm^{-2}
AB$_2$	$2.5\ 10^9$ cm^{-2}	$<10^6$ cm^{-2}	$0.2 10^9$ cm^{-2}	$10.8 10^{13}$ cm^{-3}	$0.9\ 10^{13}$ cm^{-3}	-

(iii) The number of screw dislocations in the sample type LB is by a factor 2-3 higher than in sample AB (see table1).
Comparing our TEM results to the CL and PL results we can state: In the AB samples the intensity of the SFE line increases close to the interface of layer and substrate as does the density of stacking faults

EXCITONS BOUND TO STACKING FAULTS IN WURTZITE GaN

The following discussion will argue after ruling out atomic defects, inversion domain boundaries and a class of perfect dislocations, that in fact the presence of Shockley partial dislocations ($b=1/3(1-100)$) bounding the stacking faults gives rise to the additional UV line at 364nm/3.40eV.

In principle the additional UV line of samples AB may be attributed to a difference in the densities of atomic defects in both samples. Although atomic defects cannot be directly imaged by transmission electron microscopy we exclude such a difference to cause the additional line. The occurence of the additional line is independent on the growth method (GSMBE, OMVPE) and the substrate (SiC, sapphire). Also inversion domain boundaries can be ruled out to cause the UV line. Experimental investigations of samples with a density of inversion domain boundaries of 10^9 cm^{-2} show no line besides that of the bound exciton. Recent theoretical work by Northrup et al. [22] shows that inversion domain boundaries do not give rise to gap states and thus are in good accordance with our experimental findings. We now can proceed to discuss dislocations as the possible source for the additional UV line. According to Shreter et al. [6], misfit dislocations in the GaN(0001) plane (60°-dislocations with $b=1/3<11-20>$, $b=1/3<11-23>$ are expected to be highly charged (10^9 e/m) due the strong ionic character of the bond. The respective inserted half planes that correspond to these dislocations end each with one row of identical atoms (Ga or N) [23,24]. According to Shreter et al. [6] the strong electric field of such a dislocation destroys the dislocation excitons, which usually are responsible for the dislocation luminescence.

In the following we therefore based on our TEM/CL observations discuss a model for excitons bound to stacking faults [25]: While wurtzite GaN has an AaBbAaBbaAbB stacking sequence, the stacking faults observed in our AB smples have stacking sequence aBbAaBbCcBbCc (I_1), aBbAaBbCcAaCcAa (I_2) and aBbAaBbCcAaBbCcBb (E) and corresponding Burgers vectors of the surrounding dislocations are $b(I_1) = 1/6<2-203>$, $b(I_2) = 1/3<1-100>$ and $b(E)=1/2<0001>$ [24]. These stacking faults represent layers of cubic phase with a width $L(I_1)=1.5c_0$, $L(I_2)= 2 c_0$ and $L(E)=2.5 c_0$ inserted into the host wurtzite lattice. The wurtzite lattice can be considered as a cubic crystal that is strained along the (111) axis with a corresponding strain of ($e_{zz}=0.612 c_0/a_0-1 =-0.005$ for the lattice constants $c_0=0.5185$ nm and $a_0=0.3189$ nm.

This strain shifts the edges of the conduction band by values $\Delta E_c = \Xi e_{zz}$ and of the valence Band by $\Delta E_v= (a-b) e_{zz}$ respectively, where Ξ, a, and b are the deformation potential constants for cubic GaN. Based on estimations of Ξ, a, and b [25] we can find $\Delta E_c=122$ meV and $\Delta:E_v=62$ meV The difference in band gaps for wurtzite and cubic phases estimated in this model is $\Delta E_g =\Delta E_c-\Delta E_v =60$ meV. The experimental values of ΔE_g are in the range of 90-190 meV [26]. Thus the theoretical estimate is slightly lower than the experimental value. It is important to note that both ΔE_c and ΔE_v are positive, i.e. the electrons are attracted to the stacking faults but the holes are repelled from them (cf.Fig. 4). Thus the interface between wurtzite and cubic GaN is similar to the type II heterojunction. Since the stacking faults contain a few atomic layers ($L=1.5c_0$ -2.5c_0) the quantum effects should be taken into account in calculation of the electron binding energy E_e. For the case $E_e << \Delta E_c$, E_e can be found in the δ-potential well approximation as $E_e =m_e (\Delta E_c L)^2/\hbar^2 =25$ meV for $L=1$ nm and $m_e= 0.2 m_0$ [27]. The holes are attracted to the electrons bound to the stacking faults via Coulomb force, forming excitons bound to the stacking faults. The binding energy of the stacking fault exciton can be estimated in the approximation that the electron at the stacking fault is immobile. An account for the electron motion in the stacking fault plane should reduce the energy to some extent. Since the stacking fault strongly repels the hole its wave function should go to zero at the stacking fault plane. Therefore the binding energy for the hole is given by its lowest p-state in the Coulomb potential: The energy of this level is $E_0^{J=3/2}$ $= 0.23$ Ry*, where Ry$^* =m_{hh}e^4/2 \varepsilon^2\hbar^2$ and m_{hh} is the mass of the heavy hole, ε is the dielectric constant. For GaN with $m_{hh}=1.76 m_0$ [28] and $\varepsilon=9$ [29], Ry$^*=295$ meV and the upper estimate for SFE binding energy is $E_{SFE}=60$ meV. The lower estimate is 30 meV. On the base of this model we can attribute the shift of the line with respect to the hexagonal GaN band gap $E_g - \hbar\omega=3.5$ eV-3.4 eV =0.1 eV with the sum of binding energies of carriers to the stacking fault $E_{SFE} + E_e \sim 0.08$ eV. It can be seen that a reasonable agreement exits between the model and the experimental data.

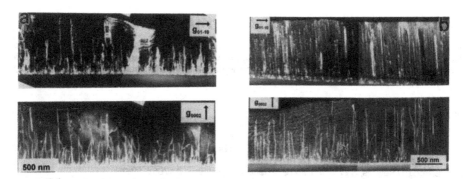

Fig.3. Defect distribution in GaN/Al₂O₃(0001). Cross-sectional transmission electron micrograph, weak beam dark field conditions. **(a)** Sample AB₁. Besides threading dislocations a high density of planar defects can be seen. **(b)** Sample LB₁. A high density of threading dislocations can be seen, no planar defects are present.

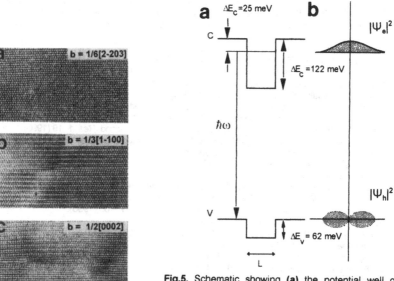

Fig.5. Schematic showing **(a)** the potential well of a stacking fault embedded in a hexagonal wurtzite lattice; **(b)** the scheme for wave functions of electrons and holes bound to the stacking fault.

Fig.4. Different types of planar defects in samples AB. High resolution transmission electron microscopy. The incident electron beam is parallel to the [2-1-10] direction. **(a)** Stacking fault bound by a Frank partial with **b**=1/6<2-203>. **(b)** Stacking fault bound by a Sockley partial with **b**=1/3<1-100>. **(c)** Frank loop bound by a Frank partial dislocation with **b**=1/2<0001>.

CONCLUSION

In summary we have shown by correlation of TEM and CL that the UV line at 364nm/3.40eV can be related to the presence of stacking faults. We explain this line by an exciton bound to a stacking fault. The stacking fault is a quantum well of cubic material embedded in the wurtzite lattice that attracts electrons to the stacking faults and repels holes.

REFERENCES

1. W.Götz, L.T.Romano, B.S.Krusor, N.M.Johnson, and R.J.Molnar, Appl.Phys.Lett. **69**, 242 (1996).
2. L.Eckey, J.-Ch. Holst, P.Maxim, A.Hoffmann, I.Broser, B.K.Meyer, C.Wetzel, E.N.Mokhov, and P.G.Baranov, Appl.Phys.Lett. **68**, 415 (1996).
3. C.Hong, D.Pavlidis, S.W.Brown, and S.C.Rand, J.Appl.Phys. **77**, 1705 (1996).
4.. F.A.Ponce, D.P.Bour, W.Götz, N.M.Johnson, H.I.Helava, I.Grzegory, J.Jun, and S.Porowski, Appl.Phys.Lett. **68**, 917 (1996).
5. A.Gassmann, T.Suski, N.Newman, C.Kisielowski, E.Jones, E.R.Weber, Z.Liliental-Weber, M.D.Rubin, H.I.Helena, I.Grzegory, M.Bockowski, J.Jun, and S.Porowski, J.Appl.Phys. **80**, 2195 (1996).
6. Shreter and Y.T.Rebane, J.Physique III to be published (1997).
7. W.Rieger, R.Dimitrov, D.Brunner, E.Rohrer and M.Stutzmann, Phys.Rev. B, to be published (1997)
8. Z.Lilienthal-Weber, H.Sohn, N.Newman, and J.Washburn, J.Vac.Sci.Technol. B **13**, 1578 (1995).
9. X.J.Ning, F.R.Chien, P.Pirouz, J.W.Yang, and M.Asif Khan, J.Mater.Res. **11**, 580 (1996).
10. J.I.Pankove, J.E.Berkeyheiser, H.P.Maruska, and J.Wittke, Solid State Commun. **8**, 1051 (1970).
11. R.Dingle and M.Ilegems, Solid State Commun. **9**, 175 (1971).
12. B.Monemar, Phys.Rev. B **10**, 676 (1973).
13. M.R.Kahn, Y.Oshita, N.Sawaki, L.Akasaki, Sol.State. Commun. **57**, 405 (1986).
14. S.Logothedidis, J.Petalas, M.Cardona, T.D.Moustakas, Phys.Rev.B **50**, 18017 (1994).
15. K.Hiramatsu, H.Amano, and I.Akasaki, J.Cryst. Growth **99**, 375 (1990).
16. J.Menninger, U.Jahn, O.Brandt, H.Yang, and K.Ploog, Phys.Rev. B **53**, 1881 (1996).
17. F.A.Ponce, D.P.Bour, W.Götz, and P.J.Wright, Appl. Phys. Lett. **68**, 57 (1996).
18. S.Christiansen, M.Albrecht, W.Dorsch, H.P.Strunk, C.Zanotti-Fregonara, G.Salviati, A.Pelzmann, M.Mayer, M.Kamp, and K.J.Ebeling, MRS Internet J. Nitride Semicond. Res. **1**, 19 (1996).
19. M.Kamp, M.Meyer, A.Pelzmann, S.Menzel, H.Y.Chung, H.Sternschulte, and K.J.Ebeling, in: Proc. of Topical workshop on III/V nitrides, Nagoya, Japan (1995).
20. T.W.Weeks, M.Bremser, A.K.Shawn, E.Carlsson, W.G.Perry, E.L.Piner, N.A.El-Masry, and R.F.Davis, J.Mater.Res. **11**, 1011 (1996).
21. Y.G.Shreter, Y.T.Rebane, T.J.Davis, J.Barnard, M.Darbyshire, J.W.Steeds, W.G.Perry, M.D.Bremser, R.F.Davis, and J.W.Steeds, Mat.Res.Soc.Symp.Proc. **449**, 683 (1997).
22. J.E.Northrup, J.Neugebauer, and L.T.Romano, Phys.Rev.Lett. **77**, 103 (1996).
23. Yu.A.Ossip'yan and I.S.Smirnova, phys.stat.sol. **30**, 19 (1968).
24. Yu.A.Ossip'yan and I.S.Smirnova, J.Phys.Chem.Solids **32**, 1521 (1971).
25. Y.T.Rebane, Y.G.Shreter, M.Albrecht, this conference.
26. S.Strite and H.Morkoç, J.Vac.Sci.Technol. B **10**, 1237 (1992).
27. A.S.Baker and M.Illegems, Phys.Rev.B **7** 743 (1973).
28. M.Suzuki and Tuenoyama, J.Appl.Phys. **80** 6868 (1996).
29. H.Morkoc, S.Strite, G.B.Gao, M.E.Lin, B.Sverdlov, and M.Burns, J.Appl.Phys. 76, 1363 (1994).

PHOTOLUMINESCENCE OF STRAIN-ENGINEERED MBE-GROWN GaN AND InGaN QUANTUM WELL STRUCTURES

Joachim Krüger[*,**,1], Christian Kisielowski[*,**], Ralf Klockenbrink[*,**], Sudhir G.S. [*,**], Yihwan Kim [*,**], Michael Rubin[**], and Eicke R. Weber[*,**]

[*] Department of Materials Science and Mineral Engineering, UC Berkeley, Berkeley, CA 94720
[**] Lawrence Berkeley National Laboratory, Materials Science Division, Berkeley, CA 94720
[1] joachim@socrates.berkeley.edu

ABSTRACT

The paper describes the influence of strain on the optical quality of GaN films grown by MBE on c-plane sapphire. The photoluminescence (PL) line width of the donor-bound exciton can be designed to be as narrow as 1.2 meV by actively utilizing hydrostatic and biaxial stress components. Unstrained p-type Mg-doped GaN films exhibit comparably narrow near band edge transitions. A sharp PL line at 3.261 eV in some of our films is identified as the donor bound exciton of the cubic phase. The formation of these cubic inclusions can be stimulated by a high III/V flux ratio at the growth temperature of T = 725°C. The PL spectrum of an InGaN multi quantum well structure is significantly broadened compared with the spectra of single quantum well structures. Combination of PL and TEM indicates that this effect relates to a progressive increase of the quantum well widths and their spacing along the growth direction. It is argued that strain affects the growth rate and the incorporation of Indium into the quantum well structures.

INTRODUCTION

In recent years, the growth of GaN has been developed within an unprecedented short time [1]. Metal-organic-chemical-vapor-deposition (MOCVD) and related methods were established to be the major crystal growth technologies for the large scale production of optoelectronic devices on GaN, such as LED's and lasers. Drawbacks of these rapid developments relate to the poor understanding of physical processes that determine and limit the material quality. Strain was recently found to be a key issue that influences many of the materials properties, such as the film morphology, the bandgap, electrical or optical properties.

It is the purpose of this paper to focus on selected optical properties of GaN that can be altered by the presence of strain. In case of GaN thin film growth, the strain originates from the growth on lattice mismatched substrates (15%) with a thermal expansion coefficient that differs from GaN. It is pointed out that the strain can be engineered by the growth of suitable buffer layers.

In case of InN/GaN quantum well structures, an additional strain component comes from the 11% lattice mismatch between GaN and InN. Experimental results indicate that this strain affects the growth of multi quantum well structures.

EXPERIMENTAL

The GaN films were grown using a rebuilt Riber 1000 chamber. Gallium was evaporated from a Knudsen effusion cell. The activated nitrogen species were provided by a Constricted Glow Discharge Plasma Source (for details see [2]). Prior to the growth, the sapphire substrate was nitridated for 10 minutes. Subsequently, a buffer layer of typically 25 nm thickness was grown. Typical growth conditions are: Ga source temperature: 1210 K, nitrogen flow rate: 35 sccm, buffer layer growth temperature: 775K, main layer growth temperature: 1000K, growth rate ≈ 0.5 µm/hr. For p-doping, the Magnesium source temperature was 280°C. The main GaN epilayers had a typical thickness of 2 µm.

Photoluminescence was excited using the 325nm line of a 50mW HeCd laser. The luminescence light was then dispersed by a 0.85m double monochromator and detected with a photomultiplier using a lock-in technique. The sample temperature was varied between 4 K and 300 K. Hall effect measurements were carried out at room temperature using the van der Pauw configuration.

RESULTS and DISCUSSION

Strain engineered n-type GaN

Thin films of hetero-epitaxially grown GaN are known to incorporate strain of up to 1 GPa, due to lattice mismatch and different thermal expansion coefficient of substrate and GaN epilayer. The amount and sign of strain are determined by the thickness and growth temperature of the buffer layers as well as by the film stoichiometry [3]. In particular, it is be noted that films grown on sapphire were found to be either under compressive or tensile stress depending on the growth conditions.

From the inspection of a variety of samples grown under different conditions strain is concluded to be the main broadening mechanism for PL lines at helium temperatures. Consequently, crystals being strain free at cryogenic temperatures exhibit PL line widths of the donor bound exciton (DX) as low as 1.2 meV (FWHM), figure 1. To the best of our knowledge, this represents the smallest line width detected so far for a hetero-epitaxially grown GaN film. This value has only been surpassed by MOCVD-films grown on bulk GaN (line width of donor bound exciton = 0.55 meV), [8]. Further indications of high optical quality of our films are the low yellow-to-exciton intensity ratio of about

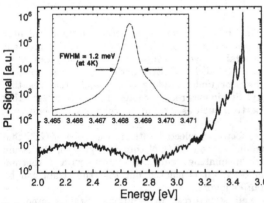

Figure 1: *PL spectrum of a strain free grown n-type GaN film on sapphire (4K)*

10^{-5} as well as the fact that phonon replica of the excitonic transitions up to the third order are visible.

To date, a free p-type carrier density of [p] = (1-2) * 10^{17} cm^{-3} along with a carrier mobility of $10 \leq \mu \leq 41$ cm^2/Vs could be achieved for Mg-doped p-type MBE-grown materials. The PL spectrum is dominated by either the donor-acceptor-transition at about 3.26 eV (ZPL, see figure 2) or the broad luminescence centered at 3.0 eV which has been assigned to the Mg acceptor. The barely visible excitonic transitions (line width = 3 meV) consist of the DX at 3.4687 eV and the acceptor bound exciton (AX) at 3.4637 eV. In agreement with [8] we assign this line to the Mg acceptor.

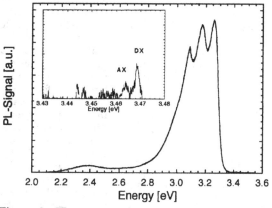

Figure 2: *PL spectrum of a strain free grown Mg-doped p-type GaN film on sapphire (4K)*

Cubic inclusions in the predominant hexagonal matrix

Figure 3 shows a PL spectrum which was frequently detected for our MBE-grown samples. The donor bound exciton (DX) of the hexagonal phase at 3.468 eV (FWHM=2.2meV) and a 9.0 meV broad line at 3.261 eV are clearly visible. In the following paragraph we present arguments assigning this transition to the DX of the cubic phase. Also, we find a structure consisting of three lines (ZPL = 3.157 eV). Temperature variation proved this to be the donor-acceptor-pair (plus two phonon replica) in the cubic phase, in agreement with cathodo-luminescence results taken from cubic GaN grown on GaAs [7].

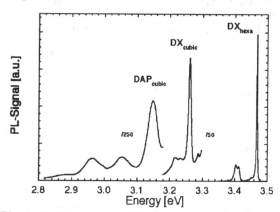

Figure 3: *PL spectrum (4K) of the hexagonal and cubic phase*

The following signals of different origin are commonly observed in the energetic region between 3.25 eV and 3.28 eV and can overlap with the DX signal of the cubic phase: the second phonon replica of the acceptor bound exciton of the hexagonal phase, the zero-phonon-transition of the donor-acceptor of the hexagonal phase [4], and the 6. replica of the resonant Raman line [5]. By temperature variation, one can clearly distinguish these signals from the excitonic transition of the cubic phase.

TEM confirmed the presence of cubic inclusions in the MBE-grown materials [10]. They account for about 1% of the hexagonal matrix and have been found throughout the whole epilayer volume. This indicates a spontaneous nucleation, with no preference for the substrate interface. In a series grown with different nitrogen flux we found the existence of cubic inclusions to be stoichiometry related, crystals grown under *gallium*-rich conditions show the strongest signal.

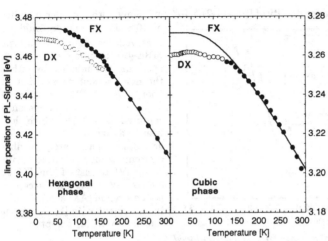

Figure 4: *Temperature dependence of the excitonic transitions in the hexagonal and cubic phase. Lines are fitted according to eq. (1)*

In order to identify the exact nature of the luminescent transition of the cubic phase the temperature dependence of the line positions of the hexagonal and cubic phase transitions was taken, figure 4. The DX of the cubic phase is visible up to room temperature, exhibiting a line width of 45.0 meV (Lorentzian). The line shape and width is identical with the free excitonic transition (FX) of the hexagonal phase. This finding additionally underlines the excitonic character of this line.

Generally, the bandgap temperature dependence of a semiconductor can be described by Cody's formula [6]:

$$E_{gap}(T) = E_{gap}(0) - \frac{\lambda}{\exp(\Theta/T) - 1)} \tag{1}$$

with λ being a constant and Θ the Einstein temperature.

We have fitted this equation to both the hexagonal and the cubic excitonic PL signals. Due to the higher line width of 9 meV, the transition from the DX to the FX signal is not resolved for the cubic phase. Apparently, the excitonic line shifts to higher energies for temperatures up to 50K. We believe this to be caused by the relative

	FX (T=4K) [eV]	DX (T=4K) [eV]	λ [meV]	Θ [K]
hexagonal phase	3.4741	3.4683	140 * ± 11	340 * ± 25
cubic phase	3.272 * ± 0.002	3.2605	135 * ± 14	340

Table I: *Energetic positions of excitonic transitions of the hexagonal and cubic phase. (Parameters derived by fitting Cody's equation (1) are marked with *)*

intensity increase of the free exciton. For the same reason, we were only able to fit Cody's relation for temperatures above 150 K, figure 4. This restriction doesn't allow to derive the Einstein temperature from the experimental data, so that we assumed it to be identical with the value of the hexagonal phase which has been determined to be Θ = 340 K. Such a value was also reported by other authors ([9] and references therein). Consequently, we get the energetic position of the free exciton at 4K as (3.272 ± 0.002) eV, table I. This is also in agreement with CL investigations of cubic GaN [7].

Photoluminescence of Single/Multi quantum well structures

Figure 5 shows the low temperature PL spectra of a two single quantum well (SQW) and one multi quantum well structure (MQW). The MQW structure consists of 10 wells grown under identical conditions.

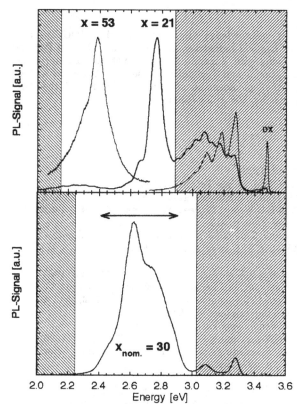

Two effects are apparent: The PL line width increases with increasing Indium content for the InGaN single quantum well structure. More strikingly, for the multi quantum well structure we observe an apparent line width of about 400 meV. Closer inspection reveals the line as being composed of several transitions. This finding can not be explained by quantum effects within the well. Therefore, a structural analysis of the well structures was conducted using high resolution TEM. The Indium distribution was spatially resolved probed [11, 12]. One realizes that the Indium content varies by up to 20% within the quantum wells. In fact, it was found for the MQW structure that the well width as well as the well spacing continuously increase with progressive growth. However, the PL results also indicate that the Indium concentration varies from well to well. Since strain

Figure 5: *PL spectra (4K) of In$_x$Ga$_{1-x}$N quantum well structures*
a) two single quantum well structures
b) multi quantum well structure

increases with the growth of each well, we feel that it is strain that effect both the growth rate and the Indium incorporation into the wells.

SUMMARY

In this paper we show that strain engineering plus the usage of a constricted glow discharge plasma source providing reactive species with low kinetic energy results in GaN films of high optical quality. We present PL spectra of p-type GaN with clearly resolved Mg-related acceptor bound exciton transition. Also, we identify the donor bound exciton of the cubic phase and analyze the PL line width broadening mechanism of multi quantum well structures. The selective examples depict how strain impacts optical properties of GaN thin films.

ACKNOWLEDGMENTS

We would like to gratefully acknowledge the supply of the MOCVD-grown quantum well structures by Nichia Chemical Industries, APA Optics, and Hewlett Packard Company. We acknowledge discussions with Zuzanna Liliental-Weber. This work was supported by the Office of Computational and Technology Research, Advanced Energy Projects and the Laboratory Technology Research Program (ERLTR) of the U.S. Department of Energy under Contract No. DE-AC03-76SF00098.

REFERENCES

[1] S.Nakamura; SPIE, Proc. **2693**, 43 (1996)

[2] A.Anders and S.Anders; Plasma Sources Sci. Technol. **4**, 571 (1995)

[3] C.Kisielowski, J.Krüger, M.S.H.Leung, R.Klockenbrink, H.Fujii, T.Suski, Sudhir G.S., J.W.Ager III, M.Rubin, and E.R.Weber; Proceedings of the 23th ICPS, Berlin 1996 (World Scientific, Singapore) 1996, p. 513
C.Kisielowski, J.Krüger, S.Ruvimov, T.Suski, J.W.Ager III, E.Jones, Z.Liliental-Weber, H.Fujii, M.Rubin, E.R.Weber, M.D.Bremser, and R.F.Davis; Phys.Rev.B II **54**, 17745 (1996)

[4] S.Fischer, C.Wetzel, E.E.Haller, and B.K.Meyer; Appl.Phys.Lett. **67**, 1298 (1995)

[5] D.J.Dewsnip, A.V.Andrianov, I.Harrison, D.E.Lacklison, J.W.Orton, J.Morgan, G.B. Renz, T.S.Cheng, S.E.Hooperz, and C.T.Foxon; Semicond.Sci.Technol.**12**, 55 (1997)

[6] G.D.Cody, in Semiconductors and Semimetals; Vol. **21B**, edited by J.I.Pankove (Academic Press, New York, 1984), chapt. 2, 11 - 79

[7] J.Menniger, U.Jahn, O.Brandt, H.Yang, and K.Ploog; Phys.Rev.B **53**, 1881 (1996)

[8] J.M.Baranowski and S.Porowski; Proceedings of the ICPS 23, Berlin 1996, (World Scientific Publishing, Singapore), p. 497 (1996)

[9] M.O.Manasreh; Phys.Rev.B **53**, 16425 (1996)

[10] M.S.H.Leung, R.Klockenbrink, C.Kisielowski, H.Fujii, J.Krüger, Sudhir G.S., A.Anders, Z.Liliental-Weber, M.Rubin, and E.R.Weber, Mat.Res.Soc.Symp. Vol. **449**, p. 221 (1997)

[11] Christian Kisielowski, Zuzanna Liliental-Weber, and Shuji Nakamura; submitted to Jap.J.Appl.Phys.

[12] Christian Kisielowski, Joachim Krüger, Zuzanna Liliental-Weber, E.R.Weber, Jinwei Yang, Asif Khan, and Chihping Kuo; to be publ.

CHARACTERISATION OF Al$_x$Ga$_{1-x}$N FILMS PREPARED BY PLASMA INDUCED MOLECULAR BEAM EPITAXY ON C-PLANE SAPPHIRE

H. ANGERER*, O. AMBACHER*, M. STUTZMANN*, T. METZGER**, R. HÖPLER**, E. BORN**,A. BERGMAIER***,G. DOLLINGER.***

*Walter Schottky Institute, Technical University Munich, Am Coulomwall, 85748 Garching, Germany, ambacher@wsi.tu-muenchen.de
**Angewandte Mineralogie und Geochemie, Technical University Munich, Lichtenbergstr. 4, 85748 Garching, Germany
***Physik Departement E12, Technical University Munich, James Frank Str., 85748 Garching, Germany

ABSTRACT

Al$_x$Ga$_{1-x}$N films were grown on c-plane sapphire by plasma induced molecular beam epitaxy with $0 \leq x \leq 1$. The composition and purity of the Al$_x$Ga$_{1-x}$N layers was determined by elastic recoil detection analysis with a relative error of 5% for the Al content. Both X-ray diffraction and atomic force microscopy indicate only a slight decrease in epitaxial quality of the Al$_x$Ga$_{1-x}$N films with increasing Al content up to x = 0.65. X-ray diffraction is used to separate the effects of thermally induced biaxial compressive stress and the alloy composition on the shift of interplanar spacings by measuring both lattice constants. The deviation of the c/a ratio from that of fully relaxed films is a quantitative measure of the biaxial compressive stress leading to a distortion of the unit cell.Values up to 0.5 GPa were observed. By the method proposed, the determination of alloy composition can be corrected for this effect. The results obtained by this method are in very good agreement with the elastic recoil detection measurements substantiating the validity of Vegard's law. These results, compared with optical measurements, indicate that the bowing parameter of the optical bandgap is 1.3 eV within the experimental error.

INTRODUCTION

The group III-nitrides with direct bandgaps ranging from 1.9 eV for InN, 3.4 eV for GaN to 6.2 eV for AlN are interesting candidates for optoelectronic devices working in the whole visible to the near UV part of the spectrum. In the meantime high efficient blue and green diodes are commercial available [1, 2, 3] The chemical, thermal and mechanical stabilities allows the use of the nitrides in aggressive environments. Especially the (Al,Ga)N system, with its high breakdown field strength [4] and high saturation drift velocity of the electrons [5], is suitable for high power, high temperature and high frequency applications. A negative electron affinity is expected [6] for AlGaN with a bandgap above 5.5 eV, a pre-condition for cold emitters. For all these applications an exact and reproducable alignment of the bandgap is necessary. We present here a method to determine the composition and the biaxial strain with X-ray diffraction (XRD). The obtained concentrations were verified with elastic recoil detection analysis (ERDA). The dependence of the bandgap on the Al content was investigated.

EXPERIMENT

The (Al, Ga)N films were grown on c-plane sapphire by plasma induced molecular beam epitaxy (PIMBE). The metallic components were supplied by conventional effusion cells, while the activated nitrogen was produced in a rf-plasma source (Oxford Applied Research CARS-25). The inductively coupled power was 400 W and the pressure during the growth was 4 x 10^{-5} mbar. The deposition temperature was varied from 840 °C for GaN up to 1000 °C for AlN. The total flux of Al and Ga was maintained at 8 x 10^{14} $s^{-1}cm^{-2}$ leading to a constant deposition rate of 0.6 μm/h. We have covered the whole composition range from GaN to AlN.

The XRD-measurements were performed using a standard triple-axis diffractometer (Philips MRD), equipped with a 4-bounce Ge 220 monochromator for the primary beam and a 3-bounce channel-cut analyser crystal for the diffracted beam. All measurements were realized in the $\Omega/2\Theta$ mode. The lattice parameter c was determined directly from the symmetric 002 reflection using Bragg's equation. For the second lattice constant a, the asymmetric 205 reflection was measured and with the relationship

$$\frac{1}{d_{hkl}^2} = \frac{4}{3} \times \frac{h^2 + k^2 + hk}{a^2} + \frac{l^2}{c^2},\qquad (1)$$

a can be calculated at known c.

The chemical composition was directly obtained by ERDA as described in detail elsewhere [7, 8]. The absorption coefficient α of the epitaxial $Al_xGa_{1-x}N$ films in the visible and ultraviolet range was determined by transmission measurements. The transmission data were corrected for the reflectivity of a multilayer system.

RESULTS

In Fig. 1 the 002 reflection of different $Al_xGa_{1-x}N$ films with $0 \leq x \leq 1$ is shown. The sharp interference maxima for alloys with $x \leq 0.6$ indicated the excellent structural quality of the samples (FWHM $\leq$ 100 arcsec). For higher Al contents the broadening of the reflection suggests an inhomogenous distribution of the cations.

Assuming the validty of Vegard's law [9],

$$c_0(x) = c_0^{GaN}(1-x) + c_0^{AlN}x,$$

$$a_0(x) = a_0^{GaN}(1-x) + a_0^{AlN}x.\qquad (2)$$

the relaxed lattice parameter a_0 and c_0 should behave linearly with the composition. Due to thermally induced biaxial homogeneous strain and the resulting deformation of the unit cell, the values of a and c obtained from XRD deviate from their relaxed parameters. The composition calculated using equation (2) yields different values for x. The Al mole fraction obtained from c(x) is significant smaller than that obtained from a(x), which is evidence for biaxial strain.

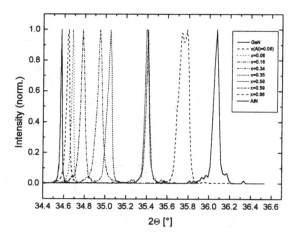

Fig. 1 Ω/2Θ scan of the 002 reflection of epitaxial Al$_x$Ga$_{1-x}$N films with $0 \leq x \leq 1$.

Additionally we compared these values with ERDA measurements, as shown in Fig. 2. The homogeneous lateral strain can be calculated from the ratio of the longitudinal expansion to the cross contraction (Poisson's ratio ν) and the modulus of elasticity (Young's modulus), and we obtain lateral compressive strain in the range of 0.3 - 0.5 GPa independent of the alloy composition. ν was determined to be 0.38 by Detchprohm et al. for GaN[10].

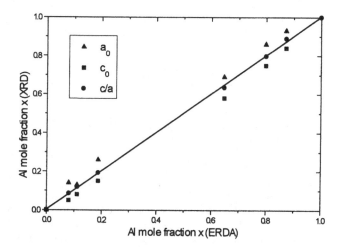

Fig. 2 Al mole fraction x determined by XRD using equation (2) for a$_0$ (triangles), c$_0$ (rectangles) and calculated (circles) using equation (3) and (4).

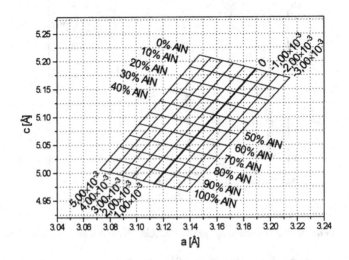

Fig. 3 Diagram for the graphical determination of the alloy composition and the homogeneous strain ε.

With

$$\nu \frac{a - a_0}{a_0} = - \frac{c - c_0}{c_0} = \varepsilon \qquad (3)$$

and

$$x = \frac{ac(1 + \nu) - ac_0^{GaN} - a_0^{GaN}c_0\nu}{ac_0^{GaN} - ac_0^{AlN} - a_0^{AlN}c\nu + a_0^{GaN}c\nu}, \qquad (4)$$

the Al mole fraction x can be calculated, eliminating the influence of biaxial stress. These values for x compared with the ERDA results (Fig. 2) shows that the linear Vegard's law is valid in the (Al, Ga)N system. From Fig. 3, the results for different compositions and several values for the homogeneous strain ε can be obtained graphically.

The absorption edge of several (Al, Ga)N alloys is shown in Fig 4. Especially at higher Al-contents ($x \geq 0.6$), due to structural disorder in these alloys, the absorption edge becomes more broad. In this case a more practical method is to assign the energy gap to a fixed absorption coefficient, as applied for disturbed and amorphous semiconductors for example. We took $\alpha = 74000$ cm^{-1}, because this is a region where all alloys have a clear absorption due to the direct band.

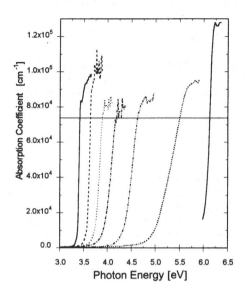

Fig. 4 The absorption coefficient as a function of energy for various alloys with $0 \leq x \leq 1$ determined by transmission data. Their band edge was determined at $\alpha = 7.4 \times 10^4$ cm^{-1}.

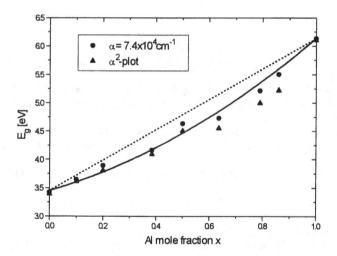

Fig. 5   Bandgap E_g of $Al_xGa_{1-x}N$ as a function of x (circles). The solid line indicates the course of the graph for b = 1.3 eV. E_g determined from the α^2 plot is shown for comparison (triangles).

The nonlinear dependence of the energy gap on the chemical composition is described by

$$E_g^{Al_xGa_{1-x}N}(x) = E_g^{GaN}(1-x) + E_g^{AlN}x - bx(1-x). \qquad (5)$$

where b is the gap bowing parameter. The bandgap as a function of the composition is shown in Fig. 5. We found a downward bowing with a bowing parameter b = 1.3 eV.
This value is consistent with Koide et al. and Khan et al. who found b = 1.0 ± 0.3 eV for alloys with x ≤ 0.4.

SUMMARY

We have grown epitaxial (Al, Ga)N films on c-plane sapphire by PIMBE, covering the whole range of composition. The lattice parameters were found to have a linear dependence on the chemical composition, if the deformation of the unit cell due to biaxial homogeneous strain is considered. A method to determine the Al mole fraction and the strain with XRD was presented. The bowing of the optical bandgap as a function of the composition was found to be 1.3 eV.

ACKNOWLEDGEMENT

This work was supported by the Deutsche Forschungsgemeinschaft (Stu 139/3) and the Bayerisches Kultusministerium (IX/2-52C(12)-61/27/87107 and FOROPTO II).

REFERENCES

[1]   S. Nakamura, T. Mukai, and M. Senoh, Appl. Phys. Lett. **64**, 1687 (1994).
[2]   S. Nakamura, M. Senoh, N. Iwasa, and S. Nagahama, Jpn. J. Appl. Phys. **34**, L797 (1995).
[3]   S. Nakamura, M. Senoh, N. Iwasa, S. Nagahama, T. Yamada, and T. Mukai, *ibid.* L1332.
[4]   V. A. Dmitiev, K. G. Irvine, C. H. Carter Jr., N. J. Kuznetsov, and E. V. Kalinina, Appl. Phys. Lett. **68**, 229 (1996)
[5]   M. S. Shur, and M. A. Kahn, Electrochem. Soc. Proc., **95-21**, 129 (1996)
[6]   M. C. Benjamin, C. Wang, R. F. Davis, and R. J. Nemanich, Appl. Phys. Lett., **64**, 3288 (1994)
[7]   G. Dollinger, S. Faestermann, and P. Maier-Komor, Nucl. Instr. and Meth. in Phys. Res., **B64**, 422 (1992).
[8]   G. Dollinger, T. Faestermann, C. M. Frey, A. Bergmaier, E. Schwabedissen, Th. Fischer, and R. Schwarz, Nucl. Instr. and Meth. in Phys. Res., **B85**, 786 (1994).
[9]   L. Vegard, Zeit. f. Physik **5**, 17 (1921).
[10] T. Detchprohm, K. Hiramatsu, K. Itoh, and I. Akasaki, Jpn. J. Appl. Phys. **31**, L1454 (1992).
[11] Y. Koide, H. Ithoh, M. R. H. Khan, K. Hiramatu, N. Sawaki, and I. Akasaki, J. Appl. Phys. **61**, 4540 (1987).
[12] M. R. H. Khan, Y. Koide, H. Itoh, N. Sawaki and I. Akasaki, Solid State Commun. **60**, 509 (1986).

INFLUENCE OF FREE CHARGE ON THE LATTICE PARAMETERS OF GaN AND OTHER SEMICONDUCTORS

M. Leszczyński[1], J. Bąk-Misiuk[2], J. Domagała[2] and T. Suski[1]

[1]High Pressure Research Center, PAS, Sokolowska 29/37, 01-142 Warsaw, Poland, mike@iris.unipress.waw.pl, [2]Institute of Physics, PAS, Al. Lotnikow 32/46, 02-668 Warsaw, Poland, bakmi@ifpan.edu.pl

ABSTRACT

Lattice parameters of semiconductors depend on the concentration of free electrons via the deformation potentials of the occupied minima of the conduction bands. In the presented work we examined the lattice parameters of variously doped GaN samples (epitaxial layers on sapphire and on SiC, bulk crystals grown at high hydrostatic pressure and homoepitaxial layers). The following dopants were used: Si, Mg and O. The measurements were performed using high resolution X-ray diffractometry. The results indicate that free electrons expand the lattice what confirms a negative value of the deformation potential of the Γ minimum of the conduction band. However, for Mg-doping (acceptor) we observed the lattice expansion as well. This violates the Vegard's law, as Mg ions are smaller than Ga ions.

INTRODUCTION

Small changes of the lattice parameters provide important information on microstructure and physical properties of semiconductors. One of the most interesting factors inducing lattice expansion or contraction is the free charge acting via the deformation potential of the energy-band extremum occupied by this charge. However, lattice parameters can be changed by other factors as well. Below, we will discuss all such factors that have been examined for the most popular semiconductors. This introduction will give us a base for discussing our experimental results for gallium nitride, a wide-band gap semiconductor which is becoming more and more popular due to its role in optoelectronics in green/blue/UV range.

Factors influencing lattice parameters

i) Size effect. The dopant ions change the lattice constant proportionally to the doping-level and to the difference between ionic radii of the substituting and substituted ions (Vegard's law). The same happens if we deal with native defects: antisites, vacancies or interstitials. Therefore, if we examine the influence of doping on lattice parameters it is also important to know a concentration of native defects. Antisites expand the lattice not only because of their ionic size, but also because their charge is of the same sign as the adjacent atoms. Additionally, real changes of the lattice parameters can be different, as the antisites may be accompanied by other point defects and/or the neighboring atoms may be re-arranged. Experimentally, lattice expansion by antisites was measured for GaAs [1] and InP [2] grown at very low temperature. For other native point defects, interstitials and vacancies, there is no reliable evidence (neither theoretical nor experimental) how they influence lattice parameters.

ii) Electronic effect. The influence of free electrons on lattice constants was examined by a number of authors [3-12]. Cargill et al. [6] examined silicon crystals doped with arsenic. The implanted and annealed layers possessed lattice constant smaller than the substrate, whereas arsenic ions (bigger than of silicon) locally increase bond length, as it was checked by the extended X-ray absorption spectroscopy (EXAFS). The decrease of the lattice constant was explained by an influence of free electrons, as the deformation potential of the conduction-band

311

for silicon is positive. In the other work [7], Cargill et al. observed the lattice constant increase for AlGaAs layers doped with silicon and tin after cooling in the dark and subsequent illumination at low temperature. In such conditions the free-electron concentration increased by about 10^{18} cm^{-3} via emptying the DX centers. The observed increase of the lattice constant could be related to a negative sign of the conduction-band deformation potential and/or to the shift of the dopant atom from the interstitial to the substitutional position [13]. In our previous works [5,10,12] we reported the changes of lattice parameters of doped GaAs versus Si , Te, S and Se concentrations. The experimental results agreed with the calculations which took into account the size effect and the electronic effect assuming deformation potential of the Γ- minimum of about -8.5 eV. This value corresponded to the values determined by various authors using other methods (-7 eV to - 16 eV [14-19]).

iii) Change of thermal expansion. If lattice parameters are measured at non-zero temperatures it is necessary to take into account a change of thermal expansion by doping (free electrons, in particular). For doped $Al_xGa_{1-x}As$ layers the thermal expansion coefficients were found to be larger relatively to the undoped samples of the same Al-content. This phenomenon is attributed to the change of anharmonic part of lattice vibrations by free electrons or/and ionized tellurium atoms. An increase of thermal expansion caused by doping is a factor which should be taken into account in lattice constant measurements at 295 K.

Those three factors can be described by the formula:

$$\Delta a/a = (\beta s + \beta e + \beta t)n \qquad (1)$$

where:

a-lattice parameter

$\beta_s = (4/\sqrt{3})*(R_d - R_s)/(a*N)$

R_d - radius of the dopant ion,

R_s - radius of the host lattice ion replaced by the dopant

N - number of host atoms per cm^3

N_i - dopant concentration

$\beta_e = -D_n/(3*B)$

n - electron concentration in the conduction-band minimum

D_n- deformation potential of this minimum

B - bulk modulus,

The parameter β_t describes the lattice constant change by free electrons (or doping) when the temperature rises from 0 K up to 295 K.

For example, for GaAs:Te at 295 K, $\beta s = 2.9 \times 10^{-24}$ cm^{-3}, $\beta e = 6.9 \times 10^{-24}$ cm^{-3},

$\beta t = 6 \times 10^{-24}$ cm^{-3} [5]

iv) Mismatch strain. For heteroepitaxial layers, a small difference between the lattice parameters of the substrate and of the thin layer can be accommodated by a tetragonal strain. For larger mismatches and layer thicknesses the strain is relaxed by misfit dislocations. Critical parameters for such lattice relaxation depend on many factors, for example, growth temperature, concentration of defects in the substrate, surface preparation. However, for needs of this paper we can recall two critical thicknesses for InGaAs layers on GaAs: 0.005µm for 1.8 % lattice mismatch [20] and 4 µm for GaAs layers on InGaAs [21] for 0.09 % lattice mismatch.

v) Thermal strain. Mismatched epitaxial layers can relax at high temperatures. When cooling, if the substrate and layer have various thermal expansion, thermal strain can be created. Such situation was observed for GaAs layer on silicon substrate [22] and for some other semiconductors.

EXPERIMENT

The following samples were examined:

i) Bulk crystals grown at extreme conditions of temperature and pressure (about 1800 K and 15 kbar) from liquid gallium supersaturated with nitrogen [23]. The crystals were grown as hexagonal (wurtzite structure) plates with surfaces perpendicular to c axis. In this paper, we report on the samples which have a high free-electron concentration (about 5×10^{19} cm^{-3}) caused presumably by the presence of oxygen and/or nitrogen vacancies. The properties of these crystals were described in Ref. [23].

ii) Homoepitaxial layers (1-2 μm) grown by metalorganic chemical vapor deposition (MOCVD) [24] and molecular beam epitaxy (MBE) [25].The free-electron concentration for these layers was of about 10^{17} cm^{-3}.

iii) Heteroepitaxial layers (0.5 -3 μm) grown by MBE or MOCVD on sapphire or silicon carbide. The concentration of free electrons for these layers depended on the level of Si-doping. For Mg-doped layers, the hole concentration was lower than 10^{18} cm^{-3}.

The free-charge concentration was established by the Hall effect and optical methods (Raman scattering, Burstein-Moss effect and others).

The lattice parameters were measured using the Bond method [26] or by direct measurements of the doubled Bragg angle. The latter method could be realized using the channel-cut analyzer of 12 arcsec acceptance in the triple axis setup [27]. In order to determine the absolute values of lattice parameters a and c we used 004, 014, 024 and 016 reflections. The measurements of lattice parameters were performed at temperature range 40 -750 K.

The values of relaxed lattice parameters were calculated accordingly to the formula:

$$c = c_o + c_o \, \nu \, (a_o - a)/a_o \qquad (2)$$

ν - Poisson ratio, $\nu = 0.38$ [28], c_o a_o- relaxed lattice parameters and using the relation $c_o/a_o = 1.6265$ measured for the bulk crystals.

The experimental results are given in Table I.

Table I. Lattice parameters for undoped and doped GaN samples (all samples, except the one Mg-doped, were of n-type). The concentrations of donor atoms are similar to the concentrations of electrons. For Mg-doped layer, the concentration of Mg-atoms was about two orders of magnitude larger than the concentration of holes.

free charge.conc.	sample	thickness [μm]	a[Å] ±0.0002	c[Å] ±0.0001	a_{relax}[Å] ±0.0003	c_{relax} [Å] ±0.0003
3×10^{19} cm^{-3}(O)	bulk		3.1881	5.1856	3.1881	5.1856
6×10^{19} cm^{-3}(O)	bulk		3.1890	5.1864	3.1890	5.1864
1×10^{17} cm^{-3}	homoepit	2	3.1881	5.1844	3.1876	5.1847
1×10^{17} cm^{-3}	on sapph.	3	3.1886	5.1849	3.1880	5.1853
1×10^{19} cm^{-3}(Si)	on sapph.	3	3.1872	5.1858	3.1880	5.1853
5×10^{17} cm^{-3}(Mg)	on sapph	3	3.1937	5.1932	3.1931	5.1936
1×10^{16} cm^{-3}	on sapph.	1.7	3.1837	5.1884	3.1882	5.1856
1×10^{17} cm^{-3}	on SiC	2	3.1920	5.1821	3.1877	5.1848

In Table I we inserted the most significant results of our research. In particular, we show that for undoped layers on sapphire the relaxed lattice parameters can differ from sample to sample. The results presented above were obtained at room temperature. Fig. 1 shows how the lattice parameters change with temperature for the bulk crystals and homoepitaxial layer (relaxed

lattice parameters change with temperature for the bulk crystals and homoepitaxial layer (relaxed values). It should be added that the lattice parameters a of the bulk crystal and homoepitaxial layer were the same for the whole temperature range.

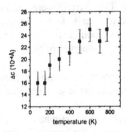

Fig.1

Difference between lattice parameter c for the homoepitaxial layer and the bulk crystal

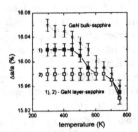

Fig.2

Thermal strain for two different layers of GaN on sapphire. The upper curve shows a difference between lattice parameters a of the bulk crystal and the sapphire.

More complicated situation is for heteroepitaxial layers. Fig. 2 shows the temperature dependence of the parallel lattice parameters with respect to the sapphire substrate for two layers grown in various laboratories. Both layers were undoped and of similar thickness. Additionally, for a comparison, a corresponding difference of lattice parameters a for the sapphire and the bulk crystal is shown.

Based on the experimental data we can draw the following conclusions:

i) The homoepitaxial layers possess the same lattice parameters (*a- in the wurtzite structure*) parallel to the interface as the substrates have. It means that the layers are fully strained. The parameters c are smaller by about 0.02%.

ii) The heteroepitaxial layers are partially relaxed. Thermal strain and relaxed lattice parameters for the undoped heteroepitaxial layers differ from sample to sample (the same observation was reported in our previous works [29, 30]).

iii) Doping with O and Mg increases the lattice parameters.

iv) Doping with Si did not change the relaxed lattice parameters with respect to the sample grown in identical conditions.

v) Thermal strain for the Si-doped layer appeared to be different with respect to the sample grown in identical conditions.

vi) Doping with oxygen increses the thermal-expansion coefficient.

DISCUSSION

i) Lattice mismatch.

There is no much information on critical conditions for lattice relaxation in heteroepitaxial layers of the nitrides. The only data published so far is for gallium nitride on aluminum nitride buffer layer [31]. The authors of this Reference found the critical thickness of 29±4 Å starting to relax 2.5% mismatch between the AlN and GaN. However, the process of relaxation can be a very special one, because the buffer layer contains a very high density of misfit dislocations relaxing 16% mismatch to the sapphire substrate. Recently, we found that critical thickness for AlN deposition on single crystal GaN substrate is just a few Angstrom [32]. For the purpose of this work we can state that the homoepitaxial layers were fully strained and heteroepitaxial layers were fully relaxed at elevated temperatures. At room temperature, the heteroepitaxial layers were strained due to their different thermal expansions with respect to the substrates.

ii) Thermal strain.

According to the published results by other authors [28] and to the results of our measurements [30] there is no clear explanation why the thermal strain is different for various samples. Different layer thickness is not the only cause as it was proposed by Detchprohm et al [30]. Most probably, a different microstructure of layers (different concentration of native defects) grown in various laboratories causes different relaxation processes. Moreover, we should have in mind that doping may influence the relaxation process and Poisson coefficient used in Equation 2.

iii) Size effect

All examined dopants expand the lattice. In the case of silicon, which is a typical substitutional donor for III-V compounds, we can apply Vegard's law. Using the ionic radius of gallium of 1.24 Å and 1.17 Å of silicon we would expect that the lattice parameters of the doped layer to about 10^{19} cm^{-3} should be decreased by about 0.0015 %. For oxygen replacing nitrogen we may assume that the size effect is negligible. For magnesium, replacing gallium ions, situation may be different. Magnesium forms with nitrogen a compound of different stoichiometry (Mg_3N_2) and the size effect is difficult to be predicted.

iv) Change of thermal expansion by doping

As it is shown in Fig. 1 free electrons increase the thermal expansion, as it is so for doped GaAs samples. The effect is not very strong but will be considered for evaluating the electronic effect.

v) Electronic effect

In conclusion, we create Table 2 where we extract electronic effect from the other factors. Using Formula 1 we calculated the values of deformation potential of the conduction band. At the present stage of research these values should be treated as semi-quantitative, but they correspond to the value of -10 eV given in the theory of Ref.[33]

Table 2 .Lattice constant change for GaN vs different concentration of free electrons

sample	electron concentration 10^{19}cm^{-3}	size effect	electron. effect	thermal	size + electron +thermal	experimental
				$\Delta c/c (10^{-5})$		
on sapph.	1(Si)	-0.3	2.6	1.7	4	0
bulk	3(O)	0	7.8	5	12.8	16
bulk	6(O)	0	16	10	26	32

CONCLUSIONS

For oxygen-rich samples of GaN, the results indicate that free electrons expand the lattice by electronic effect and by the increase of thermal expansion. This case is similar to gallium arsenide. For heteroepitaxial layers, the scattering of experimental data is too large to draw conclusions on effect caused by doping with silicon. For Mg-doped samples (p-type), the increase of the lattice parameters is unexpectedly large and suggests a complex nature of this dopant.

ACKNOWLEDGEMENTS

This work was supported by grants of Polish Committee for Scientific Research (KBN) number 175/PO3/96/10.

REFERENCES

1 M.Leszczyński, P.Franzosi, J.Cryst. Growth, **134**, 151 (1993).
2. M.Leszczyński,unpublished.
3. T.Figielski, Phys. Stat. Sol. **1**, 306 (1961).
4 W.Keyes, Solid State Phys. **20**, 37 (1967).
5. J.Bąk-Misiuk, M.Leszczyński, J.Domagała and Z.R.Zytkiewicz, J. Appl. Phys. **78**, 6994 (1995).
6. Cargill III, J.Angillelo and K.J.Kavanagh, Phys. Rev. Lett. **61**, 1748 (1988).
7. G.S.Cargill III, A.Segmüller, T. F.Kuech and T.N.Theis, Phys. Rev. **B46**, 10078 (1992).
8. U.Pietsch, J.Bak-Misiuk and V.Gottschalch, Phys.Stat. Sol. (a) **82**, K137 (1984).
9. J.Bąk-Misiuk, H.G.Brühl, W.Paszkowicz and U.Pietsch, Phys. Stat. Sol. (a) **106**, 451 (1988).
10. M.Leszczyński, J.Bąk-Misiuk, J.Domagała, J.Muszalski, M.Kaniewska and J.Marczewski, Appl. Phys. Lett. **67**, 539 (1995).
11. G.Contreras, L.Tapfer, A. K. Sood and M.Cardona, Phys. Stat. Sol. (b) **131**, 475 (1985).
12. J.Bąk-Misiuk, M.Leszczyński, W.Paszkowicz and J.Domagała, J. Appl. Phys.**69**, 3366 (1996).
13. D.J.Chandi and K.J.Chang, Phys.Rev.Lett. **61**, 873 (1988)
14. M.Cardona and N.E.Christensen, Phys. Rev. **B35**, 6182 (1987).
15. C.G.Van de Walle, Phys. Rev. **B39**, 1871 (1989).
16. D.D.Nolte, W. Walukiewicz and E. E. Haller, Phys. Rev. Lett. **59**, 501 (1987).
17. L.Samuelson and S. Nielsson, J. Lumin. **40&41**, 127 (1988).
18. P.Pfeffer, I.Gorczyca and W.Zawadzki, Solid State Commun. **51**, 178 (1984).
19. W.Walukiewicz, H.E.Ruda, J.Lagowski and H.C.Gatos, Phys. Rev. **B32**, 2645 (1985).
20. G.L.Price, Phys.Rev.Lett. **66**, 469 (1991).
21. B.H.Yang, Z.G.Wang, H.J.Heand and L.Y.Liu, J.Cryst.Growth **103**, 371 (1990).
22. N.Lucas, H.Zabel, H.Morkoc and H.UniuAppl.Phys.Lett. **52**, 2117 (1988).
23. S.Porowski, I.Gregory and J.Jun in High Pressure Chemical Synthesis, eddited by J.Jurczak and B.Baranowski (Elsevier, Amsterdam,1989), p.21
24. K.Pakula, A.Wysmolek, K.P.Korona, J.M.Baranowski, R.Stepniewski, I.Gregory, M.Bockowski, J.Jun, S.Krukowski, M.Wroblewski and S.Porowski, Proccedings of NATO Advanced Research Workshop, HEAD95 Smolenice , Slovakia (in press).
25. H.Teisseyre, G.Nowak, M.Leszczyński, I.Grzegory, M.Bockowski, S.Krukowski, S.Porowski, M.Mayer, A.Peltzman, M.Karp, K.J.Ebeling, G.Karczewski, Nitride Semiconductor Researh, MRS Internet Journal, http://nsr.mij.mrs.org/1.1/13.
26. W.L.Bond, Acta Crystalogr.**13**, 814 (1960).
27. P.F.Fewster and N.L.Andrew. J.Appl. Crystallogr. **28**, 451 (1995).
28. T.Detchprohm, K.Hiramatsu, K.Itoh and J.Akasaki, Japan, J.Appl.Phys. **31**, L1454 (1992).
29. M.Leszczyński, H.Teisseyre, T.Suski, I.Grzegory, M.Bockowski, J.Jun, K.J.M.Baranowski, C.T.Foxon and T.S.Cheng, Appl. Phys.Lett. **69**, 73 (1996).
30. M.Leszczyński, T.Suski, P.Perlin, H.Teisseyre I.Grzegory, M.Bockowski, J.Jun, S.Porowski and Major, J.Phys.D:Appl.Phys. **28**, A149 (1995).
31. Chinkyo Kim, I.K.Robinson, Jeamin Myoung, Kyuhwam Shim, Myung-Cheol Yoo and Kyekyoon Kim, Appl. Phys.Lett. **49**, 2358 (1996).
32. M.Leszczyński, Barski, I.Grzegory, S.Porowski (unpublished).
33. P.Perlin, I.Gorczyca, N.E.Christtansen, I.Grzegory, H.Teisseyre and T.Suski, Phys.Rev. **B45**, 13307 (1992).

HREM AND CBED STUDIES OF POLARITY OF NITRIDE LAYERS WITH PRISMATIC DEFECTS GROWN OVER SiC

P. Vermaut*, P. Ruterana*, G. Nouet*, A. Salvador**, H. Morkoç**
*Laboratoire d'Études et de Recherches sur les Matériaux, Unité associée CNRS 6004, Institut des Sciences de la Matière et du Rayonnement, 6 Blvd Maréchal Juin, 14050 Caen Cedex , France. (gerard@leriris1.ismra.fr)
** University of Illinois-Urbana, Coordinated Science Laboratory, Urbana, Illinois, IL61801 USA (morkoç@uiuc.edu)

ABSTRACT

The polarity of GaN films and their AlN buffer layer grown on $(0001)_{Si}$ 6H-SiC by an electron cyclotron resonance plasma enhanced molecular beam epitaxy has been investigated by convergent beam electron diffraction (CBED) and high resolution electron microscopy (HREM). The experimental results are in good agreement with the simulations and allow to determine that the free surfaces of the GaN and AlN layers are Ga and Al-terminated respectively. Moreover, $(\bar{1}2\bar{1}0)$ prismatic planar defects observed in the AlN layers have been identified as stacking faults and observations in different areas of the specimens have shown that the layers are unipolar.

INTRODUCTION

III-nitride semiconductors are very promising materials for optoelectronic applications with their direct bandgap ranging from 1.89 to 6.2 eV [1]. Optically efficient GaN layers have been observed to contain a high density of defects as threading dislocations and planar defects [2]. The influence of the chemical nature of the substrate surface i.e. $(0001)_{Si}$ and $(000\bar{1})_C$ on the surface morphology and photoluminescence properties of GaN layers grown by Metal Organic Chemical Vapour Deposition (MOCVD) without an AlN buffer layer has been analyzed [3]. It was shown that layers grown on $(0001)_{Si}$ are featureless in contrast to those grown on the $(000\bar{1})_C$ surface. In addition, from X-ray photoelectron spectroscopy measurements, they concluded that GaN epitaxial layers on $(0001)_{Si}$ of SiC are terminated with N. For AlN, HREM has recently shown that they are Al-terminated free surfaces [4].

These results would mean that GaN and AlN layers grown on $(0001)_{Si}$ SiC present opposite polarity; thus, showing that polarity of nitride layers on a polar substrate as SiC is not obvious. In the present work, the polarity of GaN and AlN layers grown on $(0001)_{Si}$ 6H-SiC by Electron Cyclotron Resonance Plasma enhanced Molecular Beam Epitaxy has been studied by CBED and HREM.

EXPERIMENTAL

6H-SiC wafers were cut 3.5° off the basal plane toward $[11\bar{2}0]$. The $(0001)_{Si}$ surface was chemically cleaned using the classical way followed by a hydrogen plasma step to reduce the amount of oxygen-carbon bonds down to below the X-ray photoemission detection limit. The details of this procedure were reported [5]. For some samples, prior to GaN, an AlN buffer layer was grown on SiC. The thickness of the AlN buffer layer was 50 nm. The growth was performed at a rate of 40 nm/h with a substrate temperature between 750 and 800° C.

Cross section samples were thinned down to 100 µm by mechanical grinding and dimpled down to 10 µm. Electron transparency was achieved by ion milling with a LN_2 cold stage at 5 kV. A final step at 3 kV was used to decrease ion beam damage. HREM observations were made

317

Mat. Res. Soc. Symp. Proc. Vol. 468 ● 1997 Materials Research Society along a <11 $\bar{2}$ 0> direction on a Topcon 002B electron microscope operating at 200 kV with a point to point resolution of 0.18 nm (Cs=0.4 mm, spread of focus and beam divergence values are estimated to be of 8 nm and 0.8 mrad, respectively). CBED experiments were carried out on a Jeol EM 2010 microscope operating at 200 kV with a probe diameter of 25 nm. HREM images and CBED patterns were simulated using the EMS software [6].

STRUCTURES

The nitride layers crystallise in the wurtzite structure which is, similarly to the 6H-SiC polytype, tetrahedrally coordinated and exhibit the $P6_3mc$ space group symmetry. These structures present similar lattices but different atomic patterns. Their stacking sequences are $T_2T'_1T_2T'_1$ and $T'_2T'_1T'_3T_1T_2T_3$ for the wurtzite and the 6H polytype, respectively [7]. Both structures are non-centrosymmetric and present an anti-mirror m' perpendicular to the stacking direction. Then, the (0001) and (000$\bar{1}$) terminated surfaces are chemically different. Conventionally, the (0001) surface is chosen to be group-III terminated for the nitride materials and Si terminated for SiC.

CBED RESULTS

Convergent beam electron diffraction is a powerful technique for structure identification, thickness and residual stress measurements [8]. CBED patterns are characteristic of the material but are also function of accelerating voltage and crystal thickness. In the following, simulation of series of zone axis patterns has been carried out for GaN and SiC using the Bloch wave method and for crystal thickness values ranging from 50 to 250 nm with a step of 5 nm.

As the absolute orientation of 6H-SiC substrates is known, determined before growth by polarity dependence of the oxidation rate [9], it was been confirmed by CBED in a first stage. The same analysis was carried out in the GaN films grown directly, or with an AlN buffer layer, on top of the SiC substrate. CBED patterns recorded at different crystal thicknesses along the [10$\bar{1}$0] zone axis are shown on figure 1 as well as the corresponding simulation.

A good match is obtained between the experimental and simulated patterns. This agreement shows without ambiguity that the [0001] directions in the GaN films and in the 6H-SiC

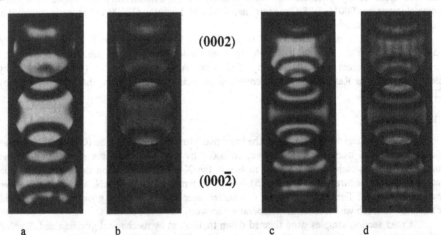

Figure 1: CBED patterns of GaN, experimental (a and c) and simulated (b and d) for a thickness of 90 and 150 nm.

substrates are parallel, meaning that the GaN films grown on $(0001)_{Si}$ surface of SiC, with and without an AlN buffer layer, exhibit a Ga-terminated free surface.

In contrast to GaN films, the AlN buffer layers present a high density of defects. Due to the small crystal thickness deposited and the density of defects, no CBED pattern could be recorded in reliable conditions in such layers.

HREM AS A TOOL FOR POLARITY DETERMINATION OF AlN

A systematic study of lattice images of compound semiconductors and particularly for wurtzite ones has been reported [10]. They have shown that the shift of the spot positions with the atomic columns is related to the phase difference between the 0002 and 000$\bar{2}$ Bijvoet-related reflections.

A series of simulated images of AlN at 6.5 nm thick, as a function of defocus is shown (Fig. 2). Two main maxima of contrast, separated by complex images, are present. These maxima are approximately visible on a wide range of defocus values: -20/-36 nm and -50/-65 nm, and correspond to black and white atomic columns, respectively. The contrast at the two maxima remains fairly constant when the thickness increases until 12 nm and then it becomes complex and non-linear. Therefore, defocus and crystal thickness parameters can be relatively easily determined. At the second maximum, white spots have an elongation which reproduces the dumbbell direction. The alternate direction of the elongation describes the stacking sequence of the structure. In addition, due to the difference between the atomic number of the species, the spots are shifted towards the heaviest species (Al). In contrast to the elongation direction of the spots which is retained for all the defocus range of this maximum, the shift of the spots decreases when the defocus value increases and disappears at -61 nm (Fig. 2). The spot shift at -52 nm defocus is independent of the crystal thickness for the range 0-12 nm. Then, for such lattice images, spots elongated upwards right are shifted on the right if the [0001] direction points upwards but are shifted on the left if [0001] points downwards. Conversely, spots elongated upwards left are shifted on

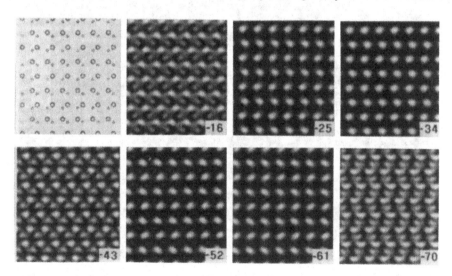

Figure 2: Simulated lattice images of AlN (O Al, o N) as a function of defocus for a 6.5 nm thick crystal along <11$\bar{2}$0>.

the left if [0001] points upwards, and on the right if [0001] points downwards. Therefore, these two features, elongation and shift of the spots around -52 nm defocus and in the range 0-12 nm, can allow polarity determination.

The second maximum of contrast (- 52 nm defocus) and the simulated image are shown in figure 3 for a thickness of about 7 nm. The left simulated image is oriented with [0001] pointing upwards whereas the right hand side one is oriented with [0001] pointing downwards. Only the left hand side simulated image presents a good match with the experimental one, with the same shift and elongation. On the right hand side image, the shift and elongation of the spots could not match the experimental image simultaneously. If the simulated image is compared to the experimental one for spot elongation fitting, it has to be translated by c/2 in order to have a match for the shift of the spots, in which case the spot elongation does not agree anymore.

The wurtzite structure is very sensitive to misalignment effects because of the presence of forbidden reflections [10], so the influence of beam tilt and crystal tilt effects on lattice images of AlN has been taken into account. The spot elongation changes and a (0001) periodicity can appear: spots become more or less elongated, but keep always the same direction. Similarly, the relative position of the spots underlining the stacking sequence is not modified which means that no confusion could be introduced on polarity by such small misalignments.

Then, using spot elongation and shift in images at -52 nm defocus, and with a careful study by image simulation of the misalignment effects, one can conclude unambiguously that AlN grows on the $(0001)_{Si}$ surface of SiC with an Al-terminated free surface.

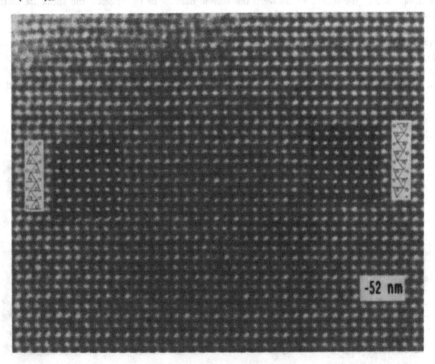

Figure 3: HREM image of AlN (- 52 nm defocus) for a thickness of 7 nm: the left simulated image is oriented with [0001] pointing upwards, the right one is [0001] downwards.

APPLICATION TO PRISMATIC DEFECTS

Prismatic planar defects are present in the AlN layers grown on SiC. They have been first reported as DPBs generated by steps at the interface with the substrate [11]. Sometimes, these defects have been observed to form closed domains in the vicinity of the interface [12]. In that case, an inversion character can be suspected and may be looked for. Similarly, if the defects cross entirely the AlN layer, an inversion is still possible. Four models of the $(\bar{1}2\bar{1}0)$ prismatic defects have been reported in the literature [13, 14]. They are two SFs and two IDBs with $\frac{1}{2}[\bar{1}011]$ and $\frac{1}{2}[\bar{1}010]$ translation vectors. An example is given for the $\frac{1}{2}[\bar{1}011]$ SF and the $\frac{1}{2}[\bar{1}010]$ IDB models, for which, projections of the structure are shown in figure 4a and b respectively. Al-N bonds are maintained at the boundary, either by an additional c/2 translation vector in the first model, or by an inversion operation in the second one. The elongation direction of the spots of one basal plane changes when the defect is crossed in the SF model in accordance with that is observed on the experimental image, whereas it does not in the IDB one (Fig. 5) [15]. Several similar defects observed in the AlN buffer layers grown on $(0001)_{Si}$ of SiC have been identified to the $\frac{1}{2}[\bar{1}011]$ SF model previously described and no IDB could be found.

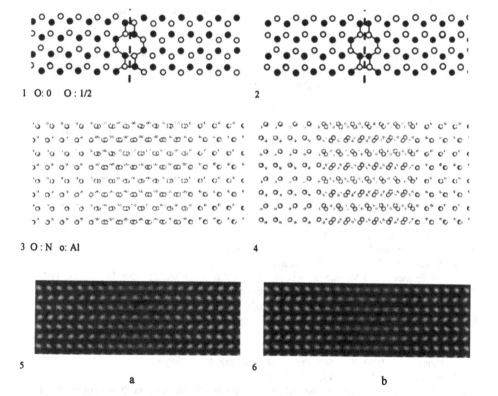

Figure 4: Models and simulated images of the stacking fault R = $\frac{1}{2}[\bar{1}011]$ (a) and of the inversion boundary R = $\frac{1}{2}[\bar{1}010]$ (b) viewed along [0001] (1-2) and [11$\bar{2}$0] (3-6) (defocus: - 52 nm, thickness: 6,5 nm).

Figure 5 : HREM image (- 52 nm defocus, thickness: 6.5 nm) of ($\overline{1}2\overline{1}0$) prismatic defect crossing the AlN layer, showing the same polarity on each side of the inclined defect.

CONCLUSION

The polarity of GaN and AlN layers grown on $(0001)_{Si}$ surface of 6H-SiC has been identified. The polarity of the substrate was shown to be transmitted to the deposited layers as if Si and C were species of group III and group V respectively.

REFERENCES

1. H. Morkoç, S. Strite, G. B. Gao, M. E. Lin, B. Sverdlov and M. Burns, J. Appl. Phys., **76**, 1363 (1994).
2. S.D. Lester, F.A. Ponce, M.G. Craford and D.A. Steigerwald, Appl. Phys. Lett., **66**, 1249 (1995).
3. T. Sasaki and T. Matsuoka, J. Appl. Phys., **64**, 4531 (1988).
4. F.A. Ponce, M.A. O'Keefe and E.C. Nelson, Phil. Mag. A, **74**, 777 (1996).
5. M.E. Lin, S. Strite, A. Agarwal, A. Salvador, G.L. Zhou, N. Teraguchi, A. Rockett and H. Morkoç, Appl. Phys. Lett. **62**, 702 (1993).
6. P. A. Stadelmann, Ultramicroscopy, **21**, 131 (1987).
7. P. Pirouz and J.W. Yang, Ultramicroscopy, **51**, 189 (1993).
8. J.C.H. Spence, and J.M. Zuo in Electron Microdiffraction, (Plenum Press, New York, 1992).
9. W. Von Müench, and I. Pfaffeneder, J. Electrochem. Soc., **122**, 642 (1975).
10. R.W. Glaisher, A.E.C. Spargo and D.J. Smith, Ultramicroscopy, **27**, 117 (1989).
11. S. Tanaka, R.S. Kern and R.F. Davis, Appl. Phys. Lett. **66**, 37 (1994).
12. P. Vermaut, P. Ruterana, G. Nouet, A. Salvador and H. Morkoç in III-Nitride, SiC and Diamond Materials for Electronic Devices, edited by D.K. Gaskill, C.D. Brandt and R.J. Nemanich (Mater. Res. Soc. Proc. **423**, Pittsburg, PA, 1996) p. 551.
13. C.M. Drum, Phil. Mag. A, **11**, 313 (1965).
14. J.L. Rouvière, M. Arlery, A. Bourret, R. Niebuhr and K. Bachem, in Proceeding of the IXth Conference on Microscopy of Semiconducting Materials, Inst. of Phys. Conf. Series N° **146**, 285 (1995).
15. P. Vermaut, P. Ruterana, G. Nouet and H. Morkoç, Phil. Mag A, **75**,239 (1997).

THE ATOMIC STRUCTURE OF THE {10$\bar{1}$0} INVERSION DOMAINS IN GaN/SAPPHIRE LAYERS

V. Potin*, P. Ruterana*, G. Nouet*, A. Salvador**, H. Morkoç**
*Laboratoire d'Études et de Recherches sur les Matériaux, Unité associée CNRS 6004, Institut des Sciences de la Matière et du Rayonnement, 6 Blvd Maréchal Juin 14050 Caen Cedex , France.
** University of Illinois-Urbana, Coordinated Science Laboratory, Urbana, Illinois, IL61801 USA

ABSTRACT

Nanometric inversion domains in GaN/Al$_2$O$_3$ layers have been investigated using HREM. They were found to be limited by {10$\bar{1}$0} planes and to cross the entire epitaxial layer. It has been possible, using extensive image simulation and matching to discriminate between possible atomic models for the boundary plane. It is shown that the inversion domain boundaries correspond to a Holt type model containing wrong bonds (Ga-Ga, N-N), and in that plane, each atom exhibits two such bonds.This probably can explain the small size of the domains (5-20 nm).

INTRODUCTION

Because of their large band gap, the nitride semiconductors are good candidates for many optoelectronic applications like short wavelength light emitting diodes and blue laser diodes [1]. In spite of large progress on the growth process, many problems are still present. The most common is the presence of large densities of defects in the active layers. In order to improve the good quality of the epitaxial films, understanding the structure and origin of these defects is necessary. These defects are mainly dislocations, basal stacking faults and planar defects [2, 3].

Inversion boundaries in prismatic planes may cross the whole active layer from the interface with the substrate to the surface. They are found in noncentrosymmetric crystals and have already been studied in GaAs [4], ZnO [5] and AlN [6].

In GaN, they were reported by Rouvière et al. [7], Wu et al [8] and Romano et al.[9]. The atomic structure of these inversion domains is still subject of controversy. Two models can be found in the literature : the Holt model consists of a pure inversion [10] whereas in the Austerman model, an inversion and a translation operate [11].

In this work, we have analyzed the atomic structure of the {10$\bar{1}$0} boundaries in GaN layers grown on sapphire using experimental high resolution electron microscopy (HREM) and extensive image simulations using the EMS software [12].

EXPERIMENTAL

The GaN layers were grown on the (0001) sapphire substrate by electron cyclotron resonance assisted molecular beam epitaxy (ECR-MBE). Prior to growth, the substrates were chemically cleaned. The GaN layers were directly grown at 800°C with a 40 nm/h deposition rate. Details about this procedure were reported by Lin et al [13].

The TEM cross section samples were thinned down to 100 μm by mechanical grinding and dimpled down to 10 μm. Electron transparency was obtained by ion milling with a LN2 cold stage at 5 KeV. The samples were analyzed using the conventional and HREM modes. HREM observations were carried out along the GaN <11$\overline{2}$0> direction on a Topcon 002B electron microscope operating at 200 kV with a point to point resolution of 0.18 nm.

RESULTS AND DISCUSSION

GaN is of wurtzite type and noncentrosymmetric. Serneels et al. [14] have demonstrated that two parts of a crystal related by an inversion present a strong contrast in multibeam dark field conditions along a zone axis which reveals the noncentrosymmetry of the crystal. Figures 1a and 1b show dark-field images recorded with g = 0002 and g = 000$\overline{2}$, respectively. The observed contrast is complementary, which implies that the small vertical domains (arrow) are related to the bulk by an inversion operation.

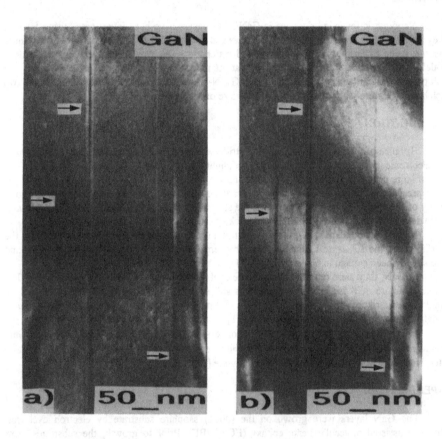

Fig. 1 Multiple dark field images along a <10$\overline{1}$0> zone axis with g = 0002 (a) and g = 000$\overline{2}$ (b). The opposite black and white contrasts show inversion domains.

An inversion domain can result from a pure inversion or from an inversion followed by a translation. In order to determine the possible translation, images were recorded in two beam conditions with g = 11$\overline{2}$0 and g = 0002 (Fig. 2a and 2b). For g = 11$\overline{2}$0, no contrast could be

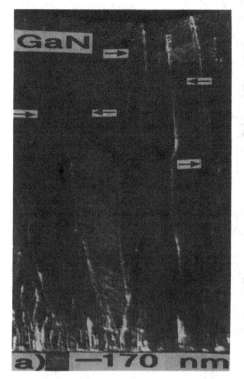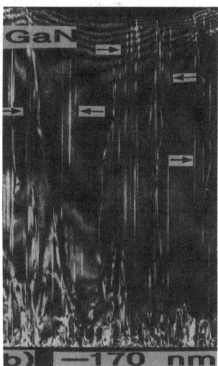

Fig. 2 Dark field images in two beam conditions with g = 11$\overline{2}$0 (domains shown by arrows are out of contrast) and g = 0002 (domains are in contrast).

observed whereas for g = 0002 a strong contrast it exhibited. So, if a translation exists, it has to be along the c-axis.

Therefore, we have constructed models using either inversion operation or a combination with a translation along the c-axis (Fig. 3). The GaN crystal was first cut into two halves along a {10$\overline{1}$0} plane and glued back after a symmetry operation on one of them. The {10$\overline{1}$0} boundary plane can be located along the in-plane bonds (which are in the {11$\overline{2}$0} projection plane) or along the out-of-plane bonds (where two bonds per atom are cut). The first type of model was proposed by Austerman and Gehman [11], and consists of an inversion followed by a translation of 1/3 <0001>. The anion sublattice is undeviated across the boundary, switching only the cation sublattice in the inversion operation (A1-A2). In the Holt model [10], only an inversion is applied (H1-H2) whereas the last type of model is defined by an inversion and a translation of ½ <0001> (V1-V2). In the A and V inversion domain boundary models, Ga-N bonds are present through the boundary plane in contrast to the H pure inversion models for which only wrong bonds (N-N, Ga-Ga) in equal number are present.

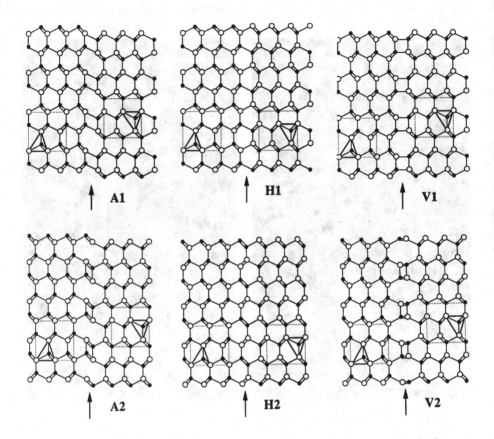

Fig. 3 Geometrical models for $\{10\bar{1}0\}$ inversion boundaries in GaN shown in the $<11\bar{2}0>$ projection. White circles represent nitrogen and black ones gallium. Single lines represent bonds which are in the $\{11\bar{2}0\}$ projection plane and double lines correspond to a projection of two bonds, one pointing behind and the other pointing in front of the projection plane. The lattice is underlined by dots and the tetrahedra show the inversion.

In this way, six models are generated; they were used in a comparison between simulation and HREM experimental results. Fig. 4a is a HREM image obtained at a defocus of 59 nm and figures 4b-g are a map of the six simulated images at the same defocus, the thickness is taken to be ~ 5 nm, the white spots correspond to atomic columns. In the experimental image, the boundary plane is formed by alternated strong and weak spots.

For A2, V1 and H1 models, the boundary plane is seen to be located between atomic columns, so that they are clearly at variance with the experimental micrograph. Among the remaining three models, only H2 exhibits a contrast similar to that of the experimental image at these defocus and thickness. Therefore, in order to assess its validity, we have carried out through

326

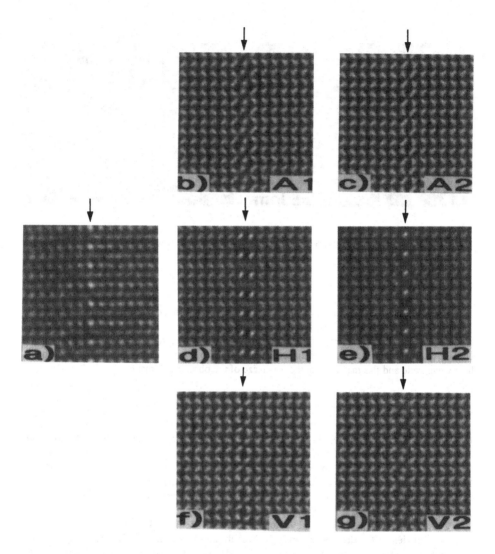

Fig. 4 Comparison of the experimental image (a) to the simulated images for the six models (b-g) at 59 nm defocus and 5 nm thickness. The incident beam is along the $<11\overline{2}0>_{GaN}$ zone axis and the arrows indicate the $\{10\overline{1}0\}$ boundary plane.

focus and thickness comparisons. As shown in fig 5a and b, simulated images using the H2 model match the experimental images. It is found to be the only model which reproduces fairly well the fine structure of the experimental image contrast.

This result is at variance with that recently published by Romano et al [9] who have chosen a model comparable to V1 for the $\{10\overline{1}0\}$ IDB in GaN in order to avoid the presence of wrong bonds in the boundary.

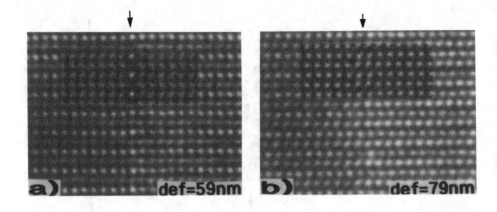

Fig. 5 HREM experimental images at 59 and 79 nm defocus. The insets show simulated images for the H2 model (thickness : 5 nm).

CONCLUSION

The examined GaN layers contain nanometric inversion domains (5-20 nm) limited by $\{10\bar{1}0\}$ planes . HREM images recorded in very thin areas ($\leq$12 nm) allow to discriminate between possible models for the boundary atomic structure. It is found that the boundary plane contain wrong bonds and two of them are present on each atom. Such configuration is probably highly energetic and this may explain the small size of the observed domains.

REFERENCES

1. S. Nakamura, M. Senoh, S. Nagahama, N. Iwasa, T. Yamada, T. Matsushita, K. Kiyoku, and Y. Sugimoto, Jpn. J. Appl. Phys. **35**, L 74 (1996).
2. F.A. Ponce, J.S. Major Jr., W.E. Plano and D.F. Welch, Appl. Phys. Lett. **65**, 2302 (1994).
3. P. Vermaut , P. Ruterana , G. Nouet and H. Morkoç, Phil. Mag A **75**, 239 (1997).
4. Z. Liliental-Weber, M. A. O'Keefe and J. Washburn, Ultramicroscopy **30**, 20 (1989).
5. J. C. Kim and E. Goo, J. Am. Ceram. Soc. **73**, 877 (1990).
6. A. D. Westwood and M. R. Notis, J. Am. Ceram. Soc. **74**, 1226 (1991).
7. J. L. Rouvière, M. Arlery, A. Bourret, R. Niebuhr and K. Bachem, in Proceeding of the IXth Conference on Microscopy of Semiconducting Materials, Inst. of Phys. Conf. Series N° **146**, (1995), p. 285.
8. X. U. Wu, L. M. Brown, D. Kapolnek, S. Keller, B. Keller, S.P. DenBaars and J. S. Speck, J. Appl. Phys. **80**, 3228 (1996).
9. L. T. Romano, J. E. Northrup and M.A. O'Keefe, Appl. Phys. Lett. **69**, 2394 (1996).
10. D. B. Holt, J. Phys. Chem. Solids **30**, 1297 (1969).
11. S. B. Austerman and W.G. Gehman, J. Mater. Sci. **1**, 249 (1966).
12. P. A. Stadelmann, Ultramicroscopy **21**, 131 (1987).
13. M.E. Lin, B.N. Sverdlov and H. Morkoç, J. Appl. Phys. **74**, 5038 (1993).
14. R. Serneels, M. Snykers, P. Delavignette, R. Gevers, and S. Amelinckx, Phys. Stat. Sol. B **58**, 277 (1973).

Part IV

Processing

PROCESSING CHALLENGES FOR GaN-BASED PHOTONIC AND ELECTRONIC DEVICES

S.J. Pearton *, F. Ren **, R.J. Shul ***, J.C. Zolper *** and A. Katz ****
* Department of Materials Science and Engineering, University of Florida, Gainesville, 32611
** Bell Laboratories, Lucent Technologies, Murray Hill, NJ 07974
*** Sandia National Laboratories, Albuquerque, NM 87185
**** EPRI, Palo Alto, CA 94304

ABSTRACT

The wide gap materials SiC, GaN and to a lesser extent diamond are attracting great interest for high power/high temperature electronics. There are a host of device processing challenges presented by these materials because of their physical and chemical stability, including difficulty in achieving stable, low contact resistances, especially for one conductivity type, absence of convenient wet etch recipes, generally slow dry etch rates, the high temperatures needed for implant activation, control of suitable gate dielectrics and the lack of cheap, large diameter conducting and semi-insulating substrates. The relatively deep ionization levels of some of the common dopants (Mg in GaN; B, Al in SiC; P in diamond) means that carrier densities may be low at room temperature even if the impurity is electrically active - this problem will be reduced at elevated temperature, and thus contact resistances will be greatly improved provided the metallization is stable and reliable. Some recent work with $CoSi_x$ on SiC and W-alloys on GaN show promise for improved ohmic contacts. The issue of unintentional hydrogen passivation of dopants will also be covered - this leads to strong increases in resistivity of p-SiC and GaN, but to large decreases in resistivity of diamond. Recent work on development of wet etches has found recipes for AlN (KOH), while photochemical etching of SiC and GaN has been reported. In the latter cases p-type materials is not etched, which can be a major liability in some devices. The dry etch results obtained with various novel reactors, including ICP, ECR and LE4 will be compared - the high ion densities in the former techniques produce the highest etch rates for strongly-bonded materials, but can lead to preferential loss of N from the nitrides and therefore to a highly conducting surface. This is potentially a major problem for fabrication of dry etched, recessed gate FET structures.

INTRODUCTION

There is an increasing interest in use of compound semiconductors for several high power/high temperature solid state devices for applications in power electronics, control and distribution circuits, hybrid drive-train automobiles, "more electric" aircraft (avionics) and next generation battleships [1-10]. The three major commercial markets for these devices are automotive, industrial factories and electric utility, while numerous tri-service defense applications also require high power heat tolerant devices. Typical device characteristic requirements for many of the existing and emerging applications are: high voltage (600-1000 V), low switching losses, high current densities (1000 A/cm^2), and operating temperatures up to 350°C. While silicon, and to a much lesser extent GaAs have been used for power devices, SiC and emerging materials such as GaN have significant advantages because of wider bandgaps (higher operating temperature), larger breakdown fields (higher operating voltage), higher electron saturated drift velocity (higher operating current) and better thermal conductivity (higher power density). Some of the properties important for power device applications for GaN, SiC

331

and Si are listed in Table I. The basic building blocks of SiC and GaN power technology are gate-turn off thyristors (GTOs), insulated gate bipolar transistors (IGBTs), and metal-oxide semiconductor controlled thyristors (MTOs). The superior device performance of compound semiconductors can be exemplified by a SiC MOS turn-off thyristor (MTO) which can carry three times higher current, and possesses eight times higher breakdown voltage than a comparable Si MTO [1]. In addition, SiC MOS MTOs can operate at much higher temperatures (250°C for SiC vs. 125°C for Si).

There are several key issues that must be addressed to fully exploit the thyristor type devices in both SiC and GaN [9]. These are:

(i)   Improvement of gate oxide quality;
(ii) Better edge termination and passivation process;
(iii) Improved ohmic contacts;
(iv) Higher implant activation efficiencies and damage removal;

For GaN-based photonic devices there are also a number of critical advances necessary in:

(i)   p-contact technology;
(ii) Mesa facet quality and yield;
(iii) Layer structure design and control.

In this paper we will show some of the recent advances in process technology for wide bandgap materials and suggest directions for future research.

Table I: High Temperature Power Devices: Potential Candidates

	Material		
Property	Si	3C SiC (6C SiC)	GaN
Bandgap	1.1	2.2 (2.9)	3.4
Maximum Operating temperature (K)	600	1200 (1580)	?
Melting Point (K)	1690	sublimes>2100?	>2200
Physical Stability	Good	Excellent	Good
Hole Mobility (RT, cm^2/Vs)	600	40	150
Electron Mobility (RT, cm^2/Vs)	1400	1000 (600)	900
Breakdown Voltage (Eb, 10^6 V/cm	0.3	4	5
Thermal Conductivity (C$_T$, W/cm)	1.5	5	1.3
Saturation Electron Drift Velocity (cm/s)	1×10^7	2×10^7	2.7×10^7
Dielectric Constant (K)	11.8	9.7	9

OMIC CONTACTS

For SiC, recent advances in epilayer growth have provided more highly-doped (Al) p-type layers for improved ohmic contacts. Much of the work involved Al-based metallization with contact resistivities of 10^{-4}-$10^{-5}\Omega cm^2$, with relatively poor thermal stability [12]. The best contact properties were obtained for samples annealed at 800-1000°C for 5 mins. Lundberg and Ostling [13] reported CoSi$_2$ ohmic contacts with ρ_c <$4\times10^{-6}\Omega cm^2$ to p-type SiC, fabricated using sequential evaporation and a 2-step anneal at 500/900°C. The silicidation process of simple Co/SiC contacts reduced the sheet resistance under the contact pads significantly. On lightly-doped SiC, the CoSi$_2$ produced barrier heights of 1.05 eV (n-type) and 1.90 eV (p-type), but on heavily doped material (doping $\geq10^{19}cm^{-3}$), ρ_c values of $3\times10^{-5}\Omega cm^2$ (n-type) and $4\times10^{-6}\Omega cm^2$ (p-type) were obtained [14].

Some novel W-based Schottky contacts based on W and WC on SiC have also been reported [14]. Chemically vapor deposited (400°C) W produced ϕ_B values of 0.79 eV on n-type material, and 1.89 eV for p-type SiC. WC films deposited by CVD from $WF_6/C_3H_8/H_2$ mixtures at 900°C produced ϕ_B values of 0.89 eV (n-type) and 1.81 eV (p-type), with little deterioration after 6 hr at 500°C for the latter.

Ren [15] has recently reviewed ohmic contracts to III-nitrides. The use of degenerately n-type InN [16] or creation of InN by N^+ ion bombardment of InP [17] produces specific contact resistivities $<10^{-6}\Omega cm^2$ with nonalloyed TiPtAu metallization.

Lin et al. [18] have obtained extremely good ohmic contacts on n-type GaN layers grown on sapphire substrates. Using Ti/Al metallization scheme they were able to obtain specific contact resistivies as low as $8x10^{-6}\Omega cm^2$ after annealing at 900°C for 30 sec This investigation also involved a materials study which linked the low contact resistance to the formation of a TiN_x interfacial phase.

It is reported that Ti/Al metallization on unintentional doped n-GaN ($1x10^{17}cm^{-3}$) showed linear I-V characteristics for small current levels (300μA). After the 500°C RTA anneal, the contact resistance of $10^{-3}\Omega$-cm^2 was obtained.[18] Upon further annealing at 900°C for 30 sec. The contact resistance was reduced to $8x10^{-6}\Omega cm^2$. The annealing time for 900°C annealing also played an important role. After annealing at 900°C for 20 sec, the Al and Ti diffraction peaks were absent from the XRD data as compared to the as-deposited sample. New peaks were observed and identified as face-center-tetragonal TiAl. AES depth profile of the annealed sample also showed that the metal/GaN interface (150 Å) was not completely abrupt. It is believed that this interface is essential for the formation low resistance ohmic contacts. During the 900°C anneal, the N diffused out from GaN and reacted with Ti to form TiN. Thus, N vacancies were created right at the metal GaN interface. Since N vacancies in GaN act as donors, this interface region would become a heavily doped n-GaN layer and provide low resistance contact formation. However, TEM analysis will be needed to confirm the reactions between TiAl and GaN. The reason for high contact resistance after annealing for 60 sec was speculated as the oxygen incorporated into the Al layer and formed a thin insulating Al_2O_3 layer on the surface of metallization. This caused the error for the contact resistance measurement.

Au/Ti metallization on GaN was also studied along with Al/Ti metallization. The as-deposited samples showed higher specific contact resistivity around 10-$10^{-1}\Omega cm^2$. After annealing at 700°C, it improved to 10^{-2}-$10^{-3}\Omega cm^2$ range. However, after further annealing at 900°C for 30 sec, the contact resistance increased substantially. The cause of this increase was not explored.

Lin et al. demonstrated a novel ohmic contact scheme to GaN using an InN/GaN short-period superlattice (SPS) and an InN cap layer [19]. A ten-period 0.5 nm/0.5 nm InN/GaN SPS structure was grown on a 0.6μm GaN with a 5 nm thick InN cap layer. The doping level for the n-GaN in the channel and SPS is around $5x10^{18}cm^{-3}$. The doping level of n-InN cap layer and SPS was about $1x10^{19}cm^{-3}$. Contact metallization consisted of 20 nm Ti and 100 nm Al. From TLM data, the specific contact resistance was $6x10^{-5}\Omega$-cm^2. Thermal annealing at temperature below 500°C led to no significant change in the contact resistivity. In this case, electrons tunnel through the SPS conduction band, effectively reducing the potential barrier formed by the InN/GaN heterostructure leading to lower contact resistances.

Generally Au/Ti contacts suffer from the problem of spiking into the underlying semiconductor upon annealing and have large contact resistance after thermal annealing, as described in previous section. To mitigate this problem, Pt which is a very good diffusion barrier has been used between Ti and Au to prevent Au spiking [20].

W was found to produce low specific contact resistance ($\rho_c \sim 8.0 \times 10^{-5} \Omega\text{-cm}^2$) ohmic contacts to n^+-GaN ($n = 1.5 \times 10^{19} \text{cm}^{-3}$) with limited reaction between the metal and semiconductor up to 1000°C. The formation of the β-W_2N and W-N interfacial phases were deemed responsible for the electrical integrity observed at these annealing temperatures. No Ga out-diffusion was observed on the surface of thin (500Å) W contacts even after 1000°C, 1 min anneals. Thus, W appears to be a stable contact to n^+-GaN for high temperature applications [21].

The unintentional doping levels of MOMBE grown $In_xGa_{1-x}N$ and $In_xAl_{1-x}N$ are very dependent on the In composition [22]. For the case of $In_xGa_{1-x}N$, the doping level of $In_xGa_{1-x}N$ is as high as 10^{20}cm^{-3} for a wide range with x (In ratio) larger than 0.37. For $In_xGa_{1-x}N$ with such high doping levels, nonalloyed ohmic contacts can be achieved. With the increase of In concentration in $In_xGa_{1-x}N$, it will also lower the bandgap of InGaN which will further reduce the contact resistance. These $In_xGa_{1-x}N$ layers were proposed for W contacts as ohmic contact layer on GaN and specific contact resistivities as a function of annealing temperature are shown in Figure 1. The contact resistance of as deposited samples is realized as low as $7 \times 10^{-6} \Omega\text{cm}^2$.

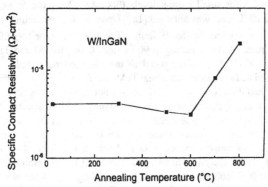

Figure 1.   Specific contact resistivity of W/InGaN as a function of annealing temperature [15].

Processing of implanted devices involves a high temperature annealing step for implant activation, typically >700°C. The stability of the WSi_x/InGaN contacts are essential to allow the high temperature process for dopant activation. The contact degradation at higher annealing temperature was related to increases in the sheet resistance, which resulted from the degradation of the metal-semiconductor interface.

From SEM studies, the as-deposited sample exhibited a very smooth surface and there was no change in the surface morphology of samples annealed at temperatures of 400 and 700°C [20]. The surface morphology of the samples annealed at 900°C showed only a small amount of surface roughness. The maximum annealing temperature to obtain good surface morphology WSi_x contacts on InGaN samples would therefore be in the range of 700-800°C. The AES studies generally confirmed the SEM observation regarding the inert nature of the metal-semiconductor interface, but indicated interdiffusion of various elements as a result of RTA at temperature of 900°C.

As shown in Figure 2, the specific contact resistivities of as-deposited W on $In_{0.6}Al_{0.4}N$ in which the unintentional doping level is 10^{18}cm^{-3} is in the high $10^{-4} \Omega\text{-cm}^2$ range [23]. Although the contact resistance reduces to $7 \times 10^{-4} \Omega\text{-cm}^2$ after annealing up to 500°C, it is still quite high for device applications. Since the unintentional doping level of InN is two orders of magnitude higher than that of $In_{0.6}Al_{0.4}N$, InN with a graded $In_xAl_{1-x}N$ layer can be used as a contact layer for InAlN devices. As illustrated in Figure 3, the contact resistance of W/InN/graded-In_xAl_{1-x}N/InAlN is half of that for W/InAlN, and the thermal stability is also improved. The contact morphology and resistance show no degradation up to 500°C. AES depth profiles of W/InN/graded-In_xAl_{1-x}N/InAlN samples showed there was only slight differences between as-deposited and 500°C annealed material. It was suggested that nitrogen diffused out into the contact metallization and formed on interfacial WN_2 phase, which improved contact resistance. Both morphology and ρ_c degraded at higher annealing temperature.

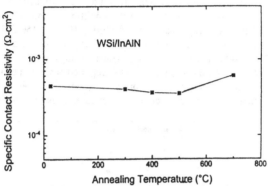

Figure 2. Specific contact resistivity of WSi_x/InAlN as a function of annealing temperature [15].

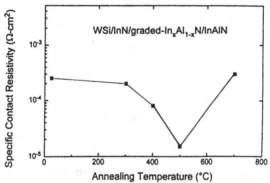

Figure 3. Specific contract resistivity of WSi_x/InN/graded In_xAl_{1-x}N/InAlN as a function of annealing temperature.

Several groups have reported improved ohmic contact properties for Ti/Al [26] or TiAlNiAu [27] on n-GaAs on which the surface was made more conducting by loss of N, either by annealing [26] or reactive ion etching [27]. Depth profiles of Ti/Al contacts annealed at 400°C showed that low contact resistance was only achieved after Al diffused to the GaN interface, suggesting a Al-Ti intermetallic is responsible for the improved properties [28]. Another intermetallic compound, $PtIn_2$, has also been used for ohmic contacts to n-GaN, and it was suggested that formation of InGaN at the interface was necessary [29].

Far less work has been done with contacts to p-GaN, where Ni/Au is the standard metallization for laser and light-emitting diodes.[30] No metallization has been found that produces the desirably low ρ_c of $<10^{-6}\Omega$-cm^2.

WET ETCHING

Much progress has recently been made in the areas of growth, dry etching, implant isolation and doping of the III-V nitrides and their ternary alloys. This has resulted in nitride-based blue/UV light emitting and electronic devices [31-39]. There has been less success in developing wet etch solutions for these materials, due to their excellent chemical stability. High etch rates have been achieved in dry etch chemistries [40-49], but damage may be introduced by ion bombardment, and controlled undercutting is difficult to attain. In addition, since dry etching has a physical component to the etch, selectivities between different materials is generally reduced.

Amorphous AlN has been reported to etch in 100°C HF/H_2O [50-52], HF/HNO_3 [53], and NaOH [54], and polycrystalline AlN in hot (≤85°C) H_3PO_4 at rates less than 500Å/min [55,56]. Mileham et al. [57] reported the etching of AlN defective single crystals in KOH based solutions at etch temperatures ranging from 23 to 80°C. They reported decreased etch rates with increasing crystal quality, as the reactions occurred favorably at grain boundaries and defect sites. InN in aqueous KOH solutions was reported to etch a few hundred angstroms per minute at 60°C [58].

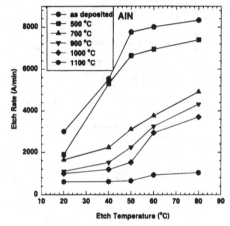

Figure 4.   Etch rate of AlN in KOH solutions as a function of etch temperature for samples as-deposited or annealed at 500, 700, 900, 1000, and 1100°C.

Wet chemical etching of AlN and $In_xAl_{1-x}N$ was investigated in KOH-based solutions as a function of etch temperature and material quality [59]. The etch rates for both materials increased with increasing etch temperatures, which was varied from 20 to 80°C. The crystal quality of AlN prepared by reactive sputtering was improved by rapid thermal annealing at temperatures to 1100°C with a decreased wet etch rate of the material measured with increasing anneal temperature [Figure 4]. The etch rate decreased approximately an order of magnitude at 80°C etch temperature after an 1100°C anneal. The etch rate for $In_{0.19}Al_{0.81}N$ grown by metallorgainc molecular beam epitaxy was approximately three times higher for material on Si than on GaAs. This corresponds to the superior crystalline quality of the material grown on GaAs. Etching of $In_xAl_{1-x}N$ was also examined as a function of In composition. The etch rate initially increased as the In composition changed from 0 to 36%, and then decreased to 0 Å/min for InN [Figure 5]. We also compared the effect of doping concentration on etch rate. Two

InAlN samples of similar crystal quality were also etched; one was fully depleted with n $<10^{16}$cm^{-3} (2.6% In) and the other n~5×10^{18}cm^{-3} (3.1% In). At low etch temperature, the rates were similar, but above 60°C the n-type sample etched faster, approximately three times faster at 80°C. The activation energy for these etches is very low, 2.0 ± 0.5 kcal mol^{-1} for sputtered AlN. The activation energies for InAlN were dependent on In composition and were in the range 2 to 6 kcal Mol^{-1}. GaN and InN layers did not show any etching in KOH at temperatures up to 80°C.

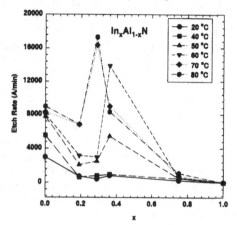

Figure 5.   Etch rate for In$_x$Al$_{1-x}$N for $0 \leq x \leq 1$ at KOH solution temperatures between 20 and 80°C.

Minsky et al. [60] demonstrated photo-electrochemical etching of GaN under illumination by a He-Cd (d=325 nm) laser using KOH and HCl solutions. Annealed (900°C) Ti contacts were used as electrical contacts and etch masks. Broad area photo-electrochemical etching of n-type β-SiC has previously been reported [61], and the same technique has been applied to n-GaN [62], using Hg lamp exposure and unannealed Ti metal contacts. Etch rates of 170-200Å ·min^{-1} were obtained for n-type material, but no etching was found for p-type GaN.

DRY ETCHING

The current status of dry etching of nitrides has recently been reviewed by Gillis et al. [63]. The baseline technique employed, reactive ion etching (RIE) produces GaN etch rates of up to 1,000Å ·min^{-1} at high dc self-biases (-300 to -400 V). Many plasma chemistries have been used, including those common for III-V semiconductors such as Cl$_2$, Cl$_3$, SiCl$_4$, CCl$_2$F$_2$, HBr/H$_2$, CH$_4$/H$_2$ and CHF$_3$. Magnetic enhancement of the discharge, as in magnetron RIE, leads to much larger rates (3000Å ·min^{-1} at -100 V self-bias) [64]. The highest etch rates are obtained with high ion density plasma sources such as Electron Cyclotron Resonance (ECR) and Inductively Coupled Plasma (ICP) [45,47-49,65]. Rates up to 1.3μm ·min^{-1} for GaN have been obtained with ICl/Ar ECR discharges. The ion energy threshold for the onset of etching GaN and InN is ~75 eV. Surface roughening and N depletion can occur at high ion energies. Little work has been done a systematic damage studies, but initial reports show that nitrides are more resistant to introduction of electrically active point defect damage that other III-V's [66], but that even low ion energy conditions can induce N-deficient surfaces [67] and degradation of luminescence [68].

A promising low damage technique is low energy, electron-enhanced etching (LE4), which avoids ion bombardment altogether [63,69]. LE4 of GaN/Si and GaN/SiC in direct current H_2 and H_2/Cl_2 plasmas has been reported at rates up to 2500Å·min^{-1} [63,69]. Another technique that avoids ion bombardment is photo-assisted vapor etching [70], where GaN has been etched in HCl vapor while being irradiated with an excimer laser and held at 200-400°C.

For SiC, basically all of the pattern steps during device processing must be carried out with dry etching due to the chemical stability and inertness of SiC in conventional acid or base solutions at normal temperatures [71]. Most of the dry etching processes reported to date have employed reactive ion etching (RIE) with chlorofluorocarbon (CHF$_3$ and related gases) or NF$_3$, and hydrogen is generally added to the plasma chemistry to avoid rough surfaces [72-75]. Flemish et al. [76-78] and Casady et al. [74] reported that higher ion density Electron Cyclotron Resonance (ECR) discharges of CF$_4$/O$_2$ or SF$_6$/O$_2$ produced much higher etch rates than RIE, and it was not necessary to add H$_2$ to the plasma chemistry to obtain smooth surface morphologies. Changes to the electrical quality of Schottky diodes fabricated on the dry etched surfaces were less severe with the ECR discharges [78], and the threshold ion energies for creating damage were also determined. (~100 eV in ECR and ~150 eV in RIE).

In work on other difficult-to-etch materials such as GaN, NiFeCo and SrS we have found that in many cases the etch rates are not limited by the volatility of the etch products, but by the initial bond-breaking that must precede formation of these products. The inherent advantage of ECR discharges over RIE plasmas is the high ion density ($\geq 10^{11}$cm^{-3} compared to $\leq 10^9$cm^{-3}) that aids in the initial bond breaking and subsequent product desorption. We have recently compared different plasma chemistries, including Cl$_2$, IBr, SF$_6$ and NF$_3$, for ECR etching of SiC. All of these chemistries are hydrogen-free and thus avoid any hydrogen passivation of near-surface dopants.

The expected etch products for SiC in Cl$_2$-based gas chemistries are SiCl$_4$ and CCl$_4$, both of which have high vapor pressure at room temperature and therefore on would expect reasonably good etch rates. Figure 6 shows the etch rate of SiC in 1000 W ECR discharges of 10Cl$_2$/5Ar, as a function of applied rf chuck power. The threshold for observing any etching is ~50 W, with the etch rate linearly dependent on ion energy (which is proportional to rf power). Therefore, even at high ion density (high microwave power) it is necessary to have an ion energy above the threshold for breaking bonds in the SiC. The presence of the Ar$^+$ ions is very important in this process – we found that the etch rate increased with percentage Cl$_2$ in the plasma up to ~10Cl$_2$/5Ar, but decreased rapidly thereafter (to ~500 Å/min in pure Cl$_2$ plasmas at 150 W rf and 1000 W microwave power). Lower etch rates were obtained with IBr plasma chemistries under the same conditions.

Oxygen has often been added to fluorine-based gas chemistries under RIE conditions to enhance the active fluorine concentration and increase SiC etch rate [76,79]. Under ECR conditions, however we observed little benefit for O$_2$ addition to either NF$_3$ or SF$_6$, as shown in Figure 7. Note that NF$_3$ produces etch rates roughly three times faster than SF$_6$ because it is more easily dissociated (bond strength 62.1 kCal/mol compared to 82.9 kCal/mol). There is little change in the intensity of the atomic fluorine lines in the 6000-7000Å region of the optical emission spectra of SF$_6$ with increasing O$_2$ content with the SF$_x$ lines at ~4500Å decreasing and the O lines at ~8000Å increasing as expected. The ion energy threshold for the onset of etching is absent with NF$_3$. We obtained etch rates of ~1100Å/min even at 0V self-bias – the ion energies in this case are what is associated with the ECR plasma generation itself, i.e. ~20 eV. Note again however that the etch rate increased in an almost linear fashion with self-bias, which suggests that the increased efficiency of bond-breaking is a key parameter in determining etch rate. The etched surface morphologies were typically similar to those of the unetched control samples, without the need for H$_2$ addition.

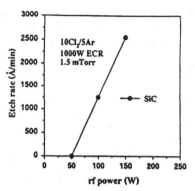

Figure 6. Etch rate of SiC in 1000 W microwave, 1.5 mTorr discharges of 10Cl$_2$/5Ar, as a function of rf power.

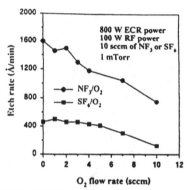

O$_2$ flow rate (sccm)

Figure 7. Etch rate of SiC in 800 W microwave, 100 W rf, 1 mTorr discharges of NF$_3$/O$_2$ or SF$_6$/O$_2$, as a function of O$_2$ flow rate. Total pressure and flow rate were held constant.

The introduction of dry etch damage into n-type SiC epilayers was measured by monitoring the sheet resistance after exposure to Ar plasmas under both RIE and ECR conditions. In these 1μm thick films, the threshold rf powers for measurable resistance changes were ~250 W (-275 V dc bias) for RIE and ~150 W (-170 W dc bias) for ECR conditions (1000 W microwave power). The SiC is much more resistant to introduction of dry etch damage than Si, as expected from its high bond strength. Significant annealing of the damage introduced by ion bombardment occurred at ~700°C, corresponding to an activation energy for damage removal of ~3.4 eV [80].

Etch rates > 1500Å/min are found for SiC in ECR Cl$_2$/Ar or NF$_3$ plasmas with moderate rf bias. There is a threshold rf power for the onset of etching in Cl$_2$/Ar (~50 W), whereas NF$_3$ is found to produce etching even with no biasing of the substrate. Addition of O$_2$ to NF$_3$ des not produce any significant etch rate enhancement, and addition of H$_2$ to Cl$_2$ plasmas greatly retards the SiC etch rate. The etched surfaces retain their original morphology in all of the chemistries we investigated, and small quantities of S-residues are detected on SF$_6$-etched samples. These

ECR processes appear quite suitable for pattern transfer into SiC at higher rates than obtainable with RIE. Photoresist is in general not a good choice as a mask material since it is readily etched in the chemistries discussed here, and indium tin oxide is a better choice [71].

ION IMPLANTATION/ANNEALING

Several recent studies have established the thermal stability limits of GaN, AlN and InN during both rapid thermal annealing [81] and vacuum annealing [82]. Figure 8 shows the nitrogen desorption rate for these binaries [82]. The effective activation energies for N_2 decomposition in vacuum were 3.48 eV (InN), 3.93 eV (GaN) and 4.29 eV (AlN). These are a factor of approximately 1.7 times higher than the binding energies of a single metal-N bond (respectively 1.93 eV, 2.2 eV and 2.88 eV for InN, GaN and AlN) [83]. The nitrogen flux from the unprotected nitride surfaces peaked at 685°C (InN), 985°C (GaN) and >1120°C (AlN) [82]. Under RTA conditions, loss of nitrogen was found to create then, degenerately n-type surfaces on the binary nitrides, with stability limits of ≤600°C (InN), 800°C ($In_{0.5}Ga_{0.5}N$, $In_{0.75}Al_{0.25}N$), ~1100°C (GaN) and ~1100°C (AlN) [81].

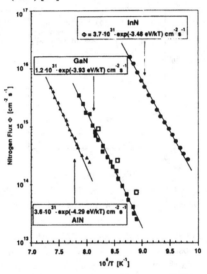

Figure 8.   Nitrogen flux or decomposition rate for InN, GaN and AlN in vacuum over the temperature range of decomposition [82].

We have compared use of GaN, InN and AlN powder for providing nitrogen partial pressure within a graphite susceptor during high temperature rapid thermal annealing of GaN, AlN, InN and InAlN. At temperatures above ~750°C vapor transport of In from InN powder produces In droplet condensation on the surface of all nitride samples being annealed. GaN powder provides better surface protection than AlN powders for temperatures up to ~1050°C when annealing GaN and AlN samples. Dissociation of nitrides from the surface is found to occur with approximate activation energies of 3.8 eV, 4.4 eV and 3.4 eV, respectively, for GaN, AlN and InN.

The first reports of the use of ion implantation to introduce impurities into GaN dates back to 1972 with work done by Pankove and coworkers on the photoluminescence of 35 implanted impurities in GaN [84]. Although luminescence data was given, no electrical data was reported. To remove the implant damage and achieve good luminescence the samples were annealed for 1 hr in flowing ammonia (NH_3). Most likely, hydrogen liberated from the ammonia ambient or hydrogen already grown into the GaN films was responsible for electrically passivating these impurities. This is likely since hydrogen is known to passivate Mg and Ca acceptors in GaN [85-87]. Once the role of hydrogen was understood with respect to passivating epitaxial dopants, primarily acceptors, it was clear that the implantation anneal sequence should also be done in a hydrogen-free ambient.

Other implantation work by Wilson et al. focused on the redistribution properties of potential dopants in GaN [88]. That work demonstrated that, at least up to 900°C, none of the implanted species studied (Be, Mg, Zn, C, Se, Ge) showed measurable redistribution with annealing for times up to 20 min. One exception was S which exhibited significant diffusion even at 600°C. This result suggests that external source diffusion will not be viable in GaN due to very low diffusivities of the dopant species. The lack of redistribution of most of these dopants was later verified up to 1100°C [89].

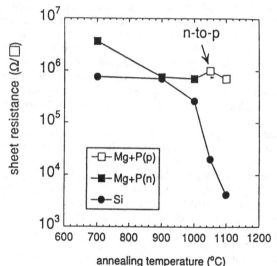

Figure 9.　Sheet resistance versus annealing temperature for Si-implanted or Mg+P implanted GaN. Significant electrical activation of the dopants, as demonstrated by the drop in sheet resistance for the Si-samples and a conversion from n-type to p-type for the Mg+P samples, starts to occur at 1050°C with increased activation at 1100°C. Unimplanted and annealed samples showed a slight decrease in sheet resistance but only from 10^6 to 10^5 Ω/sq.

Turning to the electrical activity of implanted dopants in GaN, Figure 9 shows sheet resistance data versus annealing temperature for Si or Mg+P implanted GaN. These data represent the first report of electrical activity of implanted dopants in GaN [90]. The samples were annealed in a rapid thermal annealer enclosed in a SiC coated graphite susceptor in flowing N_2. The key point from Figure 9 is that electrical activity (a sharp drop in sheet resistance for the Si-implanted samples and a conversion from n-type to p-type for the Mg+P implanted samples) does not occur until 1050°C. At this temperature the GaN film can dissociate by liberating nitrogen; a process that is accelerated in the presence of hydrogen or water vapor. Therefore, it is critical that the annealing ambient be well controlled to maintain the integrity of the semiconductor. In fact, even when bulk nitrogen loss is not detectable by sputtered Auger Electron Spectroscopy (AES), near surface loss over approximately 50Å can create a degeneratively doped n-type region since N-vacancies are believed to act as donors in GaN [91]. This layer will then enhance ohmic contact formation or degrade Schottky contact properties [92,93]. Control of this surface condition is required for fabrication of transistors incorporating ion implantation since both ohmic and Schottky contacts are required. An effective method to maintain the original surface during the anneal is to encapsulate the GaN surface with a sputter deposited AlN film which can later be removed in a selective KOH-based etch. Using this approach, Pt/Au Schottky contacts have been achieved on GaN after annealing at 1100°C while near ohmic behavior resulted on samples annealed uncapped [93].

Returning to the required implant activation anneal temperature for GaN, Table II contains typical annealing temperature and melting points for GaN and for several other semiconductors. The final column of the table shows the ratio of annealing temperature to melting point. As is the case for GaSb, InP, GaAs and Si, the implant activation temperature generally follows a two-thirds rule with respect to the melting point. For GaN (and SiC, however, the activation temperature presently employed is closer to 50% of the melting point. Therefore, although dopant activation can be achieved in GaN at 1100°C, the optimum annealing temperature may very well be closer to 1700°C to fully remove the implant damage. Since this temperature is beyond the capability of most rapid thermal annealing systems, new annealing apparatus will have to be developed if this temperature is indeed required. This point will be revisited below when the removal of implant damage is examined by channeling Rutherford Backscatterng (C-RBS).

Another important technological tool for which ion implantation is well suited is to explore doping or compensation effects of new species. In the case of GaN, one critical technological issue is the determination of an alternative acceptor species to Mg that has a smaller ionization energy (the ionization energy of Mg is ~170 meV) and therefore would yield more free holes at room temperature. Along these lines, Ca had been suggested as being a shallow acceptor in GaN [94] and ion implantation was used for the first demonstration of p-type GaN based on Ca-doping [95]. Unfortunately, the ionization energy of Ca was also shown to be equivalent to that of Mg; however, this result demonstrates the utility of ion implantation for introducing various species into the semiconductor host to study their properties.

Turning now to the build-up and removal of implant damage in GaN, Figure 10 shows channeling C-RBSA spectra for Si-implanted GaN at various doses and after an annealing treatment. Figure 10a demonstrates the GaN has a very high threshold of amorphization, on-the-order-of $2 \times 10^{16} cm^{-2}$, where amorphization is taken as the point where the implanted spectrum coincides with the random spectrum [96]. This is in contrast to GaAs where an amorphous region forms for room temperature implants near a dose of $4 \times 10^{13} cm^{-2}$ but is similar to high Al-fraction AlGaAs which is not amorphized until a similar dose is achieved [96]. Typically in other III-V semiconductors, if amorphization is avoided during the implantation process, most of the damage can be removed during the implant activation anneal with the semiconductor returned near to the pre-implant damage level. However, Figure 10b shows that even for a Si-implant dose of $5 \times 10^{15} cm^{-2}$ that does not amorphize the sample, significant damage, well above the unimplanted level, remains after a 1100°C, 15 sec anneal [97]. This supports the hypothesis that higher temperature annealing will be required to optimize the implant activation process. Recent results have shown, however, at least for Si-implantation in GaN, that even in the presence of significant residual implant damage high dopant activation efficiencies and low resistance regions can be obtained [97].

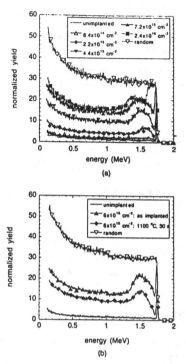

Figure 10. Channeling Rutherford Backscattering spectra for Si-implanted GaN at an energy of 100 keV a) for various Si-doses and B) for a dose of $6 \times 10^{15} \text{cm}^{-2}$ both as-implanted and after a 1100°C, 30 sec anneal. The spectra in (a) demonstrates the high threshold of amorphization of GaN during implantation while the spectra in (b) demonstrates that significant damage remains in high dose Si-implanted GaN even after annealed at 1100°C.

A final area of implantation process of compound semiconductors is the formation of select areas of high resistance material for inter-device isolation or current guiding. For both n- and p-type GaN, N-implantation is effective for introducing compensating point defects [90]. This approach yielded a maximum sheet resistance after annealing in the range of 750 to 850°C where the implantation-induced defect density is partially removed to reduce defect assisted hopping conduction but still sufficient to compensate the extrinsic epitaxial doping [90]. Additional work has shown that He is also an effective isolation species for n-type GaN while H-implantation has limited utility for isolation since the compensation anneals out below 400°C [98]. The fact that H-implantation isolation is not effective is not clearly understood but may relate to the implanted hydrogen acting to passivate the compensating point defects during the low temperature anneal. Finally, implantation isolation of In-containing III-nitride materials has shown that InGaN, as used in a LED, laser cavity, or transistor channel, can not be rendered highly resistive by F or O-implantation while InAlN can be highly compensated by O- or N-implants [99-101].

The utility of ion implantation for fabricating the variety of high power device structures, eg. MOS-controlled thyristor, MOS Turnoff thyristor, insulated gate bipolar transistor, necessary for power switching will depend largely on the ability of the crystal growers to produce thick

(>10μm), lightly doped (n<10^{15}cm^{-3}) GaN layers, and high-resistivity substrates.

Table II: Comparison of semiconductor melting points (T$_{mp}$) with the   temperature required to activate implanted dopants (T$_{act}$).

	T$_{mp}$(°C)	T$_{act}$(°C)	T$_{act}$/T$_{mp}$
GaSb	707[a]	500-600	0.77
InP	1057[a]	700-750	0.69
GaAs	1237[A]	750-900	0.69
Si	1410[a]	950	0.67
SiC	2797[a]	1300-1600	0.46-0.57
GaN	2518[b]	~1100	0.44

[a]Handbook of Chemistry and Physics, ed. Robert C. Weast, (CRC Press, Boca Raton, FL, 1983) p. E-92-93.
[B]J.A. Van Vechten, Phys. Rev. B, 7, 1479 (1973).

Casady and Johnson [7] have recently reviewed implantation in SiC technology. Boron has proven to have higher activation than Al and in most cases the implant is performed at 700-800°C to avoid amorphization. The activation annealing is performed at 1100-1650°C under Ar ambients.

SUMMARY AND CONCLUSIONS

There are still numerous technical obstacles to optimizing the performance of wide bandgap semiconductor devices, including
(i)   improved ohmic contacts to p-GaN. One potential solution here is grading to p-InGaN of the highest In concentration that will allow achievement of p-doping, and which is consistent with the requisite thermal stability of the device. The higher the In composition the lower this thermal stability will be, but the specific contact resistivity will also be improved. This solution may favor MBE and MOMBE over MOCVD, because of their lower growth temperatures and ability to incorporate higher In concentrations in InGaN.
(ii)   improved trench etching and laser mesa etching processes for SiC and GaN, respectively. Optimization of ICP and ECT plasma chemistries and conditions should be sufficient, and LE4 may play a role because of its potentially lower damage. In this respect, the availability of slow, controlled wet etch processes for damage clean-up after dry etching is also desirable.
(iii) high temperature (>500°C) stable Schottky contacts to GaN for power transistors - WN$_x$ should be explored in this context.
(iv) the availability of high quality gate oxides for MOS devices in both GaN and SiC. While SiO$_2$ works adequately in many cases on SiC, more development is needed, and there is little systematic work reported for AlN or Ga$_2$O$_3$ on GaN. The latter has produced exciting results on GaAs in recent times, but its thermal stability may be an issue on GaN.

ACKNOWLEDGMENTS

The work at UF is partially supported by NSF-DMR (DMR9421109, L.D. Hess), and DARPA (A. Husain) through AFOSR (G.L. Witt). The work at SNL is supported by USDOE (contract DE-AC04-94AL85000). SNL is a multi-program laboratory operated by Sandia Corporation, a unit of Lockheed Martin.

REFERENCES

1. L.S. Rea, Mat. Res. Soc. Vol. 423, 3 (1996).
2. K. Reihardt, J.D. Scofield, and W.C. Mitchel, Proc. Workshop on High Temperature Electronics for Vehicles, eds. G. Khalil, H. Singh, and T. Podlesak, ARL Tech. Rep., pp. 73-79, April 1995.
3. D.M. Brown, E. Downey, M. Ghezzo, J. Krechmer, V. Krishanmurty, W. Hennessy, and G. Michon, Sol. State Electronics, 39, 1543 (1996).
4. J. Palmour, J.S. King, D. Waltz, J. Edmond, and C. Carter, Trans. 1st Intl. High Temperature Conference, Albuquerque, NM, pp. 207-212 (1991).
5. A.K. Agarwal, R.R. Seirgeij, S. Seshadri, M.H. White, P.D. McMullin, A.A. Burk, L.B. Roland, C.D. Brandt, and R.H. Hopkins, Mat. Res. Soc. Symp. 423, 87 (1996).
6. M. Bhatnagar and B.J. Baliga, IEEE Trans. Electron. Dev. 40, 645 (1993).
7. J.B. Casady and R.W. Johnson, Sol. State Electronics 39, 1409 (1996).
8. See for example GaN and related materials ed. F. Ponce, R.D. Dupuis, J.A. Edmond, and S Nakamura, MRS Vol. 395 (1996) and references therein.
9. H.H. Han, J.S. Williams, J. Zou, D.J.H. Cockayne, S.J. Pearton, and R.A. Stall, Appl. Phys. Lett. 69, 2364 (1996).
10. J.C. Zolper, D.J. Reiger, A.G. Baca, S.J. Pearton, J.W. Lee, and R.A. Stall, Appl. Phys. Lett. 69, 538 (1996).
11. L.M. Porter, R.F. Davis, J.A. Bow, M.J. Kim, and R.W. Carpenter, J. Mater. Res. 10, 26 (1995).
12. J. Crofton, P.A. Barnes, J.R. Williams, and J.A. Edmond, Appl. Phys. Lett. 62, 384 (1993).
13. N. Lundberg and M. Ostling, Solid-State Electronics 39, 1559 (1996).
14. N. Lundberg, Ph.D. Thesis, Royal Inst. Technology, Sweden (1996).
15. F. Ren, in GaN and Related materials, ed. S.J. Pearton (Gabon and Breach, NY, 1997).
16. F. Ren, C.R. Abernathy, S.J. Pearton, and P.W. Wisk, Appl. Phys. Lett. 64, 1508 (1994).
17. F. Ren, C.R. Abernathy, S.N.G. Chu, J.R. Lothian, and S.J. Pearton, Appl. Phys. Lett. 66, 1503 (1995). F. Ren. S.J. Pearton, J.R. Lothian, S.N.G. Chu, W.K. Chu, R.G. Wilson, C.R. Abernathy, and S.S. Peri, Appl. Phys. Lett. 65, 2165 (1994).
18. M.E. Lin, Z. Ma, F.Y. Huang, Z.F. Fan, L.H. Allen, and H. Morkoc, Appl. Phys. Lett. 64, 1003 (1994).
19. M.E. Lin, F.Y. Huang, and H. Morkoc, Appl. Phys. Lett. 64, 2557 (1994).
20. A. Durbha, S.J. Pearton, C.R. Abernathy, J.W. Lee, P.H. Holloway, and F. Ren, J. Vac. Sci. Technol. B14, 2582 (1996).
21. M.W. Cole, D.W. Eckart, W.Y. Han, R.L. Pfeffer, T. Monahan, F. Ren, C. Yuan, R.A. Stall, S.J. Pearton, Y. Li, and Y. Lu, J. Appl. Phys. 80. 278 (1996).
22. C.R. Abernathy, J.D. MacKenzie, S.R. Bharatan, K.S. Jones, and S.J. Pearton, Appl. Phys. Lett. 66, 1632 (1995).
23. F. Ren, S.J. Pearton, S. Donovan, C.R. Abernathy, and M.W. Cole, ECS Proc. Vol. 96-11, 122 (1996).
24. C.B. Vartuli, S.J. Pearton, C.R. Abernathy, J.D. MacKenzie, R.J. Shul, J.C. Zolper, M.L. Lovejoy, A.G. Baca, and M. Hagerott-Crawford, Mat. Res. Soc. Symp. Vol. 421, 373 (1996); J. Vac. Sci. Technol. B14, 3520 (1996).
25. S.M. Donovan, J.D. MacKenzie, C.R. Abernathy, C.B. Vartuli, S.J. Pearton, F. Ren, M.W. Cole, and K. Jones, Mat. Res. Soc. Symp. Proc. 449, 771 (1997).

26. L.F. Lester, J.M. Brown, J.C. Ramer, L. Zhang, S.D. Hersee, and J.C. Zolper, Appl. Phys. Lett. 69, 2737 (1996).

27. Z. Fan, S.N. Mohammad, W. Kim, O. Aktas, A.E. Botcharev, and H. Morkoc, Appl. Phys. Lett. 68, 1672 (1996).

28. B.P. Luther, S.E. Mohney, T.N. Jackson, M.A. Khan, Q. Chen, and J.W. Wang, Appl. Phys. Lett. 70, 57 (1997).

29. D.B. Ingerly, Y.A. Chang, N.R. Perkins, and T.F. Kuech, Appl. Phys. Lett. 70, 108 (1997).

30. S. Nakamura, M. Senoh, and T. Mukai, Appl. Phys. Lett. 62, 2390 (1993).

31. S. Nakamura, M. Senoh, and T. Mukai, Jpn. J. Appl. Phys. 30, L1708 (1991).

32. S.C. Binari, L.B. Rowland, W. Kruppa, G. Kelner, K. Doverspike, and D.K. Gaskill, Electron Lett. 30, 1248 (1994).

33. M.A. Khan, M.S. Shur, and Q. Chen, ibid., 31, 2130 (1995).

34. M.A. Khan, J.N. Kuznia, A.R. Bhattarai, and D.T. Olson, Appl. Phys. Lett. 62, 1248 (1993).

35. S. Nakamura, M. Senoh, and T. Mukai, ibid., 62, 2390 (1993).

36. I. Akasaki, H. Amano, M. Kito, and K. Kiramatsu, J. Lumin. 48/49, 666 (1991).

37. S. Nakamura, M. Senoh, N. Iwasa, and S. Nagahama, Appl. Phys. Lett. 67, 1868 (1995).

38. J.C. Zolper, A.G. Baca, R.J. Shul, R.G. Wilson, S.J. Pearton, and R.A. Stall, ibid., 68, 1266 (1996).

39. S. Nakamura, M. Senoh, S. Nagahama, N. Iwasa, T. Yamada, T. Matsushita, H. Kiyoku, and Y. Sugimoto, Jpn. J. Appl. Phys. 35, L74 91996).

40. I. Adesida, A. Mahajan, E. Andideh, M. Asif Khan, D.T. Olsen, and J.N. Kuznia, Appl. Phys. Lett. 63, 2777 (1993).

41. M.E. Lin, Z.F. Zan, Z. Ma, L.H. Allen, and H. Morkoc, ibid., 64, 887 (1994).

42. A.T. Ping, I. Adesida, M. Asif-Khan, and J.N. Kuznia, Electron. Lett. 30, 1895 (1994).

43. H. Lee, D.B. Oberman, and J.S. Harris, Jr., Appl. Phys. Lett. 67, 1754 (1995).

44. S.J. Pearton, C.R. Abernathy F. Ren, J.R. Lothian, P.W. Wisk, A. Katz, and C. Constantine, Semicond. Sci Technol. 8, 310 (1993).

45. S.J. Pearton, C.R. Abernathy, and F. Ren, Appl. Phys. Lett. 64, 2294 (1994).

46. L. Zhang, J. Ramer, K. Zheng, L.F. Lester, and S.D. Hersee, Paper presented at MRS Fall Meeting, Boston, MA (1995).

47. R.J. Shul, S.P. Kilcoyne, M. Hagerott-Crawford, J.E. Parameter, C.B. Vartuli, C.R. Abernathy, and S.J. Pearton, Appl. Phys. Lett. 66, 1761 (1995).

48. R.J. Shul, S.J. Pearton, and C.R. Abernathy, Abstract 311, p. 412, The Electrochemical Society Meeting Abstracts, Vol. 96-1, Los Angeles, CA, May 5-10, 1996.

49. C.B. Vartuli, S.J. Pearton, J.W. Lee, J. Hong, J.D. MacKenzie, C.R. Abernathy, and R.J. Shul, Appl. Phys. Lett. 69, 1426 (1996).

50. K.M. Taylor and C. Lenie, J. Electrochem. Soc. 107, 308 (1960).

51. G. Long and L.M. Foster, J. Am Ceram. Soc. 42, 53 (1959).

52. N.J. Barrett, J.D. Grange, B.J. Sealy, and K.G. Stephen, J. Appl. Phys. 57, 5470 (1985).

53. C.R. Aita and C.J. Gawlak, J. Vac. Sci. Technol. A1, 403 (1983).

54. G.R. Kline and K.M. Lakin, Appl. Phys. Lett. 43, 750 (1983).

55. T. Pauleau, J. Electrochem. Soc. 129, 1045 (1982).

56. T.Y. Sheng, Z.Q. Yu, and G.J. Collins, Appl. Phys. Lett. 52, 576 91988).

57. J.R. Mileham, S.J. Pearton, C.R. Abernathy, J.D. MacKenzie, R.J. Shul, and S.P. Kilcoyne, ibid., 67, 1119 (1995).

58. Q.X. Guo, O. Kato, and Y. Yoshida, This Journal, 139, 2008 (1992).

59. C.B. Vartuli, S.J. Pearton, J.W. Lee, C.R. Abernathy, J.D. MacKenzie, J.C. Zolper, R.J. Shul, and F. Ren, J. Electrochem. Soc. 143, 3681 (1996).

60. M.S. Minsky, M. White, and E.L. Hu, Appl. Phys. Lett. 68, 1531 91996).

61. J.S. Shor and R.M. Osgood, J. Electrochem. Soc. 140, L123 (1993).

62. C. Youtsey, I. Adesida, and G. Bulman, Electronics Lett. (in press).

63. H.P. Gillis, D.A. Choutov, and K.P. Martin, J. Mater., August 1996, pp. 50-55.

64. G.F. McLane, S.J. Pearton, and C.R. Abernathy, Wide Bandgap Semiconductors and Devices, Vol. 95-21 (ECS, Pennington, NJ), pp. 204-214.

65. R.J. Shul, G.B. McClellan, S.A. Casalnuovo, D.J. Rieger, S.J. Pearton, C. Constantine, C. Barratt, and R.K. Karlicek, Appl. Phys. Lett. 69, 1119 (1996).

66. S.J. Pearton, J.W. Lee, J.D. MacKenzie, C.R. Abernathy, and R.J. Shul, Appl. Phys. Lett. 67, 2329 (1995).

67. F. Ren, J.R. Lothian, Y.K. Chen, J.D. MacKenzie, S.M. Donovan, C.B. Vartuli, C.R. Abernathy, J.W. Lee, and S.J. Pearton, J. Electrochem. Soc. 143, L217 (1996).

68. R.J. Shul, J.C. Zolper, M.H. Crawford, R.T. Hickman, R.D. Briggs, S.J. Pearton, J.W. Lee, R. Karlicek, C. Tran, C. Constantine, and C. Barratt, ECS Prod. Vol. 96-15, 232 (1996).

69. H.P. Gillis, D.A. Choutov, K.P. Martin, S.J. Pearton, and C.R. Abernathy, J. Electrochem. Soc. 143, L251 (1996).

70. R.T. Leonard and S.M. Bedair, Appl. Phys. Lett. 68, 794 (1996).

71. J.R. Flemish, K. Xie, and G.F. McLane, Mat. Res. Soc. Symp. Proc. 428, 106 (1996).

72. A.J. Steckl and P.H. Yih, Appl. Phys. Lett. 60, 1966 (1992).

73. J.W. Palmour, R.F. Davis, T.M. Wallett, and K.B. Bhashin, J. Vac. Sci. Technol. A. 4, 590 (1986).

74. J.B. Casady, E.D. Luckowski, M. Bozack, D. Steridan, R.W. Johnson, and J.R. Williams, Tech. Dig. Of Int. Conf. SiC Related materials, Kyoto, Japan 1995, pp. 382-383.

75. P.H. Yih and A.J. Steckl, J. Electrochem. Soc. 140, 1813 (1993).

76. J.R. Flemish, Proc. Widebandgap Semiconductors and Devices, ed. F. Ren (Electrochemical Society, Pennington, NJ 1995) Vol. 10, pp. 231-235.

77. J.R. Flemish, K. Xie, and J.H. Zhao, Appl. Phys. Lett. 64, 2315 (1994).

78. J.R. Flemish, K. Xie, W. Buchwald, L. Casas, J.H. Zhao, G.F. McLane, and M. Dubey, Mat. Res. Soc. Symp. Proc. 339, 145 (1994).

79. J.B. Casady and R.W. Johnson, Solid State Electron. 39, 1409 (1996); J.B. Casady, E.D. Luckowski, M. Bozack, D. Sheridan, R.W. Johnson, and J.A. Williams, J. Electrochem. Soc. 143, 750 (1996).

80. S.J. Pearton, J.W. Lee, J.M. Grown, M. Bhaskaran, and F. Ren, Appl. Phys. Lett. 68, 2987 (1996).

81. C.B. Vartuli, S.J. Pearton, C.R. Abernathy, J.D. MacKenzie, E.S. Lambers, and J.C. Zolper, J. Vac. Sci. Technol. B14, 3523 (1996).

82. O. Ambacher, M.S. Brandt, R. Dimitrov, T. Metzger, M. Stutzmann, R.A. Fischer, A. Miehr, A. Bergmaier, and G. Dollinger, J. Vac. Sci. Technol. B14, 3532 (1996).

83. J.H. Edgar, Properties of Group III Nitrides (INSPEC IEEE London 1994).

84. J.I. Pankove and J.A. Hutchby, J. Appl. Phys. 47, 5387 (1976).

85. H. Amano. M. Kito, K. Hiramatsu, and I. Akasaki, Jap. J. Appl. Phys. 28, L2118 (1989).

86. S. Nakamura, T. Mukai, M. Senoh, and N. Iwasa, Jap. J. Appl. Phys. 31, L139 (1992).

87. J.W. Lee, S.J. Pearton, J.C Zolper, and R.A. Stall, Appl. Phys. Lett. 68, 2102 (1996).

88. R.G. Wilson, C.B. Vartuli, C.R. Abernathy, S.J. Pearton, and J.M. Zavada, Solid-State Elec. 38, 1329 (1995).

89. J.C. Zolper, M. Hagerott-Crawford, S.J. Pearton, C.R. Abernathy, C.B. Vartuli, J. Ramer, S.D. Hersee, C. Yuan, and R.A. Stall, Mat. Res. Soc. Symp. Proc. Vol. 394, 801 (1996).

90. S.J. Pearton, C.R. Abernathy, C.B. Vartuli, J.C. Zolper, C. Yuan, and R.A. Stall, Appl. Phys Lett. 67, 1435 (1995).

91. H.P. Maruska and J.J. Tietjen, Appl. Phys. Lett. 15, 327 (1969).

92. L.F. Lester, J.M. Brown, J.C. Ramer, L. Zhang, S.D. Hersee, and J.C. Zolper, Appl. Phys. Lett. 69, 2737 (1996).

93. J.C. Zolper, D.J. Rieger, A.G. Baca, S.J. Pearton, J.W. Lee, and R.A. Stall, Appl. Phys. Lett. 69, 538 (1996).

94. S. Strite, Jpn. J. Appl. Phys. 33, L699 (1994).

95. J.C. Zolper, R.G. Wilson, S.J. Pearton, and R.A. Stall, Appl. Phys. Lett. 68, 1945 (1996).

96. H.H. Tan, J.S. Williams, J. Zou, D.J.H. Cockayne, S.J. Pearton, and R.A. Stall, Appl. Phys. Lett. 69, 2364 (1996).

97. J.C. Zolper, M.H. Crawford, J.S. Williams, H.H. Tan, and R.A. Stall, Conf. Proc. Of Ion Beam Modification of Materials, 1-6, 1996, Albuquerque, NM (in press).

98. S.C. Binari, H.B. Dietrich, G. Kelner, L.B. Rowland, K. Doverspike, D.K. Wickenden, J. Appl. Phys. 78, 3008 91995).

99. S.J. Pearton, C.R. Abernathy, P.W. Wisk, W.S. Hobson, and F. Ren, Appl. Phys. Lett. 63, 1143 (1993).

100. J.C. Zolper, S.J. Pearton, C.R. Abernathy, C.B. Vartuli, Appl. Phys. Lett. 66, 3042 (1995).

101. J.C. Zolper, M. Hagerott-Crawford, S.J. Pearton, C.R. Abernathy, C.B. Vartuli, C. Yuan, and R.A. Stall, J. Electron. Mat. 25, 839 (1996).

Photoelectrochemical etching of GaN

C. Youtsey*, G. Bulman**, and I. Adesida*

* Microelectronics Laboratory and Dept. of Electrical and Computer Engineering, University of Illinois at Urbana-Champaign, Urbana, IL 61801

** CREE Research, Inc, Durham, NC 27713

ABSTRACT

A photoelectrochemical etching process for n-type GaN using KOH solution and broad-area Hg arc lamp illumination is described. Etch rates as high as 320 nm/min are obtained. The etch rate is investigated as a function of light intensity for stirred and unstirred solutions. Preliminary results on etching of $Al_{0.1}Ga_{0.9}N$ layers are reported.

INTRODUCTION

GaN and the related III-V nitrides have rapidly emerged as very promising candidates to form the basis of a variety of new electronic devices, including short-wavelength emitters and detectors and high temperature/high power electronics. Recent breakthroughs in nitride growth techniques have made available for the first time high quality epitaxial films suitable for device fabrication. GaN is distinguished by its high chemical stability, and this characteristic has presented a challenge for the development of effective etching procedures. The most successful etching results to date have been reported using dry etching techniques such as reactive ion etching (RIE), chemically assisted ion beam etching (CAIBE), electron cyclotron resonance (ECR) RIE, inductively coupled plasma (ICP) RIE, and low energy electron enhanced etching (LE4) [1]. In contrast, however, GaN has proven highly resistant to conventional wet etchants, and the highest wet etch rates reported to date for high quality films have not exceeded 20 Å/min at room temperature [2]. Wet etching methods provide an important complement to dry etching techniques due to their simplicity and very low level of damage formation. In addition, wet etchants can often achieve high levels of selectivity between different materials not attainable using dry etching methods. It is therefore important to explore alternative wet etching techniques that may be viable.

Photoelectrochemical (PEC) wet etching of n-type GaN was recently reported by Minsky et al. [3] using KOH solution and HeCd (325 nm) laser illumination. PEC etching is a technique in which photogenerated electron-hole pairs enhance the oxidation and reduction reactions in an electrochemical cell [4,5]. The process described by Minksy et al. achieved etch rates of approximately 400 nm/min, but etching was localized. Lu et al. [6] have reported photo-assisted anodic etching of n-type GaN using a buffered aqueous solution of tartaric acid and ethylene glycol and Hg arc lamp illumination, and etch rates as high as 160 nm/min were obtained. Youtsey et al. [7] have described the use of KOH solution and Hg arc lamp illumination to etch n+ GaN ($n \sim 1 \times 10^{18}$) and unintentionally doped n-type ($n \sim 1 \times 10^{16}$) GaN samples. In this work, we examine the effect of light intensity and stirring of the solution on the GaN etch rate, and obtain etch rates as high as 320 nm/min at room temperature. In addition, preliminary results on etching of $Al_{0.1}Ga_{0.9}N$ films are described.

EXPERIMENT

The samples for this study were grown by metalorganic chemical vapor deposition (MOCVD) using an AlN buffer layer on SiC substrates. The GaN films were doped n-type ($n \sim 1 \times 10^{18}$) and had a thickness of 3.5 μm. The thickness of the $Al_{0.1}Ga_{0.9}N$ films was 1 μm. 100 nm of Ti was deposited by electron-beam evaporation and patterned by metal lift-off. The Ti metal served as an etch mask as well as an electrical contact to the sample. No annealing of the metal contacts was carried out. For the etched profiles shown in this work, the Ti mask was left remaining on the sample.

Figure 1 shows a schematic of the electrochemical cell used in this work. The samples were clipped to a Teflon base using a nickel washer. Electrical contact was made by the nickel

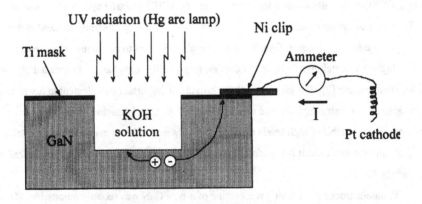

Fig. 1 Photoelectrochemical etching apparatus

washer at only one corner of the GaN sample, and was found sufficient to provide uniform etching (±10%) over the entire surface of the sample for samples as large as 0.5 x 1 cm². A Pt wire was used as the system cathode. An ammeter in series between the sample and Pt cathode was used to monitor the photocurrent flowing through the electrochemical cell. Ultraviolet illumination was provided by a Hg arc lamp, with a typical intensities of 17.4 and 19.6 mW/cm² at wavelengths of 365 and 320 nm, respectively. KOH was used as the electrolyte at a concentration of 0.04 M. All solutions were at room temperature. Etching was conducted with or without a magnetic stirrer, as denoted in the text. Sample characterization was carried out using a Tencor alpha-step profilometer and a Cambridge S360 scanning electron microscope.

RESULTS AND DISCUSSION

Figure 2 shows a plot of the time evolution of the photocurrent flowing through the electrochemical cell during the etching of a GaN sample. The instances at which illumination was begun and stopped are noted on the figure. The total duration for which the sample was illuminated was 30 min. A light intensity of 17.4 mW/cm² @ 365 nm was recorded, as described above. Less than 200 pA of current was observed without sample illumination. The "dark etch rate" was determined to be negligible, as less than 100 Å of material was removed after 12 hours of etching without sample illumination.

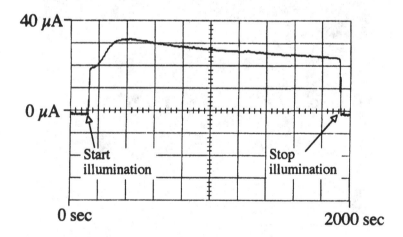

Fig. 2 Time evolution of photocurrent

Figure 3 shows an SEM micrograph of the etch profile for the sample etched in Figure 2. The etch depth is approximately 2.5 μm, indicating an etch rate of 83.3 nm/min. The etched sidewalls are seen to be fairly vertical, with minimal undercutting of the Ti etch mask. The etched surfaces are somewhat rough, which may reflect nonuniform etching resulting from the high level of dislocations present in the material. Auger electron spectroscopy (AES) surface analysis was conducted on the etched surface shown in Figure 3, and indicated that surface concentrations of Ga and N were modified by less 5 % as compared to a similar, unetched surface.

The etch rate of GaN is plotted in Figure 4 as a function of light intensity for both stirred and unstirred solutions. For the case of the stirred solution, the etch rate is seen to be linear with the light intensity for intensities as high as 50 mW/cm^2 @ 365 nm. At this intensity an etch rate of approximately 320 nm/min was recorded. A similar linear dependence of etch rate on light intensity was reported by Minsky et al. and indicates that the etch rate indeed depends on the number of carriers that are photogenerated. The unstirred solution exhibited a saturation in etch rate for intensities greater than approximately 30 mW/cm^2 @ 365 nm. This saturation indicates that the etch rate becomes diffusion limited at higher light intensities. Minsky et al. did not observe a similar saturation effect, but utilized a KOH solution at a significantly higher

Fig. 3 GaN etch profile (2.5 μm etch depth)

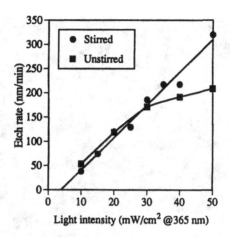

Fig. 4 Etch rate vs. light intensity for stirred and unstirred solutions

concentration.

PEC etching of $Al_{0.1}Ga_{0.9}N$ was conducted under similar conditions to that of the GaN sample described by Figure 2. An SEM micrograph of the etch profile is shown in Figure 5. The etch depth is approximately 1 μm. The etch rate is estimated at 33 nm/min, or approximately 40% of the corresponding GaN etch rate. The AlGaN layer has a bandgap approximately 8 % larger than that of the GaN, and is likely to absorb a significantly lower fraction of the 365 nm photons than the GaN film. The reduced etch rate is therefore believed to be at least in part a consequence of the reduced number of photogenerated carriers available for etching. It should be observed that careful filtering of the wavelengths incident upon the surface of GaN/AlGaN heterostructure may be a useful method of achieving bandgap-selective etching.

SUMMARY

In summary, a photoelectrochemical etching process for n-type GaN using KOH solution and Hg arc lamp illumination has been described. Etch rates as high as 320 nm/min are obtained. The etch rate has been found to depend linearly on the light intensity for stirred solutions, but saturates for unstirred solutions, indicating a diffusion limited process. The process provides high etch anisotropy, but etched surfaces tend to be somewhat rough. AES analysis of the etched surfaces indicates minimal modification of the original surface stoichiometry. Preliminary results are reported for PEC etching of $Al_{0.1}Ga_{0.9}N$ layers.

Fig. 5 $Al_{0.1}Ga_{0.9}N$ etch profile (1 μm etch depth)

REFERENCES

[1] H.P. Gillis, D.A. Choutov, and K.P. Martin, J. Mater., 48, p. 50 (1996).

[2] S.J. Pearton, C.R. Abernathy, F. Ren, J.R. Lothian, P.W. Wisk, and A. Katz, J. Vac. Sci. Technol. A, 11, p. 1772, (1993).

[3] M. Minsky, M. White, and E.L. Hu, Appl. Phys. Lett., 68, p. 1531 (1996).

[4] Ruberto, M.N, X. Zhang, R. Scarmozzino, A.E. Willner, D.V. Podlesnik, and R.M. Osgood, J. Electrochem. Soc., 138, p. 1174 (1991)

[5] R. Khare and E.L. Hu, J. Electrochem. Soc., 138, p. 1516 (1991).

[6] H. Lu, Z. Wu, and I. Bhat, Materials Research Society Fall 1996 Proceedings.

[7] C. Youtsey, G. Bulman, and I. Adesida, Electron. Lett., 33, p. 245 (1997).

PATTERNING OF GaN IN HIGH-DENSITY Cl$_2$- AND BCl$_3$-BASED PLASMAS

[a]R. J. Shul, [a]R. D. Briggs, [a]J. Han, [b]S. J. Pearton, [b]J. W. Lee, [b]C. B. Vartuli, [c]K. P. Killeen, and [c]M. J. Ludowise

[a]Sandia National Laboratories, Albuquerque, NM 87185-0603, rjshul@sandia.gov
[b]University of Florida, Department of Materials Science and Engineering, Gainesville, FL 32611
[c]Hewlett-Packard Laboratories, Palo Alto, CA 94304

ABSTRACT

Fabrication of group-III nitride electronic and photonic devices relies heavily on the ability to pattern features with anisotropic profiles, smooth surface morphologies, etch rates often exceeding 0.5 μm/min, and a low degree of plasma-induced damage. Patterning these materials has been especially difficult due to their high bond energies and their relatively inert chemical nature as compared to other compound semiconductors. However, high-density plasma etching has been an effective patterning technique due to ion fluxes which are 2 to 4 orders of magnitude higher than conventional RIE systems. GaN etch rates as high as ~1.3 μm/min have been reported in ECR generated ICl plasmas at -150 V dc-bias. In this study, we report high-density GaN etch results for ECR- and ICP-generated plasmas as a function of Cl$_2$- and BCl$_3$-based plasma chemistries.

INTRODUCTION

Interest in GaN and related group-III nitride materials continues to grow as demonstrations of blue, green, and ultraviolet (UV) light emitting diodes (LEDs), blue lasers, and high temperature electronic devices are reported.[1-16] Fabrication of many of theses devices may be attributed to improvements in material growth capabilities. However, enhanced device performance can only be realized with improved process capabilities, including plasma etch. Although many advances in plasma etch technology have transpired, the rapid development of high-performance state-of-the-art devices, sub-0.5 μm features, and complex material structures, including the group-III nitrides, has increased the requirements for etch processes. Plasma etch processes often demand highly anisotropic profiles, high etch rates, smooth morphologies, and low-damage. Etching the group-III nitrides is further complicated by their inert chemical nature and their strong bond energies as compared to other compound semiconductors. GaN has a bond energy of 8.92 eV/atom, InN 7.72 eV/atom, and AlN 11.52 eV/atom as compared to GaAs which has a bond energy of 6.52 eV/atom. Due to these high bond energies and inert chemical behavior, the group-III nitrides resist etching in standard, room temperature wet chemical etchants. Therefore, essentially all device patterning has been accomplished using plasma etching technology. For example, commercially available LEDs from Nichia are fabricated using a Cl$_2$-based RIE to expose the n-layer of the heterostructure.[1, 2, 5] The first GaN-based laser diode was also fabricated using RIE to form the laser facets.[7] Laser facet fabrication is especially dependent upon plasma etch pattern transfer since the majority of epitaxially grown group-III nitrides are on sapphire substrates which inhibits cleaving the sample with reasonable yield.

Perhaps the most significant advancement in plasma etching the group-III nitrides has been the utilization of high-density plasmas. High-density plasma etch systems typically yield higher etch rates with less damage than more conventional reactive ion etch (RIE) systems due to plasma densities which are 2 to 4 orders of magnitude greater and the ability to effectively decouple ion energies and plasma density. Etch profiles also tend to be more anisotropic due to lower process pressures which results in less collisional scattering of the plasma species. Plasma etching of GaN has been reported using several techniques including RIE, electron cyclotron resonance (ECR), inductively coupled plasma (ICP), magnetron reactive ion etch (MIE), and chemically assisted ion beam etching (CAIBE). In this paper, we compare ECR and ICP etch results for GaN in Cl$_2$- and BCl$_3$-based plasmas. Etch rates, profiles, near-surface stoichiometry, and surface and sidewall morphology will be discussed.

EXPERIMENT

The GaN films etched in this study were grown by one of three techniques; metal organic-molecular beam epitaxy (MO-MBE), radio-frequency-MBE (rf-MBE), or metal organic chemical vapor deposition (MOCVD). The ECR plasma reactor used in this study was a load-locked Plasma-Therm SLR 770 etch system with a low profile Astex 4400 ECR source in which the upper magnet was operated at 165 A. Due to the magnetic confinement of electrons within the microwave source (2.45 GHz), high-density plasmas are formed at low pressures with low plasma potentials and ion energies. Highly anisotropic etching can be achieved by superimposing an rf-bias (13.56 MHz) on the sample and employing low pressure conditions ($\leq$5 mTorr) which minimizes ion scattering and lateral etching. With rf-biasing, energetic ions are accelerated from the plasma to the sample with potential for kinetic damage to the surface. Etch gases were introduced through an annular ring into the chamber just below the quartz window. To minimize field divergence and to optimize plasma uniformity and ion density across the chamber, an external secondary collimating magnet was located on the same plane as the sample and was run at 25 A. Plasma uniformity was further enhanced by a series of external permanent rare-earth magnets located between the microwave cavity and the sample.

ICP offers an alternative high-density plasma technique where plasmas are formed in a dielectric vessel encircled by an inductive coil into which rf-power is applied.[17, 18] The electric field produced by the coils in the horizontal plane induces a strong magnetic field in the vertical plane trapping electrons in the center of the chamber and generating a high-density plasma. At low pressures ($\leq$ 20 mTorr), the plasma diffuses from the generation region and drifts to the substrate at relatively low ion energy. Thus, ICP etching is also expected to produce low damage with high etch rates. The ICP reactor used in this study was a load-locked Plasma-Therm SLR 770 with a cylindrical coil configuration and alumina vessel encircled by a three-turn inductive coil into which 2 MHz rf-power was applied. As with ECR etching, anisotropic profiles were obtained by superimposing a rf-bias (13.56 MHz) on the sample to independently control ion energy. Etch gases were introduced through an annular region at the top of the chamber. The general belief is that ICP sources are easier to scale-up than ECR sources and are more economical in terms of cost and power requirements. ICP does not require the electromagnets or waveguiding technology necessary in ECR. Additionally, automatic tuning technology is much more advanced for rf-plasmas than for microwave discharges.

All samples were mounted using vacuum grease on an anodized Al carrier that was clamped to the cathode and cooled with He gas. Samples were patterned using AZ 4330 photoresist. Etch rates were calculated from the depth of etched features measured with a Dektak stylus profilometer after the photoresist was removed with an acetone spray. Each sample was ~1 cm^2 and depth measurements were taken at a minimum of three positions. Standard deviation of the etch depth across the sample was nominally less than ± 10% with run-to-run variation less than ± 10%. The gas phase chemistry for several plasmas was studied using a quadrupole mass spectrometer (QMS) or an optical emission spectrometer (OES). Surface morphology, anisotropy, and sidewall undercutting were evaluated with a scanning electron microscope (SEM). The root-mean-square (rms) surface roughness was quantified using a Digital Instruments Dimension 3000 atomic force microscope (AFM) system operating in tapping mode with Si tips. Auger electron spectroscopy (AES) was used to investigate the near-surface stoichiometry of GaN before and after exposure to several plasmas.

RESULTS AND DISCUSSIONS

Due to their inert chemical nature and high bond strengths, the group-III nitrides resist etching in common compound semiconductor wet chemical etchants at room temperature. Very slow etching of GaN has been reported in hot alkalis or electrolitically in NaOH.[19, 20] InN etch rates of 300-600 Å/min have been reported in aqueous KOH and NaOH at 60°C.[21, 22] For amorphous or polycrystalline AlN, wet chemical etching has been reported in several different solutions including H_3PO_4, HF/H_2O, HNO_3/HF, and dilute NaOH at relatively low rates ($\leq$500 Å/min).[23-29] Mileham et al. have recently compiled etch results for binary group-III nitrides in a series of wet chemical etchants.[30] GaN and InN did not etch in either acid or base solutions below ~80°C. However, single crystal AlN etched in strong KOH- and NaOH-based solutions at room temperature. The etch was strongly dependent upon etchant concentration, temperature, and film quality. Higher etch rates were reported for lower crystalline quality material possibly due to more defects or dangling bonds with which OH⁻ ions in the etch solution can interact.

Plasma etching of GaN has been reported using several dry etch techniques including RIE, ECR, ICP, MIE, and CAIBE. Using RIE, GaN etch rates as high as 650 Å/min have been reported at dc-biases of -400 V.[31-34] Under similar dc-bias conditions, high-density plasmas typically yield higher etch rates than RIE due to ion densities which are 2 to 4 orders of magnitude greater. Etch profiles also tend to be more anisotropic due to lower process pressures which results in less collisional scattering of the plasma species. GaN etch rates have been reported up to 1.3 μm/min at -150 V dc-bias in an ECR,[35-41] 3500 Å/min at -100 V dc-bias in a MIE,[42] 2100 Å/min at -500 V in a CAIBE,[43] and 6875 Å/min at -280 V dc-bias in an ICP.[44] GaN has also been etched using low energy electron enhanced etching (LE4) at ~2500 Å/min in Cl_2 at 100°C and 0 V dc-bias.[45]

Several different plasma chemistries have been used to etch compound semiconductor materials including halogen- and hydrocarbon-based plasmas. Chlorine-based plasmas have been the dominant etch chemistries used due to the higher volatility of the group-III chlorides as compared to the other halogen-based plasmas. Table I shows boiling points for possible GaN etch products etched in halogen- and hydrocarbon-based plasma chemistries. GaN typically etches at much slower rates than GaAs. The high volatilities of the Ga- and the nitrogen-based etch products shown in Table I implies that the etch rates are not limited by desorption of the etch products. However, due to the strong bond energies of the group-III nitrides, the initial bond breaking of the group-III-N bond, which must precede the etch product formation, may be the rate limiting step.[46] Faster GaN etch rates obtained in high-density plasma etch systems as compared to RIE may be explained by a two step process directly related to the plasma flux and the bond breaking step. Initially the high-density plasmas increase the bond breaking mechanism allowing the etch products to form and then produce efficient sputter desorption of the etch products.

Etch Products	Boiling Point (°C)
$GaCl_3$	201
GaF_3	~1000
$GaBr_3$	279
GaI_3	sub 345
$(CH_3)_3Ga$	55.7
NCl_3	<71
NF_3	-129
NBr_3	na
NI_3	explodes
NH_3	-33
N_2	-196
$(CH_3)_3N$	-33

Table 1. Boiling points for possible etch products of GaN etched in halogen- or CH_4/H_2-based plasmas.

Although fast GaN etch rates have been observed in chlorine-based plasmas, the source of reactive Cl as well as the use of additive gases are important. Very often gas mixtures are used in a plasma to increase etch rate, improve anisotropy, increase selectivity, or produce smoother etch morphologies by adjusting the chemical:physical ratio of the etch mechanism. The addition of Ar,

SF$_6$, N$_2$, or H$_2$ can have a significant effect on the etch characteristics of GaN and are discussed in this paper.

Small amounts of Ar are often added to halogen-based plasmas to improve the sputter desorption efficiency of etch products from the substrate surface thus increasing etch rate and anisotropy. GaN etch rates are shown in Figure 1 as a function of %Ar in ICP- and ECR-generated Cl$_2$ plasmas. The plasma etch conditions were 2 mTorr pressure, 30 sccm total flow rate, ~-250 V dc-bias, 25°C substrate temperature, and 500 W ICP source power or 850 W ECR source power. As Ar was added to either ICP or ECR Cl$_2$ discharges, GaN etch rates decreased due to less available reactive Cl in the plasma. When BCl$_3$ was substituted for Cl$_2$, etch rates were relatively constant as function of %Ar and as much as 6 times slower than those obtained in the Cl$_2$/Ar plasma due to lower concentrations of reactive Cl.

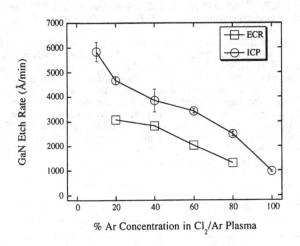

Figure 1. GaN etch rates as a function of %Ar for ECR- and ICP-generated Cl$_2$/Ar plasmas.

In Figures 2 and 3, GaN etch rates are shown for ECR- and ICP-generated Cl$_2$/SF$_6$/Ar and BCl$_3$/SF$_6$/Ar plasmas as a function of %SF$_6$. ECR plasma conditions were; 1 mTorr pressure, 850 W ECR source power, 150 W rf-cathode-power with a corresponding dc-bias of -190 ± 20 V, 25 sccm Cl$_2$/SF$_6$ or BCl$_3$/SF$_6$, and 5 sccm Ar. ICP plasma conditions were; 2 mTorr pressure, 500 W ICP source power, 95 to 115 W rf-cathode-power with a constant dc-bias of -250 ± 10 V, 25 sccm Cl$_2$/SF$_6$ or BCl$_3$/SF$_6$, and 5 sccm Ar. (The pressure increased to ~3 mTorr in the ICP reactor at >60% SF$_6$). In general, as the %SF$_6$ was increased, the etch rates decreased independent of etch technique. As the %SF$_6$ was increased from 0 to 20 in the ECR, the Cl concentration, as determined by QMS (indicated by m/e = 35 peak intensity), decreased but remained significant. Faster GaN etching at 20% SF$_6$ might be expected based on the Cl concentration alone. However, formation of SCl (m/e = 67) was observed at 20% SF$_6$ which may have been responsible for the reduced GaN etch rate due to consumption of the reactive Cl by S. At 30 and 40% SF$_6$, the Cl concentration was greatly reduced and slow GaN etch rates resulted. In Figure 3, GaN etch rates increased up to 20% SF$_6$ in the ICP and 30% SF$_6$ in the ECR and then decreased sharply in both reactors. In the ECR, the Cl concentration (m/e = 35) increased as the SF$_6$ increased to 30% and then decreased at 40%. This implied that at low SF$_6$ concentrations, the SF$_6$ enhanced the dissociation of BCl$_3$ resulting in higher concentrations of reactive Cl and faster etch rates. However above ~30% SF$_6$, the sulfur appeared to consume reactive chlorine as the SCl concentration increased and the etch rate decreased.

GaN etch rates were also obtained for Cl$_2$/N$_2$/Ar and BCl$_3$/N$_2$/Ar plasmas under the following ECR and ICP conditions; 2 mTorr pressure, 850 W ECR source power or 500 W ICP source

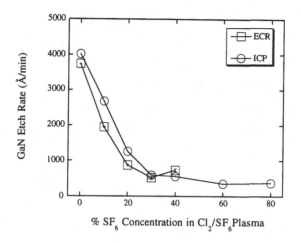

Figure 2. GaN etch rates as a function of %SF$_6$ for ECR- and ICP-generated Cl$_2$/SF$_6$/Ar plasmas.

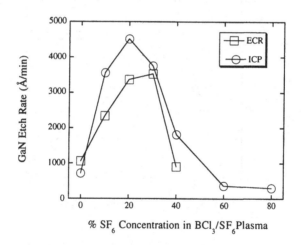

Figure 3. GaN etch rates as a function of %SF$_6$ for ECR- and ICP-generated BCl$_3$/SF$_6$/Ar plasmas.

power, 110 to 170 W rf-cathode-power with a constant dc-bias of -200 ± 10 V, 25 sccm Cl$_2$/N$_2$ or BCl$_3$/N$_2$, and 5 sccm Ar. Figure 4 shows GaN etch rates as a function of %N$_2$ concentration in both ECR and ICP Cl$_2$ plasmas. As the %N$_2$ increased in the Cl$_2$ plasma, the GaN etch rates decreased due to less available reactive Cl. However, as shown in Figure 5, GaN etch rates increased significantly as N$_2$ was added to the ECR- and ICP-generated BCl$_3$ plasma. Etch rates increased up to 40% N$_2$ and then decreased at higher N$_2$ concentrations. This trend was similar to

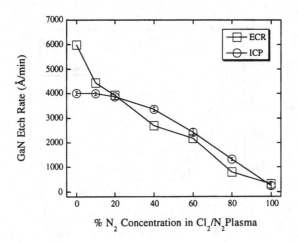

Figure 4. GaN etch rates as a function of %N₂ for ECR- and ICP-generated Cl₂/N₂/Ar plasmas.

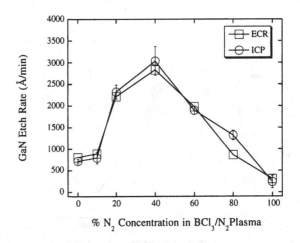

Figure 5. GaN etch rates as a function of %N₂ for ECR- and ICP-generated BCl₃/N₂/Ar plasmas.

that observed in ECR and ICP etching of GaAs, GaP, and In-containing materials.[47-49] Ren *et al.* observed peak etch rates for In-containing materials in an ECR plasma at 75% BCl₃- 25% N₂. As N₂ was added to the BCl₃ plasma, Ren observed maximum emission intensity for atomic and molecular Cl at 75% BCl₃ using OES. Correspondingly, the BCl₃ intensity decreased and a BN emission line appeared. It was suggested that at 75% BCl₃, N₂ enhanced the dissociation of BCl₃ resulting in higher concentrations of reactive Cl and Cl ions and higher etch rates. This may explain faster GaN etch rates observed as N₂ was added to the BCl₃/Ar plasmas in this study, however higher concentrations of Cl and BN emission were not observed using OES in the ICP.

In Figure 6, the rms roughness is plotted as a function of %N_2 for the ECR and ICP BCl$_3$ plasmas. The rms roughness for the as-grown GaN samples etched in the ECR was 1.53 ± 0.06 nm and 3.5 ± 0.2 nm for samples etched in the ICP. The rms roughness for GaN etched in the ECR remained relatively constant (< 2.5 nm) independent of %N_2. The rms-roughness for samples etched in the ICP were slightly higher with the roughest surface (~3.75 nm) at 20% N_2. Figure 7 shows the AFM outputs for GaN etched in ECR-generated BCl$_3$/N_2 plasmas at 100% BCl$_3$, 20% N_2/BCl$_3$, and 60% N_2/BCl$_3$.

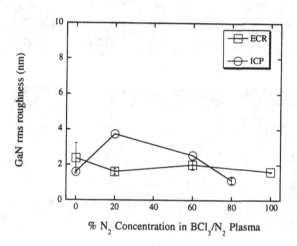

Figure 6. GaN rms-roughness as a function of %N_2 for ECR- and ICP-generated BCl$_3$/N_2/Ar plasmas.

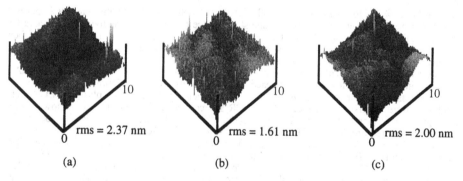

| (a) | (b) | (c) |

Figure 7. GaN rms-roughness outputs for ECR-generated a) BCl$_3$ b) 20% N_2/BCl$_3$, and c) 60% N_2/BCl$_3$ plasmas. The rms roughness was a) 2.37 ± 0.86 nm, b) 1.61 ± 0.17 nm, and c) 2.00 ± 0.27 nm.

In Figure 8, GaN etch profiles are shown as a function of %N_2 in BCl$_3$/N_2/Ar ECR- and ICP-generated plasmas. The etched surfaces were relatively smooth independent of %N_2 and etch technique. However, for the ECR-generated 60% N_2/BCl$_3$ plasma (Figure 8c), significant micromasking was observed possibly due to redeposition. The most anisotropic profiles were

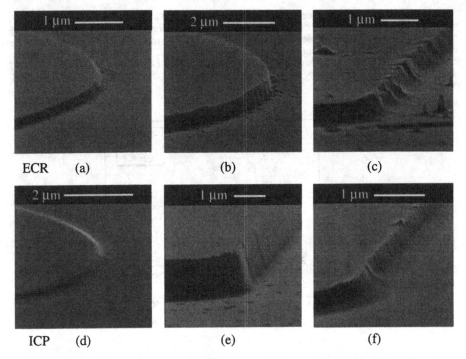

ECR   (a)   (b)   (c)

ICP   (d)   (e)   (f)

Figure 8. SEM micrographs of GaN samples etched in either an ECR or ICP $BCl_3/N_2/Ar$ plasma at (a, d) 0% N_2, (b, e) 20% N_2, and (c, f) 60% N_2. The photoresist mask has been removed.

observed at 20% N_2 for both the ECR and ICP (Figures 8b and e). In the pure BCl_3 plasma (Figures 8a and d), the etch depths were quite shallow due to the low concentration of reactive Cl and low GaN etch rate. At 20% N_2, the etch depths and rates were much higher due to higher concentrations of reactive Cl. At high N_2 concentrations (Figures 8c and f), the etch profiles were slightly overcut due possibly to breakdown of the mask-edge.

The addition of H_2 to chlorine-based plasmas typically results in slower etch rates for GaAs and GaP since H_2 acts as a scavenger of reactive Cl and forms HCl. In Figure 9, GaN etch rates are plotted as a function of %H_2 for ECR- and ICP-generated $Cl_2/H_2/Ar$ plasmas. ECR etch conditions were; 1 mTorr pressure, 850 W ECR source power, 150 W rf-cathode-power with corresponding dc-biases of -170 to -210 V, 25 sccm Cl_2/H_2, and 5 sccm Ar. ICP etch conditions were; 2 mTorr pressure, 500 W ICP source power, 95 to 115 W rf-cathode-power with a constant dc-bias of -250 ± 10 V, 25 sccm Cl_2/H_2, and 5 sccm Ar. GaN etch rates in the ECR and ICP increased slightly as H_2 was initially added to the Cl_2/Ar plasma (10% H_2). Using QMS in the ECR discharge, the Cl concentration (m/e = 35) remained relatively constant at 10% H_2. As the H_2 concentration was increased further, the Cl concentration decreased and the HCl concentration increased as the GaN etch rates decreased in both plasmas, presumably due to the consumption of reactive Cl by hydrogen.

In Figure 10, BCl_3 was substituted for Cl_2 and was used to etch GaN in both the ECR and ICP reactors. The increase in etch rate observed at 10% H_2 concentration in the ECR-generated BCl_3 plasma correlated with an increase in the reactive Cl concentration as observed by QMS. As the H_2 concentration was increased further, the Cl concentration decreased, the HCl concentration increased, and the GaN etch rates decreased due to the consumption of reactive Cl by hydrogen. In the ICP reactor, GaN etch rates were quite slow and decreased as H_2 was added to the plasma up to 80% where a slight increase was observed.

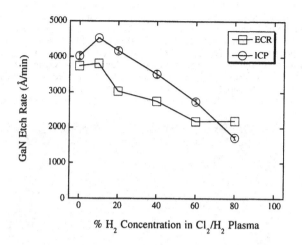

Figure 9. GaN etch rates as a function of %H₂ for ECR- and ICP-generated Cl₂/H₂/Ar plasmas.

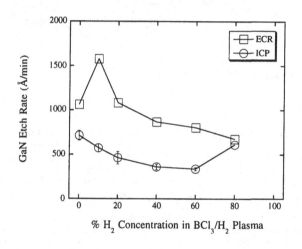

Figure 10. GaN etch rates as a function of %H₂ for ECR- and ICP-generated BCl₃/H₂/Ar plasmas.

Auger spectra for GaN samples etched under several different ICP plasma conditions were taken to determine the near-surface stoichiometry. The Auger spectrum for the as-grown GaN showed normal amounts of adventitious carbon and native oxide on the GaN surface. Following exposure to either a Cl_2- or BCl_3-based plasma, Cl and/or B were observed on the surface. The Ga:N ratios for samples exposed to a plasma were normalized to the ratio of the as-grown sample. For GaN samples etched in $Cl_2/H_2/Ar$ or $Cl_2/N_2/Ar$ plasmas, the normalized Ga:N ratio ranged from ~0.9 to 1.25. The ratios approached 1 at 60% H_2 or N_2 implying equi-rate removal of Ga and N. For the BCl_3-based plasmas, the normalized Ga:N ratios decreased at 20% H_2 and N_2 and then increased at 60%. At 20% N_2/BCl_3, the surface was N-rich, whereas all other surfaces were

Ga-rich. GaN surfaces exposed to BCl₃-based plasmas showed consistently higher ratios as compared to Cl₂-based plasmas due to the more physical nature of the etch mechanisms.

Figure 11 shows a SEM micrograph of GaN etched in a Cl₂/H₂/Ar ICP-generated plasma. The plasma conditions were; 5 mTorr pressure, 500 W ICP power, 22.5 sccm Cl₂, 2.5 sccm H₂, 5 sccm Ar, 25°C temperature, and a dc-bias of -280 ± 10V. Under these conditions the GaN etch rate was ~6880 Å/min with highly anisotropic, smooth sidewalls. The vertical striations observed in the sidewall were due to striations in the photoresist mask which were transferred into the GaN feature during the etch. The sapphire substrate was exposed during the overetch period and showed significant pitting possibly due to defects in the substrate or growth process.

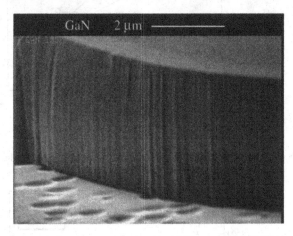

Figure 11. SEM micrograph of MOCVD GaN etched in an ICP-generated Cl₂/H₂/Ar plasma.

CONCLUSIONS

In summary, the utilization of high-density ECR and ICP chlorine-based plasmas has resulted in high rate (> 0.5 μm/min), smooth, anisotropic etching of GaN. The source of reactive Cl (Cl₂, BCl₃, ICl, etc.) and the use of additive gases (Ar, H₂, N₂, SF₆, etc.) have several effects on the etch characteristics of GaN. Cl₂-based plasmas typically resulted in faster GaN etch rates as compared to BCl₃-based plasmas due to higher concentrations of reactive Cl. The addition of Ar, SF₆, N₂, or H₂ to Cl₂- or BCl₃-based plasmas appeared to effect the concentration of reactive Cl in the plasma which directly correlated to the GaN etch rate. GaN etch rate trends were quite similar for ICP and ECR reactors independent of plasma chemistry. Very smooth pattern transfer was obtained over a wide range of plasma chemistries. The mechanism of breaking the GaN bonds appears to be critical and perhaps the rate limiting step in the etch mechanism. The use of high-density plasmas resulted in improved GaN etch results possibly due to a two step process directly related to the plasma flux. Initially the high-density plasmas increase the bond breaking mechanism allowing the etch products to form and then produce efficient sputter desorption of the etch products. ICP etching of the group-III nitrides in Cl₂/H₂ plasmas resulted in etch profiles and sidewall smoothness which may improve the yield and performance of etched facet lasers.

ACKNOWLEDGMENTS

The authors would like to thank P. L. Glarborg and L. Griego for their technical support. This work was performed at Sandia National Laboratories supported by the U.S. Department of Energy under contract # DE-AC04-94AL85000. Sandia is a multiprogram laboratory operated by Sandia Corporation, a Lockheed Martin Company, for the United States Department of Energy.

REFERENCES

1. S. Nakamura, T. Mukai, and M. Senoh, Jpn. J. Appl. Phys. **30**, L1998 (1991).
2. S. Nakamura, T. Mukai, M. Senoh, and N. Iwasa, Jpn. J. Appl. Phys. **31**, L139 (1992).
3. J. S. Foresi and T. D. Moustakas, Appl. Phys. Lett. **62**, 2859 (1993).
4. S. C. Binari, L. B. Rowland, W. Kruppa, G. Kelner, K. Doverspike, and D. K. Gaskill, Electron. Lett. **30**, 1248 (1994).
5. S. Nakamura, T. Mukai, and M. Senoh, Appl. Phys. Lett. **64**, 1687 (1994).
6. S. Nakamura, M. Senoh, N. Iwasa, and S. Nagahama, Jpn. J. Appl. Phys. **34**, L797 (1995).
7. S. Nakamura, M. Senoh, S. Nagahama, N. Iwasa, T. Yamada, T. Matsushito, H. Kiyoku, and U. Sugimoto, Jap. J. Appl. Phys. **35**, L74 (1996).
8. T. Matsuoka, T. Sasaki, and A. Katsui, Optoelectronic Devices and Technologies **5**, 53 (1990).
9. H. Amano, M. Kito, K. Hiramatsu, and I. Akasuki, Jpn. J. Appl. Phys. **28**, L2112 (1989).
10. S. Strite and H. Morkoc, J. Vac. Sci. Technol. B **10**, 1237 (1992).
11. M. A. Kahn, J. N. Kuzina, J. M. Van Hove, D. T. Olson, S. Krishnankutty, and R. M. Kolbas, Appl. Phys. Lett. **58**, 526 (1991).
12. M. A. Khan, A. Bhattarai, J. N. Kuznia, and D. T. Olson, Appl. Phys. Lett. **63**, 1214 (1993).
13. R. F. Davis. Proc. IEEE **79**, 702 (1991).
14. I. Akasaki, H. Amano, M. Kito, and K. Kiramatsu, Lumin. **48/49**, 666 (1991).
15. M. A. Khan, Q. Chen, M. S. Shur, B. T. Dermott, J. A. Higgins, J. Burm, W. Schaff, and L. F. Eastman, Electron. Lett. **32**, 257 (1996).
16. J. C. Zolper, R. J Shul, A. G. Baca, R. G. Wilson, S. J. Pearton, and R. A. Stall, Appl. Phys. Lett. **68**, 2273 (1996).
17. see for example, *High-density Plasma Sources,* ed. O. A. Popov (Noyes Publications, Park Ridge, NJ, 1996).
18. M. A. Lieberonan and R. A. Gottscho, in *Plasma Sources for Thin Film Deposition and Etching,* ed. M. H. Francombe and J. L. Vossen, Physics of Thin Films Vol. 18 (Academic Press, San Diego, 1994).
19. T. L. Chu, J. Electrochem. Soc. **119**, 1200 (1971).
20. J. I. Pankove, J. Electrochem, Soc. **119**, 1118 (1972).
21. Q. X. Guo, O. Kato, and A. Yoshida, J. Electrochem. Soc. **139**, 2008 (1992).
22. S. J. Pearton, C. R. Abernathy, F. Ren, J. R. Lothian, P. W. Wisk, and A. Katz, J. Vac. Sci. Technol. **A11**, 1772 (1993).
23. T Y. Sheng, Z. Q. Yu, and G. J. Collins, Appl. Phys. Lett. **52**, 576 (1988).
24. T. Pauleau, J. Electrochem. Soc. **129**, 1045 (1982).
25. K. M. Taylor, and C. Lenie, J. Electrochem. Soc. **107**, 308 (1960).
26. G. Long and L. M. Fuster, J. Am. Ceram. Soc. **42**, 53 (1959).
27. N. J. Barrett, J. D. Grange, B. J. Sealy, and K. G. Stephens, J. Appl. Phys. **57**, 5470 (1985).
28. C. R. Aita and C. J. Gawlak, J. Vac. Sci. Technol. **A1**, 403 (1983).
29. G. R. Kline and K. M. Lakin, Appl. Phys. Lett. **43**, 750 (1983).
30. J. R. Mileham, S. J. Pearton, C. R. Abernathy, J. D. MacKenzie, R. J. Shul, and S. P. Kilcoyne, J. Vac. Sci. Technol. **A14**, 836 (1996).
31. I. Adesida, A. Mahajan, E. Andideh, M. Asif Khan, D. T. Olsen, and J. N. Kuznia, Appl. Phys. Lett. **63**, 2777.
32. M. E. Lin, Z. F. Zan, Z. Ma. L. H. Allen, and H. Morkoc, Appl. Phys. Lett. **64**, 887 (1994).
33. A. T. Ping, I. Adesida, M. Asif Khan, and J. N. Kuznia, Electron. Lett. **30**, 1895 (1994).
34. H. Lee, D. B. Oberman, and J. S. Harris, Jr., J. Electron. Mat. **25**, 835 (1996).
35. C. B. Vartuli, S. J. Pearton, J. W. Lee, J. Hong, J. D. MacKenzie, C. R. Abernathy, and R. J. Shul, Appl. Phys. Lett. **69**, 1426 (1996).
36. C. B. Vartuli, J. D. MacKenzie, J. W. Lee, C. R. Abernathy, S. J. Pearton, and R. J. Shul, J. Appl. Phys. **80**, 3705 (1996).
37. S. J. Pearton, C. R. Abernathy, and F. Ren, Appl. Phys. Lett. **64**, 2294 (1994).

38. S. J. Pearton, C. R. Abernathy, and F. Ren, Appl. Phys. Lett. **64**, 3643 (1994).
39. R. J. Shul, S. P Kilcoyne, M. Hagerott Crawford, J. E. Parmeter, C. B. Vartuli, C. R. Abernathy, and S. J. Pearton, Appl. Phys. Lett. **66**, 1761 (1995).
40. L. Zhang, J. Ramer, J. Brown, K. Zheng, L. F. Lester, and S. D. Hersee, Appl. Phys. Lett. **68**, 367 (1996).
41. R. J. Shul, A. J. Howard, S. J. Pearton, C. R. Abernathy, C. B. Vartuli, P. A. Barnes, and M. J. Bozack, J. Vac. Sci. Technol. **B13**, 2016 (1995).
42. G. F. McLane, L. Casas, S. J. Pearton, and C. R. Abernathy, Appl. Phys. Lett. **66**, 3328 (1995).
43. I. Adesida, A. T. Ping, C. Youtsey, T. Dow, M. Asif Khan, D. T. Olson, and J. N. Kuzina, Appl. Phys. Lett **65**, 889 (1994).
44. R. J. Shul, G. B. McClellan, S. A. Casalnuovo, D. J. Rieger, S. J. Pearton, C. Constantine, C. Barratt, R. F. Karlicek, Jr., C. Tran, and M. Schurman, Appl. Phys. Lett. **69**, 1119 (1996).
45. H. P. Gillis, D. A. Choutov, and K. P. Marlin, JOM, 50 (1996).
46. S. J. Pearton and R. J. Shul, "III-Nitrides", Academic Press, in press.
47. F. Ren, J. R. Lothian, J. M. Kuo, W. S. Hobson, J. Lopata, J. A. Caballero, S. J. Pearton, and M. W. Cole, J. Vac. Sci. Technol. **B14**, 1 (1995).
48. F. Ren, W. S. Hobson, J. R. Lothian, J. Lopata, J. A. Caballero, S. J. Pearton, and M. W. Cole, Appl. Phys. Lett. **67**, 2497 (1995).
49. R. J. Shul, G. B. McClellan, R. D. Briggs, D. J. Rieger, S. J. Pearton, C. R. Abernathy, J. W. Lee, C. Constantine, and C. Barratt, J. Vac. Sci. and Technol. **A**, in press, (1996).

ETCH CHARACTERISTICS OF GaN USING INDUCTIVELY COUPLED
Cl₂/HBr and Cl₂/Ar PLASMAS

H. S. Kim*, Y. J. Lee*, Y. H. Lee*, J.W. Lee**, M. C. Yoo**, T.I. Kim** , and G. Y. Yeom*
*Department of Materials Engineering, Sung Kyun Kwan University, Suwon, 440-746, Korea
**Photonics Lab., Samsung Advanced Institute of Technology, Suwon, 440-600, Korea

ABSTRACT

To fabricate GaN-based optoelectronic devices successfully, a reproducible etch process with high etch rates, vertical etch profile, and damage-free surface is required. In this study, GaN etching was performed using planar inductively coupled plasmas and the effects of process parameters such as inductive power, bias voltage, pressures and, gas combination on the characteristics of the plasmas were studied using Langmuir probe, optical emission spectroscope (OES), and quadrupole mass spectrometer (QMS). Gas combinations of Cl₂/Ar and Cl₂/HBr were used to etch GaN. GaN etch rates increased with the increase of chlorine radical and ion energy (bias voltage), and the etch rates close to 4000 Å/min could be obtained without substrate heating over 100 ℃. The addition of HBr and Ar (more than 20%) generally reduced GaN etch rates. In our experimental condition, it appears that the chemical reaction between Cl and Ga in GaN affects more significantly to GaN etching compared to physical sputtering, and it was partially confirmed by the data measured by Langmuir probe and OES. Angle resolved XPS data showed the variation of surface Ga/N ratio depending on the process parameters, which influence the formation of low resistance ohmic contact.

INTRODUCTION

III-nitride semiconductors such as GaN have great potential for the fabrication of opto-electronic devices such as light emitting diodes and laser diode, and electronic devices operating at high temperatures. Despite some materials problems, high crystalline quality GaN thin films have been grown by various techniques such as metalorganic chemical vapor deposition (MOCVD) and molecular beam epitaxy (MBE). Recently, dry etching techniques using high density plasmas or chemically assisted ion beam have been employed to define device features with controlled profiles and etch depths[1-6]. The first III-nitride based laser diode was fabricated using chlorine-based reactive ion etching(RIE) to form laser facets but the facets were not smooth enough and the roughness of the facet surfaces was approximately 500 Å[7]. Also, even though GaN LEDs are fabricated commercially, some technical problems related to the GaN etching still remain and reproducible etching processes with high etch rate and vertical etch profile are required to fabricate GaN optoelectronic devices successfully.

In this study, planar inductively coupled Cl₂/Ar and Cl₂/HBr plasmas were used to etch GaN and the effects of process parameters such as inductive rf power, bias voltage, gas combination, and substrate temperature on the etch characteristics of GaN were studied. Also, the characteristics of chlorine-based plasmas and the variations in surface composition of GaN depending on the process parameters were also studied to investigate GaN etch mechanism.

EXPERIMENT

The 2 μm-thick GaN epitaxial layer was grown by MOCVD on the (0001) sapphire wafer and photoresist (PR) or SiO$_2$ were used as a mask layer. To generate inductively coupled plasmas, 13.56MHz rf power was applied to a planar spiral Cu coil separated by a 1cm-thick quartz window located on the top of the process chamber after the introduction of process gas combinations. Separate 13.56MHz rf power was also applied to the bottom electrode, where the substrates are located, to generate dc self- bias voltages. Inductive power was varied from 200 to 600 Watts, dc self-bias voltage from 0 to 120volts, and substrate temperature from 3 to 70 ℃. The dc-self bias voltage was measured using a high voltage probe(Tektronix P6015A). Cl$_2$ was used as the main etch gas to etch GaN and HBr or Ar was used as additive gas while the operating pressure was kept less than 20 mTorr. Etch rates were calculated from the depth of etched features measured with a Detak stylus profilometer after removing the photoresist mask.

A single Langmuir probe, quadrupole mass spectrometry (QMS : Balzers QMG/E 125), and optical emission spectroscopy (OES: SC Tech. PCM402) were used to characterize chlorine-based inductively coupled plasmas. A single Langmuir probe was inserted in the center of the chamber to estimate ion densities of the plasmas and was biased at -40 volts to collect ion currents. Because it is difficult to measure ion density for molecular gases, ion current density was used as a measure of ion density. During the etch processes, QMS and OES were used to monitor dissociated radicals and etch products such as Ga, N, and GaCl$_x$ in the plasmas. QMS was located at the sidewall of the process chamber. Optical emission was also measured from the sidewall viewport of the chamber using a fused silica lense focused on the GaN surface being etched and Ar actinometry was used to measure the relative radical densities of reactive gas[8]. For Ar actinometry, small amount of Ar(5%) was added to the plasmas and the emission intensities of 774nm from chlorine atoms was normalized by the emission intensities of 707.4nm from Ar. Surface composition of the GaN was investigated using angle-resolved XPS (Fisons Instruments Surface Systems ESCALAB 220i) before or after plasma etching.

RESULTS AND DISSCUSSION

GaN is not spontaneously etched in chlorine-containing gases at temperatures below 200 ℃. Therefore, highly energetic ion bombardment is required to break the strong chemical bonding of GaN [5]. Figure 1 shows GaN etch rates and its related plasma parameters as a function of inductive power for -60 volts and -120 volts of bias voltages at 25 ℃ of substrate temperature in the pure Cl$_2$, 60Cl$_2$/40Ar, and 60Cl$_2$/40HBr plasmas. As shown in Figure 1(a), GaN etch rates increased with inductive power possibly due to the increase of ion density and the reactive radical in the plasma as shown in Figure 1(b) and (c). As the dc bias voltage increased from -60 volts to -120 volts in the fixed inductive power of 400 Watts, GaN etch rate also increased. The increase of the GaN etch rates by the increase of bias voltage was ascribed to the increased energetic ion bombardment because the changes of the ion density and Cl radical by bias voltage were negligible(not shown). On the other hand, GaN etch selectivities over mask layers decreased with the increase of inductive power and bias voltage for all the plasma chemistries used in the experiment. The etch selectivity over photoresist was below 3:1 due to the higher ion density and energetic ion bombardment, therefore SiO$_2$ was used as the mask material for the most of GaN etching experiment.

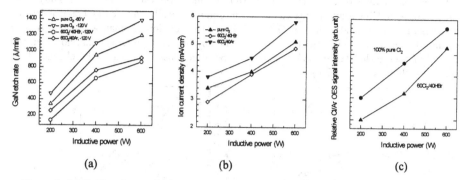

(a)   (b)   (c)

Figure 1. (a) GaN etch rate, (b) ion current density, and (c) relative Cl/Ar OES signal intensity as a function of inductive power from 200Watts to 600Watts in the Cl_2, $60Cl_2/40Ar$, and $60Cl_2/40HBr$ plasmas at 25 °C of the substrate temperature and -120 volts of bias voltage.

To investigate the effects of gas combination on the etch rate of GaN, Ar or HBr gas was added to Cl_2 at a fixed pressure of 10 mTorr and at 400Watts of inductive power. The effects of HBr addition to Cl_2 are shown in Figure 2(a) for GaN etch rates and Figure 2(b) for the measured ion current densities and radical densities. As shown in Figure 2, the addition of HBr to Cl_2 plasmas reduced GaN etch rates significantly and also reduced ion current density and Cl radical density. However, the reduction of ion current density with the increase of HBr percent was small compared to the reduction of Cl radical. Therefore, the significant reduction of GaN etch rates with the increase of HBr appears to be more related to the reduction of Cl radical density. Also, Br radicals from HBr addition appear not to affect GaN etch rates significantly due to the lower chemical reactivity with GaN compared to Cl radicals in our low temperature experimental conditions.

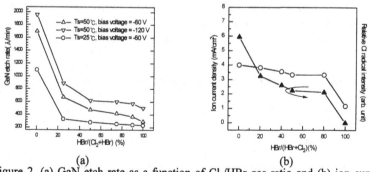

(a)   (b)

Figure 2. (a) GaN etch rate as a function of Cl_2/HBr gas ratio and (b) ion current density and relative Cl radical intensity in the Cl_2/HBr plasmas at the substrate temperature of 25 °C, 400 Watts of inductive power, and -120 volts of bias voltage.

Figure 3 shows the effects of Ar addition on GaN etch rates (a) and on ion current density and Cl radical density (b). The addition of Ar more than 20% also decreased GaN etch rates, however the addition of small amount of Ar(10%) increased GaN etch rates slightly. The addition of Ar to Cl₂ plasmas increased ion densities monotonically. In case of Cl radical density, it showed a similar trend as that of GaN etch rates. The addition of small amount of Ar can enhance the removal of etch products such as GaCl$_x$ by physical sputter etching, however, the change of GaN etch rates at 10%Ar appears to be more closely related to the change of Cl radicals as shown in Figure 3(b).

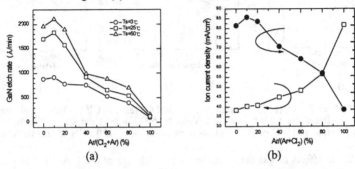

(a)   (b)

Figure. 3. (a) GaN etch rate as a function of Cl₂/Ar gas ratio at the substrate temperature from 3 ℃ to 50 ℃, and (b) ion current density and relative Cl radical intensity in the Cl₂/Ar plasmas at the substrate temperature of 25 ℃, 400 Watts of inductive power, and -120 volts of bias voltage.

Therefore, to obtain high GaN etch rates, not only sputter etching process by energetic ion bombardment but also chemical reaction forming volatile etching products is required in the GaN etching. The importance of chemical reaction in GaN etching can be also understand from the change of GaN etch rates as a function of substrate temperature as shown in Figure 4. GaN etch rates increased with the increase of the substrate temperature as shown Figure 2 and 3. For 100% Cl₂ plasmas, GaN etch rate increased linearly with the increase of substrate temperature and the etch rate was close to 4000 Å/min at 600Watts of inductive power and -120 volts of bias voltage when the substrate temperature increased to 70 ℃

Etch byproducts during the GaN etching were monitored using QMS and OES. 2 inch-blank GaN/sapphire wafers were used to increase the signals. In case of OES, the emissions from peaks such as Ga (402nm), GaCl(337nm), and N₂(358 nm) were monitored. These optical emission peaks were confirmed by the separate measurements of GaCl and Ga peaks by the GaAs etching and from N₂ peaks by N₂ plasmas. Figure 5 shows the relative optical emission intensities of the species as a function of bias voltage during the GaN etching using 100%Cl₂ and 100% Ar plasmas. The emission signals were normalized to the respective signals from no-bias conditions. As shown in Figure 5, the increase of optical emission intensities as a function of bias voltage was much higher for pure Cl₂ plasmas compared to that for pure Ar plasmas. Because the pure Ar plasma contains more ion densities and less Cl radical densities compared to the pure Cl₂ plasmas as shown in Figure 3(b), the increase of the etch byproducts is possibly related to the increased chemical reactions between Cl and GaN due to the increased GaN reactivity by the high ion energy bombardment.

Etch byproducts were also monitored using QMS while etching GaN at 400 Watts of inductive power and -120volts of bias voltage. The species related to GaN etch byproducts such

370

as Ga, N, N_2, and $GaCl_x$ (x = 1~3) were observed similar to OES experiment in Figure 5. Due to the mass range limit (200 amu) of our instrument, $GaCl_4$ can not be monitored, however, among the $GaCl_x$ (x=1~3) $GaCl_2$ (140.5 amu) was the most abundant species. As Ar or HBr is added to chlorine plasmas, the amounts of $GaCl_x$ were generally decreased with the increase of added gas percentages.

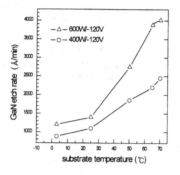

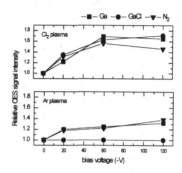

Figure 4. GaN etch rates as a function of substrate temperature for pure Cl_2 plasmas

Figure 5. Relative optical emission intensities at the temperature of 25 ℃ and 400 Watts of inductive power.

Angle resolved XPS analysis has been carried to analyze the variation in Ga/N ratio of the etched GaN surface and is shown in Figure 6. The take-off angle was varied from 30° to 90°. The higher take-off angle represents the detection of XPS signals from the deeper position of the etched GaN surface. As shown in Figure 6, as it gets closer to the surface, nitrogen-rich GaN is obtained and the GaN surface etched by pure Cl_2 shows the most nitrogen-rich surface. The addition of Ar or HBr improved the stoichiometry of the surface. The deviation from the stoichiometry could affect the resistance of the ohmic contact which is formed on the etched GaN surface.

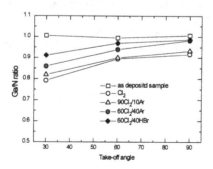

Figure 6. The variation in Ga/N ratio of the GaN surface measured by angle resolved XPS etched at 600Watts of inductive power,-120 volts of bias voltage, and 50 ℃ of substrate temperature.

SUMMARY

Planar inductively coupled Cl_2/Ar and Cl_2/HBr plasmas were used to etch GaN and the effects of process parameters such as inductive rf power, bias voltage, gas combination, and substrate temperature on the etch characteristics of GaN were studied. From the experiment, GaN etch rates close to 4000 Å/min could be obtained without substrate heating over 100 ℃ and GaN etch rate appeared to be more affected by the chemical reaction between Cl radical and GaN compared to the physical sputtering itself. This etch characteristics could be partially confirmed by Langmuir probe, OES, and QMS analysis. Angle resolved XPS analysis shows the dependence of gas combinations on the Ga/N ratio which could affect the formation of low resistance ohmic contact followed by the GaN etching. In general, nitrogen rich GaN surfaces were obtained for our experimental conditions.

ACKNOWLEDGMENTS

The authors wish to thank Dr. H. H. Park and S. H. Sa at Yousei University for XPS analysis and Dr. Y. J. Park and H. S. Park at Samsung Advanced Institute of Technology for growing GaN. This work was supported by Ministry of Information and Communication.

REFERENCES

1. J. C. Zolper and R. J. Shul, MRS BULLETIN/FEBRUARY, p36 (1997).
2. C. B. Vartuli, S. J. Pearton, C. R. Abernathy, R. J. Shul, A. J. Howard, S. P. Kilcoyne, J. E. Parmeter, and M. Hagerott-Crawford, J. Vac. Sci. Technol. A 14, 1011 (1996).
3. J. Pearton, C. B. Vartuli, R. J. Shul, and J. C. Zolper, Materials Sci. Eng. B31, 1 (1995).
4. J. Shul G. B. McClellan, S. J. Pearton, C. R. Abernathy, C. Constantine, and C. Barratt, Electron. Lett. 32, 1408 (1996).
5. T. Ping, A. C. Schmitz, M. Asif Khan, and I. Adesida, J. Electron. Mat. 25, 825 (1996).
6. I. Adesia, A. T. Ping, C. Youtsey, and T. Dow, M. Khan, D. T. Olson, and J. N. Kuznia, Appl. Phys. Lett. 65, 889 (1994).
7. Nakamura, M. Senoh, S. I. Nagahama, N. Iwasa, T. Yamada, T. Matsushita, H. Kiyoku and Y. sugimoto, Jpn. J. Appl. Phys. 35, L74 (1996).
8. M. Donnelly, J. Vac. Sci. Technol. A 14, 1076 (1996).
9. A. T. Ping, I. Adesia, M. Khan, and J. N. Kuznia, Electron. Lett. 30, 1895 (1994).

DRY ETCHING OF GaN USING REACTIVE ION BEAM ETCHING AND CHEMICALLY ASSISTED REACTIVE ION BEAM ETCHING

Jae-Won Lee*, Hyong-Soo Park*, Yong-Jo Park*, Myong-Cheol Yoo*, Tae-Il Kim*
Hyeon-Soo Kim**, Geun-Yong Yeom**
*Photonics Lab, Samsung Advanced Institute of Technology, Suwon P.O. Box 111, Korea,
jaewon@saitgw.sait.samsung.co.kr
**Department of Materials Engineering, Sung Kyun Kwan University, Suwon 440-746, Korea

ABSTRACT

Dry etching characteristics of GaN using reactive ion beam etching (RIBE) were studied. Etching profile, etching rate and etching selectivity to a photoresist (PR) mask were investigated as a function of various etching parameters. Characteristics of chemically assisted reactive ion beam etching (CARIBE) and RIBE were compared at varied mixtures of CH_4 and Cl_2. A highly anisotropic etching profile with a smooth surface was obtained for tilted RIBE with Cl_2 at room temperature. Etching selectivity to a PR was dramatically improved in RIBE and CARIBE when a volume fraction of CH_4 to the mixture of CH_4 and Cl_2 was larger than 0.83.

INTRODUCTION

Properties of GaN such as its wide band gap, chemical inertness, and high-temperature stability have attracted much attention. A great deal of interest has been paid to the development of blue GaN laser diodes (LD) since the first announcement of room temperature lasing.[1] The hexagonal structure of GaN prevents conventional cleaving for mirror facets. And wet etching is not possible due to its chemical inertness. Dry etching was used to form the mirror facets for the first room temperature operating GaN LD. A primitive dry etching study of GaN with conventional reactive ion etching (RIE) did not show good results due to low etching rate and low selectivity to mask material. Consequently almost all efforts to fabricate the mirror facets were focused on the development of dry etching using a high density plasma (HDP) or ion beam. A recent study of HDP RIE including inductively coupled plasma (ICP) RIE or electron cyclotron resonance (ECR) RIE increased the etching rate to more than 6000 Å/min[2,3]. Conventional RIE's etching rate was on the order of 500-1000 Å/min[4]. Ion beam etching techniques, including reactive ion beam etching (RIBE), chemically assisted ion beam etching (CAIBE), and chemically assisted reactive ion beam etching (CARIBE) have been widely used in compound semiconductor devices. CAIBE of GaN with chlorine based gases and a Ni/SiO_2 mask at 300 ℃ showed a vertical etching profile with a high etching rate[5]. But there was no report on RIBE or CARIBE of GaN with substrate tilting.

In this article, we report the dry etching characteristics of GaN using RIBE and CARIBE as a function of various etching parameters including beam condition, tilt angle, substrate temperature, and gas chemistry.

EXPERIMENT

The ion beam etching system used in this study had a 210mm diameter ion source, a Meissner trap and a load-lock chamber. A typical base pressure without activating the trap was 2×10^{-7} Torr. A filamentless radio frequency inductively coupled plasma (RF ICP) ion source with

3 optically aligned Mo grids was used. The beam acceleration voltage, the beam current, and the suppressor voltage were controlled independently. A plasma bridge neutralizer(PBN) was used to prevent beam spreading and to neutralize the positive charge build-up on a non-conductive substrate. Undoped 2~4 μm thick GaN layer grown by metal organic vapor phase epitaxy (MOVPE) on (0001) Al₂O₃ wafer was used for this study. A GaN sample patterned with a conventional PR was loaded on a rotational fixture which could be tilted between -70° and +70° with respect to the incident beam direction. The substrate temperature was varied from 3 °C to 300 °C by helium backside cooling or resister heating. The process pressure was maintained below 5 ×10⁻⁴ Torr throughout the experiment.

In the RIBE of GaN, chlorine based reactive gases were introduced into the ion source and a highly uniform collimated reactive ion beam was extracted through the grids. The total flow rate of the gases was fixed at 8sccm. Etching profile, etching rate, and selectivity to the PR mask were investigated as a function of beam parameters, gas chemistry, tilt angle, and substrate temperature.

Etching characteristics of CARIBE were investigated by varying the volume fraction of methane to the mixture of methane and chlorine gases. Methane was introduced into the ion source while molecular chlorine gas was distributed around the substrate through the nozzles. Etching profile, etching rate, and selectivity to the PR were investigated and compared with RIBE with the same gas mixture. Argon and oxygen plasma cleaning was done between RIBE and CARIBE to avoid potential memory effects.

RESULTS AND DISCUSSION

RIBE of GaN

Fig.1 represents angular dependency of the etching rate in RIBE with chlorine based gas. Ion energy was 500eV and ion beam current density was 0.58mA/cm². The substrate temperature was 3 °C. Fig. 2 shows a typical result of RIBE as a function of ion energy for a fixed ion beam current density of 0.87mA/cm².

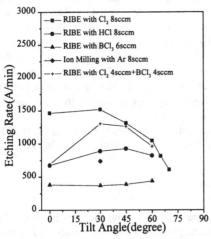

Fig.1 Angular dependency in RIBE

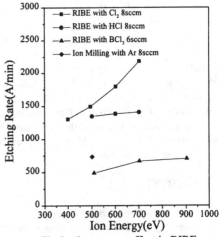

Fig.2 Ion energy effect in RIBE

In Figs. 1 and 2, we can see that the ion beam current density, the ion energy, and the angle of incidence must be optimized to obtain the best etching results, since each material has a different etching rate for the same etching conditions. The etching selectivity to the conventional PR decreased as the tilt angle increased and it was less than 3:1. The surface roughness of the etched base plane, measured by atomic force microscope (AFM), was 10 Å in RMS. A smooth and highly anisotropic profile with high etching rate was obtained in RIBE with Cl_2 at 3 ℃, while low etching rate and rough surface were obtained for BCl_3 or HCl. It was attributed to a difference in generating reactive chlorine ions under this condition. The molecular binding energy of Cl-Cl is 242.58 KJ/mol at 298K, which is much less than B-Cl's 536 KJ/mol or H-Cl's 431.62 KJ/mol at 298K[6]. This means that Cl_2 can be more easily decomposed into atomic chlorine radicals and produce more chlorine ions than BCl_3 or HCl.

Fig. 2 shows the relationship between the ion energy and the etching rate. The ion beam current density was fixed at $0.87mA/cm^2$ and the substrate temperature was 3 ℃. An increase in etching rate with the beam energy reflects an enhanced sputtering effect. This implies that GaN etching is reactant limited.[3] The relatively small increase of etching rate in HCl or BCl_3 is attributed to the limited generation of reactive ions due to the high binding energy.

Fig. 3 shows the etching profiles of RIBE with Cl_2 with respect to the tilt angle. The ion energy was 500eV and the beam current density was $0.58mA/cm^2$. The mask material was conventional non-hard baked PR. A sidewall angle of the PR pattern was about 75° and the substrate temperature was 3 ℃. The etched sidewall profile was undercut when the substrate was normally exposed to the ion beam. We could obtain a vertical profile when the tilt angle was 60°. The profile was double step with a low angle slope around the pattern edge. We define this as the slight slope area. The slight slope area was caused by a self-shadowing effect. A shadowed area was created behind the mask when the mask was obliquely exposed to a highly collimated directional beam. The slight slope area was observed when the tilt angle was higher than 45° and extended as the tilt angle increased further. An overcut profile with a slight slope was obtained when the tilt angle was 65°, even though the PR mask profile was highly undercut as shown in Fig. 3(c). This implies that we can compromise a mask profile effect by adjusting the tilt angle and can control the etching profile.

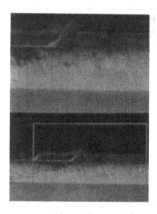 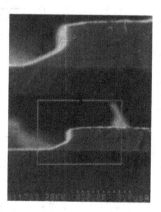 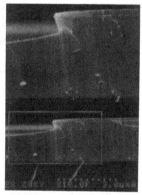

Fig.3 a) RIBE without tilting   b) RIBE with 60° tilting   c) RIBE with 65° tilting
(normal incidence)

Fig.4 shows a linear increase in etching rate with the substrate temperature. When the substrate temperature was 300 ℃, RIBE with HCl or BCl₃ showed a clean and smooth surface as did Cl₂ at 3 ℃. It was attributed to the enhanced desorption rate of volatile products. The etching rate of RIBE with HCl or BCl₃ was still slower than that of Cl₂ at 300 ℃. This result did not agree with that of CAIBE[5]. It indicated a higher etching rate for HCl at 300 ℃, while the etching rate for HCl was lower than that for Cl₂ at RT. This shows a difference between RIBE and CAIBE. RIBE of GaN is the reactant limited process and the major reactant is accelerated chlorine ions. The generation efficiency and acceleration of the reactive ions in RIBE are independent to the substrate temperature. CAIBE uses inert argon ions and molecular gases as the reactants. The reaction of the molecular gases to GaN may differ from each other as the substrate temperature varies.

CARIBE of GaN

CH₄ was introduced into the ICP source to generate artificial polymerization on the PR mask. Cl₂ gas was distributed around the substrate to enhance the desorption rate of volatile by-products. The total gas flow rate was fixed as 12sccm. This CARIBE process showed nearly the same behavior as RIBE for the same mixture with respect to etching rate and selectivity as shown in Fig.5. The etching rate and the selectivity to PR mask were strongly affected by the volume fraction of CH₄ in the total flow of reactive gases. The etching selectivity to PR was constant and was very low for a small CH₄ volume fraction, but suddenly jumped when the CH₄ volume fraction was larger than 0.83.

A scanning electron microscope (SEM) study revealed that a thin polymer layer(Fig.6) formed on the PR mask. As shown in Fig. 6(a), the PR mask remained undamaged after etching when the CH₄ volume fraction was 0.83. This is attributed to the balance between the polymerization of C-H and sputter removal on the mask. Sputter induced chemical desorption occurred on the GaN surface.[7] The thin polymer layers and the PR mask were removed easily by a conventional organic cleaning process and revealed a highly anisotropic profile with a very smooth surface.

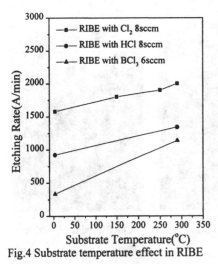

Fig.4 Substrate temperature effect in RIBE

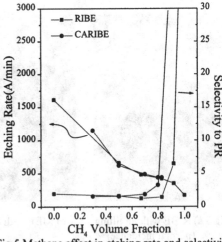

Fig.5 Methane effect in etching rate and selectivity

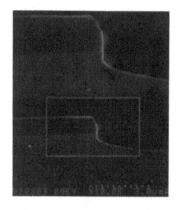

Fig. 6. a) RIBE with CH₄ and Cl₂ at 60° tilting b) CARIBE with CH₄ and Cl₂ at 30° tilting

CONCLUSIONS

Dry etching characteristics of GaN were investigated using RIBE and CARIBE with various beam parameters and gas chemistry. Etched sidewall angle could be controlled by varying etching parameters such as the tilt angle and the substrate temperature. A highly anisotropic etching profile with the smooth surface was obtained by RIBE with chlorine at 3 ℃. Extremely high etching selectivity to PR was obtained when a 0.83 volume fraction of methane was used in RIBE or CARIBE. We believe this high selectivity, low temperature RIBE or CARIBE process with a PR mask can provide a good facet formation of a GaN LD cavity.

REFERENCES

1. S.Nakamura, M.Senoh, S.Nagahama, N.Iwasa, T.Yamada, T.Matsushita, H.Kiyoku and Y. Sugimoto, Jpn. J. Appl. Phys. **35**,L74-L76(1996); Appl. Phys. Lett. **69**(26), 4056-4058(1996)
2. R.J.Shul, G.B.McClellan, S.J.Pearton, C.R.Abernathy, C.Constantine and C.Barratt, Electron. Lett. **32**(15), 1408-1409(1996)
3. C.B.Vartuli, J.D.MacKenzie, J.W.Lee, C.R.Abernathy, S.J.Pearton and R.J.Shul, J. Appl. Phys. **80**(7), 3705-3709(1996)
4. M.E.Lin, Z.F.Fan, L.H.Allen and H.Morkoc, Appl. Phys. Lett. **64**(7), 887-888(1994)
5. A.T.Ping, A.C.Schmitz, M.Asif Khan and I.Adesida, J. Electron. Mater. **25**(5), 825-829(1996)
6. R.C.Weast, CRC Handbook of Chemistry and Physics,68th Edition, F169-F170(1987-1988)
7. B.Humphreys, M.Govett, MIJ-NSR,**1**(28),1-9(1996)

DEVELOPMENT OF GaN AND InGaN GRATINGS BY DRY ETCHING

J. W. LEE[1], J. HONG[1], J. D. MACKENZIE[1], C. R. ABERNATHY[1], S. J. PEARTON[1], F. REN[2] and P. F. SCIORTINO, Jr.[2]

[1]Department of Materials Science and Engineering, University of Florida, Gainesville FL 32611
[2]Lucent Technologies, Bell Laboratories, Murray Hill NJ 07974

ABSTRACT

Sub-micron periodic gratings with pitch ~3,000Å were formed in GaN and InGaN using holographic lithography and room temperature ECR BCl_3/N_2 dry etching at moderate microwave (500W) and rf (100W) powers. The process produces uniform gratings without the need for elevated sample temperatures during the etch step.

INTRODUCTION

The GaN-InGaN laser diodes reported to date have been ridge waveguides in which the facets have been formed by dry etching,[1,2] cleaving[3] or polishing.[4] These index-guided lasers have broad bandwidth and short coherence length, and are ideal for applications such as compact-disk players. Other applications such as communications require single-mode output, and the most common method for producing a resonant cavity for single frequency output is to form a periodic grating,[5-8] either adjacent to the active region through which current flows (distributed feedback laser) or in a passive part of the cavity (distributed Bragg reflector laser). While the latter approach avoids the need for epitaxial regrowth on the grating, distributed feedback lasers are the most readily available form of semiconductor lasers.[9]

In this paper we report on the fabrication of periodic ($\leq 0.2\mu m$) gratings in GaN and InGaN using holographic lithography and Electron Cyclotron Resonance (ECR) BCl_3/N_2 dry etching for pattern transfer. The etch rates for InGaN are found to have a much stronger dependence on temperature than GaN (or AlN), and provides an additional parameter for maximizing etch selectivities in some applications.

EXPERIMENTAL

Single epitaxial layers of GaN, $In_{0.5}Ga_{0.5}N$, AlN and InN grown on c-Al_2O_3 by Metal Organic Molecular Beam Epitaxy in the system described previously [10,11] were used for these experiments. In most laser structures the active region will be InGaN or sometimes GaN, but we decided to investigate the other two binaries as well to better understand the etching differences between the various nitrides. All of the layers were ~0.5μm thick and were nominally undoped. These samples were coated with 450Å thick, imaging photoresist which was patterned by holographic exposure using the wavefront division method.[12,13]

Dry etching was performed in a Plasma-Therm SLR 770 system at substrate temperature between 100-300°C. The microwave power of the ECR source was held constant at 500W, while the rf chuck power (100W) produced a dc self-bias at the sample position of −145V. The gas chemistry employed was BCl_3/N_2 (10/5 standard cubic centimeters per minute gas flow rates), which we have found effective in preserving the surface stoichiometry of dry etched nitrides.[14] The etch rates were established by stylus profilometry on separate sections patterned with SiN_X masks, while the samples etched with the holographic resist masks were characterized by atomic force microscopy (AFM) performed in the tapping mode, and scanning electron microscopy (SEM).

RESULTS AND DISCUSSION

The following facts have been established for dry etching of nitrides :
(i) one of the rate-limiting step is the initial bond-breaking that must precede etch product formation, [15] and thus the high ion fluxes present in ECR, [16,17] magnetron-enhanced reactive ion etching [18] and Inductively Coupled Plasma [19] systems for a given input power or pressure produce much higher etch rates than conventional reactive ion etching.[20-24]
(ii) standard plasma chemistries used for etching of compound semiconductors such as Cl_2/Ar, $CH_4/H_2/Ar$ and related gases work well for the nitrides.
(iii) it is difficult to avoid preferential loss of N, leading to conducting n-type surfaces.[14]
An excellent review of dry etching of the nitrides has recently been published by Gillis et al.[25] While many different plasma chemistries, ion fluxes and ion energy ranges have been investigated, a neglected parameter has been temperature. Shul et al.[16]noted competing reactions occurring in binary nitrides etched in $Cl_2/CH_4/H_2/Ar$ up to 170°C. In this work we decided to investigate a wider temperature range and avoid the complication of the polymer-forming CH_4.
Figure 1 shows the temperature dependence of nitride etch rates at fixed microwave and rf powers. Clearly only the In-containing materials display any significant effect of temperature. In purely chemical etching one expects the etch rate to vary as $AnT^{1/2} \cdot \exp(-E_A/kT)$, where A is a rate

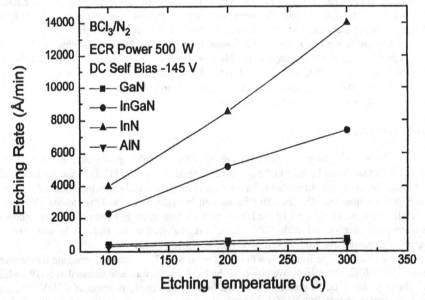

Figure 1. Etch rate of the nitrides as a function of temperature in ECR BCl_3/N_2 discharges.

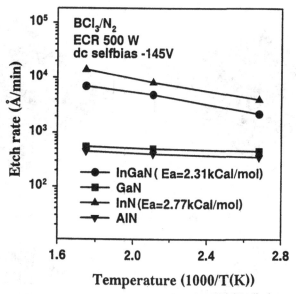

Figure 2. Arrhenius plot of nitride etch rates in ECR BCl₃/N₂ discharges.

constant, n the density of active etching species, T the absolute temperature, E_A the activation energy and k is Boltzmann's constant.[26] In Cl_2-based plasma chemistries, it is well established that the $InCl_3$ etch product with its high latent heat of vaporization (37 kCal/mol versus 12 kCal/mol for $GaCl_3$) is what limits the etch rate of InP, InAs and other III-V semiconductors.[27]

An Arrhenius plot of the data is shown in Figure 2. The activation energies calculated for InGaN and InN are in the range 2.3-2.8 kCal/mol, well below the heat of sublimation of $InCl_3$, and indicate that ion-enhanced reactions are just as important as the kinetics for the purely chemical desorption of $InCl_3$. This is expected at low pressure conditions with the high ion flux present in the ECR reactor. The etch rates of GaN and AlN are almost independent of temperature up to 300°C. Combined with the fact that the etch depth for periodic gratings is small, typically <2,000 Å, and that this is easily obtainable with the resist mask, there is no advantage in complicating the process by employing elevated sample temperatures which would require a much more difficult hard-mask technique.

The gratings were formed in both InGaN and GaN by room temperature etching with BCl₃/N₂ for short periods (30 secs - 2 mins). Figure 3 shows AFM scans of gratings in InGaN after removal of the remaining photoresist with acetone. The pattern transfer is uniform and complete, and the morphology between the gratings measured by a line scan by atomic force microscopy is similar to that of an unetched control sample (root mean square roughness 3.5 nm). Similar results are shown in Figure 4 for gratings formed in GaN. The linewidth variation is ± 8 % over a quarter of a 2 inch diameter wafer, and appears to come from the lithographic process based on measurements before and after etching.

SEM micrographs of the deepest gratings formed in InGaN are shown in Figure 5. The depth is around 2,000 Å, with a pitch of ~3,000 Å. The quality of the gratings (i.e. roughness and

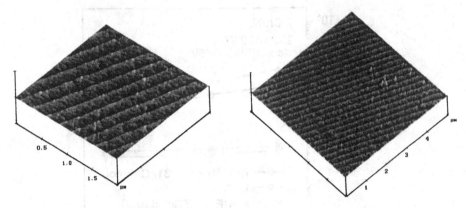

Figure 3. AFM scans of gratings formed in InGaN. The resist has been removed.

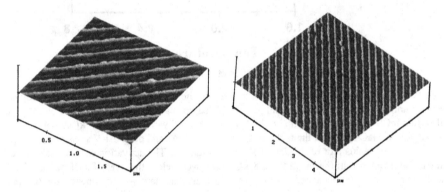

Figure 4. AFM scans of gratings formed in GaN. The resist has been removed.

linewidth uniformity) was basically constant for etch temperatures between 25 - 75 °C. We are currently performing MOMBE regrowth of GaN on such gratings to examine the conditions for overgrowth planarization.

SUMMARY AND CONCLUSION

We have formed sub-micron periodic gratings in GaN and InGaN using holographic lithography and ECR dry etching for pattern transfer. The room temperature etch process produces clean, uniform gratings in these nitride materials.

ACKNOWLEDGMENTS

The work at UF is partially supported by a DARPA grant (A. Husain), administered by AFOSR (G. L. Witt). The continued support of Y. K. Chen at Lucent Technologies is appreciated.

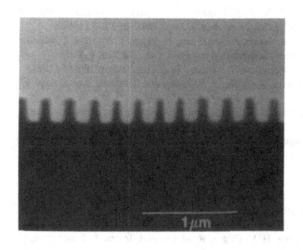

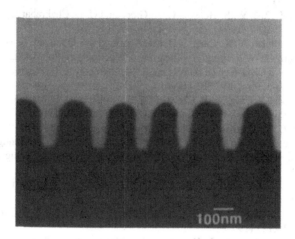

Figure 5. SEM micrographs of gratings formed in InGaN. The resist has been removed.

REFERENCES

1. S. Nakamura, M. Senoh, S. Nagahama, N. Iwasa, T. Yamada, T. Matsushita, H. Kikoyu and Y. Sugimoto, Jap. J. Appl. Phys. **35** L74 (1996).
2. I. Akasaki, S. Sota, H. Sakai, H. Amano, T. Tanaka and M. Koike, Electron. Lett. **32** 1105 (1996).

3. S. Nakamura, M. Senoh, S. Nagahama, N. Iwasa, T. Yamada, T. Matsushita, H. Kikoyu and Y. Sugimoto, Jap. J. Appl. Phys. **35** L217 (1996).

4. S. Nakamura, M. Senoh, S. Nagahama, N. Iwasa, T. Yamada, T. Matsushita, H. Kikoyu and Y. Sugimoto, Appl. Phys. Lett. **68** 2105 (1996).

5. K. Huata, U. Mikami and T. Saitoh, J. Vac. Sci. Technol. B2 45 (1984).

6. J. Yuba, K. Gamo, X. G. He, Y. S. Zhang and S. Namba, Jap. J. Appl. Phys. **22** 1211 (1983).

7. G. P. Agrawal and N. K. Dutta, Long Wavelength Semiconductor Lasers (Van Nostrand, NY 1986).

8. J. Abe, H. Sugimoto, T. Ohnishi, K. Ohtsuka, T. Matsui, H. Yoshiyasu and Y. Nomura, Proc 13th SOTAPOCS Symp., ed, H. Lee (Electrochem. Soc. Pennington NJ 1991).

9. J. Hecht, The Laser Guidebook, 2nd Ed. (McGraw-Hill NY 1992).

10. C. R. Abernathy, Mat. Sci. Eng. R **14** 203 (1995).

11. C. R. Abernathy, J. D. MacKenzie, S. R. Bharatan, K. S. Jones and S. J. Pearton, Appl. Phys. Lett. **66** 1632 (1995).

12. D. V. Podlesnik, H. H. Gilgen, R. M. Osgood, Jr. and A. Sanchez, Appl. Phys. Lett. **43** 1083 (1983).

13. S. J. Pearton, F. Ren, W. S. Hobson, C. A. Green and U. K. Chakrabarti, Semicond. Sci. Technol. 7 1217 (1992).

14. F. Ren, J. R. Lothian, Y. K. Chen, J. D. MacKenzie, S. M. Donovan, C. B. Vartuli, C. R. Abernathy, J. W. Lee and S. J. Pearton, J. Electrochem. Soc. **143** L217 (1996).

15. C. B. Vartuli, J. D. MacKenzie, J. W. Lee, C. R. Abernathy, S. J. Pearton and R. J. Shul, J. Appl Phys. **80** 2331 (1996).

16. R. J. Shul, S. P. Kilcoyne, M. H. Crawford, J. E. Parmeter, C. B. Vartuli, C. R. Abernathy and S. J. Pearton, Appl. Phys. Lett. **66** 1761 (1995).

17 S. J. Pearton, C. R. Abernathy and F. Ren, Appl. Phys. Lett. **64** 2294 (1994).

18. G. F. McLane, S. J. Pearton and C. R. Abernathy, Proc. Symp. Wide Bandgap Semiconductors and Devices, Vol. **95-21** (Electrochemical Society, Penington NJ 1995) p. 204

19. R. J. Shul, G. B. McClellan, S. J. Pearton, C. R. Abernathy, C. Constantine and C. Barratt, Electron. Lett. **32** 1408 (1996).

20. I. Adesida, A. Mahajan, E. Andideh, M. A. Khan, D. T. Olsen and J. N. Kuznia, Appl Phys. Lett. **63** 2777 (1993).

21. M. E. Lin, Z. F. Fan, Z. Ma, L. H. Allen and H. Morkoc, Appl. Phys. Lett. **64** 887 (1994).

22. A. T. Ping, I. Adesida, M. A. Khan and J. N. Kuznia, Electron. Lett. **30** 1895 (1994).

23. H. Lee, D. B. Oberman and J. S. Harris, Jr. Appl. Phys. Lett. **67** 1754 (1995).

24. W. Pletschen, R. Niegurh and K. H. Bachem, Proc. Symp. Wide Bandgap Semiconductors and Devices,, Vol. **95-21** (Electrochem. Soc. Pennington NJ 1995) p.241

25. H. P. Gillis, D. A. Choutov and K. P. Martin, J. O. M. pp 50-55 (1996).

26. D. L. Flamm, in Plasma Etching-An Introduction , ed. D. M. Manos and D. L. Flamm (Academic Press. San Diego 1989).

27. V. M. Donnelly, D. L. Flamm, C. W. Tu and D. E. Ibbottson, J. Electrochem. Soc. **129** 253 (1982).

PLASMA DAMAGE EFFECTS IN InAlN FIELD EFFECT TRANSISTORS

F. REN[1], J. R. LOTHIAN[1], Y. K. CHEN[1], J. D. MACKENZIE[2], S. M. DONOVAN[2], C. R. ABERNATHY[2], C. B. VARTURI[2], J. W. LEE[2], S. J. PEARTON[2] AND R. G. WILSON[3]

[1]Lucent Technologies, Bell Laboratories, Murray Hill, NJ 07974
[2]Department of Materials Science and Engineering, University of Florida, FL 32611
[3]Hughes Research Laboratories, Malibu, CA 90265

ABSTRACT

During gate mesa plasma etching of InN/InAlN field effect transistors the apparent conductivity in the channel can be either increased through three different mechanisms. If hydrogen is part of the plasma chemistry, hydrogen passivation of the shallow donors in the InAlN can occur, we find diffusion depths for ^2H of ≥ 0.5 micron in 30 mins at 200°C. The hydrogen remains in the material until temperatures ≥ 700°C. Energetic ion bombardment in SF_6/O_2 or BCl_3/Ar plasmas also compensates the doping in the InAlN by creation of deep acceptor states. Finally the conductivity of the immediate InAlN surface can be increased by preferential loss of N during BCl_3 plasma etching, leading to poor rectifying contact characteristics when the gate metal is deposited on this etched surface. Careful control of plasma chemistry, ion energy and stoichiometry of the etched surface are necessary for acceptable pinch-off characteristics.

INTRODUCTION

GaN-based electronics offer the promise of high breakdown voltages and high operating temperature. A number of GaN field effect transistors (FETs) and AlGaN/GaN heterostructure FETs have been reported, showing excellent speed performance [1-12]. To date there have been no studies of InAlN alloys as channel layers in electronic devices. It is expected that In-containing alloys will have good carrier transport properties and low specific contact reistance should be obtained by grading to InN which has the lowest bandgap (1.9 eV) of the III-nitrides and is degenerately n-type ($\sim10^{20}$cm^{-3}) when grown by metal organic molecular beam epitaxy (MOMBE) [13-16]. We have previously reported the use of InN as an emitter ohmic contact layer on GaAs/AlGaAs heterojunction bipolar transistor, achieving contact resistivities of 5 x 10^{-7} $\Omega \cdot$cm^2 for nonalloyed TiPtAu on the InN [17]. The $In_XAl_{1-X}N$ grown by MOMBE is also degenerately n-type for x-values above ~ 0.5 [13-16]. The source of the background n-type doping is still undetermined, but may be related to nitrogen vacancies and possibly to other impurities such as O and C. By controlling the In content it is therefore possible to grow a highly n-type InAlN channel layer, topped by a heavily doped contact layer graded to InN.

The III-nitrides are chemically very stable and few wet etching recipes exist. GaN may be etched by molten KOH or NaOH at ≥ 400 °C, while laser enhanced HCl or KOH solutions produce etch rates of a few hundreds of angstroms per minute at room temperature [18]. AlN can also be etched in KOH-based solutions at rates that are strongly dependent on the material quality [19]. Virtually all of the nitride devices reported to date have employed dry etching for pattern transfer. A variery of techniques including reactive ion etching (RIE), chemically assisted ion beam etching (CAIBE), magnetron enhanced etching and electron cyclotron resonance (ECR) etching have been employed [20-23]. Much faster rates are obtained with high ion density conditions because this enhances the initial bond breaking that must precede etch product

385

Mat. Res. Soc. Symp. Proc. Vol. 468 °1997 Materials Research Society

foundation. AlN has a bond energy of 11.5 eV, GaN 8.9 eV and InN 7.7 eV [24], campared to GaAs with only 6.5 eV and thus ion assistance is necessary to produce practical etch rates. The etch studies to date have concentrated on achieving high etch rates, vertical, smooth sidewalls and determining the optimum mask materials for fabrication of laser facets. Much less attention has been paid to etching of shallow gate mesas for transistors, especially those that contain In. In this paper we describe experiments that determine the effect of various ECR plasma treatments on InAlN FET structures.

EXPERIMENTAL

All of the samples were grown by MOMBE on 2" ϕ GaAs substrates using a wavemat ECR N_2 plasma and metalorganic group III precursors (trimethylamine alane, triethylindium) [13-16]. A low temperature (~ 400°C) AlN nucleation layer was followed by a 500 Å thick AlN buffer layer grown at 700°C. The $In_{0.3}Al_{0.7}N$ channel layer (~5x 10^{17} cm^{-3}) was ~500 Å thick and then an ohmic contact layer was produced by grading to pure InN over a distance of ~500 Å.

FET surfaces were fabricated by depositing TiPtAu source/drain ohmic contacts, which were protected by photoresist. The gate mesa was formed by dry etching down to the InAlN channel using an ECR BCl_3 or BCl_3/N_2 plasma chemistry. During this process we noticed that the total conductivity between the ohmic contacts, did not decrease under some conditions. Other potential etch chemistries include CH_4/H_2 and SF_6/O_2. To simulate the effects of these processes we exposed the FET substrates to D_2, O_2 or SF_6 plasmas, in all cases we saw strong reductions in sample conductivity. The incorporation of D_2, O_2 or F_2 into the InAlN was measured by secondary ion mass spectrometry. Changes to the surface stoichiometry were measured by Auger electron spectroscopy (AES). All plasma processes were carried out in a Plasma-Therm SLR 770 system, with an Astex 4400 low profile ECR source operating at 500 W. The samples were clamped to an rf-powered, He-backside cooled chuck, which was left at floating potential (about -30 V relative to the body of the plasma.)

RESULTS AND DISCUSSION

Figure 1 shows the I-V characteristics from an InAlN layer before and after hydrogenation at 200°C for 30 mins. The sample became much more resistive, with no current flowing for

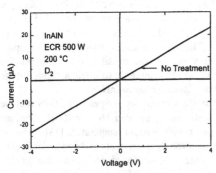

Fig. 1. Current-voltage characteristics for ohmic contacts on as-grown InAlN or on D_2-plasma exposed InAlN.

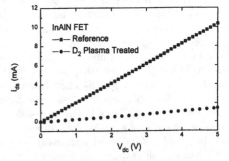

Fig. 2. I_{DS}-V_{DS} relation for as-grown on D_2-plasma treated InAlN FET.

applied voltages of ± 40 V. A similar effect was observed in the full FET structure. Figure 2 shows the drain-source current as a function of applied bias before and after a D_2 plasma treatment at 200°C for 30 mins. The loss of conductivity could be due to two different mechanism. The first is hydrogen forming neutral complexes $(D-H)^0$, with the donors extra electrons taken up in forming a bond with the hydrogen [25]. The second mechanism is creation of deep acceptor states that trap the electrons and remove them from the conduction process. These states might be formed by the energetic D^+ or D_2^+ ion bombardment from the plasma.

The decrease in I_{DS} in the InAlN FET was a strong function of both the ion energy in the plasma and of the active neutral (D^0) and ion density (D^+, D_2^+). Figure 3 shows the dependence of I_{DS} on ECR microwave power. As this power is increased both dissociation of D_2 molecules into atoms and ionization of atomic and molecular species will incrase and thus is difficult to separate out bombardment and passivation effects. At fixed ECR power, the I_{DS} values also decrease with rf power, which controls the ion energy (Fig 4.). At 150 W, the ion energy increases to ~200 eV, with a correpsonding decrease in I_{DS}. This is good evidence that creation of deep traps is playing at least some role and the fact that hydrogen is found to diffuse all the way through the sample also implicates passivation as contributing to the loss of conductivity. Figure 5 shows SIMS depth

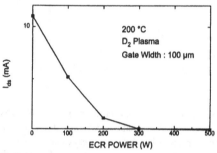

Fig. 3. I_{DS} values at 5 V bias, as a function of the ECR power used during D plasma exposure at 200 °C for 30 mins.

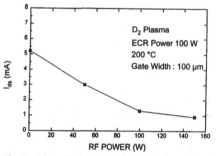

Fig. 4. I_{DS} values at 5 V bias, as a function of the rf power used during D_2 plasma exposure as 200 °C for 30 mins.

profiles of ^2H in the FET after a 500 W D_2 ECR plasma treatment at 250°C for 30 mins. A high concentration of ^2H (~ 10^{20} cm^{-3}) has permeated the entire layer structure under these conditions. Subsequent annealing at 700°C reduces the deuterium concentration by an order of magnitude, by loss of the deuterium to the surface. The fact that the ^2H profile has a plateau shape indicates it is trapped at defects or impurities with this concentration. This is a common phenomenon in highly doped semiconductors containing hydrogen [25]. Thus the results of Figs 4 and 5 lead us to believe that both mechanisms are contributing to the decrease in sample conductivity.

We observed substantial decreases in InAlN conductivity after exposure to SF_6 and O_2 plasmas as well. The background oxygen concentration was too high to allow for a quantitative measure of additional oxygen incorporation as a result of the plasma exposure. However, SIMS profiling showed that F penetration was limited to ≤ 400 Å (Fig 6) and thus could not be the cause of the loss of conductivity. In both of these cases the physical ion bombardment during the plasma exposure is the dominant carrier loss mechanism.

Upon dry etch removal of the InN capping layer, a Pt/Ti/Pt/Au gate contact was deposited on the exposed InAlN to complete the FET processing. If pure BCl_3 was employed as the plasma

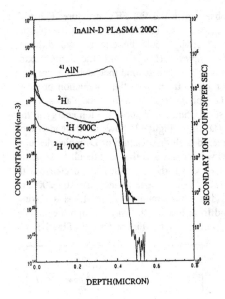

Fig. 5. SIMS profiles of ^{2}H in InAlN treated in deuterium plasma and then annealed at 500 and 700 °C.

Fig. 6. SIMS profiles of F in INAlN treated in ECR or ECR-rf SF$_6$ plasma at 250°C for 10 mins.

chemistry, we observed ohmic and not rectifying behavior for the gate contact. If BCl$_3$/N$_2$ was used, there was some improvement in the gate characteristics. A subsequent attempt at a wet-etch clean-up using either H$_2$O$_2$/HCl or H$_2$O$_2$/HI produced a reverse breakdown in excess of 2 V (Fig 7). These results suggests that the InAlN surface becomes non-stoichiometric during the dry etch step and that addition of N$_2$ retards some of this effect.

Figure 8 shows the I$_{DS}$ values obtained as a function of dry etch time in ECR discharges of either BCl$_3$ or BCl$_3$/N$_2$. In the former case the current does not decrease as material is etched

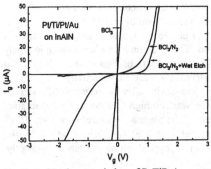
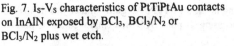

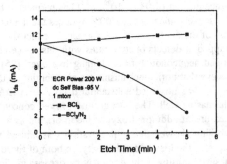

Fig. 7. I$_S$-V$_S$ characteristics of PtTiPtAu contacts on InAlN exposed by BCl$_3$, BCl$_3$/N$_2$ or BCl$_3$/N$_2$ plus wet etch.

Fig. 8. IDS values at 5 V bias for InAlN FETs etched for various times in BCl$_3$ or BCl$_3$/N$_2$.

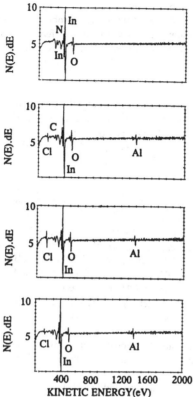

Fig. 9. AES surface scans from InAlN FET structures as grown (top), after BCl₃ etch (second from top), after BCl₃/N₂ etch (second from bottom) or after BCl₃/N₂ plus wet etch (bottom).

away, suggesting that a conducting surface layer is continually being created. By contrast BCl_3/N_2 plasma chemistry does reduce the drain-source current as expected, even though the breakdown characteristics of gate metal deposited on this surface are much poorer than would be expected

AES analysis of the etched surfaces was employed to understand these results. As shown in Fig. 9 the N/In ratio of the near-surface is decreased as a result of the dry etching from 0.35 in the as-grown sample to 0.28 for BCl_3, 0.29 for BCl_3/N_2 amd 0.31 for BCl_3/N_2 plus a wet etch. Thus it appears that nitrogen is preferentially lost from the etched surface, much as is the case with loss of P from InP in Cl_2 or CH_4/H_2 plasma chemistries. The deposition of a gate metal onto this In-rich surface produces ohmic behavior. To overcome this problem in InP it has proven necessary to heat the sample during plasma etching to enhance desorption of the $InCl_3$ etch product [26].

CONCLUSIONS AND SUMMARY

InAlN FET structures are sensitive to several effects during dry etching of the gate mesa. Firstly if hydrogen is present in the plasma there can be passivation of the doping in the InAlN channel layer. Secondly, the ion bombardment from the plasma can create deep acceptor states that compensate the material. Thirdly, even when these problems are avoided through use of H-free plasma chemistries and low ion energies and fluxes, preferential loss of N can produce poor rectifying gate characteristics for metal deposited on the etched surface.

ACKNOWLEGEMENTS

The work at UF and HRL is partially supported by grants from the U.S. Army Research Office (Dr. J. M. Zavada). The work at UF is also supported by the Solid State Microstructures program of NSF (ECS-9522887) and by Division of Materials Research of NSF (DMR-9421109) and by an ARPA grant (A. Husain) administered y AFOSR (G. L. Witt).

REFERENCES

1. S. C. Binari and H. B. Dietrich, in GaN and Related Materials (Edited by S. J. Pearton), Gordon and Breach, New York (1996).
2. T. P. Chow and R. Tayagi, IEEE Trans. Electron. Dev. 41, 1481 (1994).
3. M. S. Shur, A. Khan, B. Gelmont, R. J. Trew and M. W. Shin, In Compound Semiconductors 1994 (Edited by H. Gorokin and U. Mishra), p. 419 IOP Publishing Bristol (1995).
4. M. A. Khan, J. N. Kuznia, A. R. Bhattarai and D. T. Olson, Appl. Phys. Lett. 62 1786 (1993).
5. S. C. Binari, Proc. Electrochem. Soc. 95-21, 136 (1995).
6. S. C. Binari, L. B. Rowland, G. Kelner, W. Kruppa, H. B. Dietrich, K. Doverspike and D. K. Gaskill, Compound Semiconductors 1994 (Edited by H. Gorokin and U. Mishra), p. 459 IOP Publishing Bristol (1995).
7. S. Mohammed, A. A. Salvador and H. Morkoc, Proc. IEEE 83, 1306 (1995).
8. S. C. Binari, L. B. Rowland, W. Kruppa, G. Kelner, K. Doverspike and D. K. Gaskill, Electron. Lett. 30, 1248 (1994).
9. M. A. Khan, J. N. Kuznia, A. R. Bhattarai and D. T. Olson, W. J. Schaff, J. W. Burm and M. S. Shur, Appl. Phys. Lett. 65, 1121 (1995)
10. A. Ozgur, W. Kin, Z. Fan, A. Botchkarev, A. Salvador, S. N. Mohammed, B. Sverdlov and H. Morkoc, Electron. Lett. 31 1389 (1995).
11. J. C. Zolper, R. J. Shul, A. G. Baca, R. G. Wilson, S. J. Pearton and R. A. Stall, Appl. Phys. Lett. 66 (1996).
12. M. A. Khan, Q. Chen, C. J. Sun, J. W. Yang, M. Blasingame, M. S. Shur and H. Park, Appl. Phys. Lett. 68 514 (1996).
13. C. R. Abernathy, J. D. MacKenzie, S. R. Bharatan, K. S. Jones and S. J. Pearton, J. Vac. Sci. Technol. A. 12 843 (1995).
14. C. R. Abernathy, J. D. MacKenzie, S. R. Bharatan, K. S. Jones and S. J. Pearton, Appl Phys. Lett. 66, 1632 (1995).
15. C. R. Abernathy, S. J. Pearton, F. Ren and P. Wisk, J. Vac. Sci. Technol. B 11, 179 (1993).
16. C. R. Abernathy, Mat. Sci. Eng. Rep. R 14, 202 (1995).
17. F. Ren, C. R. Abernathy, S. J. Pearton and P. W. Wisk, Appl. Phys. Lett. 64 1508 (1994).
18. M. S. Minsky, M. White and E. L. Hu, Appl. Phys. Lett. 68 1531 (1996).
19. J. R. Mileham, S. J. Pearton, C. R. Abernathy, J. D. MacKenzie, R. J. Shul and S. P. Kilcoyne, Appl. Phys. Lett. 68, 1531 (1996).

20. S. J. Pearton, C. R. Abernathy, F. Ren, J. R. Lothian, P. Wisk and A. Katz, J. Vac. Sci. Technol. A **11**, 1772 (1993).

21. A. T. Ping, C. Youtsey, I. Adesida, M. A. Khan and J. N. Kuznia, J. Electron. Mater. **24**, 229 (1995).

22. R. J. Shul, in GaN and Related Materials (Edited by S. J. Pearton). Gordon and Breach, New York (1996).

23. H. Lee, D. B. Oberman and J. S. Harris, Appl. Phys. Lett. **67**, 1754 (1995).

24. M. E. Lin, Z. Fan, Z. Ma, L. H. Allen and H. Morkoc, Appl. Phys. Lett. **64**, 887 (1994).

25. I. Adesida, A. Mahajan, E. Andideh, M. A. Kahn, D. T. Olsen and J. N. Kuznia, Appl. Phys. Lett. **63**, 2777 (1994).

26. S. J. Pearton, C. R. Abernathy and F. Ren, Appl. Phys. Lett. **64** 2294 (1994).

ICP DRY ETCHING OF III-V NITRIDES

C.B. Vartuli, J.W. Lee, J.D. MacKenzie, S.M. Donovan, C.R. Abernathy, S.J. Pearton, University of Florida, Dept MS&E, Gainesville Fl; R.J. Shul, Sandia National Laboratories, Albuquerque NM; C. Constantine, C. Barratt, PlasmaTherm IP, St. Petersburg Fl; A. Katz, EPRI, Palo Alto, CA; A.Y.Poyakov, M.Shin, M. Skowronski, Carnegie Mellon University, Dept of MS&E, Pittsburgh. PA.

ABSTRACT

Inductively coupled plasma etching of GaN, AlN, InN, InGaN and InAlN was investigated in $CH_4/H_2/Ar$ plasmas as a function of dc bias, and ICP power. The etch rates were generally quite low, as is common for III-nitrides in CH_4 based chemistries. The etch rates increased with increasing dc bias. At low rf power (150W), the etch rates increased with increasing ICP power, while at 350W rf power, a peak was found between 500 and 750 W ICP power. The etched surfaces were found to be smooth, while selectivities of etch were ≤ 6 for InN over GaN, AlN, InGaN and InAlN under all conditions.

INTRODUCTION

GaN and related compounds are high temperature materials with strong bond strengths and excellent chemical stability. This had made it difficult to develop etch processes for these materials that have high etch rates and selectivities and still produce smooth surfaces and anisotropic features.[1] A number of investigations of dry etching of GaN and related compounds in various chemistries have been done in both reactive ion etching, (RIE),[2-5] and electron cyclotron resonance (ECR) plasma modes.[6-9] ECR plasma etching, with its higher plasma density, has proven much more efficient than RIE etching for the nitrides.[10,11] Inductively coupled plasma (ICP) sources are an alternative method for achieving high density plasmas. The plasma is contained inside the dielectric shield which is surrounded by an inductive coil powered with a 2 MHz rf source.[12] An alternating magnetic field is induced inside the chamber by the oscillating electric field, and this helps to produce a high-density plasma due to confinement of electrons. The plasma diffuses out of the generation region and drifts to the sample position with accompanying low ion energies. The ICP source is generally believed to be easier to scale-up than the ECR, and to have advantages in terms of cost of ownership and availability of truly automatic matching networks for tuning of the discharge. Shul et al. have reported ICP etching of GaN in $Cl_2/H_2/Ar$ plasmas with etch rates to ~ 7000 Å/min.[13,14] We report the results of ICP etching of GaN, AlN, InN, InGaN and InAlN in $CH_4/H_2/Ar$ plasmas as a function of dc bias and ICP power. This chemistry avoids the use of corrosive gases and thus simplifies the process considerably.

EXPERIMENTAL

The GaN, AlN, InN, $In_{0.36}Al_{0.64}N$ and $In_{0.2}Ga_{0.8}N$ samples were grown using Metal Organic Molecular Beam Epitaxy (MO-MBE) on semi-insulating, (100) GaAs and Si substrates in an Intevac Gen II system as described previously.[15,16] The group-III sources were triethylgallium, trimethylamine alane and trimethylindium, respectively, and the atomic nitrogen was derived from an Wavemat ECR source operating at 200 W forward power. The layers were single crystal with a high density of stacking faults and microtwins.

The samples were patterned with a carbon-based mask and were etched in a Plasma-Therm ICP 790 reactor. The temperature of the He back-side cooled chuck was held at 23 °C, and the process pressure at 2 mTorr. The rf power (13.56 MHz) was varied between between 150 and 450 W (dc self biases between 0 and -645V) and the ICP power between 0 and 1500 W. Step heights were obtained from Dektak stylus profilometry measurements after the removal of the mask, and used to calculate the etch rates. The error in these measurements is approximately ± 5 %. The surface morphology of selected GaN samples were examined with Atomic Force

393

Mat. Res. Soc. Symp. Proc. Vol. 468 ° 1997 Materials Research Society

Microscope (AFM) using a Si tip in tapping mode. The selectivity of etch was calculated for InN over GaN, AlN, InGaN and InAlN.

RESULTS AND DISCUSSION

Figure 1 shows the etch rate for GaN, AlN, InN, InGaN and InAlN as a function of dc bias at 500 W ICP power (top) and 1000 W ICP plasmas. The dc bias was higher at lower ICP

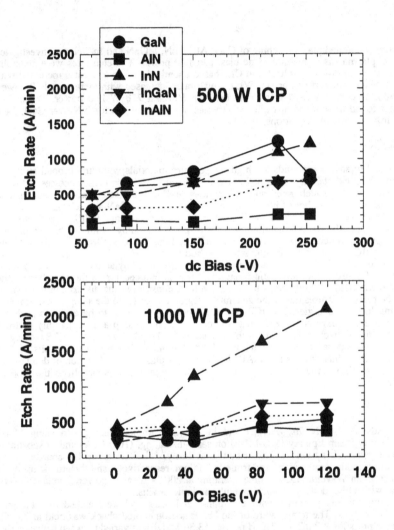

Figure 1. Etch rates for GaN, AlN, InN, InGaN and InAlN as a function of dc bias at 500 W ICP power (top) and 1000 W ICP power (bottom).

powers for the same applied rf power. This is due to the higher plasma density which suppresses the cathode dc self-bias at higher ICP powers. The etch rates were generally quite low ($\leq$ 2000Å/min) for all materials in this chemistry. This is not expected from a consideration of the volatilities of the etch products, (the boiling point of $(CH_3)_3Ga$ is 55.7 °C, $(CH_3)_3In$ is 134 °C, and $(CH_3)_3Al$ is 126 °C),[17] but other factors such as film deposition or formation on the plasma-exposed surface which may reduce the effects of the impinging ions appear to dominate the final etch rate. There was a general upward trend in the etch rates as the dc bias increased. At similar dc biases, the etch rate was in general higher at 1000 W compared to 500 W ICP power, with InN removal approximately twice as fast at 1000 W ICP power.

In Fig. 2, etch rates for GaN, AlN, InN, InGaN and InAlN are shown as a function of ICP power at 150 W rf power (top) and 350 W rf power (bottom) in $CH_4/H_2/Ar$ plasmas at 2 mTorr. The dc bias at each source power is also shown. At 150 W rf power (dc bias ranged from -405V at 0W ICP to -29V at 1500W ICP), the etch rate increased with increasing ICP power. This indicates that the etch was reaction limited under these powers (bottom), with the etch rate increasing with increasing plasma density, irrespective of the decreasing ion energy. At 350 W rf chuck power, the etch rates initially increased rapidly as the ICP power was increased from 0 W to 500 W. The dc bias was higher under these conditions, ranging from -645V at 0W ICP to -58V at 1500W ICP power. At 500 W ICP power the etch rates were up to four times faster at the higher rf power (350 W). The GaN etch rate fell sharply above 500W ICP power, while the etch rates of the In containing materials increased to 750 W ICP power and then fell off. The etch would appear to no longer be reaction-limited above the particular powers at which the etch rates are a maximum. Rather, bond breaking or removal of the reacted products may be limiting the etch, or perhaps the reactive neutrals were being sputter removed before reaction could occur on the semiconductor surface.

Figure 3 shows the RMS roughness as a function of ICP power for GaN at 150 W rf power. The etched surfaces were all smoother than the as-grown sample. This may be due to the fact that sharp features tend to be etched faster due to the angular dependence on ion milling, and as long as there is no preferential loss of nitrogen from the surface, the RMS roughness may decrease. The smoothest surface was found at 500 W ICP power. The low roughness would indicate that there was little preferential loss of the group V species from the surface, though this needs to be verified by Auger Electron Spectroscopy.

Figure 4 shows the selectivities of etch for InN over GaN, AlN, InGaN and InAlN in $CH_4/H_2/Ar$ plasmas as a function of dc bias (top) and ICP power (bottom). As the dc bias increased, the selectivity for InN over AlN increased. The bond strength of InN is much less than that of AlN, (7.7 eV and 11.5 eV respectively),[17] which indicates that the ions were able to break the bonds in the InN material more efficiently as their energy increased, but did not have sufficient energy to efficiently break the bonds in AlN. The selectivity of InN over GaN also rose initially with increasing dc bias, for the same reason. Above -55V however the ions had enough energy to efficiently remove GaN (bond strength of 8.9 eV) as well, lowering the selectivity of etch. The selectivities of InN over InGaN and InAlN were less than 3 under all conditions. Both InN/GaN and InN/AlN showed a maxima in the plot at 1000 W ICP power (Fig. 9 bottom). The dc bias decreased with increasing ICP power. As the plasma density increased, the etch rates of AlN and GaN increased, while the accompanying decrease in ion energy leads to the etch rate of InN remaining approximately constant.

SUMMARY

Inductively coupled plasma etching of GaN, AlN, InN, InGaN and InAlN was examined in $CH_4/H_2/Ar$ plasmas as a function of dc bias and ICP power. The etch rates increased with increasing dc bias. At low rf power (150W), the etch rates increased with increasing ICP power, while at 350W rf power, a peak was found between 500 and 750 W ICP power. The dc bias was found to decrease with increasing ICP power. The selectivities of etch were generally low, ≤ 6 under all conditions for InN over GaN, AlN, InGaN and InAlN. The etched surfaces were found to be smooth under most conditions.

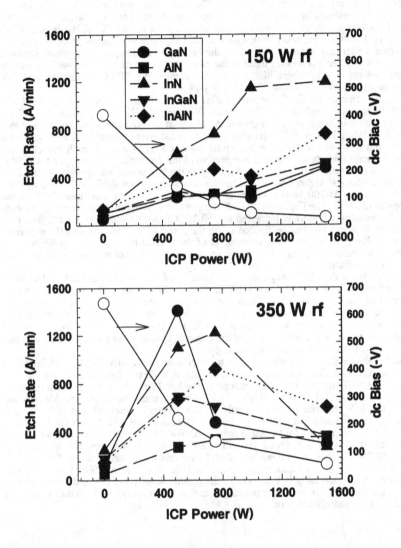

Figure 2. Etch rates for GaN, AlN, InN, InGaN and InAlN as a function of ICP power at 150 W rf power (top) and 350 W rf power (bottom).

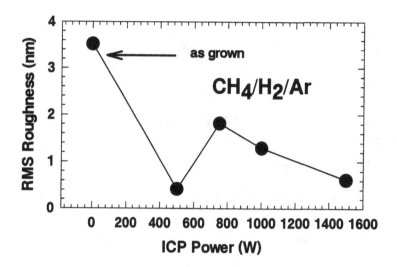

Figure 3. RMS roughness as a function of ICP power for GaN in CH$_4$/H$_2$/Ar plasma at 2 mTorr and 150W rf power.

ACKNOWLEDGMENTS

The authors would like to thank the staff of the Microfabritech Facility for their help with this work. The work at the University of Florida, PTI and SNL is supported by DARPA (A. Husain) through AFOSR (G.L. Witt), an AASERT grant through ARO (Dr. J. M. Zavada), and a University Research Initiative grant #N00014-92-J-1895 administered by ONR. The work at SNL is supported by DOE under contract DEAC04-94AL85000. Sandia is a multiprogram laboratory operated by Sandia Corporation, a Lockheed-Martin Company, for the US Department of Energy.

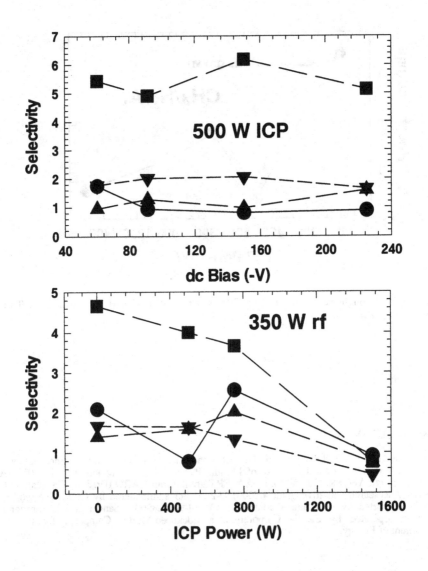

Figure 4. Selectivities of etch for InN over GaN, AlN, InGaN and InAlN in $CH_4/H_2/Ar$ plasmas as a function of dc bias (top) and ICP power (bottom).

REFERENCES

1. H.P. Gillis, D.A. Choutov and K.P. Martin, J.O.M. August 1996 pp 50-55.
2. I. Adesida, A. Mahajan, E. Andideh, M. Asif Khan, D.T. Olsen and J.N. Kuznia, Appl. Phys. Lett. **63** 2777 (1993).
3. M.E. Lin, Z.F. Zan, Z. Ma, L.H. Allen and H. Morkoc, Appl. Phys. Lett. **64** 887 (1994).
4. A.T. Ping, I Adesida, M. Asif Khan and J.N. Kuznia, Electron. Lett. **30** 1895 (1994).
5. H. Lee, D.B. Oberman and J.S. Harris, Jr., Appl. Phys. Lett. **67** 1754 (1995).
6. S.J. Pearton, C.R. Abernathy, F. Ren, J.R. Lothian, P.W. Wisk, A. Katz and C. Constantine, Semicond. Sci. Technol. **8** 310 (1993).
7. S.J. Pearton, C.R. Abernathy and F. Ren, Appl. Phys. Lett. **64** 2294 (1994).
8. L. Zhang, J. Ramer, J. Brown, K. Zheng, L.F. Lester and S.D. Hersee, Mat. Res. Soc. Symp. Proc. Vol **395** 763 (1996); Appl. Phys. Lett. **68** 367 (1996).
9. R.J. Shul, S.P. Kilcoyne, M. Hagerott Crawford, J.E. Parmeter, C.B. Vartuli, C.R. Abernathy and S.J. Pearton, Appl. Phys. Lett. **66** 1761 (1995).
10. R.J. Shul, presented at 189th ECS meeting, Los Angeles CA, May 1996; Electrochem. Soc. Proc. Vol **96-11** 159 (1996).
11. C.B. Vartuli, J.D. MacKenzie, J.W. Lee, C.R. Abernathy, S.J. Pearton and R.J. Shul, J. Appl. Phys. **80** 3264 (1996).
12. High Density Plasma Sources, ed. O.A. Pogov (Noyes Publications, Park Ridge NJ 1996).
13. R.J. Shul, G.B. McClellan, S.A. Casalnuovo, D.J. Rieger, S.J. Pearton, C. Constantine, C. Barratt, R.F. Karlicek, Jr., C. Tran and M. Schurman, Appl. Phys. Lett. **69** 1119 (1996).
14. R.J. Shul, G.B. McClellan, S.J. Pearton, C.R. Abernathy, C. Constantine and C. Barratt, Electon. Lett. **32** 1408 (1996).
15. C.R. Abernathy, J. Vac. Sci. Technol. A **11** 869 (1993).
16. C.R. Abernathy, Mat. Sci. Eng. Rep. **14**, 203 (1995).
17. CRC Handbook of Chemistry and Physics (CRC Press, Boca Raton, FL 1990).

IMPLANTATION ACTIVATION ANNEALING OF Si-IMPLANTED GALLIUM NITRIDE AT TEMPERATURES > 1100 °C

J. C. ZOLPER,[a], J. HAN,[a] R. M. BIEFELD,[a] S. B. VAN DEUSEN,[a] W. R. WAMPLER,[a] S. J. PEARTON,[b] J. S. WILLIAMS,[c] H. H. TAN,[c] R. J. KARLICEK,[d] JR., R. A. STALL[d]

[a] Sandia National Laboratories, Albuquerque, NM 87185-0603,
[b] University of Florida, Department of Materials Science and Engineering, Gainesville, FL 32611,
[c] Dept. of Electronic Materials Engineering, Australian National University, Canberra, 0200, Australia,
[d] Emcore Corp., Somerset, NJ 08873

ABSTRACT

The activation annealing of Si-implanted GaN is reported for temperatures from 1100 to 1400 °C. Although previous work has shown that Si-implanted GaN can be activated by a rapid thermal annealing at ~1100 °C, it was also shown that significant damage remained in the crystal. Therefore, both AlN-encapsulated and uncapped Si-implanted GaN samples were annealed in a metal organic chemical vapor deposition system in a N_2/NH_3 ambient to further assess the annealing process. Electrical Hall characterization shows increases in carrier density and mobility for annealing up to 1300 °C before degrading at 1400 °C due to decomposition of the GaN epilayer. Rutherford backscattering spectra show that the high annealing temperatures reduce the implantation induced damage profile but do not completely restore the as-grown crystallinity.

INTRODUCTION

With the development of GaN-based electronics for high-power and high-temperature operation the reduction of the transistor access resistance becomes a more critical issue (1-4). The two approaches taken to reduce this resistance in other III-V semiconductor transistors are recessed gate designs and self-aligned implanted structures (5). Structures based on selective area implantation may be the preferred approach for GaN-based transistors due to the present difficulty in controllable wet etching of GaN (6).

Although implantation doping of GaN has already been demonstrated, more work is needed to optimize the implant activation annealing process (7-9). In particular, recent studies on the thermal stability of implantation-induced defects in GaN suggest that the annealing temperature must be pushed significantly above 1100 °C (5). In this work, we present structural and electrical data for Si-implanted GaN annealed at temperatures up to 1400 °C. Both electrical and structural data are presented to correlate the effect of the removal of implantation-induced damage to electrical activation of implanted Si donors.

EXPERIMENTAL APPROACH

The GaN layers used in the experiments were ~1.0 μm thick grown on c-plane sapphire substrates by metalorganic chemical vapor deposition (MOCVD) in a multiwafer rotating disk reactor at 1040 °C with a ~20 nm GaN buffer layer grown at 530 °C (10). The

GaN layers were unintentionally doped, with background n-type carrier concentrations $\leq$ 1×10^{16} cm^{-3} as determined by room temperature Hall measurements. When annealed at 1100 °C for 15 s the material maintained its high resistivity. The as-grown layers had featureless surfaces and were transparent with a bandedge luminescence at 356 nm at 4 K. Additional luminescence peaks were observed near 378 and 388 nm. We speculate that these second peaks are due to carbon contamination in the film from the graphite heater in the growth reactor.

Two sets of samples were prepared for implantation and annealing. The first set of samples was encapsulated with 120 nm of sputter deposited AlN (11). The AlN was deposited in an Ar-plasma at 300 W using an Al-target and a 10 sccm flow of N$_2$. The film had an index of 2.1±0.05. Si implantation was performed through the AlN at an energy of 210 keV and a dose of 5×10^{15} cm^{-2}. Monte Carlo TRIM calculations predict that ~7% of the Si-ions come to rest in the AlN film with 4.6×10^{15} cm^{-2} being placed in the GaN (12). The Si peak range from the GaN surface is estimated to be 80 nm. The second set of samples was unencapsulated and implanted with Si at an energy of 100 keV and dose of 5×10^{15} cm^{-2}. This also gives a range from the GaN surface of 80 nm. A sample from each set was annealed under one of four conditions as shown in Table I. The samples annealed in the rapid thermal annealer (RTA) were placed in a SiC coated graphite susceptor and processed in flowing N$_2$. The remaining samples were annealed in a custom built metal organic chemical vapor deposition (MOCVD) system that employed rf-heating with the samples placed on a molybdenum holder on a SiC coated graphite susceptor. The stated temperatures were measured with a Accufiber Model-10 or a Minolta Cyclota-52 pyrometer which were calibrated by the melting of Ge at 934 °C. The pressure in the MOCVD reactor was 630 Torr with gas flows of 4 slm of N$_2$ and 3 slm of NH$_3$. The encapsulated and unencapsulated samples for a given temperature were annealed together.

TABLE I: Summary of annealing conditions.

samples	anneal temperature (°C) /time (s)	reactor	ambient
4,8	1100/15	RTA	N$_2$
3,7	1100/30	MOCVD	N$_2$/NH$_3$
2,6	1300/30	MOCVD	N$_2$/NH$_3$
1,5	1400/30	MOCVD	N$_2$/NH$_3$

Samples were characterized by channeling Rutherford Backscattering (C-RBS) with a 2 meV ^{4}He beam with a spot size of 1 mm^2 at an incident angle of 155°. Aligned spectra are taken with the beam parallel to the c-axis of the GaN film. Random spectra are the average of five off-axis, off-planar orientations. Electrical characterization was done by the Hall technique at room temperature with evaporated Ti/Au contacts deposited on the corners of each sample and annealed at 500 °C for ~15 s. The AlN encapsulated samples were analyzed by C-RBS before having the AlN stripped at ~80 °C in a KOH-based photoresist developer solution (AZ400K) (13,14). This etch has been shown to selectively etch AlN with no measurable GaN etching. The unencapsulated samples had Hall measurements performed prior to C-RBS to avoid any unintentional modification to the electrical characteristics from the He-beam. The surface of the samples was examined by optical and electron microscopy before and after annealing.

RESULTS AND DISCUSSION

Figures 1a and 1b shows the sheet electron concentration and electron Hall mobility versus the annealing conditions for the unencapsulated and AlN encapsulated samples, respectively. An unimplanted sample annealed at 1100 C for 15 s in the RTA remained highly resistive with n << 1x10¹⁵ cm⁻². The mobility of this undoped film could not be reliably measured with the Hall effect due to the low carrier concentration.

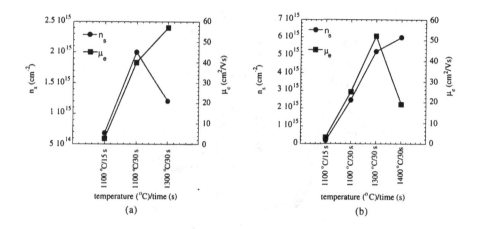

Figure 1. Sheet electron concentration and electron Hall mobility versus annealing treatment for a) unencapsulated and b) AlN encapsulated Si-implanted GaN.

First looking at the data for the unencapsulated samples (Fig 1a), the sample annealed at 1100 °C for 15 s in the RTA has a sheet electron concentration of 6.8×10^{14} cm⁻² or 13.6 % of the implanted dose. This activation percentage is in the range reported for earlier Si-implanted GaN samples annealed in this way (7,8). After the 1100 °C, 30 s MOCVD anneal the number of free electrons goes up to 40% of the implanted dose before decreasing to 24% for the 1300 °C anneal. The decrease for the 1300 °C sample was accompanied by a degradation of the surface of the sample as determined by observation under an optical microscope. (This point will be revisited when discussing the C-RBS spectra later, but it is suspected that the GaN layer has started to decompose during this anneal.) Therefore, the reduction in the electron concentration may be due to loss of material. The Hall mobility increased with increasing thermal treatment and is suggestive of improved crystalline quality.

No data is given in Fig. 1a for the unencapsulated sample annealed at 1400 °C since the GaN film completely sublimed or evaporated during this anneal. This was confirmed by C-RBS data for this sample that showed only the substrate Al and O peaks with a slight Ga surface peak.

Now turning to the data for the AlN encapsulated samples (Fig 1b). There is increasing sheet electron density with increasing thermal treatments for the encapsulated samples including the highest temperature anneal. The RTA sample has a lower activity than the comparable unencapsulated sample (3.6% versus 13.6%), however, all the other AlN encapsulated samples have higher electron concentrations than the comparable unencapsulated sample. The 1300 °C sample has a sheet electron concentration of 5.2×10^{15} cm⁻² that is 113% of the Si dose that should have been retained in the GaN layer (4.6×10^{15} cm⁻²). This may be due to indiffusion of the Si from the AlN encapsulant into the GaN substrate or to the activation of other native donor defects in the GaN layer such as N-vacancies. The sample annealed at 1400 °C has a still higher free electron concentration (6×10^{15} cm⁻²) that may also be partly due to activation of native defects. This sample had visible failures in the AlN layer (cracks and voids) that allowed some degree of decomposition of the GaN layer. This was confirmed by scanning electron microscope (SEM) images that show regions of GaN loss. Therefore, the formation of N-vacancies in this sample is very likely and may contribute to the electron concentration.

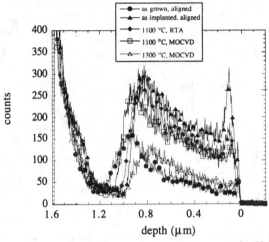

Figure 2: Channeling Rutherford Backscattering (C-RBS) spectra for Si-implanted (100 keV, 5×10^{15} cm^{-2}) GaN either as-implanted or annealed as shown in the legend.

Figure 2 shows a compilation of the C-RBS spectra for the unencapsulated samples along with the aligned spectrum for a unimplanted sample. The as-implanted sample shows the damage peak near 100 nm with an additional peak at the surface. The surface peak may be due to preferential sputtering of the surface during Si-implantation and has been reported in earlier implantation studies (15,16). The sample annealed in the RTA has improved channeling and an apparent reduction in the implant damage. It should be noted, however, that the surface peak is also diminished in the RTA sample and it has previously been reported that the change in the surface peak can account for the apparent reduction in the implantation damage peak (16). Upon annealing at higher temperatures or longer times in the MOCVD reactor the channeling continues to improve and approaches, but does not reach, the unimplanted aligned spectra. The 1300 °C sample, however, appeared to show evidence of material loss as demonstrated by the change in position and abruptness of the substrate signal as well as the observation of surface roughing. Therefore, the reduction in the implantation damage peak in this sample may be due to sublimation or evaporation of the implanted region and not recovery of the original crystal structure. The loss of at least part of the Si-implanted region is consistent will the reduction in the free electron concentration in this sample shown in Fig 1a.

Figure 3 shows a compilation of the C-RBS spectra for the AlN encapsulated samples. The as-implanted sample has a damage peak at ~80 nm with no additional surface peak as seen in the unencapsulated sample. The lack of a surface peak in this sample supports the hypothesis that this peak on the unencapsulated sample is due to preferential sputtering since the GaN surface of the encapsulated sample is protected from sputter loss during implantation. A significant reduction in the implantation-induced damage peak occurs after the RTA anneal with a further reduction with increasing thermal processing. The value for the minimum channeling yield (χ_{min}) for the GaN layer on an AlN encapsulated, unimplanted sample was estimated to be 2.5% while the 1400 °C annealed, implanted sample had a χ_{min} value of

12.6%. The as-implanted samples had χ_{min} values of 38.6% (AlN encapsulated) and 34.1% (unencapsulated). Therefore, the 1400 °C annealed sample demonstrates significant damage recovery but not complete damage removal to the virgin, unimplanted level.

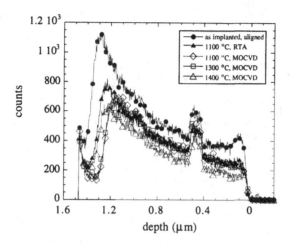

Figure 3: Channeling Rutherford Backscattering (C-RBS) spectra for Si-implanted (210 keV, 5×10^{15} cm^{-2}), GaN encapsulated with 120 nm of AlN either as-implanted or annealed as shown in the legend.

CONCLUSION

The application of ion implantation to GaN-based electronics can be expected to significantly reduced the device access resistance and associated power loss. To optimize the implantation process the activation annealing sequence must be well understood. Electrical data demonstrates that annealing up to 1400 °C increases electrical activity of Si-implanted GaN if the sample can be protected with an AlN encapsulation layer. C-RBS spectra also show that implantation-induced damage can be significantly reduced by using annealing temperatures up to 1400 °C. Further work is needed to improve the thermal stability of the AlN encapsulation; however, once this is accomplished, ion implantation doping of GaN should be a viable technology for high-power and high-temperature electronic devices.

Acknowledgments: The authors gratefully acknowledge the technical support of J. A. Avery at Sandia. Sandia is a multiprogram laboratory operated by Sandia Corporation, a Lockheed Martin Company, for the United States Department of Energy under contract #DE-ACO4-94AL85000. The work at UF is partially supported by a National Science Foundation grant (DMR-9421109) and a University Research Initiative grant from ONR (N00014-92-5-1895). Additional support for the work at Sandia, UF, and EMCORE was provided from DARPA (A. Husain) and administered by AFOSR (G. L. Witt).

References

1. M. A. Khan, A. Bhattarai, J. N. Kuznia, and D. T. Olson, Appl. Phys. Lett. **63**, 1214 (1993).
2. S. C. Binari, L. B. Rowland, W. Kruppa, G. Kelner, K. Doverspike, and D. K. Gaskill, Elect. Lett. **30**, 1248 (1994).
3. N. X. Nguyen, B. P. Keller, S. Keller, Y.-F. Wu, M. Lee, C. Nguyen, S. P. Denbaars, U. K. Mishra, and D. Grider, Electron. Lett. 33, 334 (1997).
4. J. C. Zolper, R. J. Shul, A. G. Baca, R. G. Wilson, S. J. Pearton, and R. A. Stall, Appl. Phys. Lett. **68** 2273 (1996).
5. J. C. Zolper, S. J. Pearton, J. S. Williams, H. H. Tan, and R. A. Stall, Materials Research Society, Fall 1996, Symposium N, vol. 449 (MRS, Pittsburgh, PA, in press).
6. J. C. Zolper and R. J. Shul, MRS Bulletin, 22, 36 (1997).
7. S. J. Pearton, C. R. Abernathy, C. B. Vartuli, J. C. Zolper, C. Yuan, R. A. Stall, Appl. Phys. Lett. **67**, 1435 (1995).
8. J. C. Zolper, M. Hagerott Crawford, S. J. Pearton, C. R. Abernathy, C. B. Vartuli, C. Yuan, and R. A. Stall, J. Electron. Mat. **25** 839 (1996).
9. J. C. Zolper, R. G. Wilson, S. J. Pearton, and R. A. Stall, Appl. Phys. Lett. **68** 1945 (1996).
10. C. Yuan, T. Salagaj, A. Gurary, P. Zawadzki, C. S. Chern, W. Kroll, R. A. Stall, Y. Li, M. Schurman, C.-Y. Hwang, W. E. Mayo, Y. Lu, S. J. Pearton, S. Krishnankutty, and R. M. Kolbas, J. Electrochem. Soc. **142**, L163 (1995).
11. J. C. Zolper, D. J. Rieger, A. G. Baca, S. J. Pearton, J. W. Lee, and R. A. Stall, Appl. Phys. Lett. **69**, 538 (1996).
12. J. F. Ziegler, J. P. Biersack, U. Littmark, *The Stopping and Range of Ions in Solids*, vol. 1, (Pergamon Press, New York, 1985).
13. J. R. Mileham, S. J. Pearton, C. R. Abernathy, J. D. MacKenzie, R. J. Shul, and S. P. Kilcoyne, Appl. Phys. Lett. **67**, 1119 (1995).
14. C. B. Vartuli, S. J. Pearton, J. W. Lee, C. R. Abernathy, J. D. MacKenzie, J. C. Zolper, R. J. Shul, and F. Ren, J. Electrochem. Soc. 143, 3681 (1996).
15. H. H. Tan, J. S. Williams, J. Zou, D. J. H. Cockayne, S. J. Pearton, and R. A. Stall, Appl. Phys. Lett. **69**, 2364 (1996).
16. H. H.Tan, J. S. Williams, J. Zou, D. J. H. Cockayne, S. J. Pearton, and C. Yuan, Proc. 1st Symp. on III-V Nitride Materials and Processes, Electrochemical Society, vol. 96-11, 142 (1996).

RECOVERY OF STRUCTURAL DEFECTS IN GaN
AFTER HEAVY ION IMPLANTATION

C. RONNING, M. DALMER, M. DEICHER, M. RESTLE, M.D. BREMSER*,
R.F. DAVIS*, H. HOFSÄSS
Universität Konstanz, Fakultät für Physik, Postfach 5560, D-78434 Konstanz, Germany
* North Carolina State University, Department of Materials Science and Engineering, Box 7907,
Raleigh, NC, 27695, USA

ABSTRACT

Single crystalline GaN-layers were implanted with radioactive ^{111}In ions. The lattice
location of the ions and the recovery of the implantation induced damage was studied using the
emission channeling technique and perturbed-$\gamma\gamma$-angular-correlation spectroscopy as a function of
the annealing temperature. We find the majority of indium atoms on substitutional sites even
directly after room temperature implantation, but within a heavily disturbed surrounding. During
isochronal annealing treatments in vacuum, a gradual recovery of the implantation damage takes
place between 873 K and 1173 K. After 1173 K annealing about 50 % of the In atoms occupy
substitutional lattice sites with defect free surroundings.

INTRODUCTION

Doping of semiconductors by ion implantation offers advantages in comparison to doping
during film growth or by diffusion. (i) The concentration as well as the lateral and depth
distribution of the dopants are precisely controllable, and (ii) almost all elements can be implanted
with sufficient high purity. However, doping by ion implantation is hampered by the created
radiation damage, which has to be removed by thermal treatment.

In the case of ion implantation into gallium nitride (GaN) there exists about two dozen
studies up to now. Mainly the thermal stability [1-3] and electrical activation [3-8] of the
implanted species were investigated, but little is known about the created radiation damage
[9,10]. Tan et al. [9] found that GaN is extremely resistant to amorphization and takes place only
at very high doses of about $4 \cdot 10^{16}$ cm^{-2}. Below this dose the created disorder consists mainly of
clusters, small dislocation loops, and planar defects. The amount of these defects can be reduced
by implantation at high temperatures or by subsequent annealing procedures, but a complete
recovery was not achieved [10]. To our knowledge nothing is known about the damage created
after heavy (> 70 amu) ion implantation into GaN and its annealing behavior.

In this work we report on emission channeling (EC) and perturbed-$\gamma\gamma$-angular-correlation
(PAC) measurements on GaN implanted with radioactive [111]In. Isoelectronic indium is expected
to occupy substitutional Ga sites in GaN with no attractive Coulomb interaction to the created
defects. Therefore, we have an ideal probe atom to study the annealing behavior of the
implantation damage.

EXPERIMENTAL

1-2 µm thick GaN films were grown on on-axis n-type, Si-face α(6H)-SiC(0001)
substrates at 1273 K at 45 Torr using a vertical, cold-wall, RF inductively heated MOVPE
deposition system [11,12]. A 0.1 µm high-temperature (1373 K) AlN-buffer layer was deposited

407

prior to the GaN growth. Deposition was performed using triethylaluminum (TEA) and triethylgallium (TEG) in combination with 1.5 SLM of ammonia (NH_3) and 3 SLM of H_2 diluent.

Into the single crystalline GaN-layer we have implanted radioactive ^{111}In at room temperature with an ion energy of 100 keV and a dose of $3\cdot10^{13}$ cm^{-2}. TRIM simulations gave a mean ion range of 28 nm (FWHM = 11 nm) and a peak concentration of about 10^{19} In/cm^3 [13].

The lattice location of the implanted ^{111}In has been determined with the emission channeling technique (EC). For the EC-measurements the sample was pre-oriented by Laue X-ray photographs and mounted on a three-axis goniometer in a vacuum chamber. The conversion electrons emitted in the decay of the daughter nucleus [111]Cd were detected with a silicon surface barrier detector. The implantation spot of 3 mm in diameter, the detector diameter of 5 mm and the distance between sample and detector of 360 mm lead to an angular resolution of $\sigma = 0.44°$. The emission channeling spectra were recorded for an angular range of +/- 5° around the c-axis by measuring the emitted electron intensity as a function of the tilt angle. For substitutional emitter atoms channeling effects of electrons are caused by the attractive interaction with the positively charged atom rows or planes, leading to maxima in the emission yield. Details of the EC-technique are described elsewhere [14].

The immediate neighborhood of the implanted ^{111}In and their possible interactions with other defects was monitored with perturbed-$\gamma\gamma$-angular-correlation spectroscopy (PAC) using the emitted γ-radiation. Here the electric field gradient tensor (EFG) at the site of the radioactive probe atom is measured. The EFG causes a hyperfine splitting of an excited state of the [111]Cd daughter nuclei, which is measured by PAC. The EFG is described by the quadrupol coupling constant v_Q and the asymmetry parameter η. These values are characteristic for specific defects nearby the probe atom and for the intrinsic lattice EFG in the case of hexagonal lattices. Usually one plots the time-dependent anisotropy R(t) of the emitted γ-radiation. A unique EFG gives rise to undamped periodic oscillations in the R(t) spectra and an EFG distribution, reflecting different defects surroundings of the probe atoms, a decreasing anisotropy with increasing time. Details of the PAC-technique are described elsewhere [15].

The annealing of the samples has been carried out in vacuum for 10 minutes and the EC- and PAC-measurements were done for annealing temperatures up to 1223 K.

RESULTS AND DISCUSSION

The EC-spectrum measured around the c-axis directly after room temperature implantation of ^{111}In into GaN is shown in fig.1a. Here, the normalized emission yield was obtained after correcting for background caused by Compton electrons and normalization to the off-axis random yield. Clearly visible is a higher conversion electron intensity along the c-axis as well as along 6 planar directions separated by angles of 30°. This spectrum therefore reflects the sixfold symmetry of the c-axis atom rows. The maximum yield in c-axis direction and the sixfold symmetry show that the In-atoms occupy mainly substitutional lattice sites after room temperature implantation. We can not distiguish between Ga- and N-sites from this spectrum, but it is reasonable that indium occupy Ga-sites, because Ga and In are isoelectronic.

For isochronal annealing up to 1073 K, emission channeling measurements were done in steps of 100 K. No increase of the normalized yield along the c-axis was observed as it can be seen in fig.1b in comparison with fig.1a.

In order to obtain a quantitative fraction of In atoms occupying different lattice sites, emission-channeling patterns for the conversion electrons were calculated for substitutional, interstitial, and close-to-substitutional sites. The simulations were carried out within the many-beam formalism based on the dynamical theory of electron diffraction [14,16]. The effects of

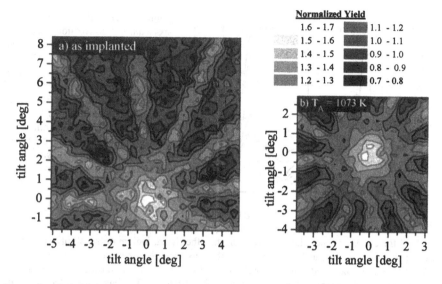

Figure 1: Emission channeling spectra of conversion electrons from the decay of ^{111}In measured for the c-axis direction in GaN (a) after implantation of ^{111}In at room temperature with 100 keV and (b) subsequent annealing at 1073 K for 10 min. in vacuum.

dechanneling and thermal vibrations of the the probe nuclei were taken into account as discussed in Ref. [14]. Corresponding to a Debye temperatures of 314 K (Ga), 745 K (N) and 247 K (In), we used thermal vibration amplitudes of $u_1 = 0.08$ Å, which lies between diamond ($u_1 = 0.044$ Å) and silicon ($u_1 = 0.1$ Å).

In fig.2 the calculated emission yield patterns for (a) interstitial and (b) substitutional indium emitters in GaN are shown. For indium at the threefold interstitial sites in the center of the hexagons, the calculations give weak channeling effects along 3 planes as well as weak blocking

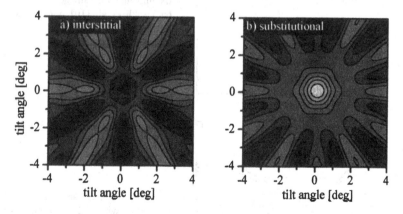

Figure 2: Calculated emission yield patterns for (a) interstitial and (b) substitutional indium emitters in GaN using the same normalization as in fig.1.

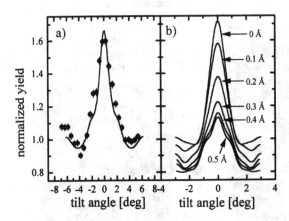

Figure 3: (a) Emission channeling spectrum of conversion electrons from the decay of ^{111}In measured for the c-axis after 1073 K annealing. The solid lines are calculated emission yield distributions (section of fig.2b) fitted to the experimental points (section of fig.1b). (b) Calculated emission channeling profile for the c-axis of GaN. The emitter position is displaced from the c-axis row position in steps of 0.1 Å.

effects along the c-axis and the other 3 planes. On the other hand, indium on substitutional sites gives rise to electron channeling effects along the c-axis and the six planes.

In fig.3a the calculated emission yield was fitted to a section through the measured 2-dimensional distribution around the c-axis of fig.1b (T_A=1073 K). The calculated spectra (solid line) fit well regarding to the maximum emission yield as well as to the width and shape of the spectrum. From the fit we obtain a substitutional fraction of 90 % and a remaining fraction of 10 % including randomly distributed emitter atoms.

In the following we want to consider the accuracy of this analysis. In fig.3b we have plotted calculated emission channeling spectra for emitters which are slightly displaced from the c-axis. A significant reduction of the channeling maxiumum yield occurs even for displacements of only 0.1 Å. A resonable fit to the experimental data can also be obtained assuming all emitters are

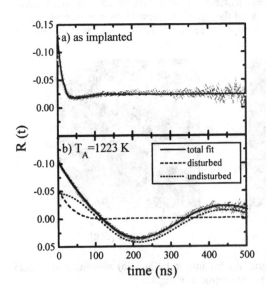

on close-to-substitutional sites, having static displacements of at most 0.05 Å. However, since a complete recovery of the implantation damage is not observed by Parikh et al. [10] nor by the PAC measurements described later on, our original analysis with 90 % In on substitutional sites appears more probable.

Figure 4 shows PAC time spectra for ^{111}In in GaN (a) directly after implantation and (b) after subsequent anneling to 1223 K. The fast drop of anisotropy seen in the R(t)

Figure 4: PAC spectra of GaN implanted with ^{111}In and 100 keV at room temperature (a) and subsequent annealing in vacuum to 1223 K (b).

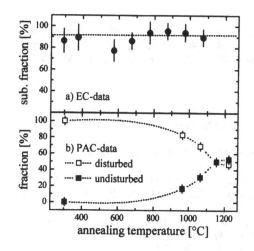

Figure 5: (a) EC- and (b) PAC-results measured on ^{111}In in GaN as a function of annealing temperature

spectrum measured direct after implantation indicates that the probe atoms are exposed to a distribution of different EFGs. This environment is characterized by an average value $\nu_Q = 45$ MHz and an associated distribution width of about $\Delta\nu_Q = 20$ MHz. Accordingly, the majority of the indium atoms is incorporated in highly distorted nonuniform environments after implantation.

After annealing to 1223 K the R(t) spectrum shows that a significant fraction of indium probe atoms occupy a unique lattice site and have the identical surrounding, because only one EFG is observed. From the amplitude of the modulation (dotted line) it is deduced that 53(5) % of the In atoms are at this site. A fit to the experimental data gives a coupling constant of $\nu_Q = 7.5(5)$ MHz and axial symmetry of the EFG tensor ($\eta = 0$). From PAC measurements with different orientations of the c-axis in respect to the γ-detectors, we found that the symmetry of the EFG tensor is parallel to the c-axis. From these and the above EC-results, it is concluded that this weak EFG is seen by indium atoms at substitutional sites with undisturbed surrounding in the non-cubic lattice, in agreement with similar results on CdS [17]. The remaining probe atoms are still exposed to a distribution of different EFGs (dashed line). But from the weaker EFG distribution with an average value of $\nu_Q = 12$ MHz ($\Delta\nu_Q = 5$ MHz) we claim that these are indium atoms on substitutional sites with defects in the second or further nearest neighborhood.

CONCLUSION

In fig.5 we have summarized the EC- and PAC-data as a function of annealing temperature. Direct after room temperature implantation a channeling effect with a maximum normalized yield of 1.6 was measured, i.e. we find 90(10) % indium atoms on substitutional sites, but with PAC we have monitored a heavily disturbed surrounding. For isochronal annealing treatments in vacuum no increase of the maximum normalized yield was observed (fig.5a), since already the majority of indium atoms occupy substitutional lattice sites. A gradual recovery of the implantation damage in the surrounding of the indium atoms takes place between 873 K and 1173 K, which was determined by PAC (dotted lines in fig.4b also include informations of PAC-measurements on GaN grown on Saphir [18]). After 1173 K annealing, 53(5) % of the probe atoms occupy undisturbed substitutional lattice sites. For higher annealing temperatures up to 1473 K no increase of this fraction is expected [18]. The remaining fraction of indium atoms occupy substitutional sites with weakly disturbed surroundings.

This experiment demonstrates that the crystalline structure of GaN is not significantly destroied by the heavy ion implantation and that the created damage can be annealed out to a large extent.

ACKNOWLEDGMENTS

This work was financially supported by the German Bundesminister für Bildung, Wissenschaft, Forschung und Technologie. The work at NCSU was supported by the Office of Naval Research on N00014-92-J-1720 and N00014-92-J-1477 and monitored by Mr. Max. Yoder and Dr. Colin Wood.

REFERENCES

1. R.G. Wilson, S.J. Pearton, C.R. Abernathy, and J.M. Zavada, Appl. Phys. Lett. **66**, 2238 (1995).
2. R.G. Wilson, C.B. Vartuli, S.J. Pearton, C.R. Abernathy, and J.M. Zavada, Sol. St. Elec. **38**, 1329 (1995).
3. J.C. Zolper, R.G. Wilson, S.J. Pearton, and R.A. Stall, Appl. Phys. Lett. **68**, 1945 (1996).
4. S.J. Pearton, C.B. Vartuli, J.C. Zolper, C. Yuan, and R.A. Stall, Appl. Phys. Lett. **67**, 1435 (1995).
5. M. Rubin, N. Newman, J.S. Chan, T.C. Fu, and J.T. Ross, Appl. Phys. Lett. **64**, 64 (1994).
6. J.S. Chan, N.W. Cheung, L. Schloss, E. Jones, W.S. Wong, N. Newman, X. Liu, E.R. Weber, A. Grassman, M.D. Rubin, Appl. Phys. Lett. **68**, 2702 (1996).
7. J.C. Zolper, R.J. Shul, A.G. Baca, R.G. Wilson, S.J. Pearton, and R.A. Stall, Appl. Phys. Lett. **68**, 2273 (1996).
8. A. Pelzmann, S. Strite, A. Dommann, C. Kirchner, M. Kamp, K. J. Ebeling, A. Nazzal, MRS Internet J. Nitride Semicond. Res. **2**, 4 (1997).
9. H.H. Tan, J.S. Williams, J. Zou, D.J.H. Cockayne, S.J. Pearton, and R.A. Stall, Appl. Phys. Lett. **69**, 2364 (1996).
10. N. Parikh, A. Suvkhanov, and M. Lioubtchenko, E. Carlson, M.D. Bremser, D. Bray, J. Hunn, and R.F. Davis, to be published in Nucl. Instr. Instr. & Meth. **B** (1997).
11. T.W. Weeks, Jr., M.D. Bremser, K.S. Ailey, E.P. Carlson, W.G. Perry, R.F. Davis, Appl. Phys. Lett. **67**, 401 (1995).
12. T.W. Weeks, Jr., M.D. Bremser, K.S. Ailey, E.P. Carlson, W.G. Perry, R.F. Davis, J. Mat. Res. **11**, 1011(1996).
13. J.F. Ziegler, J.P. Biersack, and U. Littmark, The stopping and ranges of ions in solids, (Pergamon Press, New York, 1985).
14. H. Hofsäss, Hyp. Int. **97/98**, 247 (1996); Phys. Rep. **201**, 121 (1991).
15. T. Wichert et al., Appl. Phys. **A 48**, 59 (1989); in Hyperfine Interactions of Defects in Semiconductors, ed. by G. Langouche (Elsevier, Amsterdam 1992) p. 77ff.
16. H. Hofsäss, U. Wahl, and S.G. Jahn, Hyp. Int. **84**, 123 (1991).
17. R. Magerle, M. Deicher, U. Desnica, R. Keller, W. Pfeiffer, F. Pleiter, H. Skudlik, T. Wichert, Appl. Surf. Sci. **50**, 159 (1991).
18. A. Burchard, M. Deicher, D. Forkel-Wirth, E.E. Haller, R. Magerle, A. Prospero, and A. Stötzler, to be published and submitted to ICDS 19.

CURRENT TRANSPORT IN W AND WSI$_x$ OHMIC CONTACTS TO INGAN AND INN

C.B. Vartuli, S.J. Pearton, C.R. Abernathy, J.D. MacKenzie, Dept of MSE, University of Florida, Gainesville FL; M.L. Lovejoy, R.J. Shul, J.C. Zolper, Sandia National Laboratories, Albuquerque NM; A.G. Baca Sandia National Laboratories, Compound Semiconductor Materials and Processes, Albuquerque NM; M. Hagerott-Crawford Sandia National Laboratories, Department of Photonics Research, Albuquerque NM; K.A. Jones, Army Research Laboratory, Ft. Monmouth NJ; F. Ren, Bell Laboratories, Lucent Technologies, Murray Hill NJ.

ABSTRACT

The temperature dependence of the specific contact resistance of W and WSi$_{0.44}$ contacts on n$^+$ In$_{0.65}$Ga$_{0.35}$N and InN was measured in the range -50 °C to 125 °C. The results were compared to theoretical values for different conduction mechanisms, to further elucidate the conduction mechanism in these contact structures. The data indicates the conduction mechanism is field emission for these contact schemes for all but as-deposited metal to InN where thermionic emission appears to be the dominant mechanism. The contacts were found to produce low specific resistance ohmic contacts to InGaN at room temperature, $\varrho_c \sim 10^{-7}$ Ω·cm^2 for W and ϱ_c of 4×10^{-7} Ω·cm^2 for WSi$_x$. InN metallized with W produced ohmic contacts with $\varrho_c \sim 10^{-7}$ Ω·cm^2 and $\varrho_c \sim 10^{-6}$ Ω·cm^2 for WSi$_x$ at room temperature.

INTRODUCTION

It has proven difficult to produce low resistance ohmic contacts to the III-nitride materials because of their wide bandgaps.[1-11] To date little work has been done regarding the conduction mechanism in ohmic contacts to the nitrides. We would like to establish a high temperature contact technology for the nitrides, for applications such as electronics capable of operation at ≥ 500 °C, or for power switching. Cole et. al.[7] reported that W produced contact resistivities of ~10^{-4} Ω·cm^2 on n$^+$ GaN, and was stable for annealing temperatures up to ~ 1000 °C. In particular the use of lower bandgap In-containing nitrides should be able to reduce the contact resistance on GaN, in analogy to the situation for In$_x$Ga$_{1-x}$As on GaAs.

In this paper we report the results of W and WSi$_{0.44}$ contacts deposited on n$^+$ In$_{0.65}$Ga$_{0.35}$N and n$^+$ InN. Temperature dependent transmission line measurements (TLM) in the range -50 °C to 125 °C were used to obtain information about the conduction mechanism in these contact structures. Room temperature TLM measurements were also measured as a function of annealing temperature, in order to establish the stability of the contacts. A key feature of using In-based nitrides is the trade-off between contact resistance and thermal stability.

EXPERIMENTAL

The 2000 Å thick InN and InGaN samples were grown using Metal Organic Molecular Beam Epitaxy (MO-MBE) on semi-insulating, (100) GaAs substrates in an Intevac Gen II system as described previously.[12,13] The InN and In$_{0.65}$Ga$_{0.35}$N were highly autodoped n-type (~10^{20} cm^{-3}, and ~ 10^{19} cm^{-3} respectively) due to the presence of the native defects endemic to these materials. The samples were rinsed in H$_2$O:NH$_4$OH (20:1) for 1 min just prior to deposition of the metal to remove native oxides. The metal contacts were sputter deposited to a thickness of 1000Å and then etched in SF$_6$/Ar in a Plasma Therm reactive ion etch (RIE) system to create TLM patterns.[14,15] The nitride samples were subsequently etched in Cl$_2$/CH$_4$/H$_2$/Ar in an Electron Cyclotron Resonance (ECR) etcher to produce the mesas for the TLM patterns.[16] The samples were annealed at temperatures from 300 to 900 °C for 15 sec under a nitrogen ambient in a RTA system (AG-410). Temperature dependent TLM measurements were made over the range -50 °C to 125 °C on the as-deposited and 900 °C (InGaN) and 500 °C (InN) annealed samples. These measurements make it possible to determine the dominant conduction mechanism over the barrier, and the results were compared to theoretical values. The error in

413

these measurements was estimated to be ±10 % due mainly to geometrical contact size effects. The widths of the TLM pattern spacings varied slightly due to processing, (maximum of ± 5 %) as determined by SEM measurements, which were taken into account when calculating the contact resistances.

RESULTS AND DISCUSSION

Figure 1 shows the theoretical curves for contacts to InGaN of this doping level exhibiting thermionic, thermionic field, or field emission as their dominant conduction mechanisms. The curves are shown only to give the expected temperature dependence of ϱ_c and the magnitude of the specific contact resistance is arbitrary. The theoretical values are calculated from[17]

$$\varrho_c \propto \exp(\Phi_b/E_{00}) \quad \text{for field emission} \tag{1}$$

$$\varrho_c \propto \exp[\Phi_b/E_{00}\coth(qE_{00}/kT)] \quad \text{for thermionic field emission} \tag{2}$$

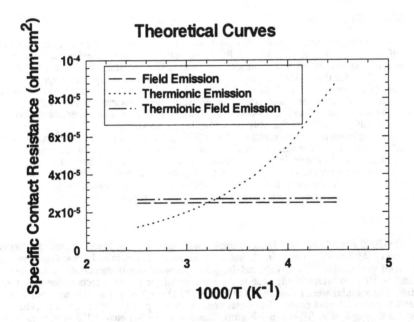

Figure 1. Theoretical curves for the temperature dependence of specific contact resistance of contacts in which thermionic emission, thermionic field emission, or field emission are the dominant conduction mechanism.

$$\varrho_c \ \alpha \ \exp(q\Phi_b/kT) \quad \text{for thermionic emission} \tag{3}$$

where

$$E_{00}= h/4\pi[N_d/m^*\varepsilon_s]^{1/2} \tag{4}$$

with Φ_b being the barrier height, N_d the donor concentration in the semiconductor, m^* the effective mass of electrons in the material and ε_s the permittivity of the semiconductor. For field emission $qE_{00}/kT \gg 1$, for thermionic field emission $qE_{00}/kT \sim 1$, and for thermionic emission $qE_{00}/kT \ll 1$, with $q/kT \cong 0.026$ eV at 300 K. A fixed barrier height (1 eV) was assumed for calculations of the three conduction mechanisms. As values have not been definitively established for m^* and ε_s for all the nitride compounds, the best available values for InN were used, ($m^*= 0.1m_e$ and $\varepsilon_s= 8\varepsilon_o$).[18]

Over the temperature range we studied there was little difference between the slope expected for the theoretical field emission and thermionic field emission plots (Figure 1). The thermionic field emission does have a slight upward slope with increasing reciprocal temperature, but it is less than the error found in the experimental measurements on the samples. By contrast, the thermionic emission case shows an obvious trend over the temperature range.

Temperature dependent contact resistance values for InGaN contacted with W and WSi$_x$are shown in Fig. 2. The specific contact resistance is very low ($< 10^{-5}$ $\Omega \cdot cm^2$) for both

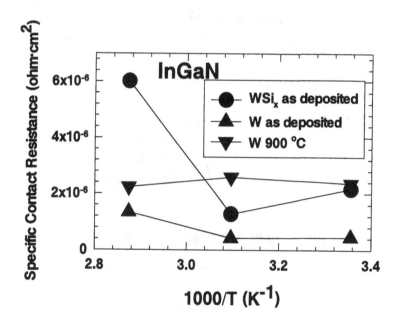

Figure 2. Experimentally measured, temperature-dependent specific contact resistance values for InGaN contacted with W and WSi$_x$.

metals. There is no clear pattern to the data over this temperature range. There is however no upward trend that would indicate thermionic emission. For this material, the value of E_{00} was estimated to be 0.63 eV based on doping levels. This gives a value of $qE_{00}/kT \sim 77$ indicating field emission conduction is expected to be dominant.

Figure 3 (top) shows the temperature dependent contact resistance data for InN contacted with WSi_x. The 500 °C annealed contact has approximately constant contact resistance over this temperature range, as is expected for InN with this doping level ($qE_{00}/kT \sim 24$).

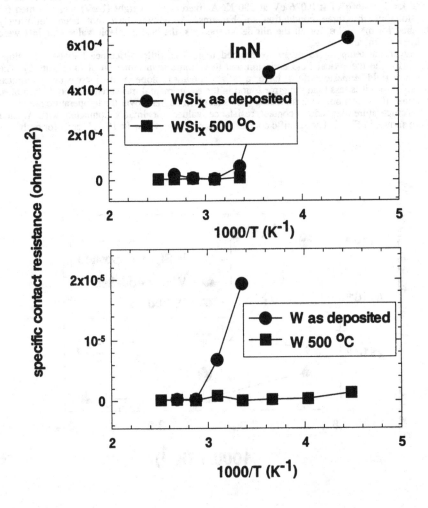

Figure 3. Experimentally determined, temperature-dependent specific contact resistance values for InN contacted with WSi_x and W.

The contact resistance for the as-deposited contact, however, rises with temperature, characteristic of thermionic emission. This may be a result of changing doping levels in the InN because of the sputter deposition of the contact, as is the case for GaAs. In comparing the data in Fig. 2 and 3 it is seen that contacts to InN are more sensitive to temperature than InGaN. The specific contact resistance of InN contacted with W as deposited and after a 500 °C anneal was also measured (Fig. 3, bottom). Again the annealed contact shows a relatively constant contact resistance over the range while the as-deposited contact shows an upward trend.

The contact resistance for W and WSi_x on InGaN as a function of subsequent annealing temperature is shown in Figure 4 (top). Both contacts had similar contact resistance as-deposited, ~ 2-4x10^{-7} $\Omega \cdot cm^2$. Above 600 °C the WSi_x showed signs of degradation, with ϱ_c ~ 10^{-5} $\Omega \cdot cm^2$ at 900 °C. ϱ_c for the W contact sample dropped to ~ 6x10^{-8} $\Omega \cdot cm^2$ at 600 °C and then increased slightly above that temperature. The contact resistances for ohmic contacts of W and WSi_x to InN as a function of annealing temperature are shown in Fig. 4 (bottom). As-deposited samples had similar contact resistances to InGaN, indicating a similar conduction mechanism. WSi_x contacts showed the most degradation at low annealing temperatures, with the resistance rising a factor of 5 after 300 °C annealing and then remaining constant. The W contacts began to degrade at 500 °C.

SUMMARY AND CONCLUSIONS

In summary, theoretical calculations based on the doping levels of InGaN and InN indicate that the dominant conduction mechanism in W-based ohmic contacts to these materials should be field emission. The experimental data fit curves for field emission or thermionic field emission for InGaN contacted with WSi_x and W. InN samples contacted with both W and WSi_x showed similar behavior after annealing at 500 °C, while for as-deposited the curves fit better to the thermionic emission case. This may indicate that the deposition of the contact metal lowered the doping levels in the InN, while annealing returned them to a higher level. W and WSi_x were found to produce low resistance ohmic contacts on n$^+$ InGaN and InN. W contacts proved to be the most stable, and also gave the lowest resistance to InGaN and InN, $\varrho_c < 10^{-7}$ $\Omega \cdot cm^2$ after 600 °C anneal, and 1x10^{-7} $\Omega \cdot cm^2$ after 300 °C anneal, respectively.

ACKNOWLEDGMENTS

The work at Sandia is supported by DOE contract DE-AC04-94AL85000. The technical help of J. Escobedo, M.A. Cavaliere, D. Tibbets, G.M. Lopez, A.T. Ongstad, J. Eng and P.G. Glarborg at SNL is appreciated. The work at the UF is supported by DARPA (monitored by AFOSR, G.L. Witt), an AASERT grant through ARO (Dr. J. M. Zavada), and a University Research Initiative grant #N00014-92-J-1895 administered by AFOSR.

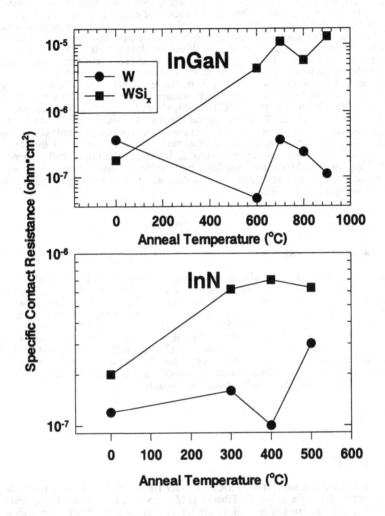

Figure 4. Specific contact resistance for W and WSi$_x$ ohmic contacts to InGaN (top) and InN (bottom) as a function of annealing temperature.

REFERENCES

1. J.S. Foresi and T.D. Moustakas, Appl. Phys. Lett. **62** 2859 (1993).
2. M.A. Khan, T.N. Kuznia, A.R. Bhattaraia and D.T. Olson, Appl. Phys. Lett. **62** 1786 (1993).
3. S. Nakamura, T. Mukai and M. Senoh, Jpn. J. Appl. Phys. **30** L1998 (1991).

4. S.C. Binari, L.B. Rowland, W. Kruppa, G. Kelner, K. Doverspike and D.K Gaskill, Electron. Lett. **30** 1248 (1994).
5. S. Nakamura, M. Senoh and T. Mukai, Appl. Phys. Lett. **62** 2390 (1993).
6. M.E. Lin, Z. Ma, F.Y. Huang, Z.F. Fan, L.A. Allen and H. Morkoc, Appl. Phys. Lett. **64** 1003 (1994).
7. M.W. Cole, D.W. Eckart, T. Monahan, R.L. Pfeffer, W.Y. Han, F. Ren, C.Yuan, R.A. Stall, S.J. Pearton, Y. Li and Y. Lu, J. Appl. Phys. **80** 278 (1996).
8. M.E. Lin, F.Y. Huang and H. Morkoc, Appl. Phys. Lett. **64** 2557 (1994).
9. F. Ren, C.R. Abernathy, S.N.G. Chu, J.R. Lothian and S.J. Pearton, Appl. Phys. Lett. **66** 1503 (1995).
10. F. Ren, C.R. Abernathy, S.J. Pearton and P.W. Wisk, Appl. Phys. Lett. **64** 1508 (1994).
11. F. Ren, in GaN and Related Materials, ed. S.J. Pearton (Gordon and Breach, NY 1996).
12. C.R. Abernathy, J. Vac. Sci. Technol. A **11** 869 (1993).
13 C.R. Abernathy, Mat. Sci. Eng. Rep. **14**, 203 (1995).
14. R.J. Shul, D.J. Rieger, A.G. Baca, C. Constantine and C. Barratt, Electron. Lett. **30** 85 (1994).
15. R.J. Shul, M.E. Sherwin, A.G. Baca and D.J. Rieger, Electron. Lett. **32** 70 (1996).
16. R.J. Shul, S.P. Kilcoyne, M. Hagerott-Crawford, J.E. Parmeter, C.B. Vartuli, C.R. Abernathy and S.J. Pearton, Appl. Phys. Lett. **66** 1761 (1995).
17. A.Y.C. Yu, Solid State Electron, 13, 239 (1970).
18. L.L. Smith and R.F. Davis, in Properties of Group III Nitrides, ed. J.H. Edgar, EMIS Datareview (INSPEC, London 1994).

TEMPERATURE BEHAVIOR OF Pt/Au OHMIC CONTACTS TO p-GaN

D. J. King, L. Zhang, J. C. Ramer, S. D. Hersee, L. F. Lester
University of New Mexico, Center for High Technology Materials, Albuquerque, NM

ABSTRACT

Ohmic contacts to Mg-doped p-GaN grown by MOCVD [1] are studied using a circular transmission line model (TLM) to avoid the need for isolation. For samples which use a p-dopant activation anneal before metallization, no appreciable difference in the specific contact resistance, r_c, as a function of different capping options is observed. However, a lower r_c is obtained when no pre-metallization anneal is employed, and the post-metallization anneal simultaneously activates the p-dopant and anneals the contact. This trend is shown for Pt/Au, Pt, Pd/Pt/Au, and Ni/Au contacts to p-GaN. The r_c's for these metal contacts are in the range of 1.4-7.6 x 10^{-3} ohm-cm^2 at room temperature at a bias of 10mA. No particular metallization formula clearly yields a consistently superior contact. Instead, the temperature of the contact has the strongest influence.

Detailed studies of the electrical properties of the Pt/Au contacts reveal that the I-V linearity improves significantly with increasing temperature. At room temperature, a slightly rectified I-V characteristic curve is obtained, while at 200°C and above, the I-V curve is linear. For all the p-GaN samples, it is also found that the sheet resistance decreases by an order of magnitude with increasing temperature from 25°C to 350°C. The specific contact resistance is also found to decrease by nearly an order of magnitude for a temperature increase of the same range. A minimum r_c of 4.2 x 10^{-4} ohm-cm^2 was obtained at a temperature of 350°C for a Pt/Au contact. This result is the lowest reported r_c for ohmic contacts to p-GaN.

INTRODUCTION

The III-V nitride semiconductors are of current interest in making blue/UV light emitting diodes (LEDs), laser diodes (LDs) and for high-temperature electronic devices because of its large direct bandgap energy (3.39 eV at room temperature). With the realization of p-GaN doped with Mg followed by rapid thermal annealing [2], high brightness blue/green LEDs have been fabricated on InGaN/GaN quantum well structure [3], and recently, pulsed and CW blue LDs have been demonstrated [4]. However, the performance of these devices such as operating voltage and quantum efficiency are restricted by the high resistance of ohmic contacts to p-GaN. Therefore, the development of low resistance ohmic contacts to p-GaN is essential for making high performance junction devices on GaN. In this work, the r_c's of various metal contacts to MOCVD-grown Mg-doped p-GaN are analyzed at room temperature as a function of different p-dopant activation anneal techniques. It is observed that activating Mg (drive out hydrogen) after the metallization step is preferable in order to avoid compensating the surface due to nitrogen desorption. Since the room temperature r_c is not particularly sensitive to the type of metallization, Pt/Au is chosen primarily for its thermal stability to study the change in r_c with temperature.

CIRCULAR TLM MODEL

The r_c is measured using a circular transmission line model (TLM). The circular contact design, shown in Fig. 1, avoids the need for isolation of the contact structures by implantation or etching. This method for patterning the metal contacts is particularly useful for GaN which, because of its chemical inertness, requires plasma etching to achieve significant etch rates [5,6,7]. For the circular TLM pattern, the resistance, R, between contacts is,

$$R = \frac{R_{sh}}{2\pi}\left[\ln\frac{r_2}{r_2-d} + L_T\left(\frac{1}{r_2-d} + \frac{1}{r_2}\right)\right], \qquad (1)$$

421

Mat. Res. Soc. Symp. Proc. Vol. 468 ° 1997 Materials Research Society

where R_{sh} is the sheet resistance of the material, r_2 is the radius of the outer circular contact, d is the gap spacing $(r_2 - r_1)$, and L_T is the transfer length. When the ring radius to gap ratio is large, the ring contact geometry reduces to the standard TLM structure,

$$R \approx 2R_c + \frac{R_{sh}d}{W},$$ (2)

where,

$$R_c = \frac{R_{sh}\,L_T}{W} \text{ and } r_c = R_{sh}\,L_T^2$$ (3)

W is the circumference of the outer ring, and r_c is the specific contact resistance. For the ring radius of 200 μm and gaps from 5 to 45 μm used in this work, small correction factors are necessary to compensate for the difference between the standard TLM and ring layouts to obtain a linear fit to the data. The calculated correction factors are given in Table 1 for a constant outer radius of 200 μm and a varying gap distance. Without these correction factors, the specific contact resistance, r_c, would be underestimated [8].

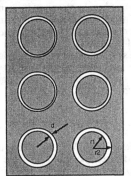

Figure 1: Circular TLM contact pattern

Table I: Correction factors.
($r_2 = 200$mm)

Gap(mm)	Correction Factors
5	1.013
10	1.026
15	1.040
25	1.070
35	1.103
45	1.139

ROOM TEMPERATURE RESULTS

Metal layers of Ni/Au, Pd/Pt/Au, Pt and Pt/Au were deposited on MOCVD grown p-GaN by e-beam evaporation below 2×10^{-6} Torr. Of Ni, Pd, and Pt, it is expected that Pt will most easily form an ohmic contact due to its large metal work function. All samples are taken from the same 2" Mg-doped GaN wafer (308) whose hole concentration and mobility at room temperature are 5-6×10^{16} cm^{-3} and 7 cm^2/V-s, respectively. Previous studies of ohmic contacts to n-GaN have shown that a surface treatment of NH$_4$OH:DI immediately prior to metallization is necessary to prevent an interfacial oxide layer from forming [9]. Therefore all samples were given this surface treatment immediately before being placed in the metallization chamber. Also, all samples were given a dopant activation anneal by RTA to drive out compensating hydrogen which is incorporated into the material during growth [10]. Pd, Pt, and Ni were chosen as contact metals because of their ability to extract hydrogen [11]. As a previous study of ohmic contacts to n-GaN shows, this high temperature anneal can cause nitrogen to desorb from the surface of the GaN [9]. The resulting nitrogen vacancies can cause dopant compensation at the surface. To inhibit nitrogen desorption during the activation anneal, a second piece of GaN that provides an overpressure of nitrogen is used to cap the study sample. In order to determine the effect of the p-dopant activation anneal on the contact characteristics, each sample was given one of the following pre-metallization anneal treatments:
a. No pre-metallization anneal.
b. A 20 minute 700°C rapid thermal anneal (RTA) with the wafer uncapped.
c. A 20 minute 700°C rapid thermal anneal (RTA) with the wafer capped with undoped GaN.

d. A 20 minute 700°C rapid thermal anneal (RTA) with the wafer capped with p-doped GaN. (This wafer will provide an overpressure of Mg.)

All samples were also given a second, post-metallization alloying anneal for 10 minutes at 750°C. For the samples that were not pre-metallization annealed, the 750°C post-metallization heat treatment performed on all wafers simultaneously activated the dopant and annealed the metal. It was found that the hydrogen had no difficulty in escaping through any of the metallizations at 750°C.

The differential resistance, R, between contacts as a function of operating current and gap spacing is measured using an HP 4145A Semiconductor Parameter Analyzer. R_{sh} and r_c are calculated from the R vs. d curve using Eqn.(2) & (3) for a fixed operating current level. Although annealing the contacts at 750°C improves their ohmic behavior, all exhibit a slight bend in their room-temperature I-V characteristic near the origin, indicating a Schottky-type barrier. It is therefore necessary to define the current at which r_c is measured. Table II compares the r_c of metal contacts to GaN that use the different pre-metallization anneal treatments. Because these contacts are not truly ohmic, r_c was measured at 1 and 10 mA. At 10 mA, which is a typical operating current for a GaN LED there is no significant difference in the r_c of samples that are pre-metallization-annealed with the different capping options. However, a lower r_c is obtained for samples that are annealed only after metallization. Presumably, the good contact between the GaN and the metal prevents nitrogen desorption and the resultant compensation at the surface of GaN. For the samples which are activated before metallization, the GaN samples are probably not in good contact with the capping wafer, therefore the dissociation of GaN at the surface occurs for all these cases. Note that no particular metallization formula is superior in r_c to another.

Table II: Comparison of specific contact resistance for different pre-anneal treatments. The specific contact resistances were measured at a current of 10 mA and 1 mA. All samples were given a 750°C post-metallization anneal for 10 minutes.

Sample	Contact Metal	Pre-Anneal	r_c (W-cm²) 10 mA	r_c (W-cm²) 1 mA
308C	Pt/Au	No	1.5×10^{-3}	1.6×10^{-2}
308A	Pt/Au	Capped with Undoped	4.8×10^{-3}	5.7×10^{-3}
308I	Pt/Au	Capped with p-Doped	3.3×10^{-3}	1.0×10^{-2}
308K	Pt/Au	Uncapped	3.4×10^{-3}	1.8×10^{-2}
308G	Pt	No	1.4×10^{-3}	3.4×10^{-2}
308F	Pt	Capped with Undoped	7.6×10^{-3}	3.6×10^{-2}
308J	Ni/Au	No	3.3×10^{-3}	2.5×10^{-2}
308B	Ni/Au	Capped with Undoped	5.5×10^{-3}	1.2×10^{-2}
308E	Pd/Pt/Au	No	2.4×10^{-3}	1.7×10^{-2}
308D	Pd/Pt/Au	Capped with Undoped	7.4×10^{-3}	1.0×10^{-2}

When r_c is measured at 1mA there is no evident trend in the data. At 1 mA bias, it is believed that the current is limited by thermionic emission which is not dependent on the resistivity of the sample, and, therefore, the N_2 desorption effect is not observed. It is also believed that, due to a graded metal-semiconductor interface, the barrier height has a voltage dependence. Thus, at 10 mA, the barrier height is significantly reduced and does not inhibit current flow so that now the resistivity of the sample is the more dominant factor in determining the contact resistance. Therefore, it is possible to observe the effect of N_2 desorption which causes a change in the resistivity of the sample. Fig. (2) shows a diagram of the barrier height decreasing with increasing applied bias for a graded metal-semiconductor interface. It is also interesting to note that sample 308A has a much lower r_c and experiences only a minor change in going from 1 mA to 10 mA. This leads us to believe that 308A has a much smaller barrier height and a more abrupt junction, therefore being less voltage dependent.

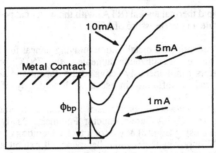

Figure 2: For a graded metal-semiconductor interface, the barrier height decreases as bias is applied.

HIGH TEMPERATURE RESULTS

Studies of the electrical properties of the Pt/Au contacts after fabrication reveal that the I-V linearity improves significantly at high temperature. Fig. 3 shows the I-V curves and their slopes, i.e. the differential resistances, measured at different temperatures. At room temperature, the I-V curve shows a bend near the origin and R decreases with increasing current. As temperature increases, the I-V linearity improves and R decreases significantly. At temperatures above 200°C, the I-V curve exhibits ideal ohmic behavior and the differential resistance is constant with current. It is believed that the current is governed by thermionic emission and the improvement in I-V linearity at high temperature is attributed to the carriers increase in thermal energy which enables them to overcome the barrier.

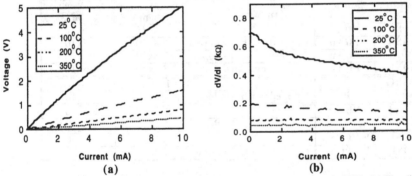

Figure 3: (a) I-V curves of the contact measured at different temperatures on sample 308A. (b) The differential resistances, $R=dV/dI$, as functions of current and temperature.

In order to gain a better understanding of this temperature dependence, the sheet resistance and r_c are measured for a range of temperatures between 25°C and 350°C. This is accomplished by placing the samples on a heated stage and measuring the temperature with a thermocouple. Table IV shows that for all samples the sheet resistance decreases significantly with increasing temperature. This is attributed to an increased hole concentration due to the relatively large acceptor ionization energy. Fig. (4a) shows the sheet resistance of sample 308A decreasing by an order of magnitude from 53,000 ohm/square at 25°C to 4,455 ohm/square at 350°C. For all the wafer 308 samples, the sheet resistance decreased by an order of magnitude.

As shown in Table III, the specific contact resistances also decrease by nearly an order of magnitude from 25°C to 350°C when measurements are taken at 1mA. Again this is attributed to the carries ability to overcome the barrier at high temperatures. If the measurements are taken at

10mA, the specific contact resistance of three samples (308C, 308I, 308K) show a very small dependence on temperature indicating that at 10 mA the barrier was almost completely reduced. Also, as Table IV shows, the current has a minimal dependence on hole concentration due to the fact that a large decrease in sheet resistance only results in a slight change in specific contact resistance. When measurements are taken at 10 mA for sample 308A, a large decrease in the specific contact resistance is seen with the increasing temperature as shown in Fig. (4b). This would indicate that at 25°C and 10 mA the barrier was not significantly reduced therefore requiring increased thermal energy for the carriers to overcome the barrier. At 350°C, a specific contact resistance of 4.2×10^{-4} ohm-cm^2 was measured for sample 308A. This is the lowest reported r_c for ohmic contacts to p-GaN. It is believed that the differences between sample 308A and the other samples is thought to be an effect of wafer non-uniformity and/or processing.

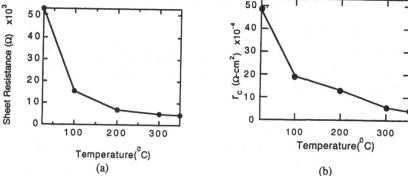

(a)   (b)

Figure 4: (a) Sheet resistance and (b) specific contact resistance as a function of temperature for a Pt/Au contact on 308A. The measurement was taken at 10 mA. For all Pt/Au samples the sheet resistance decreased by nearly an order of magnitude.

Table III: Comparison of the specific contact resistance measured at 25°C and 350°C for Pt/Au contacts.

Sample	r_c (Ω-cm^2) 1 mA 25°C	r_c (Ω-cm^2) 10 mA 25°C	r_c (Ω-cm^2) 1 and 10mA 350°C
308A	5.7×10^{-3}	4.8×10^{-3}	4.2×10^{-4}
308C	1.6×10^{-2}	1.5×10^{-3}	1.6×10^{-3}
308I	1.0×10^{-2}	3.3×10^{-3}	1.8×10^{-3}
308K	1.8×10^{-2}	3.4×10^{-3}	2.0×10^{-3}

Table IV: Comparison of the sheet resistance measured at 25°C and 350°C for Pt/Au contacts.

Sample	R_{sh} (Ω) 1 mA 25°C	R_{sh} (Ω) 10 mA 25°C	R_{sh} (Ω) 1 and 10mA 350°C
308A	71900	53100	4500
308C	52800	41300	3800
308I	42200	26700	2900
308K	52200	35000	3400

CONCLUSION

In conclusion, metal semiconductor ohmic contacts to p-GaN were studied. It was found that for samples that were pre-metallization annealed, no appreciable difference in the specific contact resistance is seen for different capping options. However, a lower r_c is obtained when no pre-metallization anneal is given and the post-metallization anneal simultaneously activates the p-dopant and anneals the contact. For Pt/Au contacts, it is found that the I-V linearity improves significantly with increasing temperature, and that at temperatures greater than 200°C, the I-V exhibits ideal ohmic behavior. Finally, the sheet resistance and the specific contact resistance decrease significantly with increasing temperature when measurements are taken at 1 mA. The sheet resistance decrease is thought to be caused by an increased hole concentration, whereas the decrease in specific contact resistance is attributed to more efficient thermionic emission across the barrier. The lowest reported r_c of 4.2×10^{-4} ohm-cm^2 was measured for a Pt/Au contact at 350°C.

ACKNOWLEDGMENTS

This work was supported by the ARPA Optoelectronics Materials Center under grant # MDA972-94-1-0003, NSF under a CAREER grant # ECS-9501785, and Sandia National Laboratories. One of the authors (DJK) is supported by a National Defense Science and Engineering Graduate Fellowship.

REFERENCES

1. S. D. Hersee, J. Ramer, K. Zheng, C. Kranenberg, K. Malloy, M. Banas, and M. Goorsky, J. Electron. Mat. **24**, 1519 (1995).

2. S. Nakamura, N. Iwasa, M. Senoh, and T. Mukai, Jpn J. Appl. Phys. **31**, 1258 (1992).

3. S. Nakamura, T. Mukai, and M. Senoh, Appl. Phys. Lett. **64**, 1678 (1994).

4. S. Nakamura, M. Senoh, S. Nagahama, N. Naruhito, T. Yamada, T. Matsushita, Y. Sugimoto, and H. Kiyoku, Appl. Phys. Lett. **69**, 1477 (1996).

5. L. Zhang. J. Ramer, J. Brown, K. Zheng, L. F. Lester, and S. D. Hersee, Appl. Phys. Lett., **68** 367 (1996).

6. C. B. Vartuli, J. D. MacKenzie, J. W. Lee, C. R. Abernathy, S. J. Pearton, and R. J. Shul, J. Appl. Phys. **80** 3264 (1996).

7. R. J. Shul, G. B. McClellan, S. A. Casalnuovo, D. J. Rieger, S. J. Pearton, C. Constantine, C. Barratt, R. F. Karlicek, C. Tran, and M. Schurman, Appl. Phys. Lett. **69** 1119 (1996).

8. Simon S. Cohen, Gennady Sh. Gildenblat, VLSI Electronics Microstucture Science Volume 13, edited by Norman G. Einspruch (Academic Press, Orlando, 1986), pp. 97-117.

9. L. F. Lester, J. M. Brown, J. C. Ramer, L. Zhang, S. D. Hersee, and J. C. Zolper, Appl. Phys. Lett., **69** 2737 (1996).

10. S.J. Pearton, R.J. Shul, R.G. Wilson, F. Ren, J.M. Zavada, C.R. Abernathy, C.B. Vartuli, J.W. Lee, J.R. Mileham, and J.D. Mackenzie, J of Electronic Materials, **25**, 845 (1996).

11. J.D. Fast, Interaction of Metals and Gases, (MacMillan, New York, 1971) Chapter 1.

LOW RESISTANCE CONTACTS TO P-TYPE GaN

Taek Kim, Jinseok Khim, Suhee Chae, and Taeil Kim
Photonics Semiconductor Lab., Samsung Advanced Institute of Technology, P.O. Box 111,
Suwon 440-600, Korea, taek@saitgw.sait.samsung.co.kr

ABSTRACT

We report the low resistance ohmic contacts to p-GaN using a Pd/Au bimetal scheme. A specific contact resistivity of 9.1 x 10^{-3} $\Omega \cdot cm^2$ was obtained after annealing. The metallization was e-beam evaporated on 2 μm-thick p-GaN (~ 9 x 10^{16}/cm^3) layers grown on c-plane sapphire substrates by metalorganic chemical vapor deposition (MOCVD). The comparison with other contacts showed that the contact resistivity of the Pd/Au contacts was at least one order smaller than those of Pt/Au and Ni/Au contacts.

INTRODUCTION

The group III-nitrides are very promising materials for optoelectronic devices such as light emitting diodes (LEDs) [1], [2] and laser diodes (LDs) [3] in the UV to the visible wavelength region. Ohmic contacts to GaN is a very challenging issue because achieving low resistance and thermally stable ohmic contacts remain one of the main obstacles for successful implementation of the devices. Since the first demonstration of contact resistance in the range of ~10^{-4} $\Omega \cdot cm^2$ using the Al metallization by Moustakas [4], relatively many results have been reported on n-GaN ohmic contacts. However, low resistance p-GaN ohmic contacts are much more difficult to obtain than n-type contacts. While the Ni/Au contacts are widely used by many groups including Nichia chemical, there have been few reports about p-GaN ohmic contacts. Recently, low resistance p-GaN contacts using the Ni/Cr/Au was reported by the authors [5]. The reported specific contact resistivity was as low as 8.3 x 10^{-2} $\Omega \cdot cm^2$. T. Mori *et al* [6] reported the schottky barrier heights and the ohmic characteristics of Pt, Ni, Au, and Ti. They obtained a specific contact resistivity of 1.3 x 10^{-2} $\Omega \cdot cm^2$ using Pt.

In this study, we investigated three high work function metal contacts : Pt/Au; Ni/Au; and Pd/Au. We found that the Pd/Au contacts exhibited the lowest contact resistance. We also compared the Pd/Au contacts with multi-level metallization schemes such as Pd/Au/At/Au and Pd/Au/Pd/Au contacts.

EXPERIMENT

The p-GaN films used in this study were grown by MOCVD on c-plane sapphire substrates. A GaN buffer layer of 250 Å was grown at 500 °C , followed by the growth of 2 μm thick p-GaN. The carrier concentration was found to be 9 x 10^{16}/cm^3 by Hall measurements. The samples were patterned and then etched to make mesa areas for transmission line method (TLM) measurements. The mesas were isolated by chemically-assisted ion beam etching (CAIBE) using Cl_2. The rectangular pads were 200 μm wide and 100 μm long. The separations between the contact pads were increased from 5 to 30 μm in 5 μm increments. All metal contacts were deposited by e-beam evaporation with a base pressure of ~ 8 x 10^{-7} torr. The contact patterns were defined by a standard lift-off process. Before metal deposition, the samples were immersed

Mat. Res. Soc. Symp. Proc. Vol. 468 © 1997 Materials Research Society

in a HCl:H₂O solution for 30 seconds followed by deionized water rinse to remove possible native oxides. The deposition sequences and the layer thickness were as follows: (1) Pd - 200 Å / Au - 5000 Å, (2) Ni - 200 Å / Au - 5000 Å, and (3) Pt - 200 Å / Au - 5000 Å. The I-V characteristics of the metal contacts were measured by a HP 4145B analyzer. The measurements were performed on the as-deposited contacts and annealed contacts at 500 °C for 30 seconds in a N₂ ambient. The interface between metal contacts and GaN was investigated by Auger depth profiles.

RESULTS & DISCUSSION

Figure 1 shows I-V characteristics of the as-deposited and annealed contacts at 500 °C for 30 seconds in a N₂ ambient. The I-V characterizations were carried out by injecting currents up to ± 7 mA and measuring the voltages across the contacts. All metal contacts showed non-linear I-V characteristics in this current range. The I-V curve of the as-deposited Ni/Au contacts was not shown in this figure because of its very rectifying behavior. The I-V curves showed that the contact resistances of all the samples used in this study were decreased after annealing. It was expected that the Pt/Au showed the lowest contact resistance since the work function of Pt (5.65 eV) is higher than those of Ni (5.15 eV) and Pd (5.12 eV). Contrary to the expectation, the Pd/Au contacts exhibited the lowest resistance leading to the lowest operation voltage in real devices. We attribute the lowest contact resistance of the Pd/Au to the fact that the Pd could act as an acceptor in GaN causing the near interface region to be highly doped [7].

The plot shown in Figure 1 is very useful for relative comparison of contact characteristics of various metal contacts. However the specific contact resistivity has to be derived from linear I-V curves to investigate the ohmic characteristics quantitatively. T. Mori et al [6] calculated specific contact resistivity using the resistance derived from non-linear I-V curves. However, this method could overestimate the real values. The I-V curves which were measured within ± 10 μA range are shown in Figure 2 The Pd/Au and the Pt/Au showed linear I-V characteristics while the I-V curve of the Ni/Au still remained slightly curved. The specific contact resistivity of the

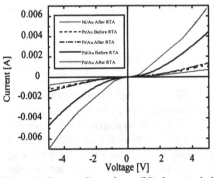

Figure 1. Current (I) - voltage (V) characteristics of various contact metals on p-GaN.

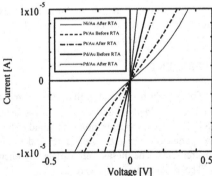

Figure 2. I-V characteristics of various contact metals measured within small current range.

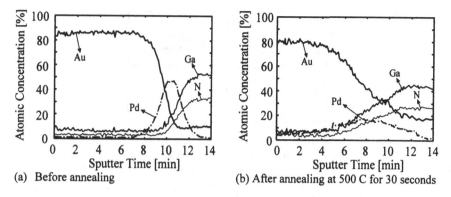

(a) Before annealing

(b) After annealing at 500 C for 30 seconds

Figure 3. Auger depth profiles of the Pd/Au contacts.

annealed Pd/Au contacts calculated from the resistance derived from the I-V curves in Figure 2 was 9.1×10^{-3} $\Omega \cdot cm^2$. The specific contact resistivities of the as-deposited Pd/Au contacts and the annealed Pt/Au were distributed around middle $\sim 10^{-2}$ $\Omega \cdot cm^2$ and high $\sim 10^{-3}$ $\Omega \cdot cm^2$, respectively. For comparison, the specific contact resistivity of the Pd/Au calculated from Figure 1 was 1×10^{-3} $\Omega \cdot cm^2$. Although the low operating voltage of Nichia laser diode [8] was a clear evidence of a reasonable contact resistance of the Ni/Au contacts, the specific contact resistivity of the Ni/Au contacts in this study was higher than those of other contacts.

Auger depth profiling was performed to investigate the interfacial reaction between the Pd/Au contacts and the p-GaN. Before the annealing, no inter-diffusion was observed and clear interfaces between Au, Pd, and GaN could be seen (Figure3-a). However, it was observed that Pd and Au diffused into the GaN in large amounts after annealing (Figure3-b). In the case of the Pd/Au contacts, we varied the annealing temperature from 400 °C to 900 °C. No significant difference in contact resistance was found in the annealed samples in the range of 400 °C ~ 700 °C. We could attribute the lower contact resistance after annealing to the diffusion of the Pd and Au cross a contamination layer on the GaN surface. A contamination layer consisting of GaO_x and adsorbed carbons was found on GaN films grown by MOCVD and this layer could not be completely removed even by sputtering the surface with Ar and N ions [9]. On the other hand, the annealing at 900 °C increased the contact resistance than that of the as-deposited sample. Nitrogen-out diffusion leaving the surface N vacancy which act as a shallow donor could explain the increase of contact resistance.

Figure 4 shows that the Pd signal was detected at 500 Å into the surface indicating Pd out-diffusion toward the surface after annealing. This result suggests that insertion of a diffusion barrier could lower the contact resistance. From this point of view, we compared Pd and Pt as a diffusion barrier. The metallization schemes were as follows : Pd/Au/Pt/Au = 200 Å/ 200 Å/ 200 Å/ 5000 Å and Pd/Au/Pd/Au = 200 Å/ 200 Å/ 200 Å/ 5000 Å. It was found that the contact resistance of as-deposited above contacts was comparable to those of the annealed Pd/Au contacts. However, the contact resistance of the annealed Pd/Au/Pt/Au contacts decreased to the value of the annealed Pt/Au contacts while contact resistance of the annealed Pd/Au/Pd/Au contacts remained unchanged. More systematic study is necessary to verify these multi-layer metallization schemes based on Pd/Au contacts.

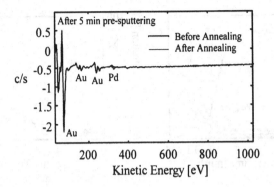

Figure 4. Auger electron spectra from the surfaces of as-deposited
and annealed at 500 C for 30 seconds Pd/Au contacts.

CONCLUSIONS

In summary, we studied ohmic characteristics of the metal contacts to p-GaN using three high work function metals. The Pd/Au exhibited a lower contact resistance than those of a Ni/Au and a Pt/Au. The specific contact resistivity was calculated to be 9.1×10^{-3} $\Omega \cdot cm^2$. The contact resistance of the Pd/Au contacts was decreased after annealing but independent on the temperature in the range of 400 °C ~ 700 °C. We attributed this to the diffusion of the metal components through the contamination layer. The auger depth profiles study showed a significant inter-diffusion of Au and Pd.

REFERNCES

[1] Shuji Nakamura, Takashi Mukai, and Masayuki Senoh, Appl. Phys. Lett. **64**, 1687 (1994)
[2] Shuji Nakamura, Masayuki Senoh, Naruhito Iwasa, and Shin-ichi Nagahama, Jpn. J. Appl.Phys. Vol. 34 (1995) pp.L797 - L799
[3] Shuji Nakamura, Masayuki Senoh, Shin-ichi Nagahama, Naruhito Iwasa, Takao Yamada, Toshio Matsushita, Hiroyuki Kiyoku, and Yasunobu sugimoto, Jpn. J. Appl.Phys. Vol. 35 (1996) pp.L74 - L76
[4] J.S. Foresi and T.D. Moustakas, Appl. Phys. Lett. **62**, 1786 (1993)
[5] Taek Kim, MyungC. Yoo, and Taeil Kim, 1996 MRS Fall Meeting, "III-V Nitrides" (in press)
[6] T. Mori, T. Kozawa, T. Ohwaki, and Y. Taga, Appl. Phys. Lett. **69**, 3537 (1996)
[7] J.T. Trexler, S.J Pearton, P.H Holloway, M.G Mier, K.R. Evans, and R.F. Kalicek. 1996 MRS Fall Meeting, "III-V Nitrides" (in press)
[8] Shuji Nakamura, Masayuki Senoh, Shin-ichi Nagahama, Naruhito Iwasa, Takao Yamada, Toshio Matsushita, Yasunobu sugimoto, and Hiroyuki Kiyoku, Appl. Phys. Lett. **69**, 4056 (1996)
[9] Hidenori Ishikawa, Setsuko Kobayashi, Y. Koide, S. Yamasaki, S. Nagai, J. Umezaki, M, Koike, and Masanori Murakami. J. Appl. Phys. **81**, 1315 (1997)

PHASE FORMATION AND MORPHOLOGY IN NICKEL AND NICKEL/GOLD CONTACTS TO GALLIUM NITRIDE

H. S. Venugopalan, S. E. Mohney, B. P. Luther*, J. M. DeLucca, S. D. Wolter, J. M. Redwing**, and G. E. Bulman***
Department of Materials Science and Engineering, The Pennsylvania State University, University Park, PA 16802
* Department of Electrical Engineering, The Pennsylvania State University, University Park, PA 16802
** ATMI, 7 Commerce Drive, Danbury, CT 06810
*** Cree Research Inc., Durham, NC 27713

ABSTRACT

Metallurgical reactions between Ni and GaN have been explored at temperatures between 400 and 900 °C in N_2, Ar, and forming gas. A trend of increasing Ga content in the reacted films was observed with increasing temperature. The reactions are consistent with the thermodynamics of the Ni-Ga-N system. Changes in the film morphology on annealing were also examined. Metal island formation and a corresponding deep, non-uniform metal penetration into GaN were observed at high temperatures. The relevance of the observed nature of phase formation and morphology in these thin films to electrical properties of Ni/GaN and Au/Ni/GaN contacts is also considered.

INTRODUCTION

The III-V nitrides are receiving considerable attention for optoelectronics in the blue and UV wavelengths as well as for their potential for high temperature electronics [1]. Therefore, ohmic contacts and Schottky barriers to GaN are of considerable interest. Since Ni has been used in Ni/Au ohmic contacts to p-type GaN [2, 3] and has also been examined as a Schottky barrier to n-type GaN [4], an understanding of the metallurgy of the Ni/GaN system is clearly needed.

Bermudez et al. [5] examined the growth of thin Ni films on GaN and noted that a pronounced interfacial reaction occurred upon annealing above 600 °C in vacuum. Guo et al. [4] studied the thermal stability of Ni Schottky contacts on n-type GaN. According to these investigators, Ga_4Ni_3 was identified by XRD along with Ni in the as-deposited film. However, no intermixing between Ni and Ga was observed in the corresponding SIMS spectra. On subsequent annealing in a N_2 atmosphere at 200 °C and 400 °C, nickel nitrides were identified by XRD. The reported reaction between Ni and GaN at room temperature is surprising from a kinetic standpoint and also given that thermodynamic estimates [6] suggest that Ni could actually be thermodynamically stable on GaN at room temperature. Thus, an improved understanding of the metallurgy of the reactions between Ni thin films and GaN is clearly desirable.

Changes in film morphology of M/GaN thin films (M= Pt, Pd, Ni) on annealing have been investigated using Rutherford backscattering spectroscopy (RBS) and scanning electron microscopy by Duxstad et al. [7]. These authors observed that for Ni on GaN, islands began to form above 700 °C. However, the effect of this metal non-uniformity on the uniformity of the metal/GaN reaction front have not been reported. If the metal in localized regions has reacted through the GaN layer in a device, the device could be rendered inoperable. In addition, it has not been reported whether this island formation is seen upon annealing Ni/Au films on GaN. This

431

Mat. Res. Soc. Symp. Proc. Vol. 468 ° 1997 Materials Research Society

system is important for ohmic contacts to p-GaN [2, 3]. Anneal temperatures of 400 °C and 600 °C have been reported for these contacts [8, 9].

In this study, metallurgical reactions between Ni and GaN have been explored at temperatures between 400 °C and 900 °C in N_2, Ar, and forming gas. Glancing angle x-ray diffraction and Auger depth profiling were employed to determine the extent of interdiffusion between Ni and GaN and identify the phases that form upon annealing. Scanning electron microscopy (SEM) of the annealed films was performed, followed by atomic force microscopy (AFM) of the GaN surface after the metal contact was etched away.

EXPERIMENTAL PROCEDURE

Gallium nitride films (ATMI) with $n = 10^{17}$ cm^{-3} were grown by MOCVD on (0001) Al_2O_3 substrates. The GaN surface was degreased in methanol, etched in 1:1 HCl:DI, rinsed in DI, and dried in N_2 gas immediately before the sample was loaded into a vacuum deposition system. Ni films of 500 Å thickness were sputtered onto the GaN at a rate of 2.0 Å/s. For Ni/Au contacts, Ni film of 500 Å thickness was first sputtered onto GaN followed by a Au film of 1000 Å.

The Ni/GaN specimens were then annealed at various temperatures between 400 °C and 900 °C in N_2 (99.99 % purity), 95 % N_2+ 5 % H_2 (forming gas), or Ar (99.99 % purity). A few of the specimens were annealed in a conventional tube furnace while others were annealed in an AG 610 rapid thermal annealing (RTA) furnace, in each case under flowing gas atmospheres. In order to determine if more extensive reaction would occur during prolonged annealing, a few samples were sealed in quartz tubes under vacuum and annealed in a box furnace for a week.

Glancing angle XRD, using Cu K_α radiation, was performed with a Philips diffractometer. Auger electron spectroscopy (AES) was performed using a Kratos scanning Auger microprobe with a 3 keV electron beam. The sample was simultaneously argon ion sputter-etched to obtain a depth profile.

Scanning electron microscopy of the annealed films was performed using an ISI SX-40A SEM at an acceleration voltage of 25 kV. Elemental analysis was obtained by energy dispersive spectroscopy (EDS) of the individual regions. For the AFM study, GaN films supplied by Cree Research were used. Nickel and Ni/Au films were deposited on the GaN by sputtering and subsequently annealed at various temperatures. The annealed Ni/GaN and Au/Ni/GaN thin films were then etched with 3:1 HNO_3:HCl and 1:5 HNO_3:HCl to dissolve Au and Ni, respectively. Subsequently, AFM was performed on the GaN surface to characterize the uniformity of reaction between the metal films and GaN.

RESULTS

Phase formation in Ni/GaN

XRD of the as-deposited Ni film on GaN indicates the presence of polycrystalline Ni on GaN. Our result is in contrast to that of *Guo et al.* [4], who reported Ga_4Ni_3 to be present in the as-deposited Ni film on GaN. To investigate the extent of reaction between Ni and GaN at 400 °C in forming gas (95% N_2 + 5% H_2), a Ni/GaN film was annealed by RTA for 10 min. in this atmosphere. No reaction between Ni and GaN was detected by XRD. The conditions were chosen to simulate those used by Trexler *et al.* [9] for the anneal of Au/Ni/p-GaN ohmic contacts. To investigate whether a reaction could be detected after more prolonged annealing, a sample was then subjected to flowing N_2 gas for 17 hours at 400 °C. The interface between Ni and GaN is as sharp after annealing for 17 hours at 400 °C as it was in the profile of the as-deposited sample. The XRD scan also showed no indication of any new phase formation or of any alteration in the lattice parameter of Ni. Our results are consistent with a recent report by Ishikawa *et al.* [10] who did not observe a reaction between Ni and GaN when the contacts were annealed at 400 °C for 10 min.

Nickel/GaN films were annealed in a tube furnace for periods of 1 and 24 hours in flowing Ar and N_2 atmospheres at 600 °C. After annealing in flowing Ar, the diffraction pattern was consistent with a face-centered cubic phase. The lattice parameter varied from 3.53 to 3.55 Å as the annealing time was increased from 1 to 24 hours. In flowing N_2, the corresponding variation was from 3.52 to 3.55 Å. The observations of an increasing lattice parameter along with an increased Ga signal in the thin film (Auger depth profiles) are consistent with the significant solubility of Ga in face-centered cubic Ni (up to 15% Ga at 600 °C, with an increase in the lattice parameter to 3.552 Å [11, 12]).

The x-ray diffraction peaks observed after annealing at 600 °C were again observed after annealing for 1 hour in Ar at 750 °C, although the peaks were shifted, revealing an increase in the lattice parameter after annealing at 750 °C. Cubic lattice parameters of 3.56 Å and 3.58 Å were observed after annealing for 1 hour in Ar and N_2 respectively. Auger depth profiles indicate that the extent of Ga dissolution is greater at 750 °C than at 600 °C, and that significantly more Ga than N is present in the film. Thus, N_2 gas has been released to the annealing environment, even when this annealing environment was N_2 at 1 atm. The identity of the Ni-Ga phase present in the samples subjected to prolonged annealing at 750 °C is not completely clear. From the lattice parameter alone, the phase is either the disordered face-centered cubic solution of Ni and Ga, near the limit of Ga solubility, or the ordered Ni_3Ga phase.

The x-ray diffraction plot for a Ni/GaN sample annealed for 10 minutes in N_2 atmosphere indicates the presence of a cubic phase with a lattice parameter of 2.89 Å in both Ar and N_2 atmospheres, consistent with the B2 NiGa phase. A comparison of the Auger depth profiles at 750 and 900 °C also indicates that the Ga/Ni ratio in the reacted film is approximately three times higher at 900 °C as compared to that at 750 °C. The reaction products in the Ni/GaN system at various times and temperatures are shown in Table I. As predicted by the calculations of Mohney *et al.* [6], the trend of Ga incorporation into the reacted film, accompanied by the release of N_2 gas, was observed in this experimental study.

Table I. Reaction products in Ni/GaN contacts.

Temperature, °C	Time	Furnace	Ni-Ga phase in contact with GaN
As-deposited			Ni
400	10 mins, 17 hours	RTA, Tube	Ni
600	1 hour, 7 days	Tube, Box	FCC Ni-Ga solid solution
750	1 hour, 7 days	Tube, Box	Ni_3Ga or FCC Ni-Ga solid solution*
900	10 mins, 30 mins	RTA	NiGa

* More Ga compared to the Ni-Ga solid solution at 600 °C

Morphology of Ni films on GaN

Our SEM studies indicate that the thin film coverage on GaN is almost uniform at 600 °C and 750 °C, with Ni-rich non-uniformities present over approximately 5% of the surface of the contact annealed at 750 °C. However, upon annealing at 900 °C, the film exhibits considerable non-uniformity whereupon Ni islands form to reduce the overall interfacial energy of the system.

Since this process occurs by surface diffusion, it is faster at high temperatures. In addition, at high temperatures, a greater amount of GaN is consumed by reaction. (More Ga-rich nickel gallides are formed.) This two-fold process of enhanced island formation along with increased extent of reaction could cause metal gallides formed beneath the islands to spike deeply into the GaN. To determine the extent of metal penetration into GaN, the annealed surfaces were etched and examined using AFM (Fig. 1). Atomic force microscopy of the surface of an etched contact that was not annealed is also shown for comparison (Fig. 2). A surface roughness of 100 Å was observed. After etching the metal from the Ni(500 Å)/GaN sample annealed at 800 °C, pits as deep as 800 Å can be observed. In other words, metal in the islands reacts with as much as 800 Å of GaN. This would certainly exert an influence in the electrical properties of the Ni/GaN interface.

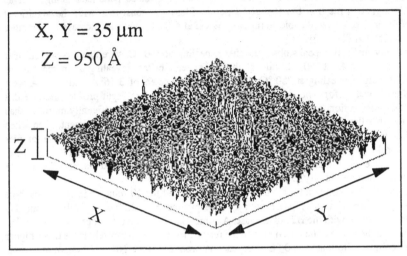

Fig 1. AFM of the GaN surface revealed after a Ni/GaN contact annealed at 800 °C for 1 min. in nitrogen was etched away.

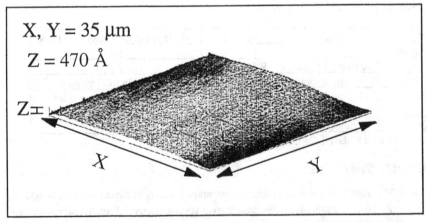

Fig 2. AFM of the GaN surface of an etched Ni/GaN contact that was not annealed.

Morphology of Ni/Au contacts to GaN

Film morphology upon annealing the Ni(500 Å)/Au(1000 Å) contact on GaN was also examined. No islands were observed to form after short anneals at 400 °C and 600 °C in N$_2$. The onset of island formation was observed by SEM for a 1 minute anneal at 700 °C in N$_2$. On etching the contact and examining the surface by AFM, pits as deep as 5000 Å were observed. On increasing the anneal temperature to 800 °C, large islands were observed to form by SEM (Fig. 3). The underlying GaN between the metal islands was clearly visible by optical microscopy. After etching the contact and examining the surface by AFM, pits as deep as 2000 Å were observed (Fig. 4). In other regions of the contact, pits as deep as 5000 Å were observed. Furthermore, the density

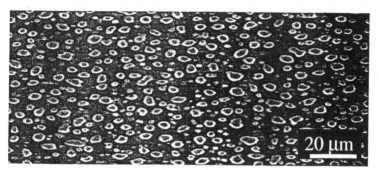

Fig 3. SEM of Au(1000 Å)/Ni(500 Å)/GaN contact after annealing at 800 °C for 1 minute in nitrogen.

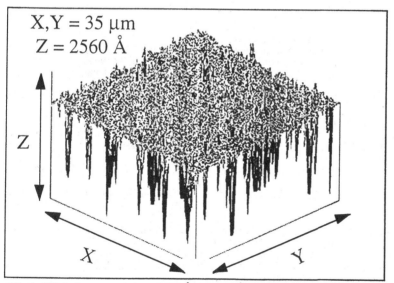

Fig 4. AFM of an annealed Au(1000 Å)/Ni (500 Å)/GaN contact after etching the metal away.

of pits on the GaN surface observed after the 800 °C anneal is more than an order of magnitude higher than the pit density observed after the 700 °C anneal. In the Nichia InGaN-AlGaN-GaN double heterojunction LED [1] and in the InGaN single quantum well (SQW) LD [13], the p-GaN layer has a thickness of 3000 Å and 5000 Å, respectively. A typical contact to p-GaN is Ni/Au [2-3, 13]. Our study indicates that for these p-layer thicknesses, a Ni(500 Å)/Au(1000 Å) contact could react through the entire p-GaN layer in localized regions when annealed at temperatures higher than 700 °C, rendering the device inoperable. Hence, anneals of Ni/Au contacts on GaN at temperatures as high as 700 °C may not be suitable for many device structures.

CONCLUSIONS

In this study, metallurgical reactions between Ni and GaN have been explored at various temperatures between 400 and 900 °C in N_2, Ar, and forming gas. A trend of increasing Ga content in the reacted films was observed with increasing temperature, while the formation of nickel nitrides was not observed. The observed reactions are consistent with the thermodynamics of the Ni-Ga-N system. Changes in the film morphology upon annealing were also examined. Metal island formation and corresponding deep, non-uniform penetration of metal into GaN were observed at high temperatures. These non-uniformities in film morphology could be significant in determining the electrical properties of Au/Ni/p-GaN and Ni/n-GaN contacts after high temperature annealing.

ACKNOWLEDGMENTS

This work was supported by DARPA (Anis Husain) through AFOSR grant F49620-95-1-0516.

REFERENCES

1. H. Morkoc and S. N. Mohammad, Science 267, p. 51 (1995).
2. S. Nakamura, M. Senoh, and T. Mukai, Appl. Phys. Lett. 62, p. 2390 (1993).
3. S. Nakamura, T. Mukai, and M. Senoh, Appl. Phys. Lett. 64, p. 2557 (1994).
4. J. D. Guo, F. M. Pan, M. S. Feng, R. J. Guo, P. F. Chou, and C. Y. Chang, J. Appl. Phys. 80, p. 1623 (1996).
5. V. M. Bermudez, R. Kaplan, M. A. Khan and J. N. Kuznia, Phys. Rev. B. 48, p. 2436 (1993).
6. S. E. Mohney and X. Lin, J. Electron. Mater. 25, p. 811 (1996).
7. K. J. Duxstad, E. E. Haller, W. R. Imler, L. T. Romano, K. M. Yu, F. A. Ponce, and D. A. Steigerwald, in III-V Nitrides, F. A. Ponce, T. D. Moustakas, I. Akasaki, and B. A. Monemar (Eds.) (Mater. Res. Soc. Symp. Proc. 449, 1997).
8. J. T. Trexler, S. J. Pearton, P. H. Holloway, M. G. Mier, K. R. Evans and R. F. Karlicek, ibid.
9. J. T. Trexler, S. J. Miller, P. H. Holloway, and M. A. Khan, in Gallium Nitride and Related Materials, F. A. Ponce, R. D. Dupuis, S. Nakamura, and J. A. Edmond (Eds.) (Mater. Res. Soc. Symp. Proc. 395, 1996), p. 819.
10. H. Ishikawa, S. Kobayashi, Y. Koide, S. Yamasaki, S. Nagai, J. Umezaki, M. Koike, and M. Murakami, J. Appl. Phys. 81 (3) p. 1315 (1997).
11. P. Feschotte and P. Eggimann, J. Less-Common. Met. 63, p. 5 (1979).
12. S. Y. Lee and P. Nash in Phase Diagrams of Binary-Nickel Alloys, P. Nash (Ed.), ASTM International, Materials Park, OH, 1991, p. 133.
13. S. Nakamura, in Gallium Nitride and Related Materials, F. A. Ponce, R. D. Dupuis, S. Nakamura, and J. A. Edmond (Eds.) (Mater. Res. Soc. Symp. Proc. 395, 1996), p. 879.

ELECTRON FIELD EMISSION FROM ALUMINUM NITRIDE

D.P. Malta, G.G. Fountain, J.B. Posthill, T.P. Humphreys, C. Pettenkofer*, and R.J. Markunas

Research Triangle Institute, Research Triangle Park, NC 27709 USA
*Hahn-Meitner-Institut, Postfach, 390128, D-1000 Berlin, Germany

ABSTRACT

AlN has been identified as a candidate material for cold cathode field emitters due to its purported negative electron affinity (NEA) surface. Recent studies by our group on AlN(0001) using angle-resolved ultraviolet photoelectron spectroscopy (ARUPS) and scanning electron microscopy (SEM) have indicated that AlN(0001) is a positive electron affinity surface. We have also investigated electron field emission behavior of AlN and pure Al films grown on Si. AlN and Al films were grown by molecular beam epitaxy (MBE) and transported via an ultra-high vacuum (UHV) integrated processing system (IPS) to an electron emission measurement system (EEMS). The reference Al film on Si showed characteristic Fowler-Nordheim behavior with a turn-on field of 120V/μm (defined at 10μA-cm^{-2}) and ~100μA-cm^{-2} emission at 140V/μm. The AlN film also showed Fowler-Nordheim behavior with a turn-on field of 60V/μm and ~10mA-cm^{-2} at 100V/μm. Air exposure of the AlN film caused a shift in turn-on to 90V/μm and ~0.1mA-cm^{-2} at 100V/μm. The I-V behavior of the AlN film is consistent with the ARUPS results on a different AlN sample - both indicating a positive electron affinity AlN surface.

INTRODUCTION

Group III-V nitrides have received much attention recently due to rapid advances in key growth areas and breakthroughs in the development of blue/ultraviolet optoelectronic devices. AlN has received particular attention as a candidate material for cold cathode field emitters due to its purported low [1] or negative [2] electron affinity (NEA). Diamond has also exhibited NEA [3]. It is believed that NEA materials could be used in applications such as flat panel displays where a stable electron field emission source requiring a small extraction field is desired.

In the NEA condition, the conduction band minimum lies above the vacuum level allowing barrier-free escape of electrons from the surface. The most widely used technique for measuring electron affinity of a semiconductor surface is ultraviolet photoelectron spectroscopy (UPS). This technique provides key information on the construction of the energy bands near the surface. Used in conjunction with field emission measurements, the electron emission behavior of the surface can be characterized. Field electron emission studies have been recently reported on BN [4], AlN [5] and most widely on diamond [e.g., 6].

We have investigated the AlN surface using angle-resolved ultraviolet photoelectron spectroscopy (ARUPS) and electron field emission measurements.

437

Mat. Res. Soc. Symp. Proc. Vol. 468 ° 1997 Materials Research Society

EXPERIMENT

ARUPS

An epitaxial structure comprised of a ~30 nm AlN layer on GaN on Al_2O_3 (0001) was loaded from air into the synchrotron at BESSY for ARUPS. Photoemission spectra were taken normal to the substrate at a photon energy of $h\nu = 60eV$. Spectra were taken in the as-loaded condition, N_2 sputtered, and annealed (600°C and 1000°C) conditions and all acquired ARUPS spectra were, for practical purposes, identical.

Electron Field Emission

Polycrystalline AlN was grown on vicinal Si(111) wafer (cut 6° off axis towards [011]) to a thickness of 100nm by molecular beam epitaxy (MBE). A large-area backside electrical contact to the n^+ Si substrate was achieved by adhering the wafer to a flat molybdenum puck with indium. The sample was transported in-vacuo via an ultra-high vacuum (UHV) integrated processing system (IPS) to the electron emission measurement system (EEMS) shown schematically in Fig. 1 Base pressure of the EEMS was ~10^{-9} Torr. The EEMS consisted of a sample docking station to facilitate transfer from the IPS, an anode capable of precision Z-approach as well as X-Y translation, vibration-isolation accomplished by spring supports, a 2000V variable DC power supply, and an HP 4140B pA meter. The anode probe consisted of a 0.25" diameter stainless steel sphere. Probe collection area was estimated based on the radius of curvature of the sphere to be ~$0.01cm^2$. Both sample and anode were floated with respect to ground. I-V characteristics were measured by applying a variable positive DC voltage to the anode and measuring the current at the sample. The anode/cathode separation distance was estimated by contacting the anode to the sample and then retracting the anode with a calibrated microtranslator. Following initial I-V measurements, the AlN/Si sample was removed from vacuum and exposed to air. It was then reloaded and I-V measurements were repeated. For comparison, a 300nm pure Al film was grown on a similar n^+ vicinal Si(111) substrate by MBE and transferred in-vacuo for I-V measurements.

RESULTS

Photoemission spectra are plotted in Fig. 2. Fig.2a shows a reduction in overall emission intensity following the sputter clean/anneal, however, the valence band cut-off remained constant at ~55.3eV. The valence band cutoff was determined by a conventional method of extrapolating a tangential line taken at the inflection point of the high energy side of the peak to the point of intersection with the noise level. This intersection defines the cut-off energy. Fig. 2b shows the low-energy region of the spectrum for several different values of negative sample bias. The lowest energy peak for each bias condition is attributed to the production of secondary electrons in the detector and is regarded as an artifact. The low energy cut-off observed at -3V bias is ~7.5eV giving a surface work function, $\phi \sim (7.5eV - 3.0eV) = 4.5eV$. The Fermi energy, E_F, was obtained from the Ta sample holder yielding $E_F \sim 60.3eV$. Fig. 3 shows a schematic construction of the energy bands near the surface based on these measurements. Assuming the band gap, E_G, for AlN of 6.3eV, the electron affinity, χ, is determined to be *positive* 3.2eV. The difference between this result and others [e.g. 1, 2] could possibly be attributed to differences in surface conditions.

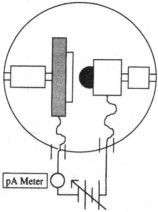

Figure 1   Schematic diagram of the electron emission measurement system (EEMS). The EEMS is part of an integrated processing system (IPS) allowing ultra-high vacuum transfer throughout. The EEMS employs a retractable spherical anode with X-Y translation. The entire system is spring supported for vibration isolation.

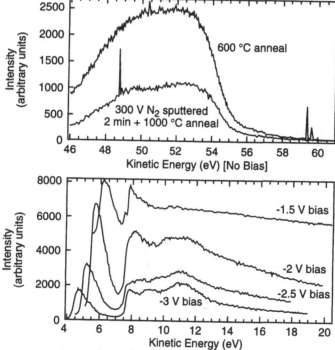

Figure 2   Synchrotron-based ARUPS spectra from air-exposed AlN/GaN/Al$_2$O$_3$(0001). The electron affinity was determined to be $\chi \sim +3.2$eV. The results were essentially unchanged following N$_2$ sputter cleaning and anneal.

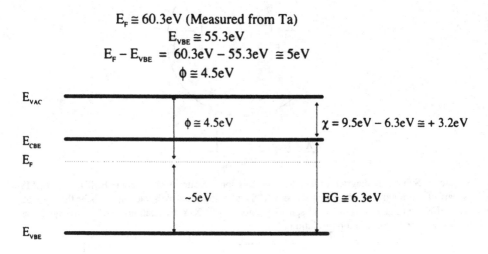

$$E_F \cong 60.3eV \text{ (Measured from Ta)}$$
$$E_{VBE} \cong 55.3eV$$
$$E_F - E_{VBE} = 60.3eV - 55.3eV \cong 5eV$$
$$\phi \cong 4.5eV$$

E_{VAC}

$\phi \cong 4.5eV$ $\chi = 9.5eV - 6.3eV \cong + 3.2eV$

E_{CBE}
E_F

~5eV $EG \cong 6.3eV$

E_{VBE}

Figure 3   Schematic energy band diagram for AlN as determined from ARUPS.   The electron affinity was ~+3.2eV.

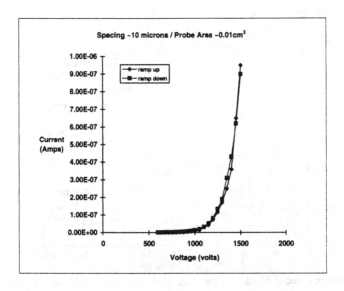

Figure 4   Electron field emission from MBE-grown Al on n+ Si.

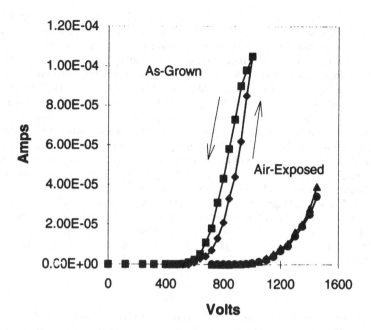

Figure 5 Electron field emission from MBE-grown AlN on n+ Si as-grown and after air exposure. Probe collection area was estimated at 0.01cm2 and the sample-probe separation distance was ~10μm.

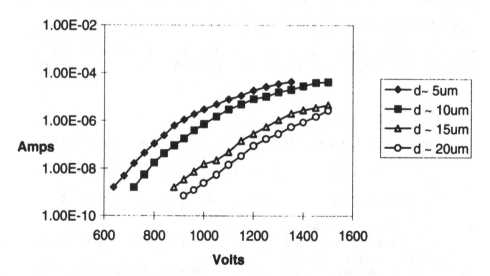

Figure 6 Electron field emission I-V curves from air-exposed AlN on n+ Si for different anode/cathode separation distances.

The Al/Si film showed very stable emission with a turn-on of 120V/μm (defined at 10μA-cm^{-2}) and ~100μA-cm^{-2} emission at 140V/μm (Fig. 4). Anode/cathode separation distance was 10μm. X-ray photoelectron spectroscopy (XPS) prior to in-vacuo transfer to the EEMS showed a clean Al surface with very little oxygen or carbon present. The relatively high turn-on and non-linear behavior was expected from a metal surface and the lack of hysteresis in the field ramp-down is consistent with a highly conductive film and substrate. Fig. 5 shows electron field emission measurements obtained from AlN/Si. The as-grown sample in-vacuo transferred to the EEMS exhibited turn-on at ~60V/μm and yielded ~10mA-cm^{-2} at 100V/μm.. Anode/cathode separation was 10μm. The non-linear behavior and relatively high turn-on field are consistent with a positive electron affinity surface. The hysteresis observed during field ramp-down is attributed to a slight charging of the AlN film. Air exposure of the AlN film caused a shift in turn-on to 90V/μm and yielded a reduced current density of ~0.1mA-cm^{-2} at 100V/μm. XPS on this surface indicated the presence of a large amount of O and C. It is interesting to note that the turn-on field increased by approximately 50% upon air exposure and the hysteresis disappeared. It is believed that the surface contamination increased the effective electron affinity of the surface and provide a conductive path to reduce charging. I-V characteristics of the air-exposed AlN/Si surface were plotted for different anode/cathode spacings (Fig. 6). The accuracy in setting anode/cathode separation distance was estimated to be within ±2μm.

CONCLUSIONS

We have measured an electron affinity of +3.2eV for the AlN surface of a AlN/GaN/Al$_2$O$_3$(0001) sample using ARUPS. Electron field emission measurements on another AlN sample - an AlN film grown on n+ Si - yielded non-linear I-V behavior with a relatively high turn-on field of 60V/μm. Air exposure of the AlN surface and re-measurement of I-V yielded an increased turn-on to 90V/μm. These observations are consistent with a positive electron affinity AlN surface.

ACKNOWLEDGMENTS

The authors gratefully acknowledge provision of one of the samples from APA Optics, Inc. and funding support of this program from BMDO/IST through ONR (Contract No. N00014-95-C-0230).

REFERENCES

1. V.M. Bermudez, T.M. Jung, K. Doverspike, and A.E. Wickenden, J. Appl. Phys. **79**, 110 (1996)
2. M.C. Benjamin, Cheng Wang, R.F. Davis, and R.J. Nemanich, Appl. Phys. Lett. **64**, 3288 (1994)
3. F.J. Himpsel, J.A. Knapp, J.A. van Vechten, and D.E. Eastman, Phys. Rev. B **20**, 624 (1979)
4. R.W. Pryor, Appl. Phys. Lett., **68**, 1802 (1996)
5. J.A. Christman, A.T. Sowers, M.D. Bremser, B.L. Ward, R.F. Davis, and R.J. Nemanich, Mat. Res. Soc. Symp. Proc., (1996), in press.
6. W. Zhu, G.P. Kochanski, S. Jin, and L. Seibles, J. Vac. Sci. Technol. B **14** 2011 (1996)

Part V

Device Performance and Design

VALENCE BAND PHYSICS IN WURTZITE GaN

T. Azuhata *, T. Sota *, S. Chichibu **, A. Kuramata †, K. Horino †, M. Yamaguchi ††,
T. Yagi ‡, and S. Nakamura ‡‡
* Department of Electrical, Electronics, and Computer Engineering, Waseda University, 3-4-1
Ohkubo, Shinjuku, Tokyo 169, JAPAN
** Faculty of Science and Technology, Science University of Tokyo, 2641 Yamazaki, Noda,
Chiba 278, JAPAN
† Optical Semiconductor Devices Laboratory, Fujitsu Laboratories Ltd., 10-1 Morinosato-
Wakamiya, Atsugi, Kanagawa 243-01, JAPAN
†† Faculty of Engineering, Hokkaido University, Sapporo 060, JAPAN
‡ Research Institute for Electronic Science, Hokkaido University, Sapporo 060, JAPAN
‡‡ Department of Research and Development, Nichia Chemical Industries Ltd., 491 Oka,
Kaminaka, Anan, Tokushima 774, JAPAN

ABSTRACT

We present a summary of recent progress towards the understanding of the valence-band physics in wurtzite GaN. Systematic studies have been performed on the strain dependence of the free exciton resonance energies by photoreflectance measurements using well-characterized samples. Analyzing the experimental data with the Hamiltonian appropriate for the valence bands, the values have been determined of the crystal field splitting, the spin-orbit splitting, the shear deformation potential constants, and the energy gap in the unstrained crystal. Discussions are given on the strain dependence of the energy gaps, of the effective masses, and of the binding energies for the free exciton ground states as well as on the valence band parameters. Using the obtained values and the generalized Elliott formula, the fundamental optical absorption spectra obtained experimentally were analyzed. The values of the elastic stiffness constants, which play a crucial role to determine the shear deformation potential constants, are also given.

INTRODUCTION

The establishment of crystal growth techniques for GaN epitaxial layers of high quality has accelerated studies on various physical properties of GaN in a past few years [1]. From a technological viewpoint, one of us (S.N.) has put the high-brightness InGaN/AlGaN double-heterostructure light-emitting diodes exhibiting from yellow to ultraviolet colors to practice [2] and has succeeded in the room-temperature pulse [3] and continuous-wave [4] operation of InGaN /InGaN multi-quantum-well laser diodes. On the other hand, from a physical viewpoint, understanding has not been fully achieved on the valence band physics, the exciton dynamics, and so on.

Wurtzite GaN grown on sapphire substrates with GaN or AlN buffer layers always suffers a certain amount of in-plane biaxial strain because of the mismatch of the lattice constants and the thermal expansion coefficients between GaN (and/or AlN) and sapphire. As is well known from a viewpoint of application, the strain-induced effects can, depending on the circumstances, rather become an advantage for achieving lower threshold current density and higher differential quantum efficiency [5]. Thus it is important to achieve a deeper understanding on the electronic structures of GaN under in-plane biaxial strain, in particular, the strain dependence of the valence band structures consisting of the three separate bands.

Experimental [6-15] and theoretical [6,16,17] values have been reported of the pressure

445

coefficient and the deformation potential constants for GaN and related nitrides, and the effects of biaxial strain on the valence band structures in zinc-blende and wurtzite GaN have been deduced from theoretical calculations [18,19]. However, there remains uncertainty arising from the fact that the values for some physical parameters such as the deformation potential constants are not yet reliable.

In this paper, we have performed photoreflectance(PR) measurements at 10 K for samples with various values of residual in-plane strain and obtained the dependence of the three exciton resonance energies on strain along the c-axis, ε_{zz}. We have also derived the expressions for them using the appropriate Hamiltonian [20] for the valence bands of GaN, whose space group is C_{6v}^4, within the cubic approximation, where the dependence of the binding energies for the exciton ground states on strain has been taken into account. By fitting the derived expressions to the experimental data, we have obtained the values for the crystal field splitting, the spin orbit splitting, the deformation potential constants, and the energy gap in the unstrained crystal. It has been found that the B- and C-valence bands show an anticrossing at $\varepsilon_{zz} \approx -0.07\%$. Discussions are given not only on the values for the physical parameters used [21] and obtained herein but also on the strain dependence of the valence band effective masses, the exciton binding energies, and the energy gaps between the conduction band and the three valence bands.

We have also measured Brillouin and fundamental optical absorption spectra of GaN. The former were used to determine the values of the elastic stiffness constants, which are used in calculating the deformation potential constants, and the latter were analyzed using the physical parameters obtained in this paper.

EXPERIMENT

Samples used herein are undoped wurtzite GaN epitaxial layers grown on sapphire(0001) substrates by the metalorganic chemical vapor deposition method [22,23], whose carrier concentrations are about 6×10^{16} cm^{-3}. We used various types of sample structures to control the lattice strain systematically. The samples are classified into three groups; (i) 1 - 10 μm GaN on low-temperature deposited thin GaN or AlGaN buffer layers, (ii) heterostructures consisting of GaN (0.1 - 0.2 μm)/Al$_{0.1}$Ga$_{0.9}$N (3 μm)/AlN buffer layer, and (iii) 0.004 - 2 μm GaN on GaN buffer layers, which are covered with Al$_{0.3}$Ga$_{0.7}$N. The quality of the samples is the same as that of the double-heterostructure light-emitting diodes [2] and laser diodes [3,4].

The values of the lattice constants, c, of wurtzite GaN were determined using the x-ray diffraction method and the values for residual strain along the c-axis were estimated by $\varepsilon_{zz}=(c-c_0)/c_0$, where c_0 is the lattice constant in the unstrained crystal and taken as 5.185 Å [24]. Note that all the values are those at room temperature because of the restriction of our equipments.

PR spectra were recorded at 10 K in a near-normal reflection configuration. A white halogen lamp as a probe light was focused onto the sample surface within 1 mm diameter and a mechanically chopped 325.0 nm line of a cw He-Cd laser was used as a modulation source. The reflected light was dispersed by a 67 cm focal length grating monochromator (McPHERSON 207) and detected using a Hamamatsu R-758 photomultiplier. To obtain the signal, $\Delta R/R$, the output of the lock-in amplifier proportional to the change in the reflectivity, ΔR, was divided by the dc-component of the output of the photomultiplier proportional to the reflectivity, R, with the use of an analog-digital converter and the signal processing program by a personal computer. The above-mentioned optical alignment enables us to record the PR spectra with low background level up to the energy of the pump laser because the monochromator acts as a filter to reject the pump laser. The accuracy and the resolution of the system were 0.5 meV and 1.2 meV, respectively. The measurements were performed in the low-field region. When necessary, polarization experiments were also made.

Typical PR spectra are shown in Fig. 1. Four transitions labeled A, B, C, and $A_{n=2}$ are observed. The transitions A, B, and C correspond to those between the three separate valence bands which belong, respectively, to the irreducible representations, Γ_9, Γ_7, and Γ_7, and the conduction band whose irreducible representation is Γ_7. As reported previously [25], these transitions are due to the free exciton resonances in wurtzite GaN of high quality. The structures labeled $A_{n=2}$ are due to the first excited states of the A exciton. To obtain the transition energies and the line widths, we used the low-field electroreflectance Lorentzian lineshape functional form [26] given by

$$
\frac{\Delta R}{R} = \mathrm{Re}\left[\sum_{j=1}^{p} C_j e^{i\theta_j} \left(E - E_{r,j} + i\Gamma_j \right)^{m_j} \right], \tag{1}
$$

where p is the number of the spectral function used in the fitting procedure, E is the photon energy, and C_j, θ_j, $E_{r,j}$, and Γ_j describe, respectively, the amplitude, the phase, the resonance energy, and the line width of the j-th spectrum. m_j are parameters representing the type of the critical point and are set to two corresponding to the exciton-type critical point throughout this paper. Eq. (1) has been fitted to the experimental data using the non-linear least-square method, to obtain the values

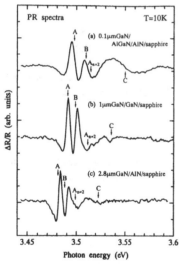

FIG. 1. Typical photoreflectance spectra.

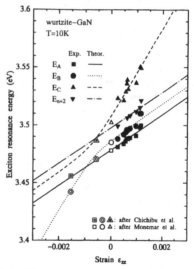

FIG. 2. The strain dependence of the free exciton resonance energies (closed plots : this work; double symbols : Ref. 29; open symbols Ref. 33).

447

for each parameter. The arrows shown in Fig. 1 indicate the resonance energies obtained from the above-mentioned procedure. Relative positions of the arrows to the peaks of the $\Delta R/R$ curves depend on phase factors, θ_j in Eq. (1).

According to the polarization selection rules for the space group C_{6v}^4, the A- and B-transitions are allowed for $E \perp c$ and the C transition for $E//c$, where E is the electric field associated with the incident light. Since the c-axis of the epitaxial layers is normal to the substrate surface, the condition $E \perp c$ is intrinsically satisfied in our experimental configuration. This is the reason why the A- and B-transitions dominate the spectra obtained experimentally. However small contribution from the C transition to the spectra is recognized. This depolarization might be due to the small but finite incident angle in our experimental configuration, the optical activity in wurtzite material [27], the polarization mixing arising from strain [28], and so on, though we do not know in detail.

To distinguish the C transition from others clearly, we tried to make quasi-polarization dependent PR measurements as follows. The incident angle was varied between 45° and 60° and for each case the suitable modulation laser intensity was adjusted. The nominal component of the light with the polarization $E \perp c$ or $E//c$ was selected from the reflected light using a Glan-Thomson prism and was again depolarized before entering the monochromator. From the results (not shown here), we assign the structures labeled C in Fig. 1 to the C exciton transitions.

The exciton resonance energies of GaN are plotted as a function of ε_{zz} in Fig. 2 , where the data measured using GaN epitaxial films with tensile strain by Chichibu et al. [29] are also shown by double symbols. It is found from Fig. 2 that the values of the resonance energies, E_A, E_B, E_C, and $E_{n=2}$, increase with increasing biaxial strain. As is shown below, E_A is a linear function of ε_{zz} and the ground state binding energy of the A exciton, $E_{ex,A}$, is not affected by strain. Therefore the least-square fit to the data of E_A and $E_{n=2}$ yields

$$E_A = 3.478 + 15.4\varepsilon_{zz} \text{ eV}, \tag{2a}$$

$$E_{n=2} = 3.498 + 15.4\varepsilon_{zz} \text{ eV}. \tag{2b}$$

From the difference between E_A and $E_{n=2}$, $E_{ex,A}$ is estimated to be 26 meV, which is simply set to four-thirds of the energy difference. The linear slope of 15.4 eV for the A-exciton is comparable to that of 12 eV reported by Amano *et al.* [8].

ANALYSIS

To analyze the strain dependence of the exciton resonance energies, we consider the valence band energy spectra near the Γ point under biaxial strain in the x-y plane. In this case the following relations hold

$$\varepsilon_{xx} = \varepsilon_{yy} = -\frac{C_{33}}{2C_{13}}\varepsilon_{zz} \text{ and } \varepsilon_{yz} = \varepsilon_{zx} = \varepsilon_{xy} = 0. \tag{3}$$

where $\varepsilon_{\alpha\beta}$ denote the components of the strain tensor and C_{ij} the components of the elastic stiffness constants.

We start from the Hamiltonian in the matrix form given by Bir and Pikus [20], which is suitable to the space group C_{6v}^4,

$$\mathcal{H} = \begin{Vmatrix} F & 0 & -H^* & 0 & K^* & 0 \\ 0 & G & \Delta & -H^* & 0 & K^* \\ -H & \Delta & \lambda & 0 & I^* & 0 \\ 0 & -H & 0 & \lambda & \Delta & I^* \\ K & 0 & I & \Delta & G & 0 \\ 0 & K & 0 & I & 0 & F \end{Vmatrix}, \tag{4}$$

where

$$\begin{aligned} &\Delta = \sqrt{2}\,\Delta_3, && H = i(A_6 k_z k_+ + D_6 \varepsilon_{z+} + A_7 k_+), \\ &F = \Delta_1 + \Delta_2 + \lambda + \theta, && I = i(A_6 k_z k_+ + D_6 \varepsilon_{z+} - A_7 k_+), \\ &G = \Delta_1 - \Delta_2 + \lambda + \theta, && \lambda = A_1 k_z^2 + A_2 k_\perp^2 + D_1 \varepsilon_{zz} + D_2 \varepsilon_\perp, \\ &K = A_5 k_+^2 + D_5 \varepsilon_+, && \theta = A_3 k_z^2 + A_4 k_\perp^2 + D_3 \varepsilon_{zz} + D_4 \varepsilon_\perp, \end{aligned} \tag{5}$$

with

$$\begin{aligned} &k_+ = k_x + i k_y, && k_\perp^2 = k_x^2 + k_y^2, \\ &\varepsilon_{z+} = \varepsilon_{xz} + i\varepsilon_{yz}, && \varepsilon_\perp = \varepsilon_{xx} + \varepsilon_{yy}, \\ &\varepsilon_+ = \varepsilon_{xx} - \varepsilon_{yy} + 2i\varepsilon_{xy}. \end{aligned}$$

Here k, ε, A_i (i=1 - 7), and D_i (i=1 - 6) denote, respectively, wavevector, strain, the Luttinger parameters, and the deformation potential constants. The quantity Δ_1 is called the crystal field splitting and the quantity $3\Delta_2$ is usually put equal to $3\Delta_3$, and called the spin-orbit splitting. For simplicity, we use the cubic approximation throughout this paper. In this approximation the constants Δ_i, A_i, and D_i satisfy the relations [20],

$$\begin{aligned} &\Delta_2 = \Delta_3, \\ &4A_5 - \sqrt{2}\,A_6 = -A_3, && 2A_4 = -A_3 = A_1 - A_2, && A_7 = 0, \\ &4D_5 - \sqrt{2}\,D_6 = -D_3, && 2D_4 = -D_3 = D_1 - D_2. \end{aligned} \tag{6}$$

It is straightforward to solve the eigenvalue problem with Eq. (4) according to Bir and Pikus [20]. The energies at the Γ point, i.e. $k = 0$, are

$$E_A^1 = E_A^0 + \left(D_1 - \frac{C_{33}}{C_{13}} D_2\right)\varepsilon_{zz} + \left(D_3 - \frac{C_{33}}{C_{13}} D_4\right)\varepsilon_{zz}, \tag{7a}$$

449

$$E^1_{B,C} = \frac{\Delta_1 - \Delta_2 + \left(D_3 - \frac{C_{33}}{C_{13}} D_4\right) \varepsilon_{zz}}{2} + \left(D_1 - \frac{C_{33}}{C_{13}} D_2\right) \varepsilon_{zz}$$

$$\pm \left[\left(\frac{\Delta_1 - \Delta_2 + \left(D_3 - \frac{C_{33}}{C_{13}} D_4\right) \varepsilon_{zz}}{2}\right)^2 + 2\Delta_2^2\right]^{1/2} \tag{7b}$$

Here $E^0_A = \Delta_1 + \Delta_2$, and the subscripts B and C correspond, respectively, to plus and minus sign in the right hand side of Eq. (7b). In Eq. (7) we have used the relation given by Eq. (3). E^1_A, E^1_B, and E^1_C denote the bands belonging to the irreducible representations, Γ_9, Γ_7, and Γ_7, respectively. Note that Eq. (7) for $\varepsilon_{zz} = 0$ is nothing but the equation derived by Hopfield [30]. The strain dependence of the effective masses for each band near the Γ point is given by the following expressions,

$$\frac{m_0}{m_{//,A}} = A_1 + A_3, \tag{8a}$$

$$\frac{m_0}{m_{//,B,C}} = A_1 + \frac{1}{2} A_3$$

$$\pm \frac{1}{4} \left[\left(\frac{\Delta_1 - \Delta_2 + \left(D_3 - \frac{C_{33}}{C_{13}} D_4\right) \varepsilon_{zz}}{2}\right)^2 + 2\Delta_2^2\right]^{-1/2} \left[\Delta_1 - \Delta_2 + \left(D_3 - \frac{C_{33}}{C_{13}} D_4\right) \varepsilon_{zz}\right] A_3, \tag{8b}$$

$$\frac{m_0}{m_{\perp,A}} = A_2 + A_4, \tag{8c}$$

$$\frac{m_0}{m_{\perp,B,C}} = A_2 + \frac{1}{2} A_4$$

$$\pm \frac{1}{4} \left[\left(\frac{\Delta_1 - \Delta_2 + \left(D_3 - \frac{C_{33}}{C_{13}} D_4\right) \varepsilon_{zz}}{2}\right)^2 + 2\Delta_2^2\right]^{-1/2} \left[\Delta_1 - \Delta_2 + \left(D_3 - \frac{C_{33}}{C_{13}} D_4\right) \varepsilon_{zz}\right] A_4, \tag{8d}$$

where $m_{//,\alpha}$ ($m_{\perp,\alpha}$) [α = A, B, and C] represents the effective mass parallel (perpendicular) to the c-axis for the α-valence band and m_0 is the free electron mass. The effective masses for the A valence band are not affected by strain.

To analyze the experimental data, we need the values for the Luttinger parameters, A_i, and the effective masses of the conduction band. According to Suzuki et al. [21], the Luttinger parameters are: $A_1 = -6.56$, $A_2 = -0.91$, $A_3 = 5.65$, $A_4 = -2.83$, $A_5 = -3.13$, and $A_6 = -4.86$. In this paper we do not use parameters A_5 and A_6, see Eqs. (7) and (8). To make our calculations consistent, we also use the values for the conduction band effective masses reported by Suzuki et al. [21] , i.e. $m_{//,e}/m_0 = 0.20$ and $m_{\perp,e}/m_0 = 0.18$, where the former (latter) is the electron effective mass parallel (perpendicular) to the c-axis. Those values tell us that the conduction band is nearly isotropic. When we use the above-mentioned effective mass parameters with the background dielectric constants $\varepsilon_{\perp,0} = 9.28$ and $\varepsilon_{//,0} = 10.1$ [31] which are typical values reported, a variational method [20,32] leads to $E_{ex,A} = 19$ meV. This does not agree with the value obtained from our experiments. To avoid treating so many parameters as adjustable ones, we have used slightly smaller values of $\varepsilon_{\perp,0} = 7.87$ and $\varepsilon_{//,0} = 8.57$, where the ratio, $\varepsilon_{\perp,0}/\varepsilon_{//,0}$, is fixed to 0.92 as obtained using the above typical values, throughout this paper to reproduce the experimentally obtained $E_{ex,A}$.

Let the conduction band minimum energy at the Γ point be E_{CB}. A concrete expression for it is written down as

$$E_{CB} = E_{gA} + E_A^0 + \Xi_D \varepsilon_{zz}, \tag{9}$$

where E_{gA} is the energy gap between the conduction and the A valence bands in the unstrained crystal and Ξ_D the combined dilational component of the deformation potential acting on the conduction band. Here the conduction band is considered as isotropic. The resonance energies observed experimentally are given by

$$E_\alpha = E_{CB} - E_{ex,\alpha} - E_\alpha^1, \quad \alpha = A, B, \text{ and } C. \tag{10}$$

Thus, E_{gA} is determined as 3.504 eV from Eq. (2a) and $E_{ex,A} = 26$ meV. Since the slope of E_A with respect to ε_{zz}, 15.4 eV, is expressed from Eqs. (7a), (9), and (10) as

$$\Xi_D - \left(D_1 - \frac{C_{33}}{C_{13}}D_2\right) - \left(D_3 - \frac{C_{33}}{C_{13}}D_4\right), \tag{11}$$

we treat Δ_1, $\Delta_2(=\Delta_3)$, and $\Xi_D - (D_1 - D_2 C_{33}/C_{13})$ as adjustable parameters. Note that we cannot determine Ξ_D and $(D_1 - D_2 C_{33}/C_{13})$ separately at this stage.

Using the non-linear least-square method, we have obtained the values for these parameters as $\Delta_1 = 22$ meV, $\Delta_2 = 5$ meV,

$$\Xi_D - \left(D_1 - \frac{C_{33}}{C_{13}}D_2\right) = 38.9 \text{ eV}, \tag{12}$$

and thus

$$D_3 - \frac{C_{33}}{C_{13}}D_4 = 23.6 \text{ eV}. \tag{13}$$

The resonance energies, E_A, E_B, and E_C, calculated with the use of the values obtained for the physical parameters are shown in Fig. 2 with the experimental data including those for unstrained

GaN measured by Monemar et al. [33]. In Fig. 2 the binding energy of the first excited state for the A exciton is simply set to be a quarter of $E_{ex,A}$. Good agreement between the experimental data and the theoretical curves has been achieved.

DISCUSSION

We first discuss the strain dependence of the effective masses, the exciton binding energies, and the energy gaps. In Figs. 3(a) and 3(b) are shown the strain dependence of the relative valence band energy levels and of the valence band effective masses calculated based on Eqs. (7) and (8), respectively. The inset in Fig. 3(a) demonstrates the strain dependence of the energy gaps between the conduction band and the three valence bands. It is found from Fig. 3 that the B- and C-valence bands show an anticrossing and exchange the band characteristics near $\varepsilon_{zz} \approx -0.07$ %. Reflecting the strain dependence of the valence band structure, the corresponding energy gaps also show an anticrossing at $\varepsilon_{zz} \approx -0.07$ %. It is found that controlling the magnitude of residual strain can modify the energy gaps by a considerable amount. Judging from Fig. 3(a), for $\varepsilon_{zz} << -0.1$ %, i.e., for large in-plane tensile strain, the valence band structure may be considered as one simple band.

The strain dependence of the reduced effective masses is shown in Fig. 4(a). In Fig. 4(b) is shown the strain dependence of the binding energies for the exciton ground states, which have been calculated based on a variational method using the reduced masses given in Fig. 4(a). Reflecting the anticrossing feature between the B- and C-valence bands, the magnitude of the exciton binding energies, $E_{ex,B}$ and $E_{ex,C}$, is interchanged continuously at $\varepsilon_{zz} \approx -0.07$ %, while $E_{ex,A}$ is not affected by strain.

Strictly speaking, the reliability of our results mentioned in the above two paragraphs depends on the Luttinger parameters given by Suzuki et al. [21] Therefore, it is desirable to reconfirm the values for the band parameters of GaN not only theoretically but also experimentally to obtain correct understanding on the electronic states and optical properties of GaN. Bearing this in mind, we discuss the values of several physical parameters obtained herein in the following.

The crystal field splitting and the spin orbit splitting are given by $\Delta_{cr} (= \Delta_1) = 22$ meV and $\Delta_{so} (=3\Delta_2) = 15$ meV. The values for Δ_{cr} and Δ_{so} obtained herein are compared with those reported previously in Table I. It is found that the parameter values obtained are similar to those reported in Refs. 34-36. The value of Δ_{cr} in Refs. 21 and 38 seems to be too large. We would like to emphasize that our value of Δ_{so} for wurtzite GaN is close to that obtained for zinc-blende GaN [39] as is the case for many other semiconductors [40].

We consider the deformation potential constants in wurtzite GaN. In order to calculate the values of D_3 and D_4 using Eqs. (6) and (13), the values of the elastic stiffness constants are needed. We have measured Brillouin spectra of 4 μm GaN on sapphire at room temperature. The results are as follows: $C_{11} = 36.5$, $C_{33} = 38.1$, $C_{44} = 10.9$, $C_{66} = 11.5$, $C_{12} = 13.5$, and $C_{13} = 11.4$ in units of 10^{11} dyn/cm^2. Thus, we have obtained the shear deformation potential constants as $D_3 = 8.82$ eV and $D_4 = -4.41$ eV. Note again that we cannot obtain the values for Ξ_D, D_1, and D_2 separately. Suzuki and Uenoyama have reported the following values for D_3 and D_4 which are obtained by *ab initio* calculations [41]; $D_3 = 3.03$ eV and $D_4 = -1.52$ eV. Those values are about one-third of ours. Here we would like to emphasize the following. It is the first time that the values of the deformation potential constants are experimentally determined based on the

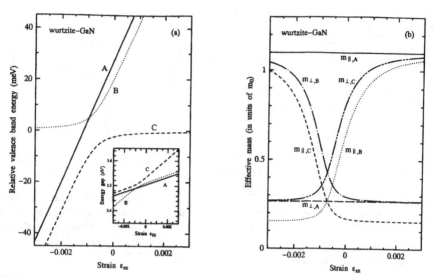

FIG. 3. The strain dependence of (a) the relative valence band energy levels and (b) the valence band effective masses.

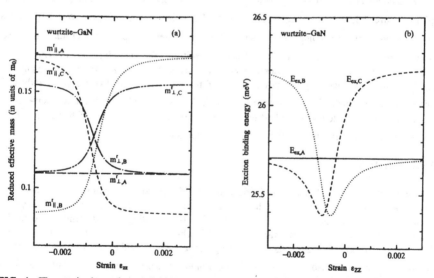

FIG. 4. The strain dependence of (a) the reduced effective masses and (b) the binding energies of the exciton ground states.

TABLE I. The values for the crystal field splitting, Δ_{cr}, and the spin orbit splitting, Δ_{so}, in wurtzite GaN under the cubic approximation.

$\Delta_{cr} (= \Delta_1)$ (meV)	$\Delta_{so} (= 3\Delta_2 = 3\Delta_3)$ (meV)	
22	11	Dingle et al. [b]
72.9	15.6	Suzuki et al. [c]
10.0 ± 0.1	17.6 ± 0.1 [a]	Gil et al. [d]
24.7	17.3	Reynolds et al [e]
19.2	10.4	Orton [f]
9.3	19.7	Korona et al [g]
42	13	Wei and Zunger [h]
22	15	Present work

[a] This value is an averaged one. In the original paper, Gil. et al. gave the values for Δ_2 and Δ_3 separately, i.e., $\Delta_2 = 6.2 \pm 0.1$ meV and $\Delta_2 = 5.5 \pm 0.1$ meV.

[b] Reference 34. [c] Reference 21. [d] Reference 11. [e] Reference 35. [f] Reference 36. [g] Reference 37. [h] Reference 38.

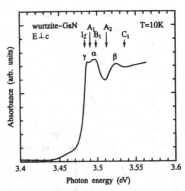

FIG. 5. Fundamental optical absorption spectra. The arrows represent the exciton resonance energy positions obtained from the photoluminescence [25] and PR measurements.

appropriate Hamiltonian for wurtzite GaN and measurements of lattice constants, both of which are important in evaluating the strain-dependence.

We have also measured fundamental optical absorption spectra of GaN using 1 μm GaN on sapphire. Figure 5 is the Absorbance spectra at 10 K, where two main peaks, α and β, and a shoulder, γ, are seen. According to the results of PR measurements presented in this paper, the peak α is due to the ground states of the A and B excitons. The shoulder γ is attributed to the transition associated with excitons bound to neutral donors. We analyzed the spectra using the generalized Elliott formula [42], which was extended for the system with two valence-bands, and physical parameters obtained above. As a result, the peak β was found to be due to a superposition of the transitions associated with the first excited A and B excitons and those with the continuum states. The peak α is clearly observed even at room temperature.

CONCLUSIONS

We have studied the biaxial strain dependence of the exciton resonance energies both experimentally and theoretically. We have determined the values of the important physical parameters, i.e., the crystal field splitting, the spin orbit splitting, the deformation potential constants, and the energy gap in the unstrained crystal, which are, we believe, the most reliable up to date. It has also been found that the anticrossing between the B- and C-valence bands occurs at $\varepsilon_{zz} \approx -0.07$ %. We have also performed Brillouin scattering measurements and determined all the values of the elastic stiffness constants. The fundamental optical absorption spectra have been measured and analyzed using the physical parameters obtained herein.

ACKNOWLEDGMENTS

The authors are grateful to K. Domen, Dr. H. Okumura, Dr. S. Yoshida, and Professor K. Suzuki for stimulating discussions. The authors wish to thank A. Takahashi and K. Haga for their help in experiments. One of the authors (S.C.) is grateful to Dr. S. Shirakata for the instructions of PR measurements. Professor H. Nakanishi is acknowledged for continuous encouragements. The work at Science University of Tokyo was supported in part by the Sasagawa Scientific Research Grant from the Japan Science Society, and the work at Waseda University was partly supported by the Izumi Science and Technology Foundation, the Japan Securities Scholarship Foundation, and the Ministry of Education, Science, Sports and Culture of Japan (High-Tech Research Center Project).

REFERENCES

1. For a review see, for example, S. Strite and H. Morkoç, J. Vac. Sci. Technol. B **10**, 1237 (1992) and *Properties of group III nitrides*, edited by J.H. Edgar, EMIS Datareviews Series No. 11 (INSPEC, the Institution of Electrical Engineers, London, United Kingdom, 1994).
2. S. Nakamura, M. Senoh, and T. Mukai, Jpn. J. Appl. Phys. **32**, L8(1993); S. Nakamura, M. Senoh, N. Iwasa, and S. Nagahama, ibid. **34**, L797(1995); S. Nakamura, T. Mukai, and M. Senoh, Appl. Phys. Lett. **64**, 1687(1994).
3. S. Nakamura, M. Senoh, S. Nagahama, N. Iwasa, T. Yamada, T. Matsushita, H. Kiyoku, and Y. Sugimoto, Jpn. J. Appl. Phys. **35**, L74(1996); **35**, L217(1996); Appl. Phys. Lett. **68**, 2105(1996); **69**, 1477(1996).
4. S. Nakamura, M. Senoh, S. Nagahama, N. Iwasa, T. Yamada, T. Matsushita, Y. Sugimoto, and H. Kiyoku, Appl. Phys. Lett. **69**, 4056(1996); **70**, 868(1997).
5. For example, see P.J.A. Thijs, L.F. Tiemeijer, P.I. Kuindersma, J.J.M. Binsma, and T.V. Dongen, IEEE J. Quantum Electron. **27**, 1426(1991) and references cited therein.
6. P. Perlin, I. Gorczyca, N.E. Christensen, I. Grzegory, H. Teisseyre, and T. Suski, Phys. Rev. B **45**, 13307(1992).
7. D.L. Camphausen and G.A.N. Connell, J. Appl. Phys. **42**, 4438(1971).
8. H. Amano, K. Hiramatsu, and I. Akasaki, Jpn. J. Appl. Phys. **27**, L1384(1988).
9. K. Naniwae, S. Itoh, H. Amano, K. Itoh, K. Hiramatsu, and I. Akasaki, J. Cryst. Growth **99**, 381(1990).
10. W. Shan, T. Schmidt, R.J. Hauenstein, J.J. Song, and B. Goldenberg, Appl. Phys. Lett. **66**, 3492(1995).
11. B. Gil, O. Briot, and R.-L. Aulombard, Phys. Rev. B **52**, R17028(1995).
12. D. Volm, K. Oettinger, T. Streibl, D. Kovalev, M. Ben-Chorin, J. Diener, B.K. Meyer, J. Majewski, L. Eckey, A. Hoffmann, H. Amano, I. Akasaki, K. Hiramatsu, and T. Detchprohm, Phys. Rev. B **53**, 16543(1996).
13. S.L. Chuang and C.S. Chang, Phys. Rev. B **54**, 2491(1996).
14. M. Tchounkeu, O. Briot, B. Gil, J.P. Alexis, and R.-L. Aulombard, J. Appl. Phys. **80**, 5352(1996).
15. W. Shan, R.J. Hauenstein, A.J. Fischer, J.J. Song, W.G. Perry, M.D. Bremser, R.F. Davis, and B. Goldenberg, Phys. Rev. B **54**, 13460(1996).
16. I. Gorczyca and N.E. Christensen, Solid State Commun. **80**, 335(1991); Physica B **185**, 410(1993).
17. N.E. Christensen and I. Gorczyca, Phys. Rev. B **50**, 4397(1994).
18. S. Kamiyama, K. Ohnaka, M. Suzuki and T. Uenoyama, Jpn. J. Appl. Phys. **34**, L821 (1995).

19. M. Nido, Jpn. J. Appl. Phys. **34**, L1513(1995).

20. G.L. Bir and G.E. Pikus, *Symmetry and Strain-Induced Effects in Semiconductors* (Wiley, New York, 1974).

21. M. Suzuki, T. Uenoyama, and A. Yanase, Phys. Rev. B **52**, 8132(1995).

22. S. Nakamura, Jpn. J. Appl. Phys. **30**, L1705(1991); S. Nakamura, M. Senoh, and T. Mukai, ibid **30**, L1708(1991); S. Nakamura, Y. Harada, and M. Senoh, Appl. Phys. Lett. **58**, 2021 (1991).

23. A. Kuramata, K. Horino, K. Domen, R. Soejima, H. Sudo, and T. Tanahashi, in *Proceedings of the International Symposium on Blue Laser and Light Emitting Diodes*, edited by A. Yoshikawa, K. Kishino, M. Kobayashi, and T. Yasuda (Ohmsha, Tokyo, 1996) p.80.

24. M. Leszczynski, H. Teisseyre, T. Suski, I. Grzegory, M. Bockowski, J. Jun, S. Porowski, K. Pakula, J.M. Baranowski, C.T. Foxon, and T.S. Cheng, Appl. Phys. Lett. **69**, 73(1996).

25. S. Chichibu, T. Azuhata, T. Sota, and S. Nakamura, J. Appl. Phys. **79**, 2784(1996).

26. D.E. Aspnes, Surf. Sci. **37**, 418(1973).

27. E.L. Ivchenko, in *Excitons*, edited by E.I. Rashba and M.D. Sturge (North-Holland, Amsterdam, 1982) Chapter 4 and references cited therein.

28. T. Koda, T. Murahashi, T. Mitani, S. Sakoda, and Y. Onodera, Phys. Rev. B **5**, 705(1972).

29. S. Chichibu, T. Azuhata, T. Sota, H. Amano, and I. Akasaki, (unpublished).

30. J.J. Hopfield, Phys. Chem. Solids **15**, 97(1960).

31. T. Azuhata, T. Sota, K. Suzuki, and S. Nakamura, J. Phys.: Condens. Matter **7**, L129(1995); T. Azuhata, T. Matsunaga, K. Shimada, K. Yoshida, T. Sota, K. Suzuki, and S. Nakamura, Physica B **219&220**, 493(1996) and references cited therein.

32. R.W. Keyes, IBM J. Res. Develop. **5**, 65(1961).

33. B. Monemar, J.P. Bergman, I.A. Buyanova, W. Li, H. Amano, and I. Akasaki, MRS Internet J. Nitride Semicond. Res. **1**, 2(1996).

34. R. Dingle, D.D. Sell, S.E. Stokowski, and M. Ilegems, Phys. Rev. B **4**, 1211(1971).

35. D.C. Reynolds, D.C. Look, W. Kim, Ö. Aktas, A. Botchkarev, A. Salvador, H. Morkoç, and D.N. Talwar, J. Appl. Phys. **80**, 594(1996).

36. J.W. Orton, Semicond. Sci. Technol. **11**, 1026(1996).

37. K.P. Korona, A. Wysmolek, K. Pakula, R. Stepniewski, J.M. Baranowski, I. Grzegory, B. Lucznik, M. Wróblewski, and S. Porowski, Appl. Phys. Lett. **69**, 788(1996).

38. S.-H. Wei and A. Zunger, Appl. Phys. Lett. **69**, 2719(1996).

39. G. Ramírez-Flores, H. Navarro-Contreras, A. Lastras-Martínez, R.C. Powell, and J.E. Greene, Phys. Rev. B **50**, 8433(1994).

40. For example , see *Landolt-Börnstein, Numerical Data and Functional Relationships in Science and Technology* (Springer-Verlag, Berlin, 1982) Vol. 17a.

41. M. Suzuki and T. Uenoyama, in *Proceedings of the International Symposium on Blue Laser and Light Emitting Diodes*, edited by A. Yoshikawa, K. Kishino, M. Kobayashi, and T. Yasuda (Ohmsha, Tokyo, 1996) p.368.

42. H. Haug and S.W. Koch, *Quantum theory of the optical and electronic properties of semiconductors* (World Scientific, Singapore, 1993) p. 240.

Comparison of Electron and Hole Initiated Impact Ionization in Zincblende and Wurtzite Phase Gallium Nitride

E. Bellotti[*], I. H. Oguzman[**], J. Kolnik[***], K. F. Brennan[*], R. Wang[†]and P. P. Ruden[†]
[*]School of ECE, Georgia Tech, Atlanta, GA, 30332, bellotti@groucho.mirc.gatech.edu
[**]National Semiconductor, Arlington, TX 76017
[***]Symbios Logic Corp., Colorado Springs, CO 80916
[†]Dept. of Electrical Engineering, University of Minnesota, MN 55455

ABSTRACT

In this paper, we present the first calculations of the electron and hole impact ionization coefficients for both wurtzite and zincblende phase GaN as a function of the applied electric field. The calculations are made using an ensemble Monte Carlo simulator including the full details of the conduction and valence bands derived from an empirical pseudopotential calculation. The interband impact ionization transition rates for both carrier species are determined by direct numerical integration including a wavevector dependent dielectric function. It is found that the electron and hole ionization coefficients are comparable in zincblende GaN at an applied field of ~ 3 MV/cm, yet vary to a slight degree at both higher and lower field strengths. In the wurtzite phase, the electron and hole coefficients are comparable at high fields but diverge at lower applied fields. The most striking result is that the ionization rates are predicted to be substantially different for both carrier species between the two phases. It is predicted that the ionization rates for both carrier species in the zincblende phase are significantly higher than in the wurtzite phase over the full range of applied fields examined.

INTRODUCTION

For most current commercial semiconductor device applications, silicon is the material of choice. However, silicon devices are incapable of operating at elevated temperatures exceeding 250 C [1]. In addition, silicon devices are of limited usage in high frequency, high power and high radiation environments [1,2]. The relatively low thermal conductivity of silicon also limits the packing density of silicon integrated circuits and requires sophisticated packaging techniques. Finally, owing to its indirect energy gap, silicon is limited in optoelectronic device applications, specifically as light emitters. There is an emerging commercial market for semiconductor devices which can operate at high temperatures, at both high power and frequencies as well as a need to increase the packing density within integrated circuits. Specific commercial applications which require high temperature operation are in aerospace (turbine engines) satellites, geothermal wells, and automobile engines, while future wireless communications systems will require high power, high frequency amplifiers. In all of these applications, silicon devices approach their fundamental limits of performance [3].

Wide bandgap semiconductors form the most attractive alternative to silicon for high temperature, and high power applications because of several inherent material advantages. Among these are the high thermal conductivity, high saturation drift velocity, small dielectric constant, high breakdown voltage, and very low thermally generated leakage current present in wide gap materials. Additionally, some of the wide band gap semiconductors, particularly GaN and the related III-nitrides, are attractive candidates for blue/ultraviolet detectors and emitters owing to the direct band gap present in these materials. For these reasons, GaN and the related III-nitride materials have attracted great interest and much work has been done on assessing their properties [4,5]. Determination of the transport properties and full device potential of GaN has been frustrated to some extent by the relative technological immaturity of GaN. Nevertheless, some electronic devices, blue light emitters and detectors have been made [6-9] from GaN and related III-nitride materials.

In many device structures, particularly in short channel field effect transistors and photodiodes, high electric fields can be present. Consequently, breakdown via carrier multiplication or tunneling can occur. The prediction of device breakdown depends critically upon knowledge of the carrier ionization coefficients in the constituent material. To date, little information is available about the high field breakdown of GaN and its related materials. The only experimental measurements that have been reported, indicated microplasmic breakdown in GaN p-n junction devices [10]. However, the critical information necessary for modeling breakdown in devices, i.e., the carrier ionization coefficients have not yet been reported. It is the purpose of this paper to present the first information about the carrier ionization coefficients in bulk GaN. Calculated values of both the electron and hole ionization coefficients in both zincblende and wurtzite polytypes of bulk GaN are presented.

MODEL DESCRIPTION

The transport calculations for the impact ionization coefficients are made using an ensemble Monte Carlo simulation. The microscopic carrier histories within the Monte Carlo simulator are traced in three dimensions in both real and k-space. The full details of the band structure are included in the Monte Carlo models for both electron and hole transport [11-13]. The carrier-phonon scattering rates are determined using the improved phonon scattering technique of Chang et al. [14] in which some quantum effects are taken into account. The magnitude of the high energy deformation potential constants chosen here and the resulting scattering rates are somewhat uncertain. This is due to the almost complete lack of experimental information about the high energy transport features of GaN. At best, only fragmentary information is presently available about the high energy transport properties of even the most studied semiconductors. As a result, some free parameters exist within the present model resulting in some loss of its predictive powers. The high energy scattering rates used in the present calculations are determined by matching to the low energy rates, calculated from Fermi's golden rule. At low energies, we assume the standard total phonon scattering rate formulations, i.e, polar optical, acoustic and intervalley deformation potential within the conduction band [11] and polar optical, acoustic and nonpolar optical within the valence band [12]. The parameters chosen for these calculations have been reported in references 15 and 16. The scattering rate parameters, of course, have some degree of uncertainty at present. Nevertheless, the values chosen are the best presently available. The full details of how the parameters were chosen and their values are reported in references 15 and 16, and for brevity will not be repeated here. For either the electrons or the holes, the magnitude of the scattering rates are adjusted so as to be comparable between the two phases. Therefore, significant differences in the calculated ionization rates between the two phases can not be solely attributed to the choice of the scattering rates.

The workings of our Monte Carlo model are essentially standard and have been reported in depth in references 11 and 12. The dielectric function, used in the calculation of the ionization transition rates, is determined directly from the band structure calculation. The details of the dielectric function calculation are reported in reference 17. The electron and hole initiated impact ionization transition rates used within the respective Monte Carlo simulators are determined numerically following the technique described in reference 18. The transition rates are determined by integrating Fermi's golden rule for a two-body, statically screened Coulomb interaction over the possible final states using the numerically generated dielectric function and pseudowavefunctions. The resulting transition rates are then averaged over energy and incorporated into the Monte Carlo simulator in the usual manner.

CALCULATED RESULTS

The Monte Carlo calculated electron and hole ionization coefficients as a function of inverse applied electric field for both phases of GaN are presented in Figure 1. As can be seen from the figure both the electron and hole ionization rates in the zincblende phase are substantially higher than in the wurtzite phase over the full range of applied electric fields examined. It is further interesting to note that the electron and hole ionization coefficients in the zincblende phase are comparable at 3 MV/cm but are slightly different from one another at both lower and higher applied field strengths.

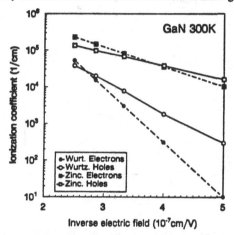

Figure 1: Calculated electron (solid) and hole (open) impact ionization coefficients as a function of inverse applied electric field in both wurtzite (circles) and zincblende (squares) bulk GaN. The rates are plotted for an electric field applied along the Γ-X and Γ-M directions for the zincblende and wurtzite phases respectively.

In the wurtzite phase of GaN, the electron and hole ionization coefficients are predicted to be comparable at the highest electric field strengths examined. At lower electric fields, the hole ionization coefficient is predicted to be larger than the electron ionization coefficient. However, it should be noted that the statistics are very poor at low field strengths in the wurtzite calculations and as such, no definitive conclusions can be drawn at these fields.

The substantial difference between the zincblende and wurtzite carrier ionization rates can be traced to the very different average carrier energies within the different phases. As can be seen from Figure 2, the average electron and hole energies are substantially higher in the zincblende GaN

from Figure 2, the average electron and hole energies are substantially higher in the zincblende GaN as opposed to the wurtzite GaN. The most striking difference between the average energies between the two phases occurs at intermediate field strengths. As can be seen from Figure 2, the average electron energy in zincblende GaN is in places nearly an order of magnitude higher than in wurtzite GaN. Though the electron-phonon scattering rates are not identical between the two phases, they have been adjusted, as mentioned above, to be comparable over the full energy range. It is therefore expected that the difference in the average electron energies and subsequently the electron initiated impact ionization coefficients between phases is not attributable solely to the slightly different scattering rates used for each phase. Instead, it is hypothesized here that the primary difference in the average electron energies between phases stems from the higher value of the density of states in wurtzite in the energy range below 3 eV. This effectively corresponds to a higher "averaged" effective mass in the wurtzite phase at these energies. As a result, the energy bands are flatter and more scatterings occur which collectively act to impede the heating of the electrons in the wurtzite conduction band.

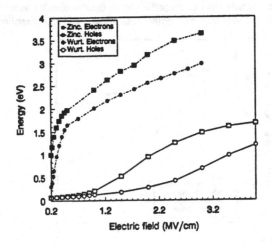

Figure 2: Calculated average electron (solid) and hole (open) energies as a function of applied electric field in zincblende (squares) and wurtzite (circles) GaN. The energies are plotted for an electric field applied along the Γ-X and Γ-M directions zincblende and wurtzite GaN respectively.

A similar argument can be made for the hole initiated impact ionization rate. As can be seen from Figure 2, there exists a substantial difference in the average hole energies between the wurtzite and zincblende phases which results in the different hole ionization coefficients within the two materials. As in the case for the electrons, the magnitudes of the phonon scattering rates were adjusted to be comparable through suitable choices of the deformation potentials. Some differences in the scattering rates remain of course, since the density of states are different between the two phases. Nevertheless, it can be reasonably conjectured that the primary difference in the average carrier energies is most likely due to the difference in the density of states and corresponding "effective" masses with their concomitant differences in the scattering rates.

Finally, in an attempt to learn more about the relative "softness" of the threshold for each carrier in each phase, the quantum yield was calculated. The quantum yield is defined as the average number of ionization events caused by a high energy particle until its kinetic energy relaxes below the ionization threshold through phonon scatterings and/or ionization events. The quantum yield for both carrier species in both phases of GaN is plotted in Figure 3. As can be seen from Figure 3, the quantum yield is substantially different between the electrons and holes. Notice that a quantum yield of 1.0 occurs at much lower energies for the holes than the electrons in both phases. This result implies that the threshold is softer for electron initiated ionization than hole initiated ionization in both phases of GaN. The relative hardness of the hole threshold as compared to the electron threshold is primarily due to the fact that the valence bands end very roughly around 7 eV in either phase, while the conduction bands extend to much higher energies.

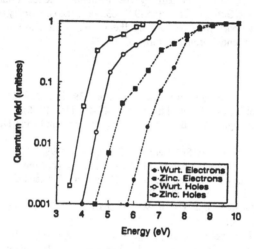

Figure 3: Calculated quantum yield for electrons and holes in zincblende and wurtzite GaN as a function of carrier energy.

CONCLUSIONS

This paper presents the first calculations of both the electron and hole initiated impact ionization coefficients for wurtzite and zincblende phase GaN. It is found that the electron and hole impact ionization coefficients are significantly higher in the zincblende phase than in the wurtzite phase over the full range of applied electric fields examined. Owing to the relatively similar phonon-scattering rates used for each different phase, it is conjectured that the primary difference between the ionization coefficients is due to the different density of states and resulting average carrier energies. It is further found that the hole ionization threshold is significantly harder than the electron ionization threshold in both phases. This is believed to be due to the fact that the valence bands end at much lower energies than the conduction bands in both phases. As such, if a carrier can achieve high energies in the valence bands, they are more likely to ionize than the electrons.

The results presented here represent a first attempt at understanding the breakdown properties and impact ionization rates in bulk GaN. It should be noted that the present level of understanding about GaN is quite limited and consequently only "educated guesses" can be made about the high energy band structure and phonon scattering rates. Since the calculations are highly sensitive to both the band structure and the phonon scattering rates, the results may vary significantly upon further refinements in the understanding of these two quantities. Further work, primarily experimental, is required in order to draw definitive conclusions about the nature of the ionization rates.

ACKNOWLEDGEMENTS

This work was sponsored in part by the National Science Foundation through a collaborative grant made to Georgia Tech (ECS-9313635) and the University of Minnesota (ECS-9408479). Additional support was received by the Georgia Tech group from the National Phosphor Center of Excellence through contract E21-Z22.

REFERENCES

1. J. B. Casady and R. W. Johnson, Solid-State Electron., 39, 1409 (1996).
2. G. Pensl and Th. Troffer, Solid State Phenomena, 47-48, 115 (1996).
3. T. P. Chow and R. Tyagi, 5th Intl. Symp. on Power Semiconductor Devices and ICs, 84 (1993).
4. S. Strite and H. Morkoc, J. Vac. Sci. Technol. B, 10, 1237 (1992).
5. S. N. Mohammad, A. A. Salvador, and H. Morkoc, Proceedings of the IEEE, 83, 1306 (1995).
6. S. Nakamura, M. Senoh, S. Nagahama, N. Iwasa, T. Yamada, T. Matsushita, Y. Sugimoto, and H. Kiyoku, Appl. Phys. Lett., 70, 868 (1997).
7. M. A. Khan, J. N. Kuznia, D. T. Olson, W. J. Schaff, J. W. Burm, and M. S. Shur, Appl. Phys. Lett., 65, 1121 (1994).
8. S. Nakamura, T. Mukai, and M. Senoh, Appl. Phys. Lett., 64, 1687 (1994).
9. M. A. Khan, J. N. Kuznia, D. T. Olson, J. M. Van Hove, M. Blasingame, and L. F. Reitz, Appl. Phys. Lett., 60, 2917 (1992).
10. V. A. Dmitriev, N. I. Kuznetsov, K. G. Irvine, and C. H. Carter, Jr., Mat. Res. Soc. Symp. Proc., 395, 909 (1996).
11. H. Shichijo and K. Hess, Phys. Rev. B, 23, 4197 (1981).
12. K. Brennan and K. Hess, Phys. Rev. B, 29, 5581 (1984).
13. M. V. Fischetti and S. E. Laux, Phys. Rev. B, 38, 9721, (1988).
14. Y. C. Chang, D. Z.-Y. Ting, J. Y. Tang, and K. Hess, Appl. Phys. Lett., 42, 26 (1983).
15. J. Kolnik, I. H. Oguzman, K. F. Brennan, R. Wang, P. P. Ruden, and Y. Wang, J. Appl. Phys., 78, 1033 (1995).
16. I. H. Oguzman, J. Kolnik, K. F. Brennan, R. Wang, T-N. Fang, and P. P. Ruden, J. Appl. Phys., 80, 4429 (1996).
17. R. Wang, P. P. Ruden, J. Kolnik, I. H. Oguzman, and K. F. Brennan, Mat. Res. Soc. Symp. Proc., 395, 601 (1995).
18. J. Kolnik, I. H. Oguzman, K. F. Brennan, R. Wang and P. P. Ruden, J. Appl. Phys., 79, 8838 (1996).

MOLECULAR DYNAMICS SIMULATION OF TRANSPORT IN DIAMOND AND GaN: ROLE OF COLLECTIVE EXCITATIONS

N. M. MISKOVSKY*, P. B. LERNER and P. H. CUTLER
Physics Department, Penn State University, University Park, PA 16802.
*nick@phys.psu.edu

ABSTRACT

In this paper we consider the role of collective excitations in transport of charge carriers in the conduction band of diamond and GaN. The present molecular dynamics simulation uses a Monte Carlo algorithm to determine the time between collisions with scattering rates calculated quantum mechanically using the Fermi Golden Rule. For very thin films (L=0.01 μm) and for low fields (F<10 V/μm), the energy spectra of both diamond and GaN contain a series of peaks which are attributed to the absorption and emission of discrete plasmons. In diamond,the intervalley LA , LO and TO scattering becomes increasingly more important at higher fields while electron-plasmon interactions decrease. For thicker diamond films (L=0.1 μm), there is hot electron transport for F~10 V/μm and quasi-ballistic transport for F~100 V/μm. In GaN with L=0.01 μm and fields F~10 V/μm, there is a discrete series of peaks in the energy spectrum corresponding to excitations associated with polar optical scattering.

INTRODUCTION

A multi-step model previously used to describe field emission from a thin film metal diamond composite cold cathode can be used to describe the field emission process in other similar wideband gap semiconductor devices involving nitride III-V compounds. The model features internal field emission at the substrate interface for the injection of electrons into the conduction band of diamond, transport through the film and subsequent emission into vacuum[1,2]. In this paper we investigate some new features of the charge transport in the conduction band in diamond and GaN.

Earlier calculations by the authors have demonstrated that both quasi-ballistic transport and hot electron transport through diamond and GaN films are possible[3]. However, the collective electron-plasmon interaction (e-pl) and the polar-optical interaction were not included. In this paper, we focus on the influence of these two scattering processes on the scattering statistics and the energy distribution. The functional form of these two different types of scattering processes is very similar despite their different physical origin[4]. Both are present in GaN: the electron-plasmon interaction due to the relatively high level of unintentional n-doping (10^{17} cm^{-3}), and the polar optical scattering due to the structure of GaN as a III-V compound. In the calculations to be described, both GaN and diamond will, to facilitate comparison, always have the same electron density in the simulation studies. Our ability to correlate the calculated energy spectra with corresponding scattering statistics provides important insights on the ability of the quasi-classical Monte Carlo algorithm to model electron energy losses due to collective excitations.

The fact that ballistic or hot-electron transport may be possible is a consequence of the following physical considerations. The emission rates for the collective mechanisms such as electron-plasmon scattering (or polar optical) scattering have a maximum which is comparable to the plasmon (or polar optical phonon) energy and decrease for faster electrons. On the other hand, optical phonon scattering roughly increases proportional to square root of energy and is expected to limit quasi-ballistic behavior for higher fields [5,6] . Thus it is important to add all possible phonon modes for a realistic simulation of electron losses at higher energies. Despite monotonically growing rates, if the electrons gain enough energy quickly from the field (this presumes a very strong field), the energy acquired from the field (E=F·L) cannot dissipate rapidly enough into the

463

lattice for values of the field above a critical value. In this regime, $E_{loss} \propto (\text{rate}) \cdot L/v \propto E^{1/2} L/v \propto$ L and the losses saturate with energy independent of the value of the field. The scattering processes for hot electrons are, of course, nonlinear and quasi-ballistic propagation is manifest only for sufficiently high fields and very thin films. The wideband gap semiconductors with high values for dielectric breakdown are the good candidates for observation of hot electron and quasi-ballistic behavior.

The present molecular dynamics simulation of electron transport through diamond and GaN uses a well-known Monte Carlo algorithm described in Ref. [7]. It combines classical molecular dynamics of conduction band electrons in the intervals between scatterings with the quantum mechanical scattering probabilities calculated using the Fermi Golden Rule. The initial distribution of electrons is taken to be an equilibrium thermal distribution at T=300 K at the bottom of the conduction band.

SCATTERING BEHAVIOR IN DIAMOND

In Fig. 1, we plot the results for the energy spectra of the conduction-band electrons for a film of thickness L=0.01 μm. For low fields, there are a series of peaks which are attributed to the absorption and emission of discrete plasmons (the density of conduction band electrons is n_e=8.3 x 10^{17} cm^{-3}). In this regime there is significant cooling of the electron gas so that the lattice and the electron gas are not in thermal equilibrium. For fields higher than 10 V/μm the central peak of the electron distribution appears at the energy comparable with the field energy FL, and the transport is

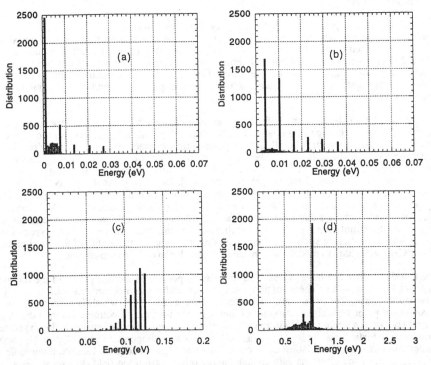

Fig. 1 Electron energy spectra for a diamond thin film with L=0.01 μm and n_e=8.3 x 10^{17} cm^{-3}: (a) F=0.1 V/μm, (b) F=1 V/μm, (c) F=10 V/μm, and (d) F=100 V/μm

quasi-ballistic. The change in behavior to quasi-ballistic propagation is reflected in the scattering statistics presented in Fig. 2. At low fields (i.e., low energies), the e-pl scattering dominates with, however, some scattering by acoustic phonons. For higher fields the number of acoustic phonon scatterings actually decreases (the scattering rates are higher, but there is less time for an electron to be scattered in the film). Eventually, for high fields the intervalley LA, LO and TO scattering begins to become increasingly important and the e-pl scattering events decrease markedly (see Fig. 2(d)). In additional simulations for a thicker film, L=0.1 μm, the energy spectrum is hot electron like for F~ 10 V/μm (with an average energy ~ 20% of the field energy) and is quasi-ballistic for fields of 100 V/μm. This same tendency was observed in earlier simulations for films as thick as 1μm with the electrons gaining a significant fraction of the field energy [8].

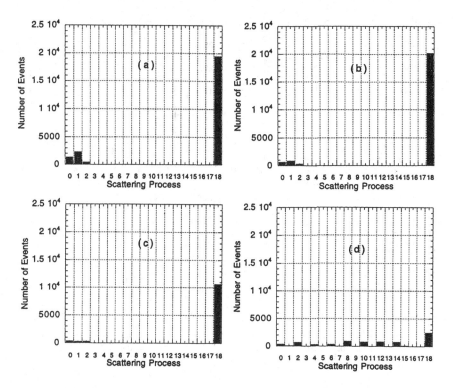

Fig. 2. Scattering statistics for conduction band electrons in diamond. The parameters are the same as in Fig. 1. Unless stated, all even(odd) processes refer to phonon emission(absorption). Process 0, elastic acoustic; 1 and 2 inelastic acoustic; 3 and 4 intervalley optical; 5 and 6 intervalley LO; 7 and 8 intervalley LA; 9 and 10 intervalley To; 11 and 12, same as 7 and 8 except to different ellipsoids; 13 and 14, same as 9 and 10 except to different ellipsoids; 15 e-e scattering; 16 e-h scattering; 17 and 18 e-pl scattering.

SCATTERING BEHAVIOR IN GaN.

For comparison we also performed simulations for GaN using the same electron density of n_e=8.3 x 10^{17} cm^{-3}. The results for the energy spectrum are shown in Fig. 3 and the scattering statistics are displayed in Fig. 4. Unlike diamond, GaN is a polar semiconductor (typical of the III-V compounds [4]) and the carriers interact with the optical mode of lattice vibrations giving rise to polar optical scattering.. For low energies a series of peaks are present which are attributed to the creation of discrete plasmons (from one to four). For realistic densities of conduction-band electrons (below 10^{21} cm^{-3}), the electron-plasmon scattering is a low-energy scattering mechanism whose rate falls off for energies greater than the plasmon energy [4]. For higher energies these discrete plasmon series become less distinct due to the increasing importance of other scattering processes (see Fig. 4). For even higher energies (F ~ 10 V/mm) we observe a discrete series with energy separation characteristic of polar optical scattering (E_p ~ 0.07 eV). For the highest field (F ~ 100 V/mm), the energy spectrum shows a majority peak at the field energy and secondary peaks at lower energies corresponding to the discrete excitations associated with polar optical scattering.

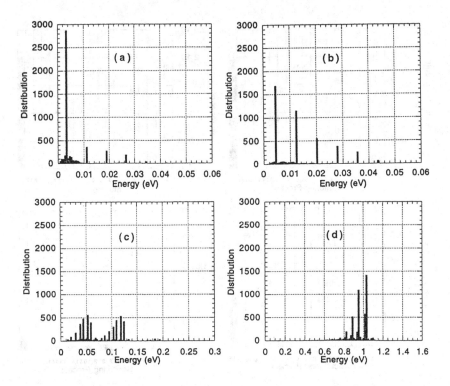

Fig. 3. The electron energy spectra for a GaN thin film with L=0.01 μm and n_e=8.3 x 10^{17} cm^{-3}: (a) F=0.1 V/μm, (b) F=1 V/μm, (c) F=10 V/μm, and (d) F=100 V/μm.

DISCUSSION

We have shown that the collective excitation modes (that is, the electron-plasmon and the polar optical modes) are manifested in a 0.01 μm film of the wideband gap semiconductors diamond and GaN. For thicker films their presence becomes less discernable because of the exchanges of much lower quanta of energy associated with acoustic-phonon processes. To verify that we indeed see the influence of collective excitations in the electron gas, we performed the same simulation with the e-e and e-pl scatterings excluded (Fig. 5). The qualitative form of the energy spectrum and the scattering statistics undergoes a dramatic change. Instead of a discrete series of peaks starting at

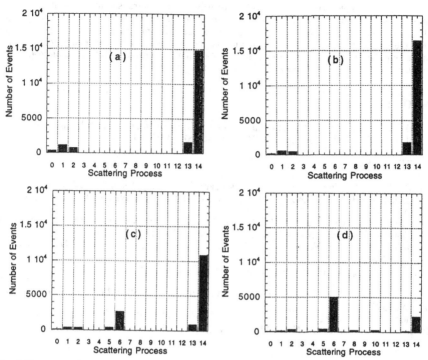

Fig. 4. The scattering statistics for conduction band electrons in GaN. The conditions correspond to those in Fig. 3. Unless stated, all even(odd) processes refer to phonon emission(absorption): Process 0, elastic acoustic; 1 and 2 inelastic acoustic; 3 and 4 polar acoustic; 5 and 6 polar optical; 7 -10 intervalley optical; 11, e-e scattering; 12, e-h scattering; 13 and 14 e-pl scattering.

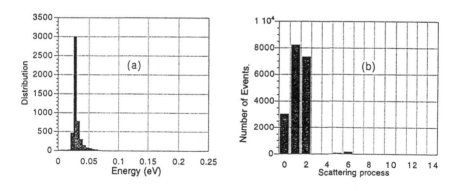

Fig. 5 (a)The energy distribution and (b)scattering statistics for a GaN film with L=0.01 μm and F=0.1 V/μm. The scattering processes are the same as those in Fig. 4. In this case, however, the e-e and e-pl interactions are excluded.

0.003 eV (see Fig. 3), we observe a Maxwellian distribution of electrons with a peak at the thermal energy ($k_B T=0.025$ eV). The dominant acoustic-phonon processes essentially restore the equilibrium of the electron gas with the crystalline lattice in the presence of a field $F \sim 0.1$ V/μm. Finally, the fact that quasi-ballistic propagation is possible when collective excitations (such as plasmons or polar optical phonons) are present implies that these processes do not compromise the use of diamond and GaN thin films in useful field emission devices.

ACKNOWLEDGEMENT

The work was supported in part by the Ballistic Missile Defense Organization administered by the Office of Naval Research through the ONR Grant No. 00014-95-0905.

REFERENCES

[1] M. W. Geis, J. W. Twichell and T. M. Lyszczarz, J. Vac. Sci. Technol. **B14**, 2060(1996).

[2] P. Lerner, P. H. Cutler and N. M. Miskovsky, Journ. De Physique III, **6**, Colloque 5-39(1996)

[3] P. Lerner, P. H. Cutler and N. M. Miskovsky, *Materials Research Society Symposium Proceedings, III-V Nitrides* (Materials Research Society, Pittsburgh, PA, 1997).

[4] Tomizawa, Kazutaka, *Numerical Simulation of Submicron Semiconductor Devices*(Artech House, Inc., Norwood, MA, 1993).

[5] Ridley, B. K., *Quantum Processes In Semiconductors*, 3rd ed.(Oxford University Press, Oxford, 1993).

[6] Seeger, Karlheinz, *Semiconductor Physics An Introduction*, 5th Ed.(Springer-Verlag, New York, NY, 1991).

[7] C. Jacoboni and L. Reggiani, Rev. Mod. Phys. **55**, 645(1983).

[8] P. Lerner, P. H. Cutler and N. M. Miskovsky, J. Vac. Sci. Technol. B15, (Mar/Apr)(1997).

ELECTRICAL AND OPTICAL PROPERTIES OF InGaN/AlGaN DOUBLE HETEROSTRUCTURE BLUE LIGHT-EMITTING DIODES

K. Yang, H.T. Shi, B. Shen, R. Zhang, Z.Z. Chen, P. Chen and Y.D. Zheng
Department of Physics, Nanjing University, Nanjing 210093, P.R. China

ABSTRACT

In this paper, we studied the electrical and optical characteristics of Nichia double heterostructure blue light-emitting diodes, with $In_{0.06}Ga_{0.94}N$:Zn, Si active layer, at 77 and 300 K. Measurement of the forward bias current-voltage behavior of the device demonstrates a departure from the Shockley model of p-n diodes, and it is observed that the dominant mechanism of carrier transport across the junction is associated with carrier tunneling. Electroluminescence experiments of the devices were performed. We obtained an emission peak located at 2.80 eV, and a relatively weaker short-wavelength peak of 3.2 eV. A significant blue shifts of the optical emission peak which is consistent with the tunneling character of electrical characteristics was observed. Furthermore, we studied the properties of electroluminescence under various pulsed currents, and a degradation in I-V characteristics and a low resistance ohmic short were observed.

INTRODUCTION

Wide band gap semiconductor InGaN, has recently attracted considerable interest because of its great potentials in electronic and optoelectronic applications[1,2]. It is being considered for fabrication of light-emitting diodes (LEDs) and lasers operating in the visible and ultraviolet (UV) [3,4]. With the substantial development of the material and device, an energy span ranging from the blue to near-UV wavelengths, which is at present inaccessible to semiconductor technology, has been opened. Recently, group-III nitride-based electronic devices are being developed for various applications, most notably for blue laser diodes (LDs) and high-temperature electronics. In addition, LEDs are the most mature amongst the nitride-based optoelectronic components, having successfully commercialized since 1994. Several papers on high-brightness InGaN/AlGaN blue LEDs from Nichia Chemical Industries have shown some behaviors of the emission spectra with various current injections[5,6,7]. While electrical characterization of the devices can provide important information about the current transport through the wide-band-gap p-n heterojunction, layer materials, and metal-semiconductor contacts[8]. So far, however, these aspects of nitride-based LEDs have received relatively little attention.

In this paper, we studied the electrical and optical properties of Nichia double heterostructure blue LEDs at various temperatures (77 K and 300 K). We observed that the main transport mechanism is associated with carrier tunneling rather than thermal diffusion. Electroluminescence (EL) experiments of the LEDs were performed, we obtained an emission peak located at 2.80 eV, and a relatively weaker short-wavelength peak of 3.2 eV which is

469

associated with interband transitions in the InGaN active region. As increasing the applied current, the light intensity of the two peaks correspondingly increases, with an obvious blue-shift of the emission peak, and the ratio of the intensity of the short-wavelength peak to that of the emission peak increases together. Furthermore, we studied the properties of EL under various pulsed currents, and a degradation in I-V characteristics and a low resistance ohmic short were observed.

EXPERIMENTS

The Nichia blue LEDs used for the I-V and injection luminescence measurements are double heterostructures (DHs) with a n-type $In_{0.06}Ga_{0.94}N$ active layer codoped with the donor Si and the acceptor Zn[9,10]. The epitaxy layers were grown by the two-flow metalorganic chemical-vapor deposition (MOCVD) method on (0001) oriented sapphire substrate. The ~30 nm thick Si-doped GaN buffer layer was grown at 510 °C and ~4 μm thick Si-doped GaN layer was grown at 1020 °C. The carrier confining Si-doped $Al_{0.15}Ga_{0.85}N$ layer was 0.15 μm thick. The growth temperature was reduced to 780 °C to grow the 50 nm thick $In_{0.06}Ga_{0.94}N$ active layer codoped with the donor Si and the acceptor Zn. The thin active layer and larger energy gap AlGaN confining layers ensure a region of uniform excitation, and the EL in the active region results from efficient recombination of carriers via Zn-related deep levels. Therefore, the spontaneous emission spectra will be a replica of the minority carrier distribution.

Prior to testing, the diodes were deencapsulated to avoid the risk of epoxy-induced stress at the cryogenic temperatures. dc I-V measurements were performed at 77 K and room temperature. For low-temperature (77 K) measurements, the diode was fixed on a cold finger in a vacuum chamber which was kept at the temperature of liquid Nitrogen. In order to understand the possible relationship between the current components and light emission, optical experiments were performed using a monochromator, chopper (or signal generator) and UV photomultiplier equipped with a lock-in amplifier (EG&G 5210).

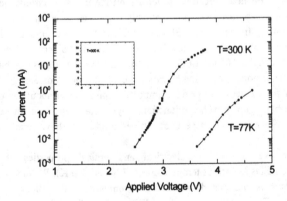

Fig. 1 Forward bias I-V characteristics of Nichia blue LEDs at 77 and 300 K.

RESULTS AND DISCUSSIONS

The forward bias I-V behaviors for the blue LEDs measured at 77 and 300 K are shown in Fig.1. Contrary to the standard Shockley model of a p-n diode, the slope of I-V characteristics in a semilogarithmic plot almost does not depend on temperature. As reported in earlier publications[8], it is observed that the low current component (from 5×10^{-6} to 5×10^{-4} A) can be approximated by an exponential function $I(A)=I_s{}^*Exp(nV)$, the diode ideality factor n was determined as large as 4.2, which indicates that the main transport mechanism is associated with carrier tunneling rather than thermal diffusion. Furthermore, the observed external quantum efficiency of ~2.7% demonstrates that the non-diffusion current does inject minority carriers from the p-type AlGaN layer into n-type InGaN active layer for the emission of light.

Similar to Ref.11, we also observed two different current components: a low voltage (0-3.1V) component and a medium voltage (3.1-3.8 V) one. A change in the I-V curve slope observed in the LEDs, indicates a change in the predominant mechanism of the current transport through the junction. As the sample temperature was reduced, carrier concentration initially decreased and then saturated at the levels usually about two orders of magnitude lower than those at room temperature. It was analyzed in Ref.12 that a significant fraction of electron transport takes place by hopping in the compensation centers leading to low mobility as compared to normal band carriers. This process becomes dominant at low temperatures. In addition, lower series resistance of the device is probably caused by a higher density of threading defects[11]. These defects can provide a parallel path for current flowing across the highly resistive p-type layers, which is consistent with the degradation mechanism under high-electrical-stress conditions as discussed below.

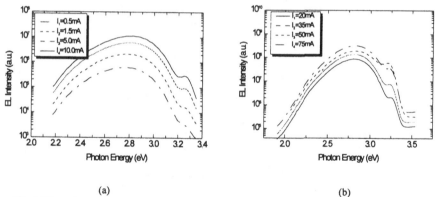

(a)   (b)

Fig.2 EL emission spectra as a function of junction current at room temperature.
(a) in small injection currents (b) in large injection currents.

EL spectra shown in Fig.2 were obtained for current ranging from 0.5 to 75 mA. At current above ~4 mA, a considerable "blue" shift of a sharp short-wavelength edge of emission band can be observed. At room temperature, in addition to the emission peak shift, a separate short-wavelength peak at ~3.2 eV appears at currents exceeding ~3 mA. The latter is obviously associated with interband transitions in the InGaN active region. A large EL spectra bandwidth is related to strong lattice coupling of the Zn state both in GaN and in InGaN.

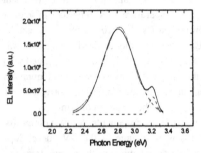

Fig.3 The EL emission peaks are expected to exhibit Gaussian broadening.

EL with shifting peak spectra has been observed and studied for GaAs p-n junctions[13]. In GaAs, the two mechanisms for shifting peak spectra are photon-assisted tunneling and band filling. Contrary to the behavior of photo-assisted tunneling, the emission spectra shown in Fig.2 are consistent with the band filling mechanism. As discussed in Ref.7, a reasonable explanation is that the shifting peak spectra are due to band filling, a progress which results from the injection of holes via tunneling into an empty acceptor impurity band and vacant valence band tails. As shown in Fig.3, the emission peaks are expected to exhibit Gaussian broadening.

The output spectra under pulsed conditions with amplitudes ranging from 0.5 to 75 mA were performed. The band-edge emission around 380 nm increases at a faster rate than the impurity-related emission around 440 nm, and becomes dominant at very large current densities, while the impurity-related emission appears to saturate. As a result of the high pulsed current measurements, several of the LEDs showed a degradation in I-V characteristics and some showed a low resistance ohmic short (~100 Ω), as shown in Fig.4. The shorts were interesting in that they could be removed by applying a modest (~5v) reverse bias and recreated by applying a

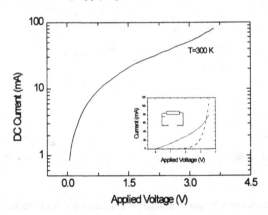

Fig.4 Typical I-V behavior of a stressed LED. As a result of the high pulsed current measurements, some LEDs showed a degradation in I-V characteristics and a low resistance ohmic short.

pulsed forward current. While the LED was in the shorted state, light output under continuous wave (cw) conditions was not possible. The shorted LED would still, however, emit light under pulsed conditions. After a reverse bias was applied to remove the short, cw operation was once again possible. The effects of the stress were not completely removable, i.e., the LED could be placed in a shorted state or in a degraded state, not back into its original, unstressed state.

Marek Osinski et al.[14] studied the performance of commercial AlGaN/InGaN/GaN blue LEDs under high current pulse conditions. They analyzed the results of DLTS, thermally stimulated capacitance, and admittance spectroscopy measurements performed on stressed devices, and the experiment data suggests that the stress did not cause a significant change in any impurity level, especially deep level impurities. The image of Electron Beam Induced current (EBIC) shows that the high forward current applied to these blue LEDs has caused metal from the p contact to migrate across the junction, and a defect density of 2.2×10^9 cm^{-2} was determined from the TEM analysis. The presence of the defect structure and the relatively high density of defects agrees with published data[15]. It is possible that the metal migration along defect tubes is the main process responsible for degradation of Nichia blue LEDs under high electrical stress conditions.

CONCLUSIONS

In summary, we have studied the electrical and optical characteristics of Nachia blue LEDs. The temperature-independent slope of I-V curves in the semilogarithmic plot suggests the involvement of tunneling transport across the junction, and associated impurity-level radiation emission is an evidence of tunneling to those levels. In EL experiments, we obtained an emission peak located at 2.80 eV, and a relatively weaker short-wavelength peak of 3.2eV which is associated with interband transitions in the InGaN active region. A significant blue shifts of the optical emission peak which is consistent with the tunneling character of electrical characteristics was observed. Furthermore, we studied the properties of EL under various pulsed currents, and a degradation in I-V characteristics and a low resistance ohmic short were observed.

REFERENCES

1. S.Strite and H.Morkoc, J.Vac.Sci.Technol. B10(4), 1237 (1992)

2. H.Morkoc, S.Strite, G.B.Gao, M.E.Lin, B.Sverdlov and M.Burns, J.Appl.Phys. 76(3), 1363(1994)

3. S.Nakamura, M.Senoh and T.Mukai, Appl. Phys. Lett. 62 , 2390(1993)

4. S.Nakamura, T.Mukai and M.Senoh, J Appl. Phys. 71, 5543 (1992)

5. S.Nakamura, T.Mukai and M.Senoh, J. Appl. Phys. 76, 8189 (1994)

6. S.D. Lester, F.A. Ponce, M.G. Craford and D.A. Steigerwald, Appl. Phys. Lett. 66, 1249 (1995)

7. H.C. Casey, Tr., J. Muth, S. Krishnankutty, and J.M. Zavada, Appl. Phys. Lett. 68, 2867 (1996)

8. L. Lee, M. Osinski and K.J. Malloy, in conf. Proc., LEOS'94 7th Annual Meeting, Boston, MA, 31 October - 3 November (IEEE, Piscataway, NJ, 1994), Vol.1, p.332

9. S.Nakamura, T.Mukai and M.Senoh, Appl. Phys. Lett. 64, 1687 (1994)

10. S.Nakamura, J. Vac. Sci. Technol. A13, 705 (1995)

11. Piotr Perlin, Marek Osinski, Petr G. Eliseev et al., Appl. Phys. Lett. 69, 1680 (1996)

12. J. Molnar, T.Lei, and T.D. Moustakas, Appl. Phys. Lett. 62, 72 (1993)

13. H.C. Casey, Jr., and R.Z. Bachrach, J. Appl. Phys. 44, 2795(1973)

14. Marek Osinski,   Joachim Zeller, Pei-Chih Chiu, B. Scott Phillips, and Daniel L. Barton, Appl. Phys. Lett. 69, 898 (1996)

15. W.Gotz, N. Johnson, H. Amano, and I. Akasaki, Appl. Phys. Lett. 65, 463 (1994)

EFFECT OF HYDROGEN CHLORIDE ON THE CAPACITANCE-VOLTAGE CHARACTERISTICS OF MOCVD-GROWN AlN/6H-SiC MIS STRUCTURES

C.C. TIN[1], A. GICHUHI[2], M.J. BOZACK[1], C.G. SHANNON[2], AND C.K. TEH[3]
[1]Department of Physics, 206 Allison Laboratory, Auburn University, AL 36849,
cctin@physics.auburn.edu
[2]Department of Chemistry, 179 Chemistry Building, Auburn University, AL 36849
[3]ATA, 2060 Evergreen Drive, Auburn, AL 36830

ABSTRACT

Aluminum nitride (AlN) is a promising material as gate insulator for 6H-SiC metal-insulator-semiconductor (MIS)-based devices. Using metalorganic chemical vapor deposition (MOCVD)-grown AlN, we have recently fabricated Au/AlN/6H-SiC MIS structures with different AlN/6H-SiC interfacial characteristics depending on the AlN growth procedures. We have also found that the use of hydrogen chloride gas is effective in improving the capacitance-voltage characteristics of the AlN/6H-SiC structure. The reason for such improvement is not well understood and several possible mechanisms for such improvement include factors such as substrate surface morphology and surface contaminants. In this paper, we will examine the relationship between surface morphology of the substrates and the capacitance-voltage characteristics of Au/AlN/6H-SiC structures.

INTRODUCTION

Intense research activities in the past several years have placed silicon carbide as a leading wide band gap semiconductor that has found applications in high temperature and high power electronic devices. A crucial component in the development of high temperature and high power electronic devices is the fabrication of a reliable MIS structure with low interfacial defects and good thermal stability. The gate dielectric material plays an important role in the electrical performance of a MIS-based device. In SiC technology, SiO_2 has been widely used and studied as the gate insulator for metal-oxide-semiconductor (MOS)-based devices. However, problems relating to high interface state density and deteriorating device performance at high temperature have plagued SiC MOS-based device development [1-2]. Although progress has been made in improving the electrical characteristics of SiC MOS structures in recent years [3-5], it is still necessary to consider alternative dielectric material for SiC MIS-based devices.

AlN [6-8] has recently emerged as a promising dielectric material for 6H-SiC due to the low lattice mismatch of <1% between the two materials. In spite of reservations about the small band offsets between AlN and 6H-SiC, AlN/6H-SiC MIS structures that were able to withstand near-theoretical breakdown field of about 2.5 MV/cm have been reported [7-8]. The electrical characteristics of the MIS structures depend on the AlN growth procedures [8]. This dependence is due to the fact that different growth procedures will produce interfaces of varying quality. We have obtained structures with low interface state density and good interfacial stability at temperature up to 300°C. However, structures with low interface state density also displayed poor high electric field characteristics. We have also observed that hydrogen chloride etching played a significant role in determining the interfacial

475

Mat. Res. Soc. Symp. Proc. Vol. 468 ° 1997 Materials Research Society

characteristics. In this paper, we will present data from atomic force microscopy (AFM) and capacitance-voltage (C-V) measurements to see if there is any connection between the morphology of the surface of the substrate before AlN deposition and the resulting electrical characteristics. Capacitance-voltage (C-V) measurement is most effective in MIS characterization because C-V characteristics are very sensitive to the presence of interface states.

EXPERIMENT

The substrates for the deposition of AlN epilayers were 5 μm thick n-type 6H-SiC epilayers on n-type 3.5° off-axis 6H-SiC substrates. The 6H-SiC substrate and epilayer have carrier concentrations of about 1×10^{18} cm^{-3} and 5×10^{16} cm^{-3} respectively. The substrates were cleaned in trichloroethylene, acetone, and methanol before being loaded into the reactor. Prior to epitaxial growth of AlN, the substrates were subjected to an *in-situ* etching with either 3% HCl/H$_2$ gas mixture or pure hydrogen at atmospheric pressure. Two different etching temperatures of 1200°C and 1350°C were used for hydrogen etching but HCl etching was done only at 1200°C. Following the etching step, the reactor was purged with pure hydrogen gas. The reactor pressure was then lowered to 300 Torr and the substrate temperature was then set at 1250°C for subsequent growth of AlN. The process gases used were 10 sccm of hydrogen bubbling through the trimethylaluminum bubbler maintained at 18°C, 2.5 slm of NH$_3$ gas, and 3.0 slm of hydrogen gas. Growth duration was about 30 minutes for all samples.

After the deposition process, ohmic contacts were fabricated on the back of the samples by sputtering about 200 nm of Ni$_{93}$V$_7$ alloy followed by annealing at 960°C for 2 minutes in vacuum. Capacitance-voltage measurements were then carried out using a Keithley Model 82 C-V system.

A second set of samples consisting of etched substrates was also prepared. These samples were subjected to the same *in-situ* etching procedures but have no AlN layer. This set of samples was used for atomic force microscopy measurements to measure the surface roughness. The AFM data were obtained in the tapping mode.

RESULTS AND DISCUSSION

Figure 1 shows the atomic force microscopy image depicting the surface morphology of a 6H-SiC sample subjected to hydrogen gas etching at 1350°C for 20 minutes. The AFM image shows undulating features similar to the macrosteps reported by Kimoto *et al.* [9]. Each of these macrosteps consists of microsteps, a phenomenon attributed to step-bunching resulting from the minimization of surface free energy [10]. Figure 2 shows the C-V curves of a Au/AlN/6H-SiC MIS structure fabricated using AlN grown on 6H-SiC that was previously etched under the same conditions as the sample shown in figure 1.

Figure 3 shows the atomic force microscopy image of a 6H-SiC sample etched with 3% HCl/H$_2$ gas mixture at 1200°C for about 4 minutes. The corresponding C-V curves for the resulting MIS structure is shown in figure 4.

The presence of interface states can be seen from the hysteresis in the C-V curves of all samples. The samples etched with 3% HCl/H$_2$ at 1200°C for 4 minutes showed negligible hysteresis whereas all other samples showed significant hysteresis. From the shift of the flatband voltage of the C-V curve, the effective charge density of the MIS structure can be computed. This effective charge density represents contributions from the mobile ionic charge, fixed charge in the dielectric material, and trapped charge in the insulator.

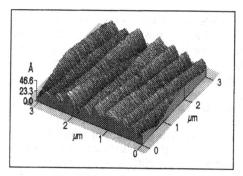

Figure 1. AFM of surface of hydrogen-etched 6H-SiC epilayer.

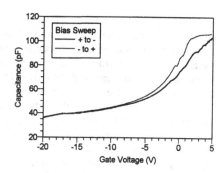

Figure 2. C-V curves of MIS structure using hydrogen-etched 6H-SiC epilayer.

Figure 3. AFM of surface of 3%HCl/H$_2$-etched 6H-SiC epilayer.

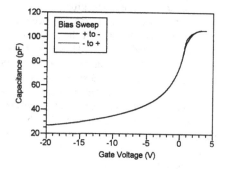

Figure 4. C-V curves of MIS structure using 3%HCl/H$_2$-etched 6H-SiC epilayer.

Table I. AFM and C-V data for 6H-SiC samples subjected to different etching conditions.

Etching Conditions	RMS Surface Roughness (Å)	Interface Effective Charge Density (cm^{-2})	
		(+ to -)	(- to +)
Hydrogen; 1200°C; 20 minutes	14.6	-1.3 x 10^{12}	1.6 x 10^{11}
Hydrogen; 1350°C; 20 minutes	9.6	-9.5 x 10^{11}	-1.7 x 10^{11}
3% HCl/H$_2$; 1200°C; 2 minutes*	1.9	-1.4 x 10^{12}	-1.9 x 10^{11}
3% HCl/H$_2$; 1200°C; 4 minutes	10.3	-5.0 x 10^{11}	-3.4 x 10^{11}

*Did not show step-bunching effect.

The resulting rms surface roughness values of the 6H-SiC epilayers subjected to different etching conditions are tabulated in Table I, together with the interface effective charge density obtained from the C-V curves of the resulting Au/AlN/ 6H-SiC MIS structures. The + and - signs denote bias sweep direction and the negative numbers indicate negative net charges. It should be noted that because of the step-bunching effect, the rms roughness is actually the rms value of the undulating feature. This fact is illustrated by the sample etched with 3% HCl/H$_2$

gas mixture for only 2 minutes. This particular sample showed a flat surface structure without any step-bunching effect and the rms surface roughness value was about 1.9 Å. The absence of step-bunching feature in this sample could be due to the presence of residual surface contaminants arising from insufficient etching which helped to reduce the surface energy without the need for step-bunching effect. This conclusion is supported by the fact that there is a large hysteresis in the C-V curves for this type of sample.

Preliminary evaluation of the data shows that samples etched with HCl for at least 4 minutes appear to be smoother or have smaller step-bunching peaks. It has been previously observed [11] in epitaxial growth of 3C, 4H, and 6H polytypes of SiC that insufficient HCl etching of the substrates played a role in determining the morphology and quality of the resulting epilayers. In the case of epitaxial growth of AlN on 6H-SiC, it is plausible that HCl etching of 6H-SiC, being more efficient than hydrogen etching, removes more contaminants from the surface thereby presenting a more prestine surface for subsequent growth of AlN.

A possible contributing reason is the presence of surface oxide considering that Al is a good getter of oxygen. The growth temperature was also high enough for both Al_2O_3 and SiO_2 formation to occur. Since small concentration of HCl may still remain after the HCl etching procedure, it is possible that the presence of residual HCl can etch away any trace deposition of both SiO_2 and Al_2O_3. However, depth profilings by x-ray photoelectron spectroscopy (XPS) and Auger electron spectroscopy (AES) showed that HCl and hydrogen-etched samples have very similar chemical structures at the interfaces with approximately the same amount of oxygen. It should be noted, however, that XPS and AES do not have the same sensitivity as C-V measurements to detect minute concentration of impurity contamination at the interface.

It must also be borne in mind that there are other reasons to be considered as well which is the subject of our ongoing study to pin-point the exact reason for the difference in C-V results of differently-etched samples. The microstruture of the interface needs to be examined by cross-sectional transmission electron microscopy to confirm whether HCl etching affects the interfacial lattice ordering.

CONCLUSION

Capacitance-voltage measurements have shown that the interface of AlN/6H-SiC seems to be critically affected by surface preparation procedure prior to deposition of AlN epilayer. AFM data showed that even though HCl-etched samples may be slightly smoother than hydrogen-etched samples, it is not conclusive that the difference in surface morphology played a role in controlling the quality of the interface. Other possible reasons could lie in the greater efficiency of HCl in etching away surface contaminants prior to AlN deposition. Ongoing work is in progress in using cross-sectional transmission electron microscopy and surface science analytical techniques to pin-point the exact reason for the difference in interfacial quality.

ACKNOWLEDGMENTS

Supported by the Center for Commercial Development of Space Power and Advanced Electronics, located at Auburn University, with funds from NASA Grant NAGW-1192-CCDS-AD, Auburn University and the Center's industrial partners; Auburn University's Faculty Research Start-up Funds; Alabama Space Grant Consortium grant under NASA Grant NGT40010; and by NSF EPSCoR Grant #95192.

REFERENCES

1. J.B. Casady, J.D. Cressler, W.C. Dillard, R.W. Johnson, A.K. Agarwal, and R.R. Siergiej, Solid-State Electronics **39**, 777 (1996).

2. A.K. Agarwal, R.R. Siergiej, S. Seshadri, M.H. White, P.G. McMullin, A.A. Burk, L.B. Rowland, C.D. Brandt, and R.H. Hopkins, Mat. Res. Soc. Symp. Proc. **423**, 87 (1996).

3. J.N. Shenoy, G.L. Chindalore, M.R. Melloch, J.A. Cooper, Jr., J.W. Palmour, and K.G. Irvine, J. Electronic Mater. **24**, 303 (1995).

4. S. Sridevan, V. Misra, P.K. McLarty, B.J. Baliga, and J.J. Wortman, Inst. Phys. Conf. Ser. **142**, 645 (1996).

5. M.R. Melloch and J.A. Cooper, Jr., MRS Bulletin **22**, 42 (1997).

6. C.I. Harris, M.O. Aboelfotoh, R.S. Kern, S. Tanaka, and R.F. Davis, Inst. Phys. Conf. Ser. **142**, 777 (1996).

7. C.-M. Zetterling, K. Wongchotigul, M.G. Spencer, C.I. Harris, S.S. Wong, and M. Östling, Mat. Res. Soc. Symp. Proc. **423**, 667 (1996).

8. C.C. Tin, Y. Song, T. Isaacs-Smith, V. Madangarli, and T.S. Sudarshan, J. Electronic Mater. **26**, 212 (1997).

9. T. Kimoto, A. Itoh, and H. Matsunami, Appl. Phys. Lett. **66**, 3645 (1995).

10. W.A. Tiller, *The Science of Crystallization: Microscopic Interfacial Phenomena* (Cambridge University Press, Cambridge, 1991), Chap. 2.

11. C.C. Tin, R. Hu, J. Liu, Y. Vohra, and Z.C. Feng, J. Cryst. Growth **158**, 509 (1996).

MOCVD GROWTH OF HIGH OUTPUT POWER INGAN MULTIPLE QUANTUM WELL LIGHT EMITTING DIODE

P. Kozodoy[*], A. Abare, R.K. Sink, M. Mack, S. Keller, S.P. DenBaars, U.K. Mishra
Electrical and Computer Engineering Department, University of California, Santa Barbara, California 93106.

D. Steigerwald
Optoelectronic Division, Hewlett-Packard, San Jose, California 95131-1008

* E-mail address: kozodoy@indy.ece.ucsb.edu

ABSTRACT

The MOCVD growth of InGaN / GaN multiple quantum well (MQW) structures for optoelectronic applications has been investigated. The structural and optical properties of the layers have been characterized by x-ray diffraction and photoluminescence. The effect of barrier and well dimensions on the optical properties have been examined; highest emission intensity and narrowest linewidth were obtained with thin wells (20-30 Å) and thick barriers (greater than 50 Å). By incorporating an MQW structure as the active region in a GaN p-n diode, high-brightness light emitting diodes (LEDs) have been produced. Under a forward current of 20 mA, these devices emit 2.2 mW of power corresponding to an external quantum efficiency of 4.5%. The emission spectrum peaks at 445 nm and exhibits a narrow linewidth of 28 nm. Under pulsed high current conditions, output power as high as 53 mW was realized and the peak emission wavelength shifted to 430 nm.

INTRODUCTION

High performance light emitting diodes (LEDs) that operate in the ultraviolet to green portion of the spectrum have been realized recently through the use of GaN and its alloys with In and Al. Early devices were constrained by difficulty obtaining p-type films, however the discovery of post-growth activation procedures for these films has led to rapid progress in nitride-based LEDs.[1,2] Devices using a Zn-doped InGaN active region produce bright but spectrally broad blue light.[3] By incorporating a single InGaN quantum well (SQW design) in the active region of the device, LEDs have been produced with increased output power and narrow emission in the blue and green.[4] For the optimization of super-bright LEDs and laser diodes an understanding of the electroluminescence of the InGaN multiple quantum well (MQW) structure is crucial. The growth and characterization of such MQW stacks has been a subject of intense research for some time,[5-7] but little data has been published on the characteristics of spontaneous emission from such structures. Yang et al have reported on MQW InGaN LEDs grown on (111) spinel substrates, emitting light at a wavelength of 385 nm with an output power of 1.8 mW at a pulsed current of 2 A.[8] Recently, Nakamura and co-workers have published impressive results on the room-temperature CW operation of a laser diode with a MQW InGaN active region, achieving a laser output power of 10 mW at a current of 100 mA.[9] In this work, we report on the growth and characterization of InGaN / GaN MQW layer structures with varying dimensions. We also report on the characteristics of an LED grown with an MQW active region, which exhibits very high output power and excellent color purity. We report for the first

481

time on measurements of MQW InGaN LED emission at very high pulsed current values, such as those relevant to laser diode operation.

EXPERIMENT

Growth Studies

InGaN / GaN MQW stacks were grown on c-plane sapphire substrates by metalorganic chemical vapor deposition. The MOCVD growth conditions and resulting material quality for GaN and InGaN have been discussed earlier.[10,11] The MQW stacks consisted of 14 periods of $In_{0.2}Ga_{0.8}N$ wells with GaN barriers and were grown on top of a 2 μm GaN buffer. Two studies were performed: in one set of growths the barrier width was varied and the well width was held constant at 20 Å. In the other set, the well width was varied and the barrier width was held constant at 50 Å.

Figure 1 shows the photoluminescence (PL) and x-ray diffraction (XRD) rocking curve data obtained from each sample in the well-width study. The PL data shows strong quantum well emission with a narrow linewidth, and very little yellow-band "deep-level" luminescence is observed. The peak emission wavelength is observed to decrease as the well thickness decreases, a trend which is to be expected from the increase in ground state energy for the thinner quantum well. Superlattice peaks are evident in the XRD data, indicating the presence of abrupt heterojunction interfaces and a high degree of vertical uniformity. As expected, the superlattice peak spacing is observed to increase as the period thickness decreases.

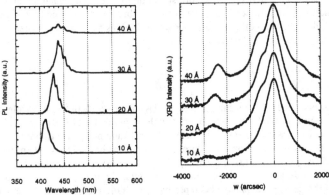

Figure 1: Photoluminescence and x-ray diffraction measurements on InGaN / GaN superlattices grown by MOCVD. Each sample contains 14 periods: the barrier is 40 Å of GaN, the wells are $In_{0.2}Ga_{0.8}N$ with width as indicated.

Two quantities of interest in the PL data—the emission peak intensity and the linewidth—are plotted in figure 2a as a function of well width. 20 - 30 Å wells are seen to give optimal luminescence qualities. In comparison the 10 Å wells have a wider emission linewidth and much reduced emission intensity; this effect is ascribed mainly to the increased importance of interface roughness in such thin wells. Increasing the well width beyond 30 Å also results in an increased linewidth and decreased emission intensity. The mechanism for this trend is believed to be degradation of the InGaN material during subsequent high-temperature growth.

The results of a similar study of barrier width are presented in figure 2b. As the barrier width is increased to about 50 Å, the emission linewidth decreases and the peak intensity improves. Reduction in wavefunction overlap between adjacent wells is believed to play an important role in these phenomena. Expanding the barrier width beyond 50 Å appears to saturate this effect. From these preliminary studies we conclude that thin InGaN quantum wells (20-30 Å) with thick GaN barriers (>50 Å) give optimal photoluminescence properties. Further studies on the growth and characterization of MQW structures are being pursued; the results will be published in a separate paper.

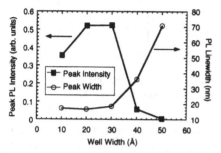 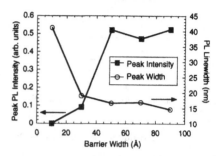

Figure 2: Summary of PL data obtained from InGaN / GaN MQW stacks. Each stack contains 14 periods. For the samples in figure 2a (left) the barrier width was held constant at 50 Å and the well width was varied. In figure 2b (right) the well width was held constant at 20 Å and the barrier width was varied. Emission peak intensity and peak width are plotted for each case.

<u>LED Device Properties</u>

The electroluminescence properties of the multi-quantum-well layers were tested in an LED structure (figure 3). 2.5 µm of Si-doped GaN with a dopant concentration of 3×10^{18} cm^{-3} was grown as the bottom layer of the device. The active region consisted of a five period MQW stack with 25 Å $In_{0.25}Ga_{0.75}N$ quantum wells and 40 Å GaN barriers. Above the active region 750Å of Mg-doped $Al_{0.07}Ga_{0.93}N$ was grown; this was capped by 0.5 µm of Mg-doped GaN. Devices were fabricated with a mesa size of 6×10^{-4} cm^2. Ni/Au was used for the p-contact and Ti/Al for the n-contact. The devices were then packaged in the standard LED lamp form.

The emission spectrum and optical power emission of the LEDs have been measured as a function of current and are shown in figures 4-6 for both low current (DC) operation and high current (pulsed). When tested under a DC current of 20 mA, the emission is seen to peak at approximately 445 nm and a narrow emission linewidth of 28 nm was obtained. The output power at this current level was 2.2 mW. The power saturates at 8 mW under a DC current of approximately 100 mA; this saturation is attributed to heating effects causing a drop in quantum efficiency. The external quantum efficiency reaches its peak value of 4.5% at a driving current of 20 mA.

By testing the LED under pulsed conditions the effect of heating can be greatly reduced. Current pulses of width 3 µs and duty ratio 3×10^{-4} were used during high current testing. As figure 5 demonstrates, output power saturation was not observed under these conditions until a driving current of 1.4 A was applied. At this current level, the output power was 53 mW, to our knowledge the highest reported for an InGaN LED. In contrast, commercially available SQW InGaN LEDs tested under the same conditions saturate at a current of 600 mA with an output power of 25 mW. We attribute the very high power level achieved in this LED to the use of a MQW structure, which reduces the degree of carrier overflow and non-radiative recombination.

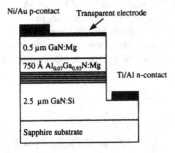

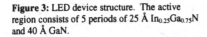

Figure 3: LED device structure. The active region consists of 5 periods of 25 Å $In_{0.25}Ga_{0.75}N$ and 40 Å GaN.

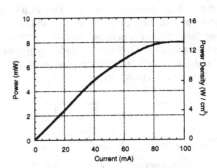

Figure 4: L-I curve for MQW LED under DC current. The quantum efficiency is 4.5%.

The electroluminescence (EL) spectrum was also measured under high current conditions (figure 6). For these measurements a pulse of width 50 ns and duty ratio of 3.0×10^{-5} was used. We note the emergence of Fabry-Perot modes at a driving current of 3A. In addition, the peak emission wavelength shifts from 445 nm to about 430 nm as the current is increased to 3A. This blue shift is tentatively attributed to band filling, although the exact mechanism is the subject of ongoing research. The saturation of an efficient donor-acceptor recombination path has reportedly been responsible for a blue shift of similar magnitude in InGaN photoluminescence under certain growth conditions,[12] however the results of the PL experiments described above point to very efficient band-to-band recombination in these quantum wells so this explanation seems unlikely. A more plausible explanation is that the long-wavelength emission originates from localized excitonic states formed by potential fluctuations in the quantum well plane caused by the inhomogenous nature of the InGaN alloy. This model has been invoked to explain a blue shift observed in the low-current (0.1 µA to 80 mA) emission of SQW LEDs and MQW lasers manufactured by Nichia Chemical Industries.[13] Gain measurements recently performed on our MQW LEDs (to be reported in a separate paper) appear to support this theory.[14]

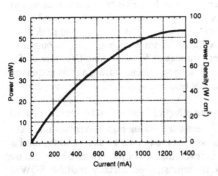

Figure 5: L-I curve taken under pulsed conditions. Pulse width = 3 µs, duty cycle is 3×10^{-4}.

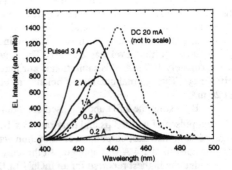

Figure 6: EL spectrum taken under pulsed current. DC spectrum shown for comparison. Pulse width = 50 ns, duty cycle is 3×10^{-5}.

CONCLUSIONS

We have investigated the luminescence properties of InGaN / GaN multi-quantum-well structures grown by MOCVD. Photoluminescence measurements performed on MQW stacks indicate that optimal emission qualities are obtained using thin quantum wells and thick barriers. The structural quality of the films is confirmed by x-ray diffraction measurements showing prominent superlattice peaks. The MQW stack has been incorporated as the active region in an LED. The resulting device demonstrates very high output power, narrow linewidth, and excellent color purity. A blue shift is observed in the emission of the LED at very high injection currents; this shift is tentatively attributed to localized states created by potential fluctuations in the quantum well plane. The output power saturation under DC conditions is attributed to heating, a problem which is exacerbated by the poor thermal conductivity of the sapphire substrate. Under pulsed conditions heating may be avoided and very high output powers are achieved.

ACKNOWLEDGMENTS

We gratefully acknowledge support for this research from the DARPA sponsored Multi-University Nitride Consortium under contract number N00014-96-1-0738. (monitored by Dr. Anis Husain). Further support was provided by the Army Research Office contract DAAH04-95-1-0329, monitored by Dr. Jon Zavada, and the National Science Foundation through Young Investigator grant number DMR-9457926, monitored by Vern Hess and Deborah Crawford.

REFERENCES

1. H. Amano, M. Kito, K. Hiramatsu, I. Akasaki, Jpn. J. Appl. Phys. **28**, L2112 (1989).

2. S. Nakamura, M. Senoh, T. Mukai, Jpn. J. Appl. Phys. **30**, L1708 (1991).

3. S. Nakamura, T. Mukai, M. Senoh, Appl. Phys. Lett. **64**, 1687 (1994).

4. S. Nakamura, M. Senoh, N. Iwasa, S. Nagahama, Jpn. J. Appl. Phys. **34**, L797 (1995).

5. K. Itoh, T. Kawamoto, H. Amano, K. Hiramatsu, I. Akasaki, Jpn. J. Appl. Phys. **30**, 1924 (1991).

6. M. Koike, S. Yamasaki, S. Nagai, N. Koide, S. Asami, H. Amano, I. Akasaki, Appl. Phys. Lett. **68**, 1403 (1996).

7. R. Singh, D. Doppalapudi, T.D. Moustakas, Appl. Phys. Let., **69**, 2388 (1996).

8. J. W. Yang, Q. Chen, C.J. Sun, B. Lim, M.Z. Anwar, M.A. Khan, H. Temkin, Appl. Phys. Lett. **69**, 369 (1996).

9. S. Nakamura, M. Senoh, S. Nagahama, N. Iwasa, T. Yamada, T. Matsushita, Y. Sugimoto, H. Kiyoku, Appl. Phys. Lett. **70**, 1417 (1997).

10. B.P. Keller, S. Keller, D. Kapolnek, W.-N. Jiang, Y.-F. Wu, H. Masui, X.H. Wu, B. Heying, J.S. Speck, U.K. Mishra, S.P. DenBaars, J. Elect. Matls. **24**, 1707 (1995).

11. S. Keller, B.P. Keller, D. Kapolnek, A.C. Abare, H. Masui, L.A. Coldren, U.K. Mishra, S.P. DenBaars, Appl. Phys. Lett. **68**, 3147 (1996).

12. C.J. Sun, J.W. Yang, Q. Chen, B.W. Lim, M.Z. Anwar, M.A. Khan, H. Temkin, D. Weismann, I. Brenner, Appl. Phys. Lett. **69**, 668 (1996).

13. S. Chichibu, T. Azuhata, T. Sota, S. Nakamura, Appl. Phys. Lett. **69**, 4188 (1996).

14. M. Kuball, E.-S. Jeon, Y.-K. Song, A.V. Nurmikko, P. Kozodoy, A. Abare, S. Keller, L.A. Coldren, U.K. Mishra, S.P. DenBaars, D. Steigerwald, (to be published in Appl. Phys. Lett.).

THEORY OF GAIN IN GROUP-III NITRIDE LASERS

W.W. CHOW*, A.F. WRIGHT**, A. GIRNDT[†], F. JAHNKE[†] and S.W. KOCH[†]
*Sandia National Laboratories, Albuquerque, NM 85718-0601, U. S. A., wwchow@sandia.gov
**Sandia National Laboratories, Albuquerque, NM 85718-0601, U. S. A.
[†]Department of Physics, Philipps University, Renthof 5, 35032 Marburg, Germany

ABSTRACT

A microscopic theory of gain in a group-III nitride quantum well laser is presented. The approach, which treats carrier correlations at the level of quantum kinetic theory, gives a consistent account of plasma and excitonic effects in an inhomogeneously broadened system.

INTRODUCTION

To analyze experimental results in group-III nitride lasers, it is helpful to be able to predict their gain spectra accurately. Both excitons and electron hole plasma play important roles in the optical properties of group-III nitride compounds, even under lasing conditions of high carrier density and temperature.[1] Also, inhomogeneous broadening is present, due to localization effects from dimensional or composition variations.[2] This paper describes a consistent treatment of the above factors.

THEORY

Our approach is based on a Hamiltonian that contains the Coulomb interaction energy among carriers.[3] Using this Hamiltonian and following a derivation similar to that resulting in the Semiconductor Bloch Equations, we get the equation of motion for the microscopic polarization, $p_{\vec{k}}$, due to an electron hole pair,[4, 5]

$$\frac{d}{dt}p_{\vec{k}} = -i\omega_{\vec{k}}p_{\vec{k}} - i\Omega_{\vec{k}}\left(n_{e,\vec{k}} + n_{h,\vec{k}} - 1\right) - \Gamma_{\vec{k}}p_{\vec{k}} + \sum_{\vec{q}}\Gamma_{\vec{k},\vec{q}}p_{\vec{k}+\vec{q}} \qquad (1)$$

The first two terms on the right hand side describe the oscillation of the polarization at the transition frequency, $\omega_{\vec{k}}$, and the stimulated emission and absorption processes. The many-body Coulomb effects appear in the form of a carrier density, N, dependence in the transition energy,

$$\hbar\omega_{\vec{k}}(N) = \varepsilon_{e,\vec{k}} + \varepsilon_{h,\vec{k}} + [\varepsilon_{g,0} + \Delta\varepsilon_{SX}(N)] \quad , \qquad (2)$$

where $\Delta\varepsilon_{SX}$ is the exchange contribution to the renormalized band gap energy. They also lead to a renormalized Rabi frequency,

$$\Omega_{\vec{k}} = \frac{1}{\hbar}\left(\mu_{\vec{k}}E + \sum_{\vec{q}}V_q p_{\vec{k}+\vec{q}}\right) \quad , \qquad (3)$$

487

where $\mu_{\vec{k}}$ is the optical dipole matrix element, E is the laser electric field, and V_q is the Fourier transform of the bare (unscreened) Coulomb potential. Carrier-carrier collisions give rise to the last two terms. The third term on the RHS describes diagonal polarization dephasing, with a dephasing rate,

$$
\begin{aligned}
\Gamma_{\vec{k}} = & \sum_{a=e,h} \sum_{b=e,h} \sum_{\vec{q}} \sum_{\vec{k}'} \frac{2\pi}{\hbar} V_q^2 D(\varepsilon_{a,\vec{k}} + \varepsilon_{b,\vec{k}'} - \varepsilon_{a,\vec{k}+\vec{q}} - \varepsilon_{b,\vec{k}'-\vec{q}}) \\
& \times \left[n_{a,\vec{k}+\vec{q}}\left(1 - n_{b,\vec{k}'}\right) n_{b,\vec{k}'-\vec{q}} + \left(1 - n_{a,\vec{k}+\vec{q}}\right) n_{b,\vec{k}'}\left(1 - n_{b,\vec{k}'-\vec{q}}\right) \right] ,
\end{aligned} \tag{4}
$$

where $D(\Delta) = \delta(\Delta) + i\pi^{-1}P(\Delta^{-1})$, and P denotes the principle value. The last term shows a nondiagonal scattering contribution that couples polarizations with different $\vec{k}$'s. The coefficient,

$$
\begin{aligned}
\Gamma_{\vec{k},\vec{q}} = & \sum_{a=e,h} \sum_{b=e,h} \sum_{\vec{k}'} \frac{2\pi}{\hbar} V_q^2 D(\varepsilon_{a,\vec{k}} + \varepsilon_{b,\vec{k}'} - \varepsilon_{a,\vec{k}+\vec{q}} - \varepsilon_{b,\vec{k}'-\vec{q}}) \\
& \times \left[\left(1 - n_{a,\vec{k}}\right)\left(1 - n_{b,\vec{k}'}\right) n_{b,\vec{k}'-\vec{q}} + n_{a,\vec{k}} n_{b,\vec{k}'}\left(1 - n_{b,\vec{k}'-\vec{q}}\right) \right] .
\end{aligned} \tag{5}
$$

In this paper, we limit the discussion to the small signal gain, where the carrier populations, $n_{e,\vec{k}}$ and $n_{h,\vec{k}}$ are inputs to the calculations.

The polarization equations are solved numerically for the steady state solution. Using a semiclassical laser theory, the intensity gain G is given by[3] (MKS units):

$$
G = -\frac{2\omega}{\varepsilon_0 ncV\mathcal{E}} \text{Im}\left(\sum_{\vec{k}} \mu_{\vec{k}}^* p_{\vec{k}} e^{i\omega t} \right) , \tag{6}
$$

where $\mathcal{E}$ is the slowly varying electric field amplitude, ω is the laser frequency, ε_0 and c are the permittivity and speed of light in vaccuum, n is the background refractive index, V is the active region volume, and the summation is over all electron and hole states.

Equation (6) gives the homogeneously broadened gain spectrum for an ideal structure, where the quantum well thickness and composition are precisely known. On the other hand, experimental data suggest that the gain region may consist of localized regions of different quantum well thicknesses or compositions. Assuming that these regions are sufficiently large so that the quantum confinement remains only along the epitaxial direction, we can treat the effects of inhomogeneous broadening by a statistical average of the homogeneous gain spectra, i.e.,

$$
G_{inh}(\omega, N, T) = \int dx \, P(x) \, G(x, \omega, N, T) , \tag{7}
$$

where $P(x)$ is a normal distribution representing the variation in x, which can either be the quantum well thickness or indium concentration.

RESULTS

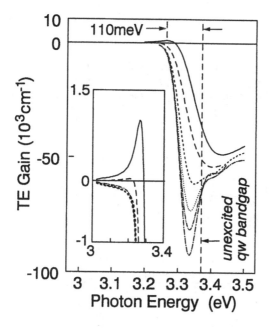

Fig. 1. Calculated TE gain spectra for a 4nm $In_{0.1}Ga_{0.9}N/Al_{0.2}Ga_{0.8}N$ quantum well at $T = 300K$ and densities N = 0.1, 0.5, 1.0, 2.0, 4.0, and $6.0 \times 10^{12} cm^{-2}$. The inset shows the gain portion of the spectra.

Figure 1 shows the computed spectra for a 4nm $In_{0.1}Ga_{0.9}N/Al_{0.2}Ga_{0.8}N$ quantum well structure and different carrier densities. The laser field polarization is in the plane of the quantum well (transverse electric or TE polarization). For the wurtzite structure considered, the orthogonal (TM) polarization has negligible gain, even at high carrier density. We use a 6×6 Luttinger-Kohn Hamiltonian and the envelope approximation[6] to compute the hole energy dispersions and optical dipole matrix elements. Input parameters are the bulk wurtzite material parameters (Table I). The ratio of the band offset (conduction/valence bands) is assumed to be 67/33. Alloy properties are obtained as composition-weighted averages of the bulk values, except for the optical bowing parameters, where we use $b = 0.53eV$ for AlGaN and $b = 1.02eV$ for InGaN.[7] The spectra are calculated assuming an inhomogeneous broadening due to a 0.01 (10%) standard deviation in the indium concentration.

Table 1. Material parameters for AlN, GaN and InN. Unless otherwise noted, the values are from density functional calculations (see Ref.[10] for details). Calculations for the crystal-field splittings and deformation potentials use the Sterne-Inkson formulation.[11].

	AlN	GaN	InN
$a(\text{Å})$[10]	3.084	3.162	3.501
$c(\text{Å})$[10]	4.948	5.142	5.669
$C_{13}(\text{GPa})$	108	103	92
$C_{33}(\text{GPa})$	373	405	224
$(a_{cz}\text{-}D_1)(\text{eV})$	-4.21	-6.11	-4.05
$(a_{ct}\text{-}D_2)(\text{eV})$	-12.04	-9.62	-6.67
$D_3(\text{eV})$	9.06	5.76	4.92
$D_4(\text{eV})$	-4.05	-3.04	-1.79
$\Delta_1(\text{eV})$	-.221	0.019	0.025
$E_g(\text{eV})$	6.28[12]	3.50[13]	1.89[14]
$\Delta_0(\text{eV})$[15]	0.019	0.013	0.001
m_c^*	0.31	0.18	0.011[16]
$m_{hh}^{\parallel}$	3.52	2.01	1.89
$m_{lh}^{\parallel}$	3.52	2.01	1.89
$m_{split}^{\parallel}$	0.25	0.15	0.024
$m_{hh}^{\perp}$	21.8	2.13	2.00
$m_{lh}^{\perp}$	0.32	0.19	0.033
$m_{split}^{\perp}$	4.38	1.45	1.60

The low density spectra show an exciton resonance. The existence of excitons at the high temperature of $T = 300K$ and carrier densities up to $N = 10^{12}cm^{-2}$ is evidence of strong Coulomb attraction between electrons and holes. It is important to note that the presence of excitonic effects in our results is not due to an *ad hoc* inclusion of excitonic transitions into a free-carrier theory, as is the case for some phenomenological models.[8] Rather the presence of excitons comes about because of the attractive Coulomb potential in the Hamiltonian we use to describe the electron-hole system. The present approach gives a consistent treatment of relaxation and screening effects in a exciton/plasma system, which is not possible with phenomenological approaches that treat the excitons and plasma as non interacting.

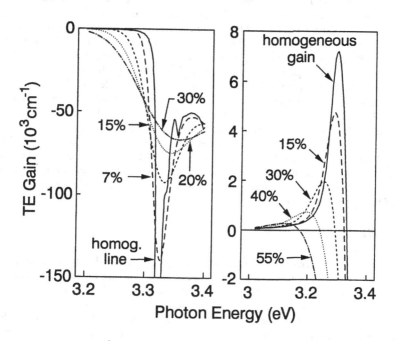

Fig. 2. TE gain/absorption spectra at $N = 10^{11}cm^{-2}$ (left) and $8 \times 10^{12}cm^{-2}$ (right). The solid curves depict the homogeneously broadened spectrum, while the other spectra are computed for increasing inhomogeneous broadening due to composition variation. All other parameters are similar to those in Fig. 1.

At high densities the exciton resonance vanishes and gain appears (see inset). Recently, there is much interest in the energy of the optical emission.[2] Our calculation shows that for the carrier density $N = 6 \times 10^{12} cm^{-2}$, which gives a local gain of $\sim 10^3 cm^{-2}$, the emission peak is over 100meV in energy lower than the unexcited quantum well bandgap energy. This red shift is the net result of the energy shifts due to several physical mechanisms. There is band filling, which leads to a blue shift that depends on the bandstructure. In addition, Coulomb interactions resulting in bandgap renormalization, Coulomb enhancement, dephasing and screening lead to a red shift, as well as reshaping of the spectrum. Finally, inhomogeneous broadening can also contribute to a significant red shift of the gain peak.

Figure 2 illustrates the effects of inhomogeneous broadening. At low densities, inhomogeneous broadening leads to broadening of the exciton resonance (Fig. 2, left). At high densities it reduces the gain, and shifts the spectrum towards lower energy (Fig. 2, right). Our calculations predict a red shift of the gain peak relative to the unexcited quantum well band gap that ranges from 70meV for the homogeneously case, to >200meV for composition variations >50% (i.e. mean indium concentration of 0.1 with standard deviation of 0.05).

Spontaneous emission spectra are more readily obtained in experiments than gain spectra. Figure 3 (top) shows the spontaneous emission spectra at carrier density $N = 5 \times 10^{12} cm^{-2}$ and different inhomogeneous broadening. The spontaneous emission spectra are obtained from the calculated gain spectra by using a relationship that is based on energy conservation arguements.[9] Comparison with the gain spectra (Fig. 3, bottom) shows that a significant energy difference can occur between spontaneous emission and gain peaks. For the homogeneously broadened spectra, the gain peak is red shifted by 40meV from the spontaneous emission. For a standard deviation of 0.03 (30%) in the indium concentration, this shift increases to >100meV, making propagation effects important in the determination of the energy of the optical emission.

CONCLUSION

In summary, we describe a theory of gain for group-III nitride quantum well lasers. The effects of excitons are integrated with those of an interacting electron-hole plasma by using a Hamiltonian for the electron-hole system that includes the many-body Coulomb interactions. The description of carrier correlation effects at the level of quantum kinetic theory gives a consistent treatment of broadening and screening effects, due to both the electron-hole plasma and the excitons. Finally, by taking into account the inhomogeneously broadening due to spatial variations in quantum well thickness or composition, we provide a realistic description of actual experimental configurations.

ACKNOWLEDGMENTS

This work was supported in parts by the U. S. Department of Energy under contract No. DE-AC04-94AL85000, the Deutsche Forschungsgemeinschaft (Germany), and the Leibniz Prize.

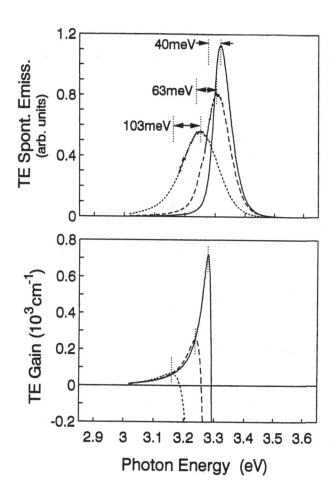

Fig. 3 Spontaneous emission (top) and gain spectra (bottom) at $N = 5 \times 10^{12} \text{cm}^{-2}$. The solid curves are the homogeneously broadened spectra, the long and short dashed curves have 15% and 30% variations in Indium concentrations, respectively. The energy values of 40, 63 and 103meV correspond to the energy differences between the gain and spontaneous emission peaks. All other parameters are similar to those in Fig. 1.

References

[1] W. W. Chow, A. Knorr and S. W. Koch, Appl. Phys. Letts. **67**, 754 (1995).

[2] S. Nakamura, SPIE Conference on Physics and Simulation of Opto-Electronic Devices, 8-14 Feb. 1997, San Jose, CA.

[3] W. W. Chow, S. W. Koch and M. Sargent III, Semiconductor-Laser Physics, (Springer Verlag, Berlin, 1994), p. 124.

[4] M. Lindberg and S. W. Koch, Phys. Rev. **B38**, 3342 (1988).

[5] F. Jahnke, et al, Phys. Rev. Lett. **77**, 5257 (1996).

[6] S. L. Chuang and C. S. Chang, Phys. Rev. B54 (1996).

[7] A. F. Wright and J. S. Nelson, Appl. Phys. Letts. **66**, 3051 (1995).

[8] F. Jain and W. Huang, IEEE J. Quantum Electron. 32, 859 (1996).

[9] C. H. Henry, R. A. Logan and F. R. Merritt, J. Appl. Phys. **51**, 3042 (1980).

[10] A. F. Wright and J. S. Nelson, Phys. Rev. **B51**, 7866 (1995).

[11] S. J. Jenkins, G. P. Srivastava and J. C. Inkson, Phys. Rev. **B48**, 4388 (1993).

[12] P. B. Perry and R. F. Rutz, Appl. Phys. Letts. **33**, 319 (1978).

[13] B. Monemar, Phys. Rev. **B10**, 676 (1974).

[14] T. L. Tansley and C. P. Foley, J. Appl. Phys. **59**, 3241 (1986).

[15] S. H. Wei and A. Zunger, Appl. Phys. Letts. **69**, 2719 (1996).

[16] V. A. Tyagai, A. M. Evstigneev, A. N. Krasiko, A. F. Andreeva and V. Y. Malakhov, Soviet Phys. Semicond. **11**, 1257 (1977).

GROWTH AND CHARACTERIZATION OF THERMAL OXIDES ON GALLIUM NITRIDE

SCOTT D. WOLTER, SUZANNE E. MOHNEY, HARI VENUGOPALAN, and DEBRA L. WALTEMYER
Department of Materials Science and Engineering
The Pennsylvania State University, University Park, PA 16802

BRIAN P. LUTHER
Department of Electrical Engineering
The Pennsylvania State University, University Park, PA 16802

ABSTRACT

Little information is available about the thermal oxidation of GaN. Since GaN is of interest for high temperature electronics, knowledge of the stability of GaN in potentially oxidizing environments would be useful. Furthermore, evaluation of the characteristics of the thermal oxide will provide information needed for assessing the potential of this oxide in processing or device applications.

In this study, thick GaN epilayers and GaN powders were exposed to dry air at 450°C, 750°C, 900°C, 925°C, 950°C, and 1000°C for periods of 1 to 25 hours. Following oxidation, the epilayers were analyzed by x-ray photoelectron spectroscopy and glancing incidence x-ray diffraction, and the powders were analyzed by conventional x-ray diffraction. For both the GaN films and powders, significant oxidation was observed at 900°C, and the oxide was identified as monoclinic β-Ga_2O_3. Oxidation in dry air resulted in roughening of the oxide/GaN interface and oxide surface. In the temperature range 900°C to 1000°C, linear kinetics were observed for times up to 10 hours indicating an interfacial reaction mechanism as the rate limiting step for oxidation. An apparent activation energy of ~72 kcal/mole was determined for this process.

INTRODUCTION

Gallium nitride possesses fundamental properties that make it well suited for application in optoelectronics. Improvements in the processing of GaN epitaxial films have resulted in the development of LEDs producing high brightness blue to UV light [1,2]. Other potential applications include LASERs and high power and high temperature electronics [3-5].

The oxidation of GaN is an important concern when processing at elevated temperatures in a potentially oxidizing ambient. A study by Johnson et al.[6] conducted in the 1930's indicated reaction of GaN with oxygen at approximately 900°C. However, very little research has been conducted in the subsequent years. Roy et al.[7] have reported the most thermodynamically stable form of gallium oxide to be β-Ga_2O_3, although many other polytypes have been observed. Recently, Wolter et al.[8] have studied the oxidation of GaN thick films grown on sapphire with results that are consistent with the previous observations. In this recent study, significant growth of β-Ga_2O_3 was observed at temperatures of 900°C or greater. An x-ray photoelectron spectroscopy (XPS) depth profile of the thermally grown oxide was also performed, providing

495

Mat. Res. Soc. Symp. Proc. Vol. 468 ° 1997 Materials Research Society

core level peak position data and information about interface roughening under the dry oxidation conditions.

This paper will provide further details of our previous research [8] on the oxidation of GaN using a dry air source gas. Emphasis will be placed on providing information concerning the oxidation kinetics using x-ray diffraction (XRD). In addition, XPS and scanning electron microscopy (SEM) will be utilized to characterize the grown oxide and assess the extent of oxidation at 900°C.

EXPERIMENTAL

GaN powder and GaN(0001) epilayers were used in this study. The GaN powder was of 99.9% purity and the average particle size was 10μm to 15μm in diameter. A quantity of 0.25g of GaN powder was placed in a quartz boat, which was in turn placed in a 1 inch diameter closed quartz tube. The source gas was flowed through a desiccant into the quartz tube at a rate of 25 sccm and allowed to purge the system for several minutes. The quartz tube containing the GaN powder was then placed in a horizontal electric furnace for various times (1 to 25 hours) and temperatures (450°C, 750°C, 900°C, 925°C, 950°C, 950°C, and 1000°C) to obtain information regarding oxide growth as a function of temperature and time. After completing each experiment, the quartz tube was removed and brought to room temperature within 5 minutes.

The GaN(0001) epilayers were oxidized in dry air in the closed quartz tube set-up under the same procedure used for the GaN powder experimentation. Details regarding the growth of these films have been described in previous work [9]. The samples were solvent cleaned in acetone, methanol, and DI water rinsed. The samples were then etched in HCl:DI water (1:1) for 1 minute in an effort to further remove surface contamination and the adventitious oxide.

The GaN powder was analyzed using bulk x-ray diffraction (XRD) employing θ–2θ scans and Cu Kα radiation. The powder was processed into a slurry using methanol and placed onto a microscope slide for analysis. The preparation of all the specimens for analysis followed a consistent procedure in order to insure that similar amounts of material were analyzed within the sampling area.

The oxidized films were characterized using x-ray photoelectron spectroscopy (XPS) (Mg Kα radiation) and glancing angle XRD (Cu Kα radiation). Details about the XPS analysis system and the procedure for assigning the peak positions relative to the C 1s spectra may be obtained in reference [8]. XPS depth profiling using argon ion sputtering was conducted to establish the Ga-O and Ga-N binding energy peak positions for the Ga 3d core level. This study will also examine the degree of interface roughening following oxidation using SEM.

RESULTS

Figure 1 reveals an XRD spectra of the GaN powder obtained after oxidation was undertaken at 900°C for various times. Similar data was collected for oxidation temperatures of 450°C, 750°C, 925°C, 950°C, and 1000°C. This graph indicates the formation of a crystalline oxide on the GaN powder which was identified as the monoclinic β-Ga$_2$O$_3$ phase by comparing the XRD peaks to those reported in literature. The oxide peak intensities shown in Figure 1 are representative of a randomly oriented oxide film and the respective oxide peak intensities

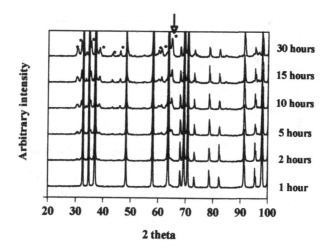

Figure 1. X-ray diffraction patterns of GaN powder oxidized at 900°C for increasing oxidation times. (* denotes the β-Ga₂O₃ XRD peaks)

increased at the same rate relative to each other for increasing oxidation duration, such as the strong β-Ga₂O₃ ($\bar{2}$ 1 7) peak at 64.73°. The intensity of this particular peak referenced to the baseline value is plotted versus oxidation time for all the respective temperatures and is shown in Figure 2. Significant oxidation was observed on the GaN powder oxidized in dry air at

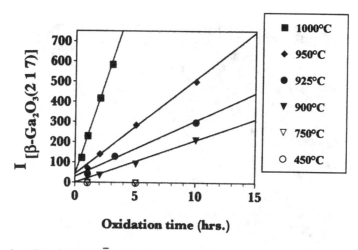

Figure 2. A plot of the β-Ga₂O₃($\bar{2}$ 1 7) peak intensity versus oxidation time reveals measurable oxidation between 750°C and 900°C.

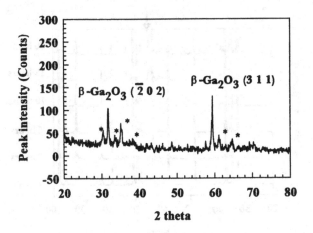

Figure 3. Glancing angle XRD of thick film material oxidized at 900°C for 25 hours. (* denotes β-Ga₂O₃ XRD peaks)

temperatures of 900°C or higher. No crystalline oxide phase(s) were detected using XRD at 450°C and 750°C for up to 25 hours in duration.

Figure 3 displays glancing angle XRD of an oxide grown at 900°C for 25 hours on the GaN epilayers. Again, the peaks were indexed as β-Ga₂O₃, although the oxide appears to be

Figure 4. A tilted SEM image of β-Ga₂O₃ on GaN grown in dry air at 900°C for 25 hours.

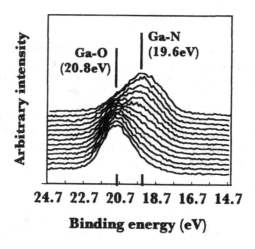

Figure 5. XPS depth profiled Ga 3d spectra for the grown oxide shown in Figure 4. This spectra was obtained after an initial 60 minutes of sputtering with each of the subsequent spectra obtained at 3 minute sputter intervals.

oriented with respect to the GaN. An SEM micrograph of this 5000Å thick overlayer is shown in Figure 4. A portion of this oxide film was removed in a solution of 10% HF acid for several hours, revealing both a roughened oxide surface and oxide/GaN interface. XPS depth profiling of this oxide is exhibited in Figure 5 for the Ga 3d core level spectra. This data reveals a gradual shift from Ga-O to Ga-N bonding of 1.2 eV, consistent with the roughening at the oxide/GaN interface.

DISCUSSION

The oxide formed by dry oxidation of the GaN powder and GaN epilayers was the monoclinic β-Ga_2O_3. Significant oxidation was observed at temperatures of 900°C or greater using XRD of the oxidized GaN powder specimen and XPS of oxidized GaN epilayers.

The oxidation kinetics for the powders were determined using a planar oxide growth model since the oxide thickness relative to the GaN particle radius was small. A thorough evaluation of this data will be published elsewhere; however, an interfacial reaction mechanism was found to be rate limiting for these oxidation conditions. In addition, an apparent activation energy within this oxidation regime was determined by extracting the β-$Ga_2O_3(\bar{2}\ 1\ 7)$ peak intensity values for each of the temperatures after 1 hour of oxidation and was determined to be ~72 kcal/mole. Interestingly, this value is similar to a reported activation energy for the decomposition of GaN in vacuum over a similar temperature range [10]. Furthermore, decomposition of GaN into liquid gallium and nitrogen gas at 1 atmosphere could occur at temperatures near 900°C based on recent thermodynamic data [11]. This temperature is similar to that at which measurable oxidation was observed to have occurred.

CONCLUSIONS

Oxidation of GaN in dry air has been investigated. The monoclinic β-Ga$_2$O$_3$ phase was observed to have formed on both the GaN powder and thick film material with measurable oxidation observed at temperatures of 900°C or greater. Linear oxidation kinetics were observed using XRD of the powder specimen by evaluating the growing oxide peak intensities. Furthermore, an activation energy of ~72 kcal/mole was calculated.

X-ray photoelectron spectroscopy of the thermal oxide formed at 900°C for 25 hours was characterized by depth profiling. A gradual peak shift from Ga-O to Ga-N bonding of 1.2 eV was revealed. This data suggested the possibility of a roughening at the oxide/GaN interface which was confirmed by scanning electron microscopy.

ACKNOWLEDGEMENTS

Study of the oxidation of GaN was performed under NSF grant DMR-9624995. The authors would like to thank R.J. Molnar (MIT Lincoln Laboratory) for providing the GaN epilayers.

REFERENCES

1. H. Amano, M. Kito, X. Hiramatsu, and I. Akasaki, Jpn. J. Appl. Phys. **28**, L2112 (1989).

2. S. Nakamura, T. Mukai, and M. Senoh, Jpn. J. Appl. Phys. **30**, L1998 (1991).

3. M. Asif Kahn, A. Bhattarai, J.N. Kuznia, and D.T. Olsen, Appl. Phys. Lett. **63**, 1214 (1993).

4. M.W. Shin and R.J. Trew, Electron. Lett. **31**, 498 (1995).

5. O. Aktas, W. Zim, Z. Fan, F. Stengel, A. Botchkarev, A. Salvador, B. Sverdlov, S.N. Mohammad, and H. Morkoc, International Electronic Device Meeting, 205 (1995).

6. W.C. Johnson, J.B. Parsons, and M.C. Crew, J. Phys. Chem. **36**, 2651 (1932).

7. R. Roy, V.G. Hill, and E.F. Osborn, Am. Chem. Soc. **74**, 719 (1952).

8. S.D. Wolter, B.P. Luther, D.L. Waltemyer, C. Önneby, S.E. Mohney, and R.J. Molnar, Appl. Phys. Lett. **70**, 2156 (1997).

9. R.J. Molnar, K.B. Nichols, P. Maki, E.R. Brown, and I. Melngailis, Mater. Res. Soc. Symp. Proc. **378**, 479 (1995).

10. R. Groh, G. Gerey, L. Bartha, and J.I. Pankove, Phys. Stat. Sol. (a) **26**, 353 (1974).

11. J. Karpinski and S. Porowski, J. Crystal Growth **66**, 11 (1984).

AUTHOR INDEX

SUBJECT INDEX